Third Edition

PREALGEBRA

James Van Dyke
James Rogers
Hollis Adams

Portland Community College

SAUNDERS COLLEGE PUBLISHING

Harcourt Brace College Publishers

Fort Worth   Philadelphia   San Diego   New York   Orlando   Austin
San Antonio   Toronto   Montreal   London   Sydney   Tokyo

Publisher: Emily Barrosse
Acquisitions Editor: Bill Hoffman
Product Manager: Nick Agnew
Developmental Editor: Carol Loyd, Terri Ward
Production Manager: Alicia Jackson
Project Editor: Linda Boyle Riley
Art Director: Lisa Caro
Text Designer: Rebecca Lloyd Lemna
Cover Credit: UPI/Corbis-Bettman

Printed in the United States of America

PREALGEBRA, third edition
0-03-019638-8

Library of Congress Catalog Card Number: 97-69267

123456   032   10 98765

To our children:

Dan, Claudia, Larry, Karla, Greg, and Ann Van Dyke

Becky and Terry Washington

Paul and Patty Hurst

Heather and John Fincher

Michelle and Jim Raible

Jessica and Ben Adams

Preface

Prealgebra provides students with a review of the basic skills of arithmetic while they are learning the basic principles of algebra at the college level. Students completing this course will be prepared to fulfill competency requirements, enter business mathematics and elementary algebra courses, and achieve satisfactory scores on placement exams.

Changes in the Third Edition

A number of changes and improvements have been made in preparing the new edition of this text. Many of these changes are in response to comments and suggestions offered by users and reviewers of the manuscript. Others have been made to bring the text in line with math reform standards.

   The following list describes the major changes in the organization and content in the third edition.

- Sections on drawing and interpreting graphs and reading and interpreting tables have been moved to the end of Chapter 1. This allows the use of graphs and tables throughout the text, giving students more exposure to these "real world" uses of mathematics.

- The chapter on measurement has been moved to Chapter 2, following whole numbers. The examples and exercises are restricted to whole numbers. Geometric formulas involving π have been deferred to the chapters on fractions and decimals. Moving this material to Chapter 2 allows for the use of geometry throughout the text, thus providing a greater variety of applications.

- Conversion of units within the English or metric system is now in Section 5.4. The placement here provides students with an immediate application of the multiplication property of one.

- Conversion of units between English and metric measures in now in Section 6.6. The conversion factors are approximations.

- Divisibility tests for 2, 3, 5, or 10 have been included in Chapter 5. This makes it easier for students to find the LCD of a group of fractions.

- Estimation of sums, differences, products, and quotients has been included for whole numbers and decimals. Students practice making quick estimates to determine whether computed answers are reasonable.

The next list of changes are pedagogical in nature and enhance the teachability of the text.

- Each chapter leads off with an **application** that gives the student a real world context for the mathematics. Exercises concerning or related to the theme problem are included in most every section of the chapter.

- Each objective is treated separately in the **How and Why,** which is followed by examples. Students can now focus on a single topic until they understand it.

- The format of the **Examples** has been changed by giving general directions and strategies at the beginning of each set. This assists the student by seeing how to attack the exercise for a given objective. The word **"application"** has been eliminated from word problems, as students need to see these as, "just another exercise," **not** as special or difficult exercises.

- Key strokes in **calculator examples** have been eliminated. Showing key strokes could cause confusion because of the wide variety of calculators available. Calculator descriptions are included in Appendix III.

- The **Exercise sets** have been reorganized. The drill exercises (both A and B groups) are separated by objective. The more difficult C exercises and the former applications mix the objectives together, giving students more practice in applying skills to real situations.

- **Chart, graph, and table readings** are included in most sections so students will gain skills in obtaining or displaying information in these formats.

- **Journal-writing exercises** are included in the State Your Understanding section of the Exercises. These exercises are identified by a **pencil icon** and allow the student to explain, in their own words, key concepts. Maintaining a journal will allow them to review concepts as they have written them.

- An optional **open-ended project** has been included at the end of each chapter. Students have an opportunity to work in groups and practice applying their new skills in a real world situation.

Format

This write-in text gives the student space to practice mathematical skills and have ready reference to worked examples with step-by-step directions. Each section contains the following pedagogical features: Objectives, Vocabulary, How and Why, Examples accompanied by Warm-Up Exercises for immediate student reinforcement, and Exercises. These pedagogical features are described in detail below. Each chapter ends with an optional Chapter Project, a True-False Concept Review, a Chapter Review, and a Chapter Test. The Chapter Project allows groups of students to apply the concepts to an open-ended real world situation. The True-False Concept Review tests the students' understanding of the mathematical theory, and the Chapter Test measures their skill level. The Chapter Review contains questions referenced to the objectives so that students can easily refer to a specific section for assistance. This format makes it possible for students to work their way through the material in a Math Lab or Learning Center. There are 3 cumulative reviews included at the end of Chapters 3, 6, and 8.

Objectives

Objectives are identified at the beginning of each section, in each example, in each exercise set, and are referenced in the Chapter Review at the end of each chapter. The objective of this text as a whole is to aid the student in making a successful move from the basic skills and concepts of arithmetic to the related skills and concepts of algebra. To enhance student success a spiral approach is used, where the new topics of algebra are revisited as the traditional topics of arithmetic are reviewed. This gradual approach to mastering the algebra gives the student ample opportunity to learn the new skills.

In **Chapter 1** the arithmetic concepts of place value and whole number exponents lead into the first three rules of exponents to prepare the students for multiplying terms in algebra in Chapter 3 and solving equations throughout the text. The skills of basic operations (addition, subtraction, multiplication, and division of whole numbers) are used to develop the algebraic skills of combining like terms and in Chapter 3, solving simple equations. Order of operations (algebraic logic) is introduced/reviewed so that two or more operations can be performed in a single exercise. A comprehensive presentation of drawing, reading, and interpreting graphs and tables is given. These are practical skills in themselves and also allow for a transition to linear graphing. Ideally, the material in Sections 1.1–1.7 is review. These sections could be treated lightly or skipped altogether, depending on the background of your students.

In **Chapter 2** measurements in both English and metric systems are discussed, and conversions are done within systems using whole numbers. Formulas for perimeter, area, and volume (arithmetic and algebra) are shown for fundamental polygons and solids. The material in Chapter 2 may be review as well. However, many students at this level do not understand the basic concepts of perimeter, area, and volume. It is our feeling that the time you spend in Chapter 2 will greatly benefit your students both for the rest of this text and in subsequent courses.

In **Chapter 3** the basic concepts of algebra are introduced as generalizations of arithmetic concepts. Students apply the arithmetic operations on algebraic terms. The arithmetic skills of Chapter 1 are expanded so the algebraic skills necessary for the solution of equations seems a natural extension. Furthermore, some of the vocabulary ("language of algebra") that is naturally a part of algebra is based on the students' familiarity with the way English is used in arithmetic.

In **Chapter 4** the arithmetic of whole numbers leads to integers. The negative integers are defined in terms of the opposites of whole numbers, using the number line, and follow as a natural extension of the set of whole numbers. The operations on integers are defined in terms of absolute values, so the student is using whole number arithmetic. To ensure competency in these skills, this introduction is immediately followed by practice with operations on terms, order of operations, evaluating expressions, and solving equations with integers.

In **Chapters 5 and 6** the arithmetic concept of rational numbers is the logical basis for the affiliated algebraic concepts of rational numbers and rational expressions. The arithmetic skills used in operations with fractions are required in the algebraic skills of adding, subtracting, multiplying, and dividing fractions involving variables. Again, proficiency in these skills is ensured by additional work with the order of operations, the evaluation of expressions, and solving equations using these numbers. The introduction of these skills provides a logical base for student success in a more intensive study of rational expressions.

Chapter 6 presents decimals as rational numbers in a different form. Square roots and the Pythagorean Formula are covered here.

Geometric formulas involving π are introduced in Chapter 5 using a fractional approximation and then in Chapter 6 using a decimal approximation. Circle graphs are also introduced in this chapter.

In **Chapter 7** the typical arithmetic concepts of proportion and percent are shown. Algebraic solutions are presented.

In **Chapter 8** the concept of (algebraic) graphing equations of two variables is introduced to provide a natural transition to the next course in algebra. Sections on slope and intercepts and variation are included here but marked as optional, to be used at the discretion of the instructor.

Vocabulary

Definitions of words that have not been previously used in the text are provided under this heading.

How and Why

Under this heading the concepts and skills are explained. Throughout the text the explanations are primarily intuitive. Rules and procedures are highlighted by bold type and color screens for quick reference and easy review. The explanations in this section, pared with the Examples that follow provide an immediate linking of theory and practice. Each objective has its own How and Why followed by examples.

Examples — Warm Ups

The examples illustrate the concept explained in the How and Why section. Each set of examples leads off with a general direction and strategy for working the examples. As each example is worked out, a step-by-step explanation is given. These explain the procedures and thinking necessary to work the example. The examples also illustrate shortcuts and include cautions about common errors and potential pitfalls where applicable.

Each example is paired with a warm-up problem of the same type and level of difficulty to reinforce the procedures used to solve the example. These Warm Ups are useful for students to check their understanding of the material before advancing to more difficult examples. The answers appear at the bottom of the page for easy reference.

The algebra topics in this text are introduced as a natural extension of the corresponding arithmetic topics. The algebra is thus integrated throughout the text, as opposed to being presented in separate sections or chapters. Here is an example of the presentation used in the text that shows the integration of arithmetic and algebra.

Examples	**Warm Ups**
Arithmetic	
A. Multiply: $\dfrac{3}{4} \cdot \dfrac{5}{8}$	A. Multiply: $\dfrac{4}{9} \cdot \dfrac{7}{5}$
$\dfrac{3}{4} \cdot \dfrac{5}{8} = \dfrac{3 \cdot 5}{4 \cdot 8} = \dfrac{15}{32}$ **There are no common factors, so write the product of the numerators over the product of the denominators.**	

Arithmetic/Algebra Bridge

B. Multiply: $\dfrac{1}{2} \cdot \left(-\dfrac{3}{4}\right) \cdot \dfrac{5}{8}$

$\dfrac{1}{2} \cdot \left(-\dfrac{3}{4}\right) \cdot \left(-\dfrac{5}{8}\right) = -\dfrac{1 \cdot 3 \cdot 5}{2 \cdot 4 \cdot 8}$ The product is negative, as the product of the first two factors is negative and that product times the third factor is negative. There are no common factors. Write the product of the numerators over the product of the denominators.

$= -\dfrac{15}{64}$

B. Multiply: $\dfrac{1}{3} \cdot \left(-\dfrac{4}{7}\right) \cdot \left(-\dfrac{8}{5}\right)$

Arithmetic/Algebra Bridge

C. Multiply: $\dfrac{3}{4}a \cdot \dfrac{6}{7}$

$\dfrac{3}{4}a \cdot \dfrac{6}{7} = \left(\dfrac{3}{4} \cdot \dfrac{6}{7}\right)a$ First use the commutative and associative properties of multiplication to group the fractions. Since 4 and 6 have a common factor of 2, simplify first, and then multiply.

$= \left(\dfrac{3 \cdot \overset{3}{\cancel{6}}}{\underset{2}{\cancel{4}} \cdot 7}\right)a$

$= \dfrac{9}{14}a$

C. Multiply: $\dfrac{7}{8} \cdot \dfrac{4}{5}x$

Algebra

D. Multiply: $\left(-\dfrac{8}{15}x\right)\left(\dfrac{5}{12}x\right)$

$\left(-\dfrac{8}{15}x\right)\left(\dfrac{5}{12}x\right) = \left(-\dfrac{8}{15} \cdot \dfrac{5}{12}\right)(x \cdot x)$ Use the commutative and associative properties of multiplication to group the coefficients and the variables.

$= \left(-\dfrac{\overset{2}{\cancel{8}}}{\underset{3}{\cancel{15}}} \cdot \dfrac{\overset{1}{\cancel{5}}}{\underset{3}{\cancel{12}}}\right)(x \cdot x)$ Reduce.

$= -\dfrac{2}{9}x^2$ Multiply.

D. Multiply: $\dfrac{2}{9}a\left(-\dfrac{2}{9}a\right)$

Answers to Warm Ups   A. $\dfrac{28}{45}$ B. $\dfrac{32}{105}$ C. $\dfrac{7}{10}x$ D. $-\dfrac{4}{81}a^2$

Exercises

The exercises are grouped according to level of difficulty. **Group A** exercises are relatively easy and can often be used in class as oral exercises. **Group B** exercises are more difficult and involve computation with larger numbers or more complex expressions. Group A and B exercises are divided by objective so the student will know the skill being practiced. Group C exercises include all objectives and a variety of real world problems. The student must decide on the strategy needed to set up or solve the problem.

The next group of exercises is **State Your Understanding,** where students are required to put their understanding of the topic down in writing using their own words. **Journal** exercises are included here. **Challenge** exercises are given to test the limits of the more capable student. **Group Activity** exercises encourage cooperative learning as recommended by **AMATYC** and **NTCM** guidelines. **Maintain Your Skills** exercises review material previously covered and in many cases review topics needed for the upcoming section.

Timetable

This text can be used in a variety of classroom situations, depending on the needs of the students. Two such possibilities are:

1. A one-quarter or one-semester course given as a review of basic arithmetic and an introduction to negative numbers, equations, and the algebra of monomials. Such a course would cover Chapters 1–7 or Sections 1.8, 1.9, and Chapters 2–7.
2. A one-quarter or one-semester course with a review of arithmetic and introduction to algebra. Such a course would cover Chapters 1–8, with Sections 8.4 and 8.5 optional.

Ancillary Materials

The following supplements to accompany *Prealgebra,* third edition, are available to enhance the presentation and understanding of the course:

Student Solutions Manual This guide contains worked-out solutions to one quarter of the problems in the exercise sets (every other odd-numbered problem) to help the student learn and practice the techniques used in solving problems.

Instructor's Manual This supplement features instructor-appropriate solutions to problems in the text. All solutions have been reviewed for accuracy.

Test Bank and Prepared Tests This resource contains written tests, including both open-ended and multiple choice questions, for each chapter of the book. In addition, it provides a printed test bank generated from a computerized test bank.

ExaMaster+™ A flexible, powerful testing system, ExaMaster+™ offers instructors a wide range of integrated testing options and features. For each chapter, test items are provided that can be selected with or without multiple-choice distractors. Teachers can select test items according to a variety of other criteria, including section, objective, focus (skill, concept, or application), and difficulty (easy, medium, or hard). Teachers can scramble the order of test items, administer tests on-line, and print objective-referenced answer keys. ExaMaster+™ can also be used to create extra practice worksheets, and includes a full-function gradebook and graphing features.

MathCue Interactive Software Available in Macintosh and Windows versions, this interactive software provides additional, self-paced support to students and is free to adopters. The new Windows version combines Tutorial and Solution Finder programs on one disk and includes new features, such as hot links to math summaries, formulas, pop-up definitions, and a searchable index. The following features apply to both platforms:

Tutorial Keyed to topics in Prealgebra, this software allows students to test their skills and pinpoint and correct weak areas. Students may choose to see step-by-step solutions or partial solutions to all problems. The software features a Missed Problem and Disk Review to enable students to review problems answered incorrectly and to summarize all topics on the disk.

Solution Finder This software allows students to input their own questions through use of an expert system, a branch of artificial intelligence. Students can check answers or receive help as if they were working with a tutor. The software will refer the student to the appropriate section of the text and will record the number of problems entered and evaluate a function at a point, graph up to four functions simultaneously, and save and retrieve function setups via disk files.

Videotapes These section-by-section videos feature on-location segments to illustrate applications. Each tape covers one chapter, with roughly 15 minutes per section.

Core Concepts Video This four-hour video tutorial covers the core concepts and works through selected examples from each section of the text. Students with access to a VCR can use the video as a take-home tutorial.

Saunders College Publishing may provide complimentary instructional aids and supplements or supplement packages to those adopters qualified under our adoption policy. Please contact your sales representative for more information. If as an adopter or potential user you receive supplements you do not need, please return them to your sales representative or send them to
Attn: Returns Department
Troy Warehouse
465 South Lincoln Drive
Troy, MO 63379

Acknowledgments

The authors appreciate the unfailing and continuous support of their spouses— Carol Van Dyke, Elinore Rogers, and Doug Adams—who made the completion of this work possible. We are grateful to Bill Hoffman, Carol Loyd, Terri Ward, Linda Boyle, Alicia Jackson, Sue Kinney, and Lisa Caro of Saunders College Publishing for their suggestions during the preparation of the text. We also want to express our thanks to the following professors and reviewers for their many excellent contributions to the development of the text:

April Allen, Hartnell College
Debra D. Bryant, Tennessee Technological University
Tom Carson, Midlands Technical College
John F. Close, Salt Lake Community College
Richard N. Dodge, Jackson Community College
Irene Doo, Austin Community College
Rebecca Easley, Rose State College

Thomas Granata, Manatee Community College (South Campus)
Mark Greenhalgh, Fullerton College
Lorette Griffy, Austin Peay State University
James W. Harris, John A. Logan College
Todd A. Hendricks, Dekalb College
Bonnie M. Hodge, Austin Peay State University
Maryann E. Justinger, Erie Community College–South
Carolyn T. Krause, Delaware Technical & Community College
Gerald R. Krusinski, College of Du Page
Mort Mattson, Lansing Community College
Loretta Palmer, Utah Valley State College
Lymeda Singleton, Abilene Christian University
Diane Thompson, University of Alaska Anchorage
Catrinus Tjeerdsma, Community College of Denver
Matrid Hurst Whidden, Edison Community College
Deborah Woods, University of Cincinnati, Raymond Walters College

Special thanks to Sharon Edgmon of Bakersfield College and Barbara Hughes of San Jacinto College for their careful reading of the text and for the accuracy review of all the problems and exercises in the text.

Jim Van Dyke
Jim Rogers
Hollis Adams

ELM Mathematical Skills

The following table lists the California ELM MATHEMATICAL SKILLS and where coverage of these skills can be found in the text. Skills not covered in this text can be found in Basic Algebra, fourth edition, or Intermediate Algebra, fourth edition. Location of the skills is indicated by chapter section or chapter.

SKILL	LOCATION IN TEXT
Cluster A: Algebra	
Real numbers and their operations	Chapters 1, 5, 6
Scientific notation	6.4
Absolute value	4.1
Applications (e.g., estimations, percents, word problems, charts and graphs)	1.2, 1.3, 1.7, 1.8, 4.4, 4.5, 7.6, and throughout the exercise sets
Evaluation of polynomials	4.8, 5.6, 6.7
Addition and subtraction of polynomials	3.3, 4.4
Multiplication and division of polynomials	3.4, 4.6
Integer exponents	1.4, 6.4
Radicals	6.9
Linear equations in one unknown with numerical or literal coefficients	3.5, 3.6, 3.7, 4.9, 5.7
Equations involving absolute values	4.1
Applications and word problems (including ratio and proportion)	7.2
Points on the number line or in the coordinate plane	8.1
Linear functions ($ax + by + c = 0$); slopes and intercepts	8.2, 8.3, 8.4

Cluster B: Geometry

Perimeter of triangles	2.2
Area of triangles	2.3
Perimeter of squares, rectangles, and parallelograms	2.2
Areas of squares, rectangles, and parallelograms	2.3
Radius and diameter of circles	5.6
Circumference of a circle	5.6
Area of a circle	5.6
Volume of rectangular solids	2.4
Volume of cylinders	5.6
Volume of spheres	5.6
Interior and exterior angles: sum of interior angles	Geometry G.5
Angles formed by parallel and perpendicular lines	Geometry G.2, G.3
Equilateral and isosceles triangles	Geometry G.3
Right triangles and the Pythagorean Theorem	6.9, Geometry G.3
45°-45°-90° and 30°-60°-90° triangles	Geometry G.3
Congruent and similar triangles	Geometry G.4, G.6

Cluster C: Data Interpretation, Counting, Probability, and Statistics

Reading data from graphs and charts	1.7
Computation with data	1.7, 1.8, and throughout text
Finite probabilities	Chapter 5
Average (arithmetic mean)	1.6

CLAST Mathematical Skills

The following table lists the Florida CLAST MATHEMATICAL SKILLS and where coverage of these skills can be found in the text. Skills not covered in this text can be found in Basic Algebra, third Edition, or Intermediate Algebra, third Edition. Location of skills is indicated by chapter section or chapter.

SKILL	LOCATION IN TEXT
1A1a — Adds and subtracts rational numbers	5.5
1A1b — Multiplies and divides rational numbers	5.3
1A2a — Adds and subtracts rational numbers in decimal form	6.2
1A2b — Multiplies and divides rational numbers in decimal form	6.3, 6.4, 6.5
1A3 — Calculates percent increase and percent decrease	7.6
2A1 — Recognizes the meaning of exponents	1.4, 1.5, 6.4
2A2 — Recognizes the role of the base number in determining place value in the base-ten numeration system and in systems that are patterned after it.	1.1, 6.1
2A3 — Identifies equivalent forms of positive rational numbers involving decimals, percents, and fractions	7.4

2A4 — Determines the order-relation between magnitudes   1.1, 4.1, 5.5, 6.1

4A1 — Solves real-world problems that do not require the use of variables and that do not require the use of percent   Ch. 1, Ch. 2, Ch.3, Ch. 4, Ch. 5, Ch. 6

4A2 — Solves real-world problems that do not require the use of variables and that do require the use of percent   Ch. 7

4A3 — Solves problems that involve the structure and logic of arithmetic   In applications throughout text

1B1 — Rounds measurements to the nearest given unit of the measuring device   1.1, 6.1

1B2a — Calculates distances   2.2, 5.6, 6.7

1B2b — Calculates areas   2.3, 5.6, 6.7

1B2c — Calculates volumes   2.4, 5.6, 6.7

2B1 — Identifies relationships between angle measures   Geometry G.2

2B2 — Classifies simple plane figures by recognizing their properties   Ch. 2, 5.6

2B3 — Recognizes similar triangles and their properties   Geometry G.6

3B1 — Infers formulas for measuring geometric figures   Ch. 2, 5.6

3B2 — Identifies applicable formulas for computing measures of geometric figures   Ch. 2, 5.6

4B1 — Solves real-world problems involving perimeters, areas, and volumes of geometric figures   2.2, 2.3, 2.4, 5.6, 6.7

4B2 — Solves real-world problems involving the Pythagorean property   6.9

1C1a — Adds and subtracts real numbers   Ch. 3, Ch. 4, Ch. 5

1C1b — Multiplies and divides real numbers   Ch. 1, Ch. 4, Ch. 5, Ch. 6

1C2 — Applies the order-of-operations agreement to computations involving numbers and variables   1.6, 4.7, 4.8, 5.6, 6.7

1C4 — Solves linear equations and inequalities   3.5, 3.6, 3.7, 4.9, 5.7, 6.8, 7.2, 7.5, 8.5

1C5 — Uses given formulas to compute results when geometric measurements are not involved   In applications throughout text

2C3 — Recognizes statements and conditions of proportionality and variation   7.2, 7.3, 8.5

1D1 — Identifies information contained in bar, line, and circle graphs   1.7, 1.8, 7.6

4D1 — Interprets real-world data from tables and charts   1.7 and in applications throughout the text

4E1 — Draws logical conclusions when facts warrant them   Geometry G.1

TASP Mathematics Skills

The following table lists the Texas TASP MATHEMATICS SKILLS and where coverage of these skills can be found in the text. Skills not covered in this text can be found in Basic Algebra, third Edition, or Intermediate Algebra, third Edition. Location of skills is indicated by chapter section or chapter.

SKILL	LOCATION IN TEXT
Use number concepts and computation skills.	Ch. 1, Ch. 2, Ch. 3, Ch. 4, Ch. 5, Ch. 6
Solve word problems involving integers, fractions, or decimals (including percents, ratios, and proportions).	Ch. 1, Ch. 2, Ch. 3, Ch. 4, Ch. 5, Ch. 6, Ch. 7
Solve one- and two-variable equations.	3.5, 3.6, 3.7, 4.9, 5.7, 6.8, 7.2, 7.5, 8.2
Solve problems involving geometric figures.	Ch. 2, and in applications throughout the text
Apply reasoning skills.	Geometry G. 1

Student Preface

"It looks so easy when you do it, but when I get home…" is a popular lament of many students studying mathematics.

The process of learning mathematics evolves in stages. For most students, the first stage is listening to and watching others. In the middle stage, students experiment, discover, and practice. In the final stage, students analyze and summarize what they have learned. Many students try to do only the middle stage because they do not realize how important the entire process is.

Here are some steps that will help you to work through all the learning stages:

1. Go to class every day. Be prepared, take notes, and most of all, think actively about what is happening. Ask questions and keep yourself focused. This is prime study time.

2. Begin your homework as soon after class as possible. Start by reviewing your class notes and then read the text. Each section is organized in the same manner to help you find information easily. The objectives tell you what concepts will be covered, and the vocabulary lists all the new technical words. There is a How and Why section for each objective that explains the basic concept, followed by worked sample problems. As you read each example, make sure you understand every step. Then work the corresponding Warm Up problem to reinforce what you have learned. You can check your answer at the bottom of the page. Continue through the whole section in this manner.

3. Now work the exercises at the end of the section. The A group of exercises can usually be done in your head. The B group is harder and will probably require pencil and paper. The C group problems are more difficult, and the objectives are mixed to give you practice at distinguishing the different solving strategies. As a general rule, do not spend more than 15 minutes on any one problem. If you cannot do a problem, mark it and ask someone (your teacher, a tutor, or a study buddy) to help you with it later. Do not skip the Maintain Your Skills problems. They are for review and will help you practice earlier procedures so you do not become "rusty." The answers to the odd exercises are in the back of the text so you can check your progress.

4. In this text, you will find State Your Understanding exercises in every section. You may do these orally or in writing. Their purpose is to encourage you to analyze or summarize a skill and put it into words. Some of these exercises are designated as journal entries. The journal entries are intended to be written. Taken as a whole, the journal entries cover *all* the basic concepts in the text. We recommend that the journal entries be kept together in a special place in your notebook. Then they are readily available as a review for chapter tests and exams.

5. When preparing for a test, work the material at the end of the chapter. The True-False Concept Review, the Chapter Review, and the Chapter Test all give you a chance to review the concepts you have learned. You may want to use the Chapter Test as a practice test. The Chapter Review is organized by objectives so you can refer back to the text if you need extra work on a particular concept.

If you have never had to write in a math class, the idea can be intimidating. Write as if you are explaining to a classmate who was absent the day the concept was discussed. Use your own words—*do not copy out of the text.* The goal is that you understand the concept, not that you can quote what the authors have said. Always use complete sentences, correct spelling, and proper punctuation. Like everything else, writing about math is a learned skill. Be patient with yourself and you will catch on.

Since we have many students who do not have a happy history with math, we have included Good Advice for Studying—a series of eight essays that address various problems that are common for students. They include advice on time organization, test taking, and reducing math anxiety. We talk about these things with our own students, and hope that you will find some useful tips.

We really want you to succeed in this course. If you go through each stage of learning and follow all the steps, you will have an excellent chance for success. But remember, you are in control of your learning. The effort that you put into this course is the single biggest factor in determining the outcome. Good luck!

Jim Van Dyke
Jim Rogers
Hollis Adams

Contents

Good Advice for Studying

Strategies for Success

Are you afraid of math? Do you panic on tests or "blank out" and forget what you have studied, only to recall the material after the test? Then you are just like many other students. In fact, research studies estimate that as many as 50% of you have some degree of math anxiety.

What is math anxiety? It is a learned fear response to math that causes disruptive, debilitating reactions to tests. It can be so encompassing that it becomes a dread of doing *anything* that involves numbers. Although some anxiety at test time is beneficial—it can motivate and energize you, for example—numerous studies show that too much anxiety results in poorer test scores. Besides performing poorly on tests, you may be distracted by worrisome thoughts, and be unable to concentrate and recall what you've learned. You may also set unrealistic performance standards for yourself and imagine catastrophic consequences for your failure to be successful in math. Your physical signs could be muscle tightness, stomach upset, sweating, headache, shortness of breath, shaking, or rapid heart beat.

The good news is that anxiety is a learned behavior and therefore can be unlearned. If you want to stop feeling anxious, the choice is up to you. You can choose to learn behaviors that are more useful to achieve success in math. You can learn and choose the ways that work best for you.

To achieve success, you can focus on two broad strategies. First, you can study math in ways *proven* to be effective in learning mathematics and taking tests. Second, you can learn to physically and mentally *relax,* to manage your anxious feelings and to think rationally and positively. Make a time commitment to practice relaxation techniques, study math, and record your thought patterns. A commitment of one or two hours a day may be necessary in the beginning. Remember, it took time to learn your present study habits and to be anxious. It will take time to unlearn these behaviors. After you become proficient with these methods, you can devote less time to them.

Begin now to learn your strategies for success. Be sure you have read the Preface to the Student in the beginning of this book. The purpose of this preface is to introduce you to the authors' plan for this text. This will help you to understand the authors' organization or "game plan" for your math experience in this course.

At the beginning of each chapter you will find more Good Advice for Studying that will help you study and take tests more effectively, and manage your anxiety. You may want to read ahead so that you can improve even more quickly. Good Luck!

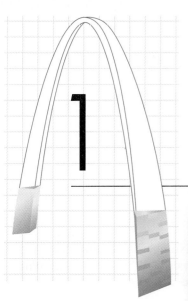

1 Whole Numbers

Terry Vine/Tony Stone Images ©

APPLICATION

Many people are concerned about the spread of the HIV virus around the world. The World Health Organization gathers statistics regularly, but it is difficult to establish the accuracy of the information. One significant problem in the gathering of statistics is that infected people can experience no symptoms and be unaware that they have the virus for many years. Testing is not universal, and the rate of false negatives (test results are negative when in fact the person is infected) is high. For these and other factors, the World Health Organization can only estimate the number of HIV infections worldwide.

The table below gives the World Health Organization's estimates of the HIV infections in adults in 1994.

Region	HIV Infections
Sub-Saharan Africa	10 million
South and Southeast Asia	3 million
Latin America/Caribbean	2 million
North America	1 million
Western Europe	500 thousand
North Africa/Middle East	100 thousand
Eastern Europe/Central Asia	50 thousand
East Asia/Pacific	50 thousand
Australia	25 thousand

What parts of the world are not included in these statistics? Give a possible explanation for this.

1

1.1 Whole Numbers: Writing, Rounding, and Inequalities

OBJECTIVES

1. Write word names from place value notation and place value notation from word names.
2. Write an inequality statement about two numbers.
3. Round off a given whole number.

VOCABULARY

The **digits** are 0, 1, 2, 3, 4, 5, 6, 7, 8, and 9.

The **natural numbers (counting numbers)** are 1, 2, 3, 4, 5, and so on.

The **whole numbers** are 0, 1, 2, 3, 4, 5, and so on. Numbers larger than 9 are written in **place value notation** by writing the digits in positions having standard **place value.** Words that name the numbers are called **word names.**

The symbols **less than,** "$<$," and **greater than,** "$>$," are used to compare two whole numbers that are not equal to each other. So, $2 < 9$, and $13 > 6$.

To **round** a whole number means to give an approximate value. The approximate value is found by rounding to an indicated place. The symbol "$\approx$" means "approximately equal to."

HOW AND WHY
Objective 1

Write word names from place value notation and place value notation from word names.

In our written whole number system (called the Hindu-Arabic system), digits and commas are the only symbols used. This system is a positional base ten (decimal) system. The location of the digit determines its value, from right to left. The first three place value names are one, ten, and hundred.

For the number 782,

2 is in the one's place, so it contributes 2 ones or 2 to the value of the number,

8 is in the ten's place, so it contributes 8 tens or 80 to the value of the number,

7 is in the hundred's place, so it contributes 7 hundreds or 700 to the value of the number.

So, 782 is 7 hundreds + 8 tens + 2 ones   or $700 + 80 + 2$. These are called expanded forms of the number.

For numbers larger than 999, we use commas to separate groups of three digits. The first four groups are unit, thousand, million, and billion (Figure 1.1). The group on the far left may have one, two, or three digits. All other groups must have three digits. Within each group the names are the same (hundred, ten, and one).

hun-dred	ten	one	hun-dred	ten	one	hun-dred	ten	one	hun-dred	ten	one
billion			million			thousand			(unit)		

Figure 1.1

For 74,896,314,555 the group names are

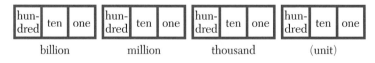

74	896	314	555
billion	million	thousand	unit

The number is read "74 billion, 896 million, 314 thousand, 555." The group name units is not read.

The word name of a three digit or smaller number is written by first writing the name of the digit in the hundreds place followed by hundred and then the name of the two digit number in the tens and ones places. The name of the two digit number in the tens and ones place is the name used to count from one to ninety-nine. So,

123 is written one hundred twenty-three,

583 is written five hundred eighty-three,

65 is written sixty-five, and

8 is written eight.

The word name of a number larger than 999 is written by writing the word name for each set of three digits followed by the group name. For example,

TABLE 1.1

Place value notation	415,	609
Word name for each set of digits	four hundred fifteen	six hundred nine
Group name	thousand	(unit)

The word name is written: four hundred fifteen thousand, six hundred nine.

To change from place value notation to the word name

1. From left to right, write the word name for each set of three digits followed by the group name (except units).
2. Insert a comma after each group name.

CAUTION **The word "and" is not used to write names of whole numbers. So write: three hundred ten, NOT three hundred and ten, also one thousand, two hundred twenty-three, NOT one thousand and two hundred twenty-three.**

To write a number in place value notation from the word name, we reverse the above process. First identify the group names and then write each group name in place value notation. Remember to write a 0 for each missing place value. Consider

Two billion, three hundred forty-five million, six thousand, one hundred thirteen. Identify the group names. (*Hint:* Look for the commas.)

Two *billion*, three hundred forty-five *million*, six *thousand*, one hundred thirteen

Write the place value notation for each group.

2 billion, 345 million, 6 thousand, 113

Drop the group value names, and make all groups (except possibly the first one) three digits. Keep all commas.

2,345,006,113   **Zeros must be inserted to show that there are no hundreds or tens in the thousands group.**

To write place value notation from a word name

1. Identify the group names.
2. Write the three digit number before each group name followed by a comma.

(The first group, on left, may have fewer than three digits.)

Numbers like 13,000,000,000, with all zeros following a single group of digits, is often written in a combination of place value notation and word name. The first set of digits on the left is written in place value notation followed by the group name. So,

13,000,000,000 is written 13 billion

| **Examples A–B** | **Warm Ups A–B** |

Direction: Write the word name.

Strategy: Write the word name of each set of three digits, from left to right, followed by the group name.

A. 17,698,453

17,	698,	453
seventeen	six hundred ninety-eight	four hundred fifty-three
million	thousand	(unit)

C A U T I O N **Do not write the word "and" between the names of the groups or in the word name of the group.**

Word name: seventeen million, six hundred ninety-eight thousand, four hundred fifty-three

A. 6,455,091

B. Write the word name for 3,189,025. Write the solution without the help of a chart.

Three million, one hundred eighty-nine thousand, twenty-five

B. Write the word name for 7,597,234.

| **Examples C–E** | **Warm Ups C–E** |

Direction: Write in place value notation.

Strategy: Write the three digit number for each group followed by a comma.

C. Two million, thirty-seven thousand, five hundred sixty-four

2,037,564

C. Thirty-two million, twenty-seven thousand, nine hundred ten

---

Answers to Warm Ups A. Six million, four hundred fifty-five thousand, ninety-one B. Seven million, five hundred ninety-seven thousand, two hundred thirty-four C. 32,027,910

D. Write 346 million in place value notation. Replace the word million with six zeros.

346,000,000

D. Write 5 thousand in place value notation.

E. The purchasing agent for the Russet Corporation received a telephone bid of twenty-three thousand eighty-one dollars as the price of a new printing press. What is the place value form of the bid that she will include in her report to her superior?

twenty-three thousand, eighty-one

23,   081

$23,081

E. The purchasing agent for the Russet Corporation also received a bid of seventeen thousand, two hundred eighteen dollars for a supply of paper. What is the place value form of the bid that she will include in her report to her superior?

HOW AND WHY
Objective 2

Write an inequality statement about two numbers.

If two whole numbers are not equal, then the first is either *less than* or *greater than* the second. This can be seen on a number line (or ruler):

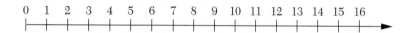

Given two numbers on a number line or ruler, the number on the right is the larger. For example,

$8 > 5$ **8 is to the right of 5, so 8 is greater than 5.**

$7 > 1$ **7 is to the right of 1, so 7 is greater than 1.**

$14 > 12$ **14 is to the right of 12, so 14 is greater than 12.**

$15 > 0$ **15 is to the right of 0, so 15 is greater than 0.**

Given two numbers on a number line or ruler, the number on the left is the smaller. For example,

$2 < 6$ **2 is to the left of 6, so 2 is less than 6.**

$7 < 10$ **7 is to the left of 10, so 7 is less than 10.**

$5 < 9$ **5 is to the left of 9, so 5 is less than 9.**

$11 < 13$ **11 is to the left of 13, so 11 is less than 13.**

For larger numbers imagine a longer number line. Notice how the points in the symbols " $<$ " and " $>$ " point to the smaller of the two numbers. For example,

$109 < 405$

$34 > 25$

$1009 > 1007$

 To write an inequality statement about two numbers

1. Insert $<$ between the numbers if the number on the left is smaller.

2. Insert $>$ between the numbers if the number on the left is larger.

Examples F–G   **Warm Ups F–G**

Directions: Insert < or > to make a true statement.

Strategy: Imagine a number line. The smaller number is on the left. Insert the symbol that points to the smaller number.

F. 67 97	F. 118 134
$\qquad$ 67 < 97	

G. 1314 1299	G. 3678 3499
$\qquad$ 1314 > 1299	

HOW AND WHY

Objective 3   **Round off a given whole number.**

Many numbers that we see in daily life are approximations. These are used to indicate the approximate value where it is felt that the exact value does not lend to the discussion. So, attendance at a political rally may be stated at 15,000 where it was actually 14,783. The amount of a deficit in the budget may be stated as $2,000,000 instead of $2,067,973. In this chapter, we use these approximations to estimate the outcome of operations with whole numbers. The symbol " $\approx$ " read "approximately equal to" is used to show the approximation. So, $2,067,973 \approx $2,000,000.

The number line can be used to see how whole numbers are rounded. Suppose we wish to round 27 to the nearest ten.

The arrow under the 27 is closer to 30 than to 20. We say "to the nearest ten, 27 rounds to 30."

We use the same idea to round any number, although we usually only make a mental image of the number line. The key question is, "Is this number closer to the smaller rounded number or closer to the larger one?" Practically, we only need to determine if the number is more or less than half the distance between the rounded numbers.

To round 34,568 to the nearest thousand without a number line, draw an arrow under the digit in the thousands place.

34,568
$\uparrow$

Since 34,568 is between 34,000 and 35,000, we must decide which number it is closer to. Since 34,500 is halfway between 34,000 and 35,000 and since 34,568 > 34,500, we conclude that 34,568 is more than halfway to 35,000. Whenever the number is halfway or closer to the larger number, we choose the larger number.

34,568 ≈ 35,000   **34,568 is closer to 35,000 than to 34,000.**

To round a number to a given place value

1. Draw an arrow under the given place value.
2. If the digit to the right of the arrow is 5, 6, 7, 8, or 9, add one to the digit above the arrow. (Round to the larger number.)
3. If the digit to the right of the arrow is 0, 1, 2, 3, or 4, do not change the digit above the arrow. (Round to the smaller number.)
4. Replace all the digits to right of the arrow with zeros.

Examples H–I   **Warm Ups H–I**

Directions: Round to the indicated place value.

Strategy: Choose the larger number if the digit to the right of the round off place is 5 or more (the number is at least halfway to the larger rounded value), otherwise choose the smaller number.

H. 127,456; ten thousand

    127,456   **Draw an arrow under the ten thousands place.**
    ↑

    130,000   **The digit to the right of the arrow is 7. Since 127,456 > 125,000, choose the larger number.**

H. 99,858; ten

I. Round to the indicated place value in the table.

Number	Ten	Hundred	Thousand
365,733	365,730	365,700	366,000
98,327	98,330	98,300	98,000

I. Round to the indicated place value in the table.

Number	Ten	Hundred	Thousand
491,356			
480,639			

Answers to Warm Ups   H. 99,860   I.

Number	Ten	Hundred	Thousand
491,356	491,360	491,400	491,000
480,639	480,640	480,600	481,000

Exercises 1.1

OBJECTIVE 1: *Write word names from place value notation and place value notation from word names.*

A.

Write the word names of each of these numbers.

1. 542

2. 391

3. 890

4. 500

5. 7015

6. 40,051

Write the place value notation.

7. Fifty-seven

8. Seventy-eight

9. Seven thousand, five hundred

10. Seven thousand, five

11. 10 million

12. 123 thousand

B.

Write the word name for each of these numbers.

13. 25,310

14. 25,031

15. 205,310

16. 250,031

17. 45,000,000

18. 750,000

Write the place value notation.

19. Two hundred forty-three thousand, seven hundred

20. Two hundred forty-three thousand, seven

21. Twenty-three thousand, four hundred seventy

22. Twenty-three thousand, four hundred seventy-seven

23. Seventeen million

24. Seven hundred thousand, seven

OBJECTIVE 2: *Write an inequality statement about two numbers.*

A.

Insert < or > between the numbers to make a true statement.

25. 18   21   **26.** 33   29

27. 51   44   **28.** 62   71

B.

29. 145   152   **30.** 212   208

31. 348   351   **32.** 275   269

OBJECTIVE 3: *Round off a given whole number.*

A.

Round to the indicated place value.

33. 694 (ten)   **34.** 786 (ten)   **35.** 1658 (hundred)

36. 3450 (hundred)

B.

Round the numbers in the table to the indicated place value.

	Number	Ten	Hundred	Thousand	Ten Thousand
37.	102,385				
38.	689,377				
39.	7,250,978				
40.	4,309,498				

C.

Write in place value notation.

41. Five hundred sixty million, three hundred fifty-three thousand, seven hundred thirty

42. Three hundred fifty million, six hundred sixty-three

For Exercises 43–44, the figure shows the population of the Portland, Oregon, area as compared to the population of the entire state.

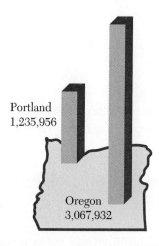

Portland
1,235,956

Oregon
3,067,932

43. Write the word name for the population in the Portland area.

44. Write the word name for the population of Oregon.

Insert < or > between the numbers to make a true statement.

45. 5634   5637

46. 10,276   10,199

47. What is the smallest 4-digit number?

48. What is the largest 5-digit number?

Round to the indicated place value.

49. 43,784,675 (ten thousand)

50. 47,078,665 (hundred thousand)

51. Round 42,749 to the nearest hundred. Round 42,749 to the nearest ten and then round your result to the nearest hundred.

     Why did you get a different result the second time?

     Which method is correct?

52. Sally bought a jet ski boat for $2075. She wrote a check to pay for it. What word name does she write on the check?

53. Jim bought a used Infinity for $18,465 and wrote a check to pay for it. What word name does he write on the check?

54. The U.S. Fish and Wildlife Department estimates the salmon runs could be as high as 154,320 fish by the year 2002 on the Rogue River if new management practices are used in logging along the river. Write the word name for the number of fish.

55. Ducks Unlimited estimated that 389,500 ducks spent the winter at the Klamath Falls refuge. Write the word name for the number of ducks.

56. The world population during 1990 exceeded 5 billion 3 hundred thousand. Write the place value notation for this world population benchmark.

57. The purchasing agent for the Upright Corporation received a telephone bid of thirty-six thousand, four hundred seven dollars as the price for a new printing press. What is the place value notation for the bid?

58. An office building in downtown Birmingham sold for $2,458,950. Give the purchase price of the building to the nearest hundred thousand dollars.

59. Ten thousand shares of Intel Corp. sold for $685,560. What is the value of the sale, to the nearest thousand dollars?

For Exercises 60–62, the figure shows the quantity of the top three industrial releases of toxic materials a few years ago. These values have been decreasing in recent years.

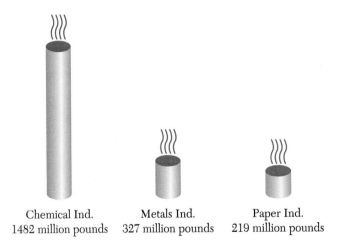

Chemical Ind.   Metals Ind.   Paper Ind.
1482 million pounds   327 million pounds   219 million pounds

60. Write the number of pounds of toxic material released by chemical industries in place value notation.

61. Write the number of pounds of toxic material released by metals industries in place value notation.

62. Write the number of pounds of toxic material released by paper industries in place value notation.

For Exercises 63–66, the per capita personal income in the New England states is given in the following table.

Massachusetts	$26,994
New Hampshire	$25,151
Maine	$20,527
Connecticut	$30,303
Rhode Island	$23,310
Vermont	$20,927

63. Write the word name for the per capita personal income in Maine.

64. Round the per capita personal income in Massachusetts to the nearest thousand.

65. Which state has the smallest per capita personal income?

66. Of Massachusetts and New Hampshire, which has the larger per capita personal income?

67. The distance from the earth to the sun was measured and determined to be 92,875,328 miles. To the nearest million miles, what is the distance?

 For Exercises 68–71, refer to Chapter 1 application, see page 1.

68. Write the place value notation for the number of HIV infections in Sub-Saharan Africa.

69. Write the place value notation for the number of HIV infections in Western Europe.

70. The numbers in the table seem to be rounded. To what place value is the number of HIV infections in Latin America/Caribbean rounded? Were all of the other entries in the table rounded to the same place value? Explain your answer.

71. Give a justification for the rounding in the table.

72. The state motor vehicle department estimated the number of licensed automobiles in the state to be 2,376,000, to the nearest thousand. A check of the records indicated that there were actually 2,376,499. Was their estimate correct?

73. The total land area of the earth is approximately 52,425,000 square miles. What is the land area to the nearest million square miles?

STATE YOUR UNDERSTANDING

74. Explain why base ten is a good name for our number system.

75. Explain why the digit 7 has two values associated with it in 175,892 and tell what the values are.

76. What is rounding? Explain how to round 87,452 to the nearest thousand and to the nearest hundred.

CHALLENGE

77. What is the place value for the digit 4 in 3,456,709,230,000?

78. Write the word name for 3,456,709,230,000.

79. Arrange the following numbers from smallest to largest: 1234, 1342, 1432, 1145, 1243, 1324, and 1229

80. What is the largest value of X that makes 2X56 > 2649 false?

81. Round 8275 to the nearest ten thousand.

82. Round 37,254 to the nearest hundred thousand.

GROUP ACTIVITY

83. Two other methods of rounding are called the "odd-even method" and "truncating." Find these methods and be prepared to explain them in class. (*Hint:* Try the library or talk to science and business instructors.)

1.2   Adding and Subtracting Whole Numbers

OBJECTIVES
1. Find the sum of two or more whole numbers.
2. Estimate the sum of a group of whole numbers.
3. Find the difference of two whole numbers.
4. Estimate the difference of two whole numbers.

VOCABULARY
Addends are the numbers that are added. In $9 + 20 + 3 = 32$, the addends are 9, 20, and 3.

The **sum** is the answer to an addition exercise. In $9 + 20 + 3 = 32$, the sum is 32.

The **commutative property of addition** permits us to change the order of the addends in addition exercises while the sum remains unchanged. For example, $63 + 75 = 75 + 63$.

Property of Addition

> **Commutative Property of Addition**
>
> $a + b = b + a$

The **associative property of addition** permits us to change the grouping of addends in addition exercises while the sum remains unchanged. For example, $(35 + 8) + 19 = 35 + (8 + 19)$.

Property of Addition

> **Associative Property of Addition**
>
> $(a + b) + c = a + (b + c)$

The **difference** is the answer in a subtraction exercise. So, in $62 - 34 = 28$, 28 is the difference.

Find the sum of two or more whole numbers.

When Pedro graduated from high school he received cash gifts of $45, $30, and $25. The total number of dollars received is found by adding the individual gifts. The total number of dollars he received is $100. In this section we will study how to add whole numbers.

The commutative and associative properties of addition are of great importance. They are the basis for the way we add. When two numbers are added, the commutative property allows addition in either order. For example,

$7 + 5 = 5 + 7$ **Since $7 + 5 = 12$ and $5 + 7 = 12$.**
$6 + 8 = 8 + 6$ **Since $6 + 8 = 14$ and $8 + 6 = 14$.**

The associative property allows for the regrouping of addends when there is more than one addition to be performed. For example,

$4 + (7 + 1) = (4 + 7) + 1$

We understand the parentheses to mean "do the operation inside first." Now we can verify that $4 + (7 + 1) = (4 + 7) + 1$.

$4 + (7 + 1) = 4 + 8$ **Add $7 + 1$ first.**
$\qquad\qquad\ = 12$

$$(4 + 7) + 1 = 11 + 1 \qquad \text{Add } 4 + 7 \text{ first.}$$
$$= 12$$

Using these properties the sum of two or more addends can be simplified to the addition of the ones, tens, hundreds, and so on. For example,

$352 + 45 = (300 + 50 + 2) + (40 + 5)$	**Write in expanded form.**
$\qquad = 300 + (50 + 40) + (2 + 5)$	**Group the hundreds, tens, and units using the associative and commutative properties.**
$\qquad = 300 + 90 + 7$	**Add.**
$\qquad = 397$	**Write in place value notation.**

The properties also allow us to use the shortcut of writing the numbers in columns so the like place values are aligned.

$$\begin{array}{r} 352 \\ + \ 45 \\ \hline 397 \end{array}$$ **Writing the numbers in columns provides a natural grouping of the ones and tens column without the tedious manipulations shown above.**

In algebra, it is very important to understand these properties when letters, called *variables*, are used for numbers. We must recognize when we can change the order and grouping of expressions involving addition. See Section 3.1.

To add $567 + 204 + 198$, write the numbers in a column.

$$\begin{array}{r} 567 \\ 204 \\ + \ 198 \end{array}$$ **Written this way the digit in the ones, tens, and hundreds places are aligned.**

$$\begin{array}{r} 1 \\ 567 \\ 204 \\ + \ 198 \\ \hline 9 \end{array}$$ **Add the digits in the ones column: $7 + 4 + 8 = 19$. Write "9" and carry the "1" (1 ten) to the tens column.**

$$\begin{array}{r} 1 \ 1 \\ 567 \\ 204 \\ + \ 198 \\ \hline 69 \end{array}$$ **Add the digits in the tens column: $1 + 6 + 0 + 9 = 16$. Write "6" and carry the "1" (10 tens = 1 hundred) to the hundreds column.**

$$\begin{array}{r} 567 \\ 204 \\ + \ 198 \\ \hline 969 \end{array}$$ **Add the digits in the hundreds column: $1 + 5 + 2 + 1 = 9$.**

▶ *To add whole numbers*

1. Write the numbers in a column so that the place values are aligned.
2. Add each column, starting with the ones (or units) column.
3. If the sum of any column is greater than nine, write the ones digit and "carry" the tens digit to the next column.

Examples A–D	**Warm Ups A–D**

Directions: Add.

Strategy: Write the numbers in a column. Add the digits in the columns starting on the right. If the sum is greater than 9, "carry" the tens digit to the next column.

A. 788 + 643	A. 864 + 657

$$\begin{array}{r} ^{11} \\ 788 \\ +\ 643 \\ \hline 1431 \end{array}$$

B. Find the sum: 1773 + 5486 + 3497	B. Find the sum: 3467 + 8912 + 4569

$$\begin{array}{r} ^{1\ 2\ 1} \\ 1773 \\ 5486 \\ +\ 3497 \\ \hline 10756 \end{array}$$

C. Add 59, 423, 5, and 1607; round the sum to the nearest ten.	C. Add 87, 6598, 47, and 3; round the sum to the nearest ten.

$$\begin{array}{r} ^{1\quad 2} \\ 59 \\ 423 \\ 5 \\ +1607 \\ \hline 2094 \end{array}$$ **When writing in a column, make sure the place values are aligned properly.**

$2094 \approx 2090$ **Round to the nearest ten.**

Calculator Example

D. Add: 4509 + 678 + 2345 + 1923 + 6789	D. Add: 5482 + 9742 + 847 + 1324 + 7321

Calculators have an internal mechanism that adds numbers just as we have been doing by hand. No special preparation is required on the part of the operator. Simply enter the exercise as it is written horizontally and the calculator will do the rest.

The sum is 16,244.

HOW AND WHY
Objective 2   **Estimate the sum of a group of whole numbers.**

The sum of a group of numbers can be estimated by rounding each member of the group to the largest place value in the group and then adding the rounded values. For instance,

4392	4000	**The largest place value is thousand, so round each**
6110	6000	**number to the nearest thousand.**
5785	6000	
7515	8000	
+ 945	+ 1000	
	25000	

Answers to Warm Ups   A. 1521   B. 16,948   C. 6740   D. 24,716

The estimate of the sum is 25,000. One use of the estimate is to see if the sum of the group of numbers is correct. If the calculated sum is not close to the estimated sum, 25,000, you should check the addition by re-adding. In this case the calculated sum, 24,747, is close to the estimate.

Example E	**Warm Up E**

Direction: Estimate the sum. Then add and compare.

Strategy: Round each number to the largest place value of all the addends. Then add and compare.

E.   356 + 7895 + 679 + 4567 + 3188 + 12 | E.   543 + 12 + 792 + 395 + 3 + 87

```
        0      Round each number to the nearest thousand.
     8000      With practice, this can be done mentally for a
     1000      quick check.
     5000
     3000
+       0
    17000
```

Now add and compare.

```
      356
     7895
      679
     4567
     3188
+      12
    16697
```

The sum, 16,697, is close to the estimation, 17,000.

HOW AND WHY
Objective 3   **Find the difference of two whole numbers.**

Felicia went shopping with \$95. She made purchases totaling \$53. How much money does she have left? Finding the difference in two quantities is called subtraction. When we subtract \$53 from \$95 we get \$42. In this section we will study the process of subtraction.

The subtraction, $16 - 7$, is usually thought of one of two ways.

"What is 7 from 16?"	This is called the "take away" version: "Take seven away from sixteen."
"16 is how much more than 7?"	This is called the "how much more?" version. "How much more is 16 than 7?" or "What do you add to 7 to get 16?"

The second question can be written $7 + ? = 16$. Since $7 + 9 = 16$, then $16 - 7 = 9$.

Answer to Warm Up   E. 1800; 1832

Observe that

$$16 - 7 + 7 = 16$$

that is,

$$16 - 7 + 7 = 9 + 7$$
$$= 16$$

and also that

$$16 + 7 - 7 = 16$$

that is,

$$16 + 7 - 7 = 23 - 7$$
$$= 16$$

We see that subtraction "undoes" addition and addition "undoes" subtraction. In algebra we say that addition and subtraction are *inverse operations*.

This leads to a method for checking subtraction. If we add the number being subtracted to the difference we will undo the subtraction. To check that $16 - 7 = 9$, we add 9 and 7. Since $9 + 7 = 16$, this verifies that 9 is the correct difference.

To find the difference $965 - 534$, write in column form and subtract in each column.

$$965 = 9 \text{ hundreds} + 6 \text{ tens} + 5 \text{ ones}$$
$$\underline{- \ 534 = 5 \text{ hundreds} + 3 \text{ tens} + 4 \text{ ones}}$$
$$431 = 4 \text{ hundreds} + 3 \text{ tens} + 1 \text{ one}$$

Check by adding.

$$\begin{array}{r} 534 \\ + \ \underline{431} \\ 965 \end{array}$$

Now consider the difference $784 - 369$. Write the numbers in column form.

$$784 = 7 \text{ hundreds} + 8 \text{ tens} + 4 \text{ ones}$$
$$\underline{- \ 369 = 3 \text{ hundreds} + 6 \text{ tens} + 9 \text{ ones}}$$

Here we cannot subtract 9 ones from 4 ones, so we rename by "borrowing" one of the tens from the 8 tens (1 ten = 10 ones) and adding the 10 ones to the 4 ones.

$$\overset{7 \ 14}{7 \cancel{8} \cancel{4}} = 7 \text{ hundreds} + 7 \text{ tens} + 14 \text{ ones}$$
$$\underline{- \ 369 = 3 \text{ hundreds} + 6 \text{ tens} + \ \ 9 \text{ ones}}$$
$$415 = 4 \text{ hundreds} + 1 \text{ ten} + \ \ \ 5 \text{ ones}$$

Check by adding.

$$\begin{array}{r} {\scriptstyle 1} \\ 369 \\ + \ \underline{415} \\ 784 \end{array}$$

We generally don't bother to write the expanded form when we subtract. We use the shortcuts in the examples.

▶ ***To subtract whole numbers***

  1. Write the numbers in a column so that the place values are aligned.
  2. Subtract in each column, starting with the ones (or units) column.
  3. When the numbers in a column cannot be subtracted, rename by borrowing.

C A U T I O N **The commutative and associative properties do not apply to subtraction.**

The following examples illustrate that the commutative and associative properties do not hold for subtraction.

Commutative

Is $8 - 3 = 3 - 8$?   **Change the order of (commute) the numbers.**

$5 = ?$ **On the left, $8 - 3 = 5$, but we have no whole number answer for $3 - 8$. In algebra, there is a new set of numbers (integers) that provide an answer for $3 - 8$ but that answer is not 5.**

Associative

Is $(9 - 5) - 2 = 9 - (5 - 2)$?   **Change the grouping (association) of the numbers.**

Does $4 - 2 = 9 - 3$?   **Subtract within the parentheses.**

$2 \neq 6$ **No, the answers are not the same. The symbol "$\neq$" is read "is not equal to."**

Since subtraction is not commutative, we must be careful not to change the order of the numbers. To emphasize the point, in this text, when asked to find the difference between two numbers we will always subtract the second number from the first number. So, the difference between 372 and 291 is 81.

Examples F–J   **Warm Ups F–J**

Directions: Subtract and check.

Strategy: Write the numbers in columns. Subtract in each column. Rename by borrowing when the numbers in a column cannot be subtracted.

F. $752 - 295$ F. $567 - 398$

$$
\begin{array}{r}
{\scriptstyle 4\ 12} \\
7\,\cancel{5}\,2 \\
-\ 2\,9\,5 \\
\end{array}
$$
   In order to subtract in the ones column we borrow 1 ten (10 ones) from the tens column and rename the ones $(10 + 2 = 12)$.

$$
\begin{array}{r}
{\scriptstyle 6\ 14} \\
{\scriptstyle 4\ 12} \\
\cancel{7}\,\cancel{5}\,2 \\
-\ 2\,9\,5 \\
\hline
4\,5\,7 \\
\end{array}
$$
   Now in order to subtract in the tens column we must borrow 1 hundred (10 tens) from the hundreds column and rename the tens $(10 + 4 = 14)$.

Check: $295 + 457 = 752$.

G.   Subtract 3700 and 948.   G.   Subtract 4500 and 891.

$$
\begin{array}{r}
3700 \\
-\ \ 948 \\
\end{array}
$$
   We cannot subtract in the ones column, and since there are 0 tens, we cannot borrow from the tens column.

Answers to Warm Ups   F. 169

$$
\begin{array}{r}
{}^{6\ 10} \\
3\cancel{7}00 \\
-\ \ 948 \\
\end{array}
$$
We borrow 1 hundred (1 hundred = 10 tens) from the hundreds place.

$$
\begin{array}{r}
{}^{9\ 10} \\
{}^{6\ 1\cancel{0}} \\
3\cancel{7}\cancel{0}\cancel{0} \\
-\ \ 948 \\
\end{array}
$$
Now borrow 1 ten (1 ten = 10 ones). We can now subtract in the ones and tens columns, but not in the hundreds column.

$$
\begin{array}{r}
{}^{2\ 16\ 9\ 10} \\
{}^{6\ 1\cancel{0}} \\
\cancel{3}\cancel{7}\cancel{0}\cancel{0} \\
-\ \ 948 \\
\hline
2752 \\
\end{array}
$$
Now borrow 1 thousand (1 thousand = 10 hundreds) and we can subtract in every column.

Check: $948 + 2752 = 3700$.

Let's try Example G again using a technique called reverse adding. Just ask yourself "What do I have to add to 948 to get 3700?"

$$
\begin{array}{r}
3700 \\
-\ 948 \\
\hline
2 \\
\end{array}
$$
Begin with the ones column. 8 is larger than 0 so ask "What do I add to 8 to make 10?"

$$
\begin{array}{r}
3700 \\
-\ 948 \\
\hline
52 \\
\end{array}
$$
Since 8 + 2 = 10, we carry the one over to the 4 to make 5. Now ask "What do I add to 5 to make 10?"

$$
\begin{array}{r}
3700 \\
-\ 948 \\
\hline
752 \\
\end{array}
$$
Carry the one to the nine in the hundreds column. Now ask "What do I add to 10 to make 17?"

$$
\begin{array}{r}
3700 \\
-\ 948 \\
\hline
2752 \\
\end{array}
$$
Finally, ask "What do I add to the carried 1 to make 3?"

The advantage of this method is that since 1 is the largest amount carried most people can do this process mentally.

H. Find the difference between 7061 and 736 and round to the nearest hundred.

$$
\begin{array}{r}
{}^{6\ 10\ 5\ 11} \\
\cancel{7}\cancel{0}\cancel{6}\cancel{1} \\
-\ \ 736 \\
\hline
6325 \\
\end{array}
$$
The "borrowing" is all shown at once.

Check: $736 + 6325 = 7061$.

$6325 \approx 6300$ **Round to the nearest hundred.**

Calculator Example:

 I. Subtract 14,691 from 33,894.

Enter: $33{,}894 - 14{,}691$

The difference is 19,203.

H. Find the difference between 6051 and 827 and round to the nearest hundred.

I. Subtract 17,358 from 44,679.

Answers to Warm Ups   G. 3609   H. 5200   I. 27,321

J. Maxwell Auto is advertising a $742 rebate on all new cars priced above $12,000. What is the cost after rebate of a car originally priced at $13,763?

Strategy: Since the price of the car is over $12,000, we subtract the amount of the rebate to find the cost.

$$
\begin{array}{r}
13763 \\
- 742 \\
\hline
13021
\end{array}
$$

The car costs $13,021.

J. Maxwell Auto is also advertising a $1438 rebate on all new cars priced above $21,000. What is the cost after rebate of a car originally priced at $27,829?

HOW AND WHY
Objective 4

Estimate the difference of two whole numbers.

The difference of two whole numbers can be estimated by rounding each number to the largest place value in the two numbers and then subtracting these rounded numbers. For instance,

$$
\begin{array}{r}
6110 \\
- 4392 \\
\end{array}
\qquad
\begin{array}{r}
6000 \\
- 4000 \\
\hline
2000
\end{array}
$$

The largest place value is thousand. Round each number to the nearest thousand and subtract.

The estimate of the difference is 2000. One use of the estimate is to see if the difference is correct. If the calculated difference is not close to 2000, you should check the subtraction. In this case the difference is 1718, which is close to the estimate.

 To estimate the difference of two whole numbers

1. Round each number to the largest place value in either number.
2. Subtract the rounded numbers.

Examples K–L

Warm Ups K–L

Directions: Estimate the difference. Then subtract and compare.

Strategy: Round each number to the largest place value in either number and then subtract.

K. 63,590 and 14,350

$$
\begin{array}{r}
60,000 \\
- 10,000 \\
\hline
50,000
\end{array}
$$
Round to the nearest ten thousand.

Subtract.

$$
\begin{array}{r}
63,590 \\
- 14,350 \\
\hline
49,240
\end{array}
$$

The difference is estimated to be 50,000; it is 49,240.

K. 73,555 and 26,956

L.   Estimate the difference of 73,425 and 48,240. Then subtract and compare.

Round each number to the nearest ten thousand.

$$70,000$$
$$-50,000$$
$$\overline{20,000}$$

Now subtract.

$$73,425$$
$$-48,240$$
$$\overline{35,185}$$

The estimated sum and the calculated sum are not close. Check to see if the subtraction was done correctly.

$$73,425$$
$$-48,240$$
$$\overline{25,185}$$
In the first subtraction, the fact that 1 was borrowed from the 7 was ignored.

Now the estimated difference, 20,000, is close to the actual difference, 25,185.

L.   Estimate the difference of 22,450 and 9,874. Then subtract and compare.

Exercises 1.2

OBJECTIVE 1: *Find the sum of two or more whole numbers.*

A.

Add.

1. 65 + 32

2. 87 + 12

3. 748 + 231

4. 533 + 254

5. 756
     +236

6. 146
     +363

7. When you add 36 and 48, the sum of the ones column is 14. You must carry the _____ to the tens column.

8. In 563 + 275 the sum is X38. The value of X is _____.

B.

9. 586 + 3492 + 321

10. 783 + 5703 + 529

11. 30,000 + 30,803 + 8740

12. 20,000 + 40,083 + 5632

13. 8 + 90 + 403 + 6070

14. 6 + 80 + 608 + 4030

15. 2795 + 3643 + 7055 + 4004 (Round sum to the nearest hundred.)

16. 6832 + 8712 + 9032 + 5111 (Round sum to the nearest hundred.)

OBJECTIVE 2: *Estimate the sum of a group of whole numbers.*

A.

Estimate the sum.

17. 345 + 782

18. 295 + 812

19. 6783 + 3599

20. 6782 + 9100

21.
$$\begin{array}{r} 3411 \\ 2001 \\ + 4561 \\ \hline \end{array}$$

22.
$$\begin{array}{r} 4567 \\ 3611 \\ + 2399 \\ \hline \end{array}$$

B.

Estimate the sum and then add.

23.
$$\begin{array}{r} 3209 \\ 7095 \\ 4444 \\ 2004 \\ + 3166 \\ \hline \end{array}$$

24.
$$\begin{array}{r} 6073 \\ 3284 \\ 1212 \\ 3593 \\ + 5606 \\ \hline \end{array}$$

25.
$$\begin{array}{r} 45,902 \\ 33,333 \\ 57,700 \\ + 23,653 \\ \hline \end{array}$$

26.
$$\begin{array}{r} 11,923 \\ 30,871 \\ 21,211 \\ + 74,486 \\ \hline \end{array}$$

OBJECTIVE 3: *Find the difference of two whole numbers.*

A.

Subtract.

27.
$$\begin{array}{r} 4 \text{ hundreds} + 9 \text{ tens} + 3 \text{ ones} \\ - 2 \text{ hundreds} + 9 \text{ tens} + 2 \text{ ones} \\ \hline \end{array}$$

28.
$$\begin{array}{r} 6 \text{ hundreds} + 3 \text{ tens} + 5 \text{ ones} \\ - 4 \text{ hundreds} + 2 \text{ tens} + 4 \text{ ones} \\ \hline \end{array}$$

29. 608 − 82

30. 642 − 80

31. 689 − 238

32. 848 − 611

33. When subtracting 34 − 19, you must "borrow" 1 from the 3. The value of the "borrowed 1" is _____ ones.

34. To subtract 524 − 462, you must borrow from the _____ column to subtract in the _____ column.

B.

Subtract.

35. 745 − 392

36. 585 − 294

37. 741 − 583

38. $932 - 857$ **39.** $800 - 378$ **40.** $600 - 338$

41. $8743 - 4078$ (Round difference to the nearest hundred.)

42. $9045 - 5786$ (Round difference to the nearest hundred.)

43.
$$\begin{array}{r} 5687 \\ -\ 3499 \\ \hline \end{array}$$

44.
$$\begin{array}{r} 9834 \\ -\ 6945 \\ \hline \end{array}$$

OBJECTIVE 4: *Estimate the difference of two whole numbers.*

A.

Estimate the difference.

45. $673 - 423$ **46.** $854 - 392$

47. $4950 - 2781$ **48.** $5693 - 2495$

49.
$$\begin{array}{r} 45,678 \\ -\ 34,722 \\ \hline \end{array}$$

50.
$$\begin{array}{r} 67,235 \\ -\ 58,991 \\ \hline \end{array}$$

B.

Estimate the difference and then subtract.

51.
$$\begin{array}{r} 875 \\ -\ 406 \\ \hline \end{array}$$

52.
$$\begin{array}{r} 455 \\ -\ 207 \\ \hline \end{array}$$

53.
$$\begin{array}{r} 6580 \\ -\ 3217 \\ \hline \end{array}$$

54.
$$\begin{array}{r} 8732 \\ -\ 5569 \\ \hline \end{array}$$

C.

A survey of car sales in Wisconsin shows the following distribution of sales among these dealers.

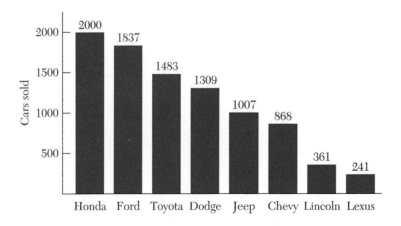

55. What is the total number of Fords, Toyotas, and Lexuses sold?

56. What is the total number of Chevys, Lincolns, Dodges, and Hondas sold?

57. How many more Hondas are sold than Fords?

58. How many more Toyotas are sold than Jeeps?

59. What is the total number sold of the three best selling cars?

60. What is the total number sold of the three worst selling cars?

61. How many fewer Lexuses are sold than Jeeps?

62. How many fewer Dodges are sold than Fords?

63. Estimate the total number of cars sold and then find the exact total.

64. What is the difference in cars sold between the best selling car and the least selling car?

65. Find the sum of 24, 9382, 5093, and 27,853. (Round the sum to the nearest hundred.)

66. Find the sum of 39, 2758, 6838, and 43,265. (Round the sum to the nearest hundred.)

67. Find the difference of 50,000 and 29,315.

68. Find the difference of 80,000 and 20,274.

69. Estimate the sum of 23,706, 34, 7561, 9346, and 236, then find the sum.

70. Estimate the sum of 96, 783, 3678, 12,555, and 8999, then find the sum.

71. Estimate the difference of 203,855 and 195,622, then find the difference.

72. Estimate the difference of 423,876 and 298,788, then find the difference.

Exercises 73–75 all refer to the Chapter 1 application problem. See page 1.

73. What is the total estimated number of HIV infections for Asia?

74. How many more estimated cases of HIV infections are there in Western Europe than in Eastern Europe/Central Asia?

75. What is the total estimated number of HIV infections in Africa?

76. The biologist at the Bonneville fish ladder counted the following number of coho salmon during a 1-week period: Monday, 895; Tuesday, 675; Wednesday, 124; Thursday, 1056; Friday, 308; Saturday, 312; and Sunday, 219. How many salmon went through the ladder that week? How many more salmon went through the ladder on Tuesday than on Saturday?

For Exercises 77–79. A new car dealership has the following sales in a given week:

Monday	$36,750
Tuesday	$46,780
Wednesday	$21,995
Thursday	$35,900
Friday	$67,950
Saturday	$212,752
Sunday	$345,720

77. What is the gross sales for the week?

78. How much less was the gross sales on Wednesday than on Friday?

79. What is the difference between the largest and the smallest daily gross sales?

80. The Stark Raving Theater had 74 patrons on Thursday night, 108 on Friday night, 114 on Saturday night, and 88 on Sunday night to see their new production. How many patrons attended the new production?

For Exercises 81–83. According to the FBI *Uniform Crime Reports,* the estimated arrests in 1993 in the United States were as follows:

Murder/manslaughter	18,856
Forcible rape	29,432
Robbery	143,877
Aggravated assault	408,148
Burglary	308,849
Larceny-theft	1,131,768
Motor vehicle theft	156,711
Arson	14,504

81. Find the total number of estimated arrests for violent crimes (murder/manslaughter, forcible rape, robbery, and aggravated assault).

82. Find the total number of estimated arrests for property crimes (burglary, larceny-theft, motor vehicle theft, and arson).

83. How many more aggravated assaults than robberies were there?

84. The attendance at three consecutive Super Bowls is 78,943, 85,782, and 103,456. What is the total attendance at the three games? How many more fans were at the game with largest attendance as opposed to the one with the least attendance?

85. Lea sold five houses last month for the following prices: $123,675, $457,000, $89,050, $312,885, and $210,560. What is the total value of her sales? (Round to the nearest hundred dollars.)

86. A forester counted 23,679 trees that are ready for harvest on a certain acreage. If Forestry Service rules require that 8543 mature trees must be left on the acreage, how many trees can be harvested?

87. The new sewer line being installed in downtown Kearney will handle 345,760 gallons of refuse per minute. The old line handled 178,550 gallons per minute. How many more gallons per minute will the new line handle?

88. The Acme warehouse has 456,893 cases of peas at the beginning of the month. During the month they ship out the following number of cases: week one, 5680; week two, 10,560; week three, 23,900; and week four, 9456. How many cases, to the nearest hundred, are left in the warehouse?

89. Fong's Grocery owes a supplier $25,875. During the month Fong's makes payments of $460, $983, $565, and $10,730. How much does Fong's still owe, to the nearest hundred dollars?

90. In the spring of 1989, an oil tanker hit a reef and spilled 10,100,000 gallons of oil off the coast of Alaska. The tanker carried a total of 45,700,000 gallons of oil. The oil that did not spill was pumped into another tanker. How many gallons of oil were pumped into the second tanker? Round to the nearest million gallons.

91. In the southwestern United States are found the Grand Canyon, Zion, and Bryce Canyon parks. Geologic changes over a billion years have created these formations and canyons. The table shows the highest and lowest elevations in each of these parks. Find the change in elevation in each park. In which park is the change greatest and by how much?

	Highest Elevation	**Lowest Elevation**
Bryce Canyon	8500 ft	6600 ft
Grand Canyon	8300 ft	2500 ft
Zion	7500 ft	4000 ft

STATE YOUR UNDERSTANDING

92. Explain to an 8-year-old child why $15 - 9 = 6$.

93. Explain to an 8-year-old child why $8 + 7 = 15$.

94. Define and give an example of a sum.

95. Define and give an example of a difference.

CHALLENGE

96. Add the following numbers, round the sum to the nearest hundred, and write the word name for the rounded sum: one hundred sixty; eighty thousand, three hundred twelve; four hundred seventy-two thousand, nine hundred fifty-two; and one hundred forty-seven thousand, five hundred twenty-three.

97. How much greater is seven million, two hundred forty-seven thousand, one hundred ninety-five than two million, eight hundred four thousand, fifty-three? Write the word name for the difference.

98. Peter sells three Honda Civics for $14,385 each, four Accords for $17,435 each, and two Acuras for $26,548 each. What is the total dollar sales for the nine cars? How many more dollars were paid for the four Accords than the three Civics?

Complete by writing in the correct digit whenever you see a letter.

99.
```
   5A68
    241
 + 10A9
   ────
   B64C
```

100.
```
   A5,30B
   34,106
       68
 +    D42
   ──────
   50,018
```

101.
```
   4A6B
 − C251
   ────
   15D1
```

102.
```
   98A,1B6
 − A35,10C
   ───────
   7D6,E98
```

GROUP ACTIVITY

103. Add and round to the nearest hundred.

```
 14,657
  3,766
123,900
    569
 54,861
346,780
```

Now, round each addend to the nearest hundred and then add. Discuss why the answers are different. Be prepared to explain why this happens.

104. If Ramon delivers 112 loaves of bread to each store on his delivery route, how many stores are on the route if he delivers a total of 4368 loaves? (*Hint:* Subtract 112 loaves for each stop from the total number of loaves.) What operation does this perform? Make up three more examples and be prepared to demonstrate them in class.

1.3   Multiplying and Dividing Whole Numbers

OBJECTIVES
1. Multiply whole numbers.
2. Estimate the product of whole numbers.
3. Divide whole numbers.
4. Estimate the quotient of whole numbers.

VOCABULARY
There are several ways to show a multiplication exercise. Here are examples of most of them, using 18 and 29.

18×29 $18 \cdot 29$ $\begin{array}{r} 29 \\ \times\, \underline{18} \end{array}$ **Used in arithmetic.**

$(18)(29)$ $18(29)$ $(18)29$ **Used in both arithmetic and algebra.**

$18 * 29$ **Use with computers and some calculators.**

The **factors** of a multiplication exercise are the numbers being multiplied. In $5(8) = 40$, 5 and 8 are the factors.

The **product** is the answer to a multiplication exercise. In $5(8) = 40$, the product is 40.

The **commutative property of multiplication** permits us to change the order of factors in multiplication while the product remains unchanged. So, $12(9) = 9(12)$.

Property of Multiplication

> **Commutative Property of Multiplication**
>
> $a \cdot b = b \cdot a$

The **associative property of multiplication** permits us to change the grouping of factors in a multiplication exercise, and the result remains unchanged. So, $(3 \cdot 2)15 = 3(2 \cdot 15)$.

Property of Multiplication

> **Associative Property of Multiplication**
>
> $(a \cdot b) \cdot c = a \cdot (b \cdot c)$

The **distributive property** permits the writing of the product of a number times a sum as the sum of two products and vice versa. For example,

$$3(2 + 4) = 3(2) + 3(4)$$
$$3(6) = 6 + 12$$
$$18 = 18$$

and

$$5(8) + 5(9) = 5(8 + 9)$$
$$40 + 45 = 5(17)$$
$$85 = 85$$

Property of Multiplication and Addition	**Distributive Property**
	$a \cdot (b + c) = a \cdot b + a \cdot c$ and $a \cdot b + a \cdot c = a \cdot (b + c)$

There are a variety of ways to show a division exercise. Here are the most commonly used ones.

$51 \div 3$ $3\overline{)51}$ **Used in arithmetic.**

$\dfrac{51}{3}$ **Used in both arithmetic and algebra.**

$51/3$ **Used with computers and some calculators.**

The **dividend** is the number being divided, so in $36 \div 4 = 9$, the dividend is 36.

The **divisor** is the number that we are dividing by, so in $36 \div 4 = 9$, the divisor is 4.

The **quotient** is the answer to a division exercise, so in $36 \div 4 = 9$, the quotient is 9.

When a division exercise does not "come out even" as in $53 \div 3$, the quotient is not a whole number.

$$\begin{array}{r} 17 \\ 3\overline{)53} \\ -\underline{51} \\ 2 \end{array}$$

We call 17 the **partial quotient** and 2 the **remainder.** The quotient is written 17 R 2.

HOW AND WHY
Objective 1

Multiply whole numbers.

Multiplying whole numbers is a shortcut for repeated addition:

$8 + 8 + 8 + 8 + 8 + 8 = 48$ or $6 \cdot 8 = 48$

6 EIGHTS

As numbers get larger, the shortcut saves time. Imagine adding 146 eights.

$8 + 8 + 8 + 8 + 8 + \cdots + 8 = ?$

146 EIGHTS

The shortcut involves the distributive property. The distributive property allows us to multiply across addends. Here are two examples.

$$\begin{aligned} 5(2 + 3) &= 5(2) + 5(3) \\ &= 10 + 15 \\ &= 25 \end{aligned} \quad \text{or} \quad \begin{aligned} (2 + 3)\,5 &= 2(5) + 3(5) \\ &= 10 + 15 \\ &= 25 \end{aligned}$$

The multiplication can be on the right or the left side. In fact, we can distribute the multiplication across any number of addends.

$$\begin{aligned} 3(2 + 4 + 7 + 9) &= 3(2) + 3(4) + 3(7) + 3(9) \\ &= 6 + 12 + 21 + 27 \\ &= 66 \end{aligned}$$

Now let's see how we use the distributive property to find the sum of the 146 eights. We multiply 8 times 146 using the expanded form of 146.

$$8(146) = 8(100 + 40 + 6)$$ **Write 146 in expanded form.**
$$= 8(100) + 8(40) + 8(6)$$ **Use the distributive property.**
$$= 800 + 320 + 48$$ **Multiply.**
$$= 1168$$ **Add.**

By writing the multiplication in column form, the distribution takes place naturally.

$146 =$ $100 + 40 + 6$	$8(6) =$ **48 = 48 ones**
$\times \underline{8} = \times \underline{8}$	$8(40) =$ **320 = 32 tens**
$=$ $800 + 320 + 48 = 1168$	$8(100) =$ **800 = 8 hundreds**

The exercise can also be performed in column form without expanding the factors.

$$
\begin{array}{r}
146 \\
\times \quad 8 \\
\hline
48 \\
320 \\
800 \\
\hline
1168
\end{array}
\qquad
\begin{array}{l}
8(6) = 48 \\
8(40) = 320 \\
8(100) = 800
\end{array}
\qquad\qquad
\begin{array}{r}
^{3\,4} \\
146 \\
\times \quad 8 \\
\hline
1168
\end{array}
$$

The form on the right shows the usual shortcut. The addition is done mentally as you go. Study this example.

$$
\begin{array}{r}
629 \\
\times \quad 46 \\
\end{array}
$$

First multiply 629 by 6.

$$
\begin{array}{r}
^{1\,5} \\
629 \\
\times \quad 46 \\
\hline
3774
\end{array}
$$

6(9) = 54. Carry the 5 to the tens column.
6(2 tens) = 12 tens. Add the 5 tens that were carried: (12 + 5) tens = 17 tens. Carry the 1 to the hundreds column.
6(6 hundreds) = 36 hundreds. Add the 1 hundred that was carried: (36 + 1) hundreds = 37 hundreds.

Now multiply 629 by 40.

$$
\begin{array}{r}
^{1\,3} \\
^{1\,\cancel{5}} \\
629 \\
\times \quad 46 \\
\hline
3774 \\
25160
\end{array}
$$

40(9) = 360 or 36 tens. Carry the 3 to the hundreds column.
40(20) = 800 or 8 hundreds. Add the 3 hundreds that were carried. (8 + 3) hundreds = 11 hundreds. Carry the 1 to the thousands column. 40(600) = 24,000 or 24 thousands. Add the 1 thousand that was carried: (24 + 1) thousands = 25 thousands. Write the 5 in the thousands column and the 2 in the ten thousands column.

$$
\begin{array}{r}
^{1\,3} \\
^{1\,\cancel{5}} \\
629 \\
\times \quad 46 \\
\hline
3774 \\
25160 \\
\hline
28934
\end{array}
$$ **Add the products.**

Two important properties of both arithmetic and algebra are the *multiplication property of zero* and the *multiplication property of one.*

Property of Multiplication	**Multiplication Property of Zero**
	$a \cdot 0 = 0 \cdot a = 0$
	Any number times zero is zero.

Therefore, as a result of the above property, we know that

$$0(19) = 19(0) = 0 \qquad \text{and} \qquad 0(165) = 165(0) = 0$$

Property of Multiplication	**Multiplication Property of One**
	$a \cdot 1 = 1 \cdot a = a$
	Any number times one is that number.

As a result of the second property, we know that

$$1(27) = 27(1) = 27 \qquad \text{and} \qquad 1(354) = 354(1) = 354$$

The commutative and associative properties of multiplication are of great importance in arithmetic because they can be used to make multiplication easier. For instance, the multiplication problem $(6 \cdot 8)(5 \cdot 5)$ becomes easier to do mentally when it is rearranged.

$$
\begin{aligned}
(6 \cdot 8)(5 \cdot 5) &= (6 \cdot 5)(8 \cdot 5) \\
&= (30)(40) \\
&= 1200
\end{aligned}
$$

Use the commutative and associative properties to change the order and the grouping of the numbers.

We shall see that these properties are also of importance in algebra.

Examples A–F   **Warm Ups A–F**

Directions: Multiply.

Strategy: Write the factors in columns. Start multiplying with the ones digit. If the product is ten or more, carry the tens digit to the next column and add it to the product in that column. Repeat the process for every digit in the second factor. When the multiplication is complete, add to find the product.

A. $0(145)$	A. $(210)(0)$
$0(145) = 0$ **Multiplication property of zero.**	
B. Find the product: $6(4592)$	B. Find the product: $8(3456)$
Write the numbers in columns, lining up the places. It is easier to multiply by the smaller number.	

$$
\begin{array}{r}
^{351} \\
4592 \\
\times \quad 6 \\
\hline
27552
\end{array}
$$
Multiply 6 times each digit, carry when necessary, and add the number carried to the next product.

Answers to Warm Ups   A. 0   B. 27,648

C. Multiply: 38 · 74

$$
\begin{array}{r}
\overset{\overset{1}{\cancel{3}}}{74} \\
\times \ \ 38 \\
\hline
592 \\
2220 \\
\hline
2812
\end{array}
$$

When multiplying by the 3 in the tens place, write a 0 in the ones column to keep the places lined up.

C. Multiply: 59 · 84

D. Find the product of 513 and 205.

Strategy: When multiplying by zero in the tens place, rather than showing a row of zeros, just put a zero in the tens column. Then multiply by the 2 in the hundreds place.

$$
\begin{array}{r}
513 \\
\times \ \ \ 205 \\
\hline
2565 \\
102600 \\
\hline
105165
\end{array}
$$

D. Find the product of 326 and 707.

Calculator Example:

 E. 346(76)

Enter the multiplication.

The product is 26,296.

E. 398(148)

F. The Sweet & Sour Company ships 62 cartons of packaged candy to the Chewum Candy Store. Each carton contains 48 packages of candy. What is the total number of packages of candy shipped?

Strategy: To find the total number of packages, multiply the number of cartons by the number of packages.

$$
\begin{array}{r}
62 \\
\times \ \ 48 \\
\hline
496 \\
2480 \\
\hline
2976
\end{array}
$$

The company shipped 2976 packages of candy.

F. The Sweet & Sour Company ships 68 cartons of chewing gum to the Sweet Shop. If each carton contains 72 packages of gum, how many packages of gum are shipped to the Sweet Shop?

HOW AND WHY

Objective 2 **Estimate the product of whole numbers.**

The product of two numbers can be estimated by rounding each factor to the largest place value in each factor and then multiplying the rounded factors. For instance,

6110	6000	Round to the nearest thousand.
× 87	× 90	Round to the nearest ten.
	540000	Multiply.

The estimate of the product is 540,000. One use of the estimate is to see if the product is correct. If the calculated product is not close to 540,000, you should check the multiplication. In this case the actual product is 531,570, which is close to the estimate.

To estimate the product of two whole numbers

1. Round each number to its largest place value.
2. Multiply the rounded numbers.

Example G	**Warm Up G**

Directions: Estimate the product. Then find the product and compare.

Strategy: Round each number to its largest place value. Multiply the rounded numbers. Multiply the original numbers.

G. 687 and 347	G. 634 and 578

$$\begin{array}{r} 700 \\ \times\ \ 300 \\ \hline 210000 \end{array}$$ **Round to the nearest hundred.**
Round to the nearest hundred.

$$\begin{array}{r} 687 \\ \times\ \ 347 \\ \hline 4809 \\ 27480 \\ 206100 \\ \hline 238389 \end{array}$$

The product is 238,389 and is close to the estimated product of 210,000.

HOW AND WHY
Objective 3

Divide whole numbers.

The division exercise $128 \div 16$ can be interpreted in one of two ways.

"How many times can 16 be subtracted from 128?"   This is called the "repeated subtraction" version.

"What number times 16 is equal to 128?"   This is called the "missing factor" version.

All division problems can be done using repeated subtraction. The process can be shortened using the traditional method of guessing the number of groups (subtractions), starting with the largest place value on the left in the dividend and then working toward the right. For example to divide 128 by 16, first try a quotient of 5.

$$\begin{array}{r} 5 \\ 16\overline{)128} \\ -\ 80 \\ \hline 48 \end{array}$$ The remainder, 48, is larger than the divisor, 16. Try a larger quotient.

$$\begin{array}{r} 7 \\ 16\overline{)128} \\ -112 \\ \hline 16 \end{array}$$

The remainder, 16, is the same as the divisor. Since the remainder and the divisor are the same, increase the quotient by 1 to 8.

$$\begin{array}{r} 8 \\ 16\overline{)128} \\ -128 \\ \hline 0 \end{array}$$

The remainder, 0, is smaller than the divisor, 16. We are done.

We see that the missing factor in $(16)(?) = 128$ is 8. So, $16(8) = 128$ or $128 \div 16 = 8$.

Note that $\dfrac{128}{16} \times 16 = 128$ and that $\dfrac{128(16)}{16} = 128$. We see that division "undoes" multiplication and vice versa. In algebra we say that multiplication and division are *inverse operations*.

This leads to a method for checking division. If we multiply the divisor times the quotient we will "undo" the division and the product will be the dividend. To check $128 \div 16 = 8$, we multiply 16 and 8.

$(16)(8) = 128$

So, 8 is correct.

To divide 17,135 by 23 $(17,135 \div 23 = ?)$ is to ask, "What number times 23 equals 17,135?"

$$17,135 \div 23 = \underset{\text{Quotient}}{?} \qquad \text{asks} \qquad 23 \times \underset{\text{Missing Factor}}{?} = 17,135$$

We find the missing factor by finding the number of groups, starting on the left in the dividend and then working to the right.

$23\overline{)17135}$ Working from left to right, we note that 23 does not divide 1, and it does not divide 17. However, 23 does divide 171 seven times. Write the 7 above the 1 in the dividend.

$$\begin{array}{r} 745 \\ 23\overline{)17135} \\ -161 \\ \hline 103 \\ -\ 92 \\ \hline 115 \\ -115 \\ \hline 0 \end{array}$$

$7(23) = 161$. Subtract 161 from 171. Since the difference is less than the divisor, no adjustment is necessary. Bring down the next digit, which is 3. Next, 23 divides 103 four times. The 4 is placed above the 3 in the dividend. $4(23) = 92$. Subtract 92 from 103. Again, no adjustment is necessary since $11 < 23$. Bring down the next digit, which is 5.

Finally, 23 divides 115 five times. Place the 5 above the 5 in the dividend. $5(23) = 115$. Subtract 115 from 115, the remainder is zero. The division is complete.

Check:

$$\begin{array}{r} 745 \\ \times\ \ \ 23 \\ \hline 2235 \\ 14900 \\ \hline 17135 \end{array}$$

Check by multiplying the quotient by the divisor.

Not all division problems have a zero remainder (see Example I).

Recall that $45 \div 0 = ?$ asks what number times 0 is 45: $0 \times ? = 45$. By the zero-product property we know that $0 \times$ (any number) $= 0$, so it cannot equal 45. Therefore no replacement for the question mark makes $0 \times ? = 45$ true.

C A U T I O N **Division by zero is not defined. It is an operation that cannot be performed.**

Since division is not commutative, we must be careful not to change the order of the numbers. To emphasize this in the text, when asked to find the quotient of two numbers, we will always divide the first number by the second number. So the quotient of 32 and 4 is 8.

C A U T I O N **The commutative and associative properties do not apply to division. That is**

$$7 \div 3 \neq 3 \div 7 \quad \text{and} \quad (48 \div 12) \div 12 \neq 48 \div (12 \div 12)$$

Examples H–K	**Warm Ups H–K**

Directions: Divide and check.

Strategy: Divide from left to right.

H. $6\overline{)5412}$

Strategy: Write the partial quotients above the dividend with the place values aligned.

$$
\begin{array}{r}
902 \\
6\overline{)5412} \\
-54 \\
\hline
1 \\
-0 \\
\hline
12 \\
-12 \\
\hline
0
\end{array}
$$

$6(9) = 54$

6 does not divide 1, so multiply by 0

$6(2) = 12$

C A U T I O N When the numbers in the quotient are not carefully aligned, errors are likely to result. A common error in this problem is to get 92 for the answer instead of 902. The zero must be placed above the tens digit for the place values of 9 and 2 to be correct.

Check: $6(902) = 5412.$

The quotient is 902.

H. $8\overline{)4064}$

I.   Find the quotient: $\dfrac{103297}{365}$

Strategy: When a division is written as a fraction, the dividend is above the fraction bar and the divisor is below.

```
        283
365)103297      365 does not divide 1.
   −730         365 does not divide 10.
    3029        365 does not divide 103.
   −2920        365 divides 1032 two times.
    1097        365 divides 3029 eight times.
   −1095        365 divides 1097 three times.
       2        The remainder is 2.
```

Check: To check, multiply the divisor by the partial quotient and add the remainder.

$$283(365) + 2 = 103{,}295 + 2$$
$$= 103{,}297$$

The answer is 283 with remainder 2. This can also be written 283 R 2.

You may recall other ways to write a remainder using fractions or decimals. These are covered later.

Calculator Example:

J.   Divide 2756 by 143.

Enter the division: 2756 ÷ 143.

$$2756 \div 143 = 19.27272727$$

The quotient is not a whole number. This means that 19 is the partial quotient and there is a remainder. To find the remainder, multiply 19 times 143. Subtract the product from 2756. The result is the remainder.

$$2756 − 19(143) = 39$$

So, $2756 \div 143 = 19$ R 39.

K.   When planting Christmas trees, the Greenfir Tree Farm allows 64 square feet per tree. If there are 43,520 square feet in 1 acre, how many trees will they plant per acre?

Strategy: Since each tree is allowed 64 square feet, we divide the number of square feet in 1 acre by 64 to find out how many trees will be planted in 1 acre.

```
       680
64)43520
  −384
    512
   −512
     00
     −0
      0
```

There will be a total of 680 trees planted per acre.

I.   Find the quotient: $\dfrac{196008}{365}$

J.   Divide 4337 by 123.

K.   The Greenfir Tree Farm allows 256 square feet per large spruce tree. If there are 43,520 square feet in 1 acre, how many trees will they plant per acre?

Answers to Warm Ups   I. 537 R 3   J. 35 R 32   K. 170 trees

HOW AND WHY

Objective 4   **Estimate the quotient of whole numbers.**

The quotient of two numbers can be estimated by rounding the divisor and the dividend to their largest place value and then dividing the rounded numbers. For instance,

$27\overline{)6345}$ $30\overline{)6000}$ **Round each to its largest place value.**

$$
\begin{array}{r}
200 \\
30\overline{)6000} \\
-6000 \\
\hline
0
\end{array}
$$ **Divide.**

The estimate of the quotient is 200. One use of the estimate is to see if the quotient is correct. If the calculated quotient is not close to 200, you should check the division. In this case the actual quotient is 235, which is close to the estimate.

The estimated quotient will not always come out even. For instance,

$741\overline{)340678}$ $700\overline{)300000}$ **Round each number to its largest place value.**

Now, 700 divides 3000 four times. Since 4(700) = 2800, there is a remainder of 200. Now multiply 700 by 5, one more than the first quotient, 700(5) = 3500. The product 3500 is 500 more than 3000. Since 4(700) is closer to 3000, we choose 4 for the estimate.

$$
\begin{array}{r}
400 \\
700\overline{)300000}
\end{array}
$$

The estimated quotient is 400.

> **To estimate the quotient of two whole numbers**
>
> 1. Round each number to its largest place value.
> 2. Divide the rounded numbers.
> 3. If the first partial quotient has a remainder, multiply the divisor by the partial quotient and one more than the partial quotient. Choose for the estimate the number whose product is closer to the dividend. Write zeros to complete the estimated quotient.

Example L   **Warm Up L**

Directions: Estimate the quotient and divide.

Strategy: Round each number to its largest place value and then divide the rounded numbers. If the first partial quotient has a remainder, choose the digit that will give the closer value when multiplied by the divisor. Then divide.

L.   58,590 and 158 | L.   78,624 and 416

$158\overline{)58590}$ $200\overline{)60000}$

$$
\begin{array}{r}
300 \\
200\overline{)60000}
\end{array}
$$ **Divide.**

$$
\begin{array}{r}
370 \\
158\overline{)58590} \\
-\underline{474} \\
1119 \\
-\underline{1106} \\
130 \\
-\underline{0} \\
130
\end{array}
$$

The quotient is estimated at 300 and is 370 R 130.

Exercises 1.3

OBJECTIVE 1: *Multiply whole numbers.*

A.

1. $(32)(3)$ **2.** $(21)(4)$ **3.** 132×5

4. 331×4 **5.** $\begin{array}{r} 45 \\ \times 11 \\ \hline \end{array}$ **6.** $\begin{array}{r} 35 \\ \times 22 \\ \hline \end{array}$

7. In 326×52 the place value of the product of "5" and "3" is _____.

8. In 326×52 the product of "5" and "6" is "30" and you must carry the 3 to the _____ column.

B.

Multiply.

9. $(42)(38)$ **10.** $(24)(65)$ **11.** $(89)(32)$

12. $(98)(23)$ **13.** $\begin{array}{r} 646 \\ \times 45 \\ \hline \end{array}$ **14.** $\begin{array}{r} 328 \\ \times 52 \\ \hline \end{array}$

15. $(92)(145)$. Round product to the nearest hundred.

16. $(78)(246)$. Round product to the nearest thousand.

OBJECTIVE 2: *Estimate the product.*

A.

Estimate the product.

17. 38(42)

18. 71(67)

19. 213(81)

20. 365(24)

21.   362
   × 19

22.   756
   × 37

B.

Estimate the product and multiply.

23.   412
   × 84

24.   277
   × 53

25.   684
   × 47

26.   823
   × 65

OBJECTIVE 3: *Divide whole numbers.*

A.

27. 497 ÷ 7

28. 639 ÷ 3

29. 355 ÷ 5

30. 488 ÷ 8

31. 51 ÷ 23

32. 97 ÷ 31

33. The division has a remainder when the last difference in the division is smaller than the _____ and is not zero.

34. For 2600 ÷ 13, in the partial division 26 ÷ 13 = 2, 2 has place value _____.

B.

Divide.

35. 12,208 ÷ 4

36. 12,324 ÷ 6

37. $\dfrac{768}{32}$

38. $\dfrac{632}{79}$

39. $43\overline{)675}$

40. $28\overline{)456}$

41. $(62)(?) = 3596$

42. $(?)(73) = 2555$

43. $46{,}113 \div 57$ (Round quotient to the nearest ten.)

44. $20{,}020 \div 65$ (Round quotient to the nearest hundred.)

OBJECTIVE 4: *Estimate the quotient.*

A.

Estimate the quotient.

45. $625 \div 57$

46. $789 \div 29$

47. $3500 \div 43$

48. $2356 \div 33$

49. $610\overline{)34{,}560}$

50. $459\overline{)55{,}923}$

51. $\dfrac{34{,}976}{712}$

52. $\dfrac{81{,}782}{198}$

B.

Estimate the quotient and divide.

53. $103\overline{)59{,}602}$

54. $108\overline{)67{,}891}$

55. $780{,}854 \div 436$

56. $560{,}999 \div 356$

C.

57. Find the product of 606 and 415.

58. Find the product of 707 and 526.

59. Find the quotient of 157,281 and 309.

60. Find the quotient of 166,242 and 9807.

61. Estimate the product and multiply: (633)(2361)

62. Estimate the product and multiply: (6004)(405)

63. Estimate the quotient and divide: 6,784,821 ÷ 423

64. Estimate the quotient and divide: 4,378,921 ÷ 241

For Exercises 65–68, the figure shows the monthly sales at Dick's Country Cars.

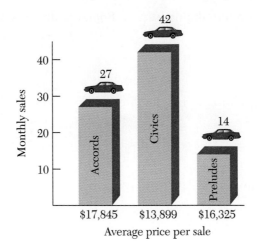

65. Estimate the gross receipts from the sale of the Accords.

66. What is the gross receipts from the sale of the Civics?

67. Estimate the gross receipts from the sale of Preludes and find the actual gross receipts.

68. Estimate the gross receipts for the month (the sum of the estimates for each model) and then find the actual gross receipts rounded to the nearest thousand dollars.

For Exercises 69–72, the revenue department of a mid-western state reported the following data for the first three weeks of March.

	Number of Returns	**Total Taxes Paid**
Week 1	4563	$24,986,988
Week 2	3981	$19,315,812
Week 3	11,765	$48,660,040

69. Estimate the taxes paid per return during week 2.

70. Find the actual taxes paid per return during week 1.

71. Find the actual taxes paid per return during week 3. Round to the nearest hundred dollars.

72. Find the actual taxes paid per return during the three weeks. Round to the nearest hundred dollars.

73. Estimate the product and multiply: $(24)(45)(36)$

74. Estimate the product and multiply: $(32)(71)(82)$

75. Estimate the quotient and divide: $\dfrac{229,367}{1216}$

76. Estimate the quotient and divide: $\dfrac{932,486}{3722}$

77. Salmon counted at the Bonneville fish ladder average 134 fish per day during a 17-day period. How many salmon are counted during the 17-day period?

78. During 1997 the population of Washington County grew at a pace of 2043 people per month. What was the total growth in population during 1997?

79. A forestry survey finds that 1664 trees are ready to harvest on a 13-acre plot. On the average, how many trees are ready to harvest per acre?

80. Rosebud Lumber Company replants 5696 seedling fir trees on a 16-acre plot of logged-over land. What is the average number of seedlings planted per acre?

81. A certain bacteria culture triples its size every hour. If the culture has a count of 375 at 10 AM, what will the count be at 2 PM the same day?

82. The starling population of the United States doubles every 3 years. If the current population of these birds is estimated to be 3,575,000, what will the population be in 9 years?

83. The estate of Ken Barker totals $347,875. It is to be shared equally by his five nephews. How much will each nephew receive?

84. Eight co-owners of the Alley Cat dress shop share equally in the proceeds when the business is sold. If the business sells for $229,928, how much will each receive?

85. The water consumption in Hebo averages 320,450 gallons per day. How many gallons of water are consumed in a 31-day month, rounded to the nearest thousand gallons?

86. The property tax on homes in Mt. Pedro averages $1536 per home. What is the total tax collected from 8347 homes, rounded to the nearest $10,000?

87. The Nippon Electronics firm assembles radios for export to the United States. Each radio is constructed using 14 resistors. How many radios can be assembled using 32,278 resistors in stock? How many resistors are left over?

88. The Nippon Electronics firm in Exercise 87 assembles a second radio containing 17 resistors. Using the 32,278 resistors in stock, how many of these radios can be assembled? How many resistors are left over?

89. Janet orders 450 radios for sale in her discount store. If she pays $78 per radio and sells them for $112, how much do the radios cost her and what is the net income from their sale? How much are her profits from the sale of the radios?

90. Jose orders 325 snow shovels for his hardware store. He pays $12 per shovel and plans to sell them for $21 each. What do the shovels cost Jose and what will be his gross income from their sale? What net income does he receive from the sale of the shovels?

91. It takes the Morris Packing Plant 8 hours to process 12 tons of lima beans. How long does it take the plant to process 780 tons of lima beans?

92. It takes the Pacific Packing Plant 6 hours to process four tons of Dungeness crab. How long does it take the plant to process 392 tons of Dungeness crab?

Exercises 93 and 94 refer to the Chapter 1 application problem. See page 1.

93. There are six instances where one region has twice the estimated number of HIV infections as another region. Find them.

94. How many times more cases of HIV are estimated to be in Sub-Saharan Africa than in South and Southeast Asia?

95. A new subway system is being built in Los Angeles at an estimated cost of $290 million per mile. It is designed to accommodate a peak-hour load of 36,000 passengers. What is the estimated cost per mile per passenger rounded to the nearest $10?

96. Portland, Oregon, is building a surface light rail system at an estimated cost of $96 million per mile. It is designed to accommodate a peak hour load of 6000 passengers. What is the estimated cost per mile per passenger?

97. Using the results of Exercises 95 and 96, is Portland or Los Angeles getting the better deal for their transit system? Justify your answer?

STATE YOUR UNDERSTANDING

98. Explain to an 8-year-old child that 3(8) = 24.

99. Explain to an 8-year-old child that 45 ÷ 9 = 5.

100. Explain the concept of remainder.

101. When 74 is multiplied by 8, we carry 3 to the tens column. Explain why this is necessary.

102. Define and give an example of a product.

103. Define and give an example of a quotient.

CHALLENGE

104. Find the product of twenty-four thousand, fifty-five and two hundred thirteen thousand, two hundred seventy-six. Write the word name for the product.

106. The Belgium Bulb Company has 171,000 tulip bulbs to market. Eight bulbs are put in a package when shipping to the United States and sold for $3 per package. Twelve bulbs are put in a package when shipping to France and sold for $5 per package. In which country will the Belgium Bulb Company get the greatest gross return? What is the difference in gross receipts?

105. Jose harvests 75 bushels of grain per acre from his 11,575 acres of grain. If Jose can sell the grain for $27 a bushel, what is the crop worth, to the nearest thousand dollars?

Complete the problems by writing in the correct digit whenever you see a letter.

107.
$$\begin{array}{r} 51A \\ \times\ \ \underline{B2} \\ 10B2 \\ \underline{154C} \\ 1A5E2 \end{array}$$

108.
$$\begin{array}{r} 1A57 \\ \times\ \ \underline{42} \\ B71C \\ \underline{D428} \\ 569E4 \end{array}$$

109.
$$\begin{array}{r} 5AB2 \\ 3\overline{)1653C} \end{array}$$

110.
$$\begin{array}{r} 21B \\ A3\overline{)4CC1} \end{array}$$

GROUP ACTIVITY

111. Multiply 23, 56, 789, 214, and 1345 by 10, 100, and 1000. What do you observe? Can you devise a rule for multiplying by 10, 100, and 1000?

Divide 23,000,000 and 140,000,000 by 10, 100, 1000, 10,000, and 100,000. What do you observe? Can you devise a rule for dividing by 10, 100, 1000, 10,000, and 100,000?

1.4   Whole Number Exponents and Powers of Ten

OBJECTIVES

1. Find the value of an expression written in exponential form.

2. Multiply or divide a whole number by a power of 10.

VOCABULARY

A **base** is a number used as a repeated factor. An **exponent** indicates the number of times the base is used as a factor.

The **values** of 2^3 and 10^5 are 8 and 100,000.

Exponents of 2 and 3 are often read "**squared**" and "**cubed**."

A **power of 10** is the value obtained when 10 is written with an exponent.

HOW AND WHY
Objective 1

Find the value of an expression written in exponential form.

Just like multiplication is repeated addition, exponents show repeated multiplication. Whole number *exponents* greater than 1 are used to write repeated multiplications in shorter form. For example,

3^4 means $3 \cdot 3 \cdot 3 \cdot 3$

and since $3 \cdot 3 \cdot 3 \cdot 3 = 81$, we write $3^4 = 81$. The number 81 is sometimes called the "fourth power of three" or "the *value* of 3^4."

Exponent
$$\downarrow$$
$$3^4 = 81$$
$$\uparrow \qquad \uparrow$$
Base   Value

Similarly, the value of 8^5 is

$8^5 = 8 \cdot 8 \cdot 8 \cdot 8 \cdot 8 = 32,768$

The base, the repeated factor, is 8. The exponent, which indicates the number of times the base is used as a factor, is 5.

Exponents of 0 and 1 are special cases.

**Property
of Exponents**

If 1 is used as an exponent, the value is equal to the base. In general,

$x^1 = x$

So, $2^1 = 2$, $11^1 = 11$, $3^1 = 3$, and $(123)^1 = 123$.

**Property
of Exponents**

If 0 is used as an exponent, the value is 1 (unless the base is also zero). In general,

$x^0 = 1,$ 	when $x \neq 0$

So, $2^0 = 1$, $11^0 = 1$, $3^0 = 1$, and $(123)^0 = 1$.

We can see a reason for the meaning of 6^1 ($6^1 = 6$) and 6^0 ($6^0 = 1$) by studying the following pattern.

$6^4 = 1 \cdot 6 \cdot 6 \cdot 6 \cdot 6$

$6^3 = 1 \cdot 6 \cdot 6 \cdot 6$

$6^2 = 1 \cdot 6 \cdot 6$

$6^1 = 1 \cdot 6$

$6^0 = 1$

> ### To find the value of an expression with a whole number exponent
>
> **1.** If the exponent is 0 and the base not 0, the value is 1.
> **2.** If the exponent is 1, the value is the same as the base.
> **3.** If the exponent is greater than 1, use the base number as a factor as many times as shown by the exponent. Multiply.

Examples A–E	**Warm Ups A–E**

Directions: Find the value.

Strategy: Identify the exponent. If it is zero, the value is 1. If it is one, the value is the base number. If it is greater than one, use it to tell how many times the base is used as a factor and then multiply.

A. 6^3

$6^3 = 6 \cdot 6 \cdot 6$

$ = 216$

A. 11^2

B.   Simplify: 29^1

$29^1 = 29$ **If the exponent is one, the value is the base number.**

B.   Simplify: 29^0

C.   Find the value of 10^7.

$10^7 = (10)(10)(10)(10)(10)(10)(10)$

$ = 10{,}000{,}000$ **Ten million. Note that the value has seven zeros.**

C.   Find the value of 10^6.

D.   Evaluate: 6^4

$6^4 = 6(6)(6)(6) = 1296$

D.   Evaluate: 7^3

 Calculator Example:

E.   Find the value of 7^9.

Calculators usually have an exponent key marked $\boxed{y^x}$ or $\boxed{\wedge}$. If your calculator doesn't have such a key you will need to multiply the repeated factors.

The value is 40,353,607.

E.   Find the value of 5^{11}.

Answers to Warm Ups   A. 121   B. 1   C. 1,000,000   D. 343   E. 48,828,125

HOW AND WHY
Objective 2

Multiply or divide a whole number by a power of 10.

It is particularly easy to multiply or divide a whole number by a power of 10. Consider the following and their products when multiplied by 10.

$$5 \times 10 = 50 \qquad 7 \times 10 = 70 \qquad 3 \times 10 = 30$$

Each product is the single digit with a zero written on the right. If the whole number is not a single-digit number (larger than nine), the place value of every digit becomes ten times larger when the number is multiplied by 10.

$$24 \times 10 = 2 \text{ tens} + 4 \text{ ones}$$
$$\underline{\times \qquad\qquad 10}$$
$$2 \text{ hundreds} + 4 \text{ tens} \qquad \textbf{Since ten} \times \textbf{ten} = \textbf{hundred}$$
$$= 240 \qquad\qquad\qquad \textbf{and one} \times \textbf{ten} = \textbf{ten.}$$

So, to multiply by 10, we need to merely write a zero on the right of the whole number. If a whole number is multiplied by 10 more than once, a zero is written on the right for each 10. So,

$$24 \times 10^4 = 240,000 \qquad \textbf{Four zeros are written on the right, one for each 10.}$$

Since division is the inverse of multiplication, dividing by 10 will eliminate the last zero on the right of a whole number. So,

$$240,000 \div 10 = 24,000 \qquad \textbf{Eliminate the final zero on the right.}$$

If we divide by 10 more than once, one zero is eliminated for each 10. So,

$$240,000 \div 10^3 = 240 \qquad \textbf{Eliminate three zeros.}$$

 To multiply a whole number by a power of 10

1. Identify the exponent of 10.
2. Write as many zeros to the right of the whole number as the exponent of 10.

 To divide a whole number by a power of 10

1. Identify the exponent of 10.
2. Eliminate the same number of zeros on the right of the whole number as the exponent of 10.

Using powers of 10, we have a third way of writing a whole number in expanded form.

$$2345 = 2000 + 300 + 40 + 5, \text{ or}$$
$$= 2 \text{ thousands} + 3 \text{ hundreds} + 4 \text{ tens} + 5 \text{ ones, or}$$
$$= 2 \times 10^3 + 3 \times 10^2 + 4 \times 10^1 + 5 \times 10^0$$

Examples F–J   **Warm Ups F–J**

Directions: Multiply or divide.

Strategy: Identify the exponent of 10. For multiplication, write the same number of zeros on the right of the whole number as the exponent of 10. For division, eliminate the same number of zeros on the right of the whole number as the exponent of 10.

F. Multiply: $12{,}748 \times 10^5$ $12{,}748 \times 10^5 = 1{,}274{,}800{,}000$ **The exponent of 10 is 5. To multiply, write 5 zeros on the right of the whole number.**	F. Multiply: 1699×10^8
G. Simplify: 346×10^2 $346 \times 10^2 = 34{,}600$	G. Simplify: 57×10^4
H. Divide: $\dfrac{975{,}000}{10^2}$ $\dfrac{975{,}000}{10^2} = 9750$ **The exponent of 10 is 2. To divide, eliminate 2 zeros on the right of the whole number.**	H. Divide: $\dfrac{1{,}860{,}000}{10^4}$
I. Simplify: $496{,}230{,}000 \div 10^4$ $496{,}230{,}000 \div 10^4 = 49{,}623$ **Eliminate 4 zeros on the right.**	I. Simplify: $281{,}000 \div 10^2$

J.   A recent fund-raising campaign raised an average of \$123 per donor. How much was raised if there were 10^4 (ten thousand) donors?
 	
 Strategy: To find the total raised, multiply the average donation by the number of donors.
 	
     $123 \times 10^4 = 1{,}230{,}000$ **The exponent is 4. To multiply, write 4 zeros on the right of the whole number.**
 	
     The campaign raised \$1,230,000.
	J. A survey of 10^5 (one hundred thousand) people indicated that they pay an average of \$3786 in federal taxes. What was the total paid in taxes?

Answers to Warm Ups   F. 169,900,000,000   G. 570,000   H. 186   I. 2810   J. \$378,600,000

Exercises 1.4

OBJECTIVE 1: *Find the value of the exponential expression.*

A.

Write in exponential form.

1. 12(12)(12)(12)(12)(12)

2. $73 \times 73 \times 73 \times 73 \times 73 \times 73 \times 73 \times 73$

Find the value.

3. 9^2

4. 8^2

5. 2^3

6. 3^3

7. 19^0

8. 17^1

9. In $7^3 = 343$, 7 is the _____ , 3 is the _____ , and 343 is the _____ .

10. In $5^4 = 625$, 625 is the _____ , 5 is the _____ , and 4 is the _____ .

B.

Find the value.

11. 6^3

12. 2^6

13. 19^2

14. 21^2

15. 10^4

16. 10^6

17. 8^3

18. 7^4

19. 3^8

20. 5^6

OBJECTIVE 2: *Multiply or divide by a power of 10.*

A.

21. 45×10^2

22. 56×10^1

23. 7×10^4

24. 13×10^3

25. $1200 \div 10^2$

26. $1600 \div 10^2$

27. $340,000 \div 10^3$

28. $4500 \div 10^1$

29. To multiply a whole number by a power of 10, write as many zeros to the right of the number as the _____ of 10.

30. To divide a whole number by a power of 10, eliminate as many _____ on the right of the number as the exponent of 10.

B.

Multiply or divide.

31. 435×10^4

32. 276×10^3

33. $1,200,000 \div 10^3$

34. $35,000,000 \div 10^4$

35. 3591×10^4

36. 6711×10^3

37. $\dfrac{30,200}{100}$

38. $\dfrac{95,500}{10^2}$

39. 705×10^8

40. 300×10^6

41. $970,000,000 \div 10^5$

42. $3,506,000,000 \div 10^6$

C.

43. Write in exponent form: $10(10)(10)(10)(10)(10)(10)(10)(10)(10)(10)$

44. Write in exponent form: $4(4)(4)(4)(4)(4)(4)(4)(4)(4)(4)(4)(4)(4)(4)(4)(4)$

Find the value.

45. 14^4

46. 16^4

47. 9^9

48. 8^8

Multiply or divide.

49. 3350×10^9

50. 420×10^{11}

51. $\dfrac{438,000,000,000}{10^8}$

52. $\dfrac{1,460,000,000,000,000}{10^7}$

53. The operating budget of a community college is approximately 73×10^6 dollars. Write this amount in place value notation.

54. A congressional committee proposes to decrease the national debt by 125×10^5 dollars. Write this amount in place value notation.

55. During one week last year approximately 32×10^6 shares of Microsoft were traded on the New York Stock Exchange. Write this amount in place value notation.

56. The distance from the earth to the nearest star outside our solar system (Alpha Centauri) is approximately 255×10^{11} miles. Write this distance in place value notation.

▼ Exercises 57–59 refer to the Chapter 1 application problem. See page 1.

57. Express the estimated number of HIV infections in Sub-Saharan Africa and in North Africa/Middle East as powers of 10.

58. Express the estimated number of HIV infections in Latin America/Caribbean and in South and Southeast Asia as the product of a number and a power of 10.

59. Express the estimated number of HIV infections in Western Europe as the product of a number and a power of 10 in three different ways.

60. The number of bacteria in a certain culture doubles every hour. If there are two bacteria at the start, zero hour, how many bacteria will be in the culture at the end of 14 hours? Express the answer as a power of 2 and as a whole number.

61. A high roller in Atlantic City places nine consecutive bets at the "Twenty-One" table. The first bet is $5 and each succeeding bet is five times the one before. How much does she wager on the ninth bet? Express the answer as a power of 5 and as a whole number.

62. The world population in 1900 was 1625 million. Write this as a product of a number and a power of ten.

STATE YOUR UNDERSTANDING

63. Explain what is meant by 4^{10}.

64. If we were using a base-five system instead of a base-ten system, what would be the contributed value of the digit 3 in 123,111? Justify your answer.

CHALLENGE

65. Mitchell's grandparents deposit $3 on his first birthday and triple that amount on each succeeding birthday until he is 12. What amount did Mitchell's grandparents deposit on his twelfth birthday? What is the total amount they have deposited in the account?

66. Find the sum of the cubes of the digits.

67. Find the difference between the sum of the cubes of 4, 8, 11, and 23 and the sum of the fourth powers of 2, 5, 6, and 7.

68. Research your local newspaper and find at least five numbers that could be written using a power of 10. Write these as a product using a power of 10.

1.5   Operations with Exponents

OBJECTIVES

1. Multiply powers with the same base.

2. Raise a power to a power.

3. Raise a product to a power.

HOW AND WHY
Objective 1

Multiply powers with the same base.

In algebra, you will study five basic properties of exponents. In prealgebra, we study three of these as they apply to whole numbers.

To discover how to multiply powers with the same base, study these examples

$$2^5 \cdot 2^4 = (2 \cdot 2 \cdot 2 \cdot 2 \cdot 2)(2 \cdot 2 \cdot 2 \cdot 2)$$

Write each of the factors.

$$= 2 \cdot 2 \cdot 2 \cdot 2 \cdot 2 \cdot 2 \cdot 2 \cdot 2 \cdot 2$$

Repeated use of the associative property of multiplication allows us to write the list of factors without the parentheses.

$$= 2^9$$

There are nine factors of 2.

and

$$17^3 \cdot 17^5 = (17 \cdot 17 \cdot 17)(17 \cdot 17 \cdot 17 \cdot 17 \cdot 17)$$

$$= 17 \cdot 17 \cdot 17 \cdot 17 \cdot 17 \cdot 17 \cdot 17 \cdot 17$$

$$= 17^8$$

The first property of exponents is a shortcut for these steps. Instead of writing out all of the factors, we can get the same result by adding the exponents.

$$2^5 \cdot 2^4 = 2^{5+4}$$

$$= 2^9$$

and

$$17^3 \cdot 17^5 = 17^{3+5}$$

$$= 17^8$$

   To multiply powers with the same base

Add the exponents and keep the common base.

$$x^a \cdot x^b = x^{a+b}$$

CAUTION   **If the bases are not the same, the property for multiplying exponents with the same base does not apply.**

$$2^5 \cdot 3^2 \neq 6^7$$

Examples A–C   **Warm Ups A–C**

Directions: Simplify.

Strategy: Use the property of exponents for multiplying powers of the same base.

A. $6^3 \cdot 6^7$

   $6^3 \cdot 6^7 = 6^{3+7}$ Add the exponents.

   $= 6^{10}$

A. $7^5 \cdot 7^9$

B. Multiply: $33^7 \cdot 33^2 \cdot 33^5$

   $33^7 \cdot 33^2 \cdot 33^5 = 33^{7+2+5}$

   $= 33^{14}$

B. Multiply: $27^4 \cdot 27^2 \cdot 27^3 \cdot 27^5$

C A U T I O N   **Do not think of this as an addition problem, this is multiplication. Adding exponents is a shortcut for performing the multiplication problem.**

C. The volume of a crate containing a new computer is given by $V = \ell wh$, where ℓ is the length, w is the width, and h is the height. If the number of feet in the length is the square of the number of feet in the width ($\ell = w^2$) and the number of feet in the height is the cube of the number of feet in the width ($h = w^3$), express the volume in terms of w, using a single exponent, and find the number of cubic feet in the volume if $w = 2$ ft.

Strategy: To express the volume of the crate in terms of the width (w), substitute w^2 for ℓ and w^3 for h in the formula and simplify.

   Formula: $V = \ell wh$

   Substitute: $V = w^2 \cdot w \cdot w^3$

   Solve: $V = w^6$ New formula.

   To find the volume when $w = 2$, substitute in the new formula and simplify.

   $V = (2)^6$

   $= 64$

   Check using the original formula.

   $w = 2, \ell = 2^2 = 4, h = 2^3 = 8$

   $V = \ell wh = 2(4)(8) = 64$

   Therefore, the volume of the crate containing the new computer is w^6 in terms of the width and 64 cubic feet when the width is 2 ft.

C. Another crate has a length that is the cube of the width (w^3) and a height that is equal to the width (w). Express the volume of the crate in terms of the width and find the number of cubic feet in the volume if the width is 3 ft.

Answers to Warm Ups   A. 7^{14} B. 27^{14} C. $V = w^5$, 243 cubic feet

HOW AND WHY
Objective 2   **Raise a power to a power.**

The first property of exponents leads us to the second property of exponents. The following table shows examples of raising a power to a power.

Problem	Factors	Result
$(3^3)^2$	$3^3 \cdot 3^3$	$3^{3+3} = 3^6 = 3^{3 \cdot 2}$
$(4^5)^3$	$4^5 \cdot 4^5 \cdot 4^5$	$4^{5+5+5} = 4^{15} = 4^{5 \cdot 3}$
$(2^2)^4$	$2^2 \cdot 2^2 \cdot 2^2 \cdot 2^2$	$2^{2+2+2+2} = 2^8 = 2^{2 \cdot 4}$

Each row of the table shows that the new exponent is the product of the exponents.

 To raise a power to a power

Multiply the exponents.

$(x^a)^b = x^{a \cdot b}$

For example, $(4^3)^5 = 4^{3 \cdot 5} = 4^{15}$.

Examples D and E   **Warm Ups D and E**

Directions: Simplify.

Strategy: Use the property of exponents for raising a power to a power.

D. $(3^2)^7$ D. $(5^4)^6$

   $(3^2)^7 = 3^{2 \cdot 7} = 3^{14}$

E. Write using a single exponent: $(8^5)^9$ E. Write using a single exponent: $(12^7)^6$

   $(8^5)^9 = 8^{5 \cdot 9} = 8^{45}$

HOW AND WHY
Objective 3   **Raise a product to a power.**

We use the definition of an exponent to arrive at the third property of exponents.

Problem	Factors	Result
$(2 \cdot 3)^2$	$(2 \cdot 3)(2 \cdot 3)$	$(2 \cdot 2)(3 \cdot 3) = 2^2 \cdot 3^2$
$(3 \cdot 4)^3$	$(3 \cdot 4)(3 \cdot 4)(3 \cdot 4)$	$(3 \cdot 3 \cdot 3)(4 \cdot 4 \cdot 4) = 3^3 \cdot 4^3$
$(2 \cdot 5)^4$	$(2 \cdot 5)(2 \cdot 5)(2 \cdot 5)(2 \cdot 5)$	$(2 \cdot 2 \cdot 2 \cdot 2)(5 \cdot 5 \cdot 5 \cdot 5) = 2^4 \cdot 5^4$

Each row of the table shows that each factor is raised to the same power.

Answers to Warm Ups   D. 5^{24} E. 12^{42}

To raise a product to a power

Raise each factor to the power.

$$(x \cdot y)^a = x^a \cdot y^a$$

For example, $(3 \cdot 6)^2 = 3^2 \cdot 6^2$.

Examples F–G   **Warm Ups F–G**

Directions: Simplify.

Strategy: Use the property of exponents for raising a product to a power.

F. $(3 \cdot 8)^7$ F. $(6 \cdot 7)^3$

  $(3 \cdot 8)^7 = 3^7 \cdot 8^7$

G.   Simplify: $(5 \cdot 7)^6$ G.   Simplify: $(8 \cdot 5)^5$

  $(5 \cdot 7)^6 = 5^6 \cdot 7^6$ **Each factor is raised to the sixth power. The place value notation for the number is 1,838,265,625.**

Here is the list of the three properties.

Properties of Exponents	Product and Power
	Product of like bases: $x^a \cdot x^b = x^{a+b}$ Power to a power: $(x^a)^b = x^{ab}$ Product to a power: $(x \cdot y)^a = x^a \cdot y^a$

Answers to Warm Ups   F. $6^3 \cdot 7^3$ G. $8^5 \cdot 5^5$

Exercises 1.5

OBJECTIVE 1: *Multiply powers with like bases.*

A.

Simplify.

1. $6^2 \cdot 6^4$

2. $7^3 \cdot 7^5$

3. $11^4 \cdot 11^6$

4. $13^5 \cdot 13^5$

5. $22^6 \cdot 22^2$

6. $23^7 \cdot 23^7$

7. Write the missing factor in exponent form: $14^3 \cdot ? = 14^{11}$

8. Write the missing factor in exponent form: $? \cdot 16^{12} = 16^{12}$

B.

Simplify.

9. $19^{11} \cdot 19^4$

10. $34^6 \cdot 34^2$

11. $58^7 \cdot 58^{19}$

12. $72^7 \cdot 72^{21}$

13. $6^2 \cdot 6^4 \cdot 6^7$

14. $12^3 \cdot 12^7 \cdot 12^4$

OBJECTIVE 2: *Raise a power to a power.*

A.

Simplify.

15. $(8^2)^2$ *mult.*

16. $(7^3)^2$

17. $(10^4)^3$

18. $(5^4)^4$

19. $(19^0)^{11}$

20. $(13^1)^{13}$

21. Find the missing exponent: $(14^5)^? = 14^{20}$

22. Find the missing exponent: $(7^?)^6 = 7^{24}$

B.

Simplify.

23. $(12^6)^3$

24. $(15^6)^4$

25. $(10^3)^7$

26. $(19^3)^{11}$

27. $(18^5)^9$

28. $(21^4)^{10}$

OBJECTIVE 3: *Raise a product to a power.*

A.

Simplify.

29. $(2 \cdot 5)^2$

30. $(4 \cdot 3)^3$

31. $(8 \cdot 9)^4$

32. $(7 \cdot 6)^5$

33. $(13 \cdot 12)^7$

34. $(15 \cdot 9)^{10}$

B.

35. $(12 \cdot 23)^{12}$

36. $(32 \cdot 13)^{20}$

37. $(14 \cdot 45)^{19}$

38. $(36 \cdot 29)^{17}$

39. $(4 \cdot 6 \cdot 8)^{11}$

40. $(9 \cdot 7 \cdot 11)^9$

C.

Simplify.

41. $(13^4)^7$

42. $(12 \cdot 17)^{19}$

43. $16^3 \cdot 16^7 \cdot 16^8$

44. $15^3 \cdot 15^7 \cdot 15^3 \cdot 15^{11}$

45. $(24^7)^9$

46. $(45 \cdot 71)^{10}$

47. $(3^3 \cdot 7^2)^4$

48. $(8^5 \cdot 9^3)^6$

49. $[(5^2)^3]^4$

50. $[(7^3)^4]^5$

51. The volume of a cube is expressed by the formula $V = s^3$. Find the number of cubic inches in the volume of a cube if $s = 5$ in.

52. The area of a triangle is given by the formula $A = \dfrac{b \cdot h}{2}$. Express the area of the triangle in terms of the height (h) if the base is the square of the height $(b = h^2)$. Find the number of square inches in the area if the height is 6 in.

53. The volume of a rectangular box is given by the formula $V = \ell \cdot w \cdot h$. Express the volume of a box in terms of the width (w) if the number of feet in the length is equal to the number of feet in the width $(\ell = w)$ and the number of feet in the height is the square of the number of feet in the width $(h = w^2)$. Find the number of cubic feet in the volume if the width is 5 ft.

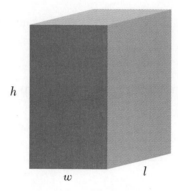

54. The area of a rectangle is given by the formula $a = \ell \cdot w$. Express the area in terms of the width (w) when the number of feet in the length (ℓ) is the cube of the number of feet in the width $(\ell = w^3)$. Find the number of square feet in the area when the width is 6 ft.

55. The volume of a right circular cylinder is given by the formula $V = \pi \cdot r^2 \cdot h$. Express the volume in terms of the radius (r) when the number of inches in the height (h) is the cube of the number of inches in the radius $(h = r^3)$. Find the number of cubic inches in the volume if the radius is 3 in. Let $\pi \approx 3$. (The symbol "$\approx$" means "is approximately equal to.")

56. The volume of a sphere is expressed by the formula $V = \dfrac{4 \cdot \pi \cdot r^3}{3}$. Find the number of cubic centimeters in the volume of a sphere if the radius (r) is 2 cm and $\pi \approx 3$.

 Exercises 57 and 58 refer to the Chapter 1 application problem. See page 1.

57. Which region has 100 times the estimated number of HIV infections as in North Africa/Middle East? Express the estimated number of HIV infections for each region as a power of 10 and then write a mathematical sentence that shows the relationship between them and the uses multiplying powers with the same base. Write a different mathematical sentence that shows the relationship between them and uses the dividing powers with the same base.

58. Which region has $\dfrac{1}{10}$ the estimated number of HIV infections as North America? Express the estimated number of HIV infections for each region as a power of 10. Write a mathematical sentence that shows the relationship between them and uses multiplying powers with the same base.

STATE YOUR UNDERSTANDING

59. Explain, in your own words, how to multiply powers with the same base.

60. Explain how to simplify $(5 \cdot 12)^8 \cdot 5^2$ using the properties for exponents.

CHALLENGE

61. Find n: $(2^n)^4 = 256$

62. Find b: $3^b \cdot 3 = 27$

63. Find b: $(2 \cdot 3)^b = 216$

GROUP ACTIVITY

64. Have each member of the group work the following problem independently.

$19 - 2 \cdot 6 - 5 + 7$

Compare your answers. If they are different, be prepared to discuss in class why they are different.

1.6   Order of Operations and Average

OBJECTIVES

1. Perform any combination of operations on whole numbers.

2. Find the average of a group of whole numbers.

VOCABULARY

Parentheses ( ) and **brackets** [ ] are used in mathematics as **grouping symbols.** These symbols indicate that the operations inside are to be performed first. Other grouping symbols that are often used are **braces** { } and the **fraction bar** —.

The **average** or **mean** of a set of numbers is the sum of the set of numbers divided by the total number of numbers in the set.

HOW AND WHY
Objective 1

Perform any combination of operations on whole numbers.

Without a rule is it possible to interpret $5 + 3 \cdot 12$ in two ways.

$$5 + 3 \cdot 12 = 5 + 36$$
$$= 41$$

or

$$5 + 3 \cdot 12 = 8 \cdot 12$$
$$= 96$$

In order to decide which answer to use, we need a standard protocol. Mathematicians agreed to multiply first, so

$$5 + 3 \cdot 12 = 41$$

The order in which the operations are performed is important since the order often determines the answer. Because of this, there is an established *order of operations.* This established order was agreed upon many years ago. In fact it is built into most of today's calculators and computers. The order of operations is frequently referred to as *algebraic logic.*

Order of Operations

To simplify an expression with more than one operation follow these steps:

Step 1. PARENTHESES—Do the operations within grouping symbols first (parentheses, fraction bar, etc.), in the order given in steps 2, 3, and 4.

Step 2. EXPONENTS—Do the operations indicated by exponents.

Step 3. MULTIPLY and DIVIDE—Do only multiplication and division as they appear from left to right.

Step 4. ADD and SUBTRACT—Do addition and subtraction as they appear from left to right.

So, we see that

$$8 - 10 \div 2 = 8 - 5 \qquad \textbf{Divide first.}$$
$$= 3 \qquad \textbf{Then subtract.}$$

$$(6 - 4)(6) = 2(6) \qquad \textbf{Subtract in parentheses first.}$$
$$= 12 \qquad \textbf{Then multiply.}$$

$$54 \div 6 \cdot 3 = 9 \cdot 3$$

Neither multiplication nor division takes preference

$$= 27$$

over the other. So do them from left to right.

Other exercises involving all of the operations are shown in the examples.

Examples A–G	**Warm Ups A–G**

Directions: Simplify.

Strategy: The operations are done in this order: operations in parentheses first, exponents next, then multiplication and division, and finally, addition and subtraction.

A. $7 \cdot 9 + 6 \cdot 2$

   $\quad 7 \cdot 9 + 6 \cdot 2 = 63 + 12$ **Multiply first.**

   $\quad\quad\quad\quad\quad = 75$ **Add.**

A. $4 \cdot 3 + 6 \cdot 5$

B. Simplify: $25 - 6 \div 3 + 8 \cdot 4$

   $\quad 25 - 6 \div 3 + 8 \cdot 4 = 25 - 2 + 32$ **Divide and multiply.**

   $\quad\quad\quad\quad\quad\quad\quad = 23 + 32$ **Subtract.**

   $\quad\quad\quad\quad\quad\quad\quad = 55$ **Add.**

B. Simplify: $4 \cdot 14 - 9 \div 3 + 6 \cdot 2$

C. Simplify: $5 \cdot 9 + 9 - 6(7 + 1)$

   $\quad 5 \cdot 9 + 9 - 6(7 + 1) = 5 \cdot 9 + 9 - 6(8)$ **Add in parentheses first.**

   $\quad\quad\quad\quad\quad\quad\quad\quad = 45 + 9 - 48$ **Multiply.**

   $\quad\quad\quad\quad\quad\quad\quad\quad = 54 - 48$ **Add.**

   $\quad\quad\quad\quad\quad\quad\quad\quad = 6$ **Subtract.**

C. Simplify: $24 \div 6 + 6 - 3(5 - 3)$

D. Simplify: $3 \cdot 4^3 - 8 \cdot 3^2 + 11$

   $\quad 3 \cdot 4^3 - 8 \cdot 3^2 + 11 = 3 \cdot 64 - 8 \cdot 9 + 11$ **Do exponents first.**

   $\quad\quad\quad\quad\quad\quad\quad\quad = 192 - 72 + 11$ **Multiply.**

   $\quad\quad\quad\quad\quad\quad\quad\quad = 120 + 11$ **Subtract.**

   $\quad\quad\quad\quad\quad\quad\quad\quad = 131$ **Add.**

D. Simplify: $5 \cdot 2^3 - 2 \cdot 4^2 + 25 - 7 \cdot 3$

E. Simplify: $(2^2 + 2 \cdot 3)^2 + 3^2$

Strategy: First do the operations in the parentheses following the proper order.

   $\quad (2^2 + 2 \cdot 3)^2 + 3^2 = (4 + 2 \cdot 3)^2 + 3^2$ **Do the exponent first.**

   $\quad\quad\quad\quad\quad\quad\quad = (4 + 6)^2 + 3^2$ **Multiply.**

   $\quad\quad\quad\quad\quad\quad\quad = (10)^2 + 3^2$ **Add.**

Now that the operations inside the parentheses are complete, continue using the order of operations.

   $\quad\quad\quad\quad\quad\quad\quad = 100 + 9$ **Do the exponents.**

   $\quad\quad\quad\quad\quad\quad\quad = 109$ **Add.**

E. Simplify: $(3^3 - 12 \div 4)^2 + 5^2$

Answers to Warm Ups   A. 42   B. 65   C. 4   D. 12   E. 601

Calculator Example:

 F. Simplify: $1845 + 165 \cdot 18 - 3798$

Enter the numbers and operations as they appear from left to right.

The answer is 1017.

F. Simplify: $1366 + 19 \cdot 372 \div 12$

G. The Lend A Helping Hand Association prepares two types of food baskets for distribution to the needy. The family pack contains nine cans of vegetables and the elderly pack contains four cans of vegetables. How many cans of vegetables are needed for 125 family packs and 50 elderly packs?

Strategy: To find the number of cans of vegetables needed for the packs, multiply the number of packs by the number of cans per pack. Then add the two amounts.

$$125(9) + 50(4) = 1125 + 200 \quad \textbf{Multiply.}$$
$$= 1325 \quad \textbf{Add.}$$

The Lend A Helping Hand Association needs 1325 cans of vegetables.

G. The Fruit of the Month Club prepares two types of boxes for shipment. Box A contains six apples and Box B contains ten apples. How many apples are needed for 96 orders of Box A and 82 orders of Box B?

HOW AND WHY
Objective 2

Find the average of a group of whole numbers.

The *average* or *mean* of a set of numbers is used in statistics. It is one of the ways to measure central tendency (like the average of a set of test grades). The average of a set of numbers is found by adding the numbers in the set and dividing the sum by the number of numbers in the set. For example, to find the average of 7, 11, and 15,

$$7 + 11 + 15 = 33 \quad \textbf{Find the sum of the numbers in the set.}$$
$$33 \div 3 = 11 \quad \textbf{Divide the sum by the number of numbers.}$$

The average is 11.

The "central" number or average does not need to be one of the members of the set. For instance find the average of 23, 32, 45, and 56.

$$23 + 32 + 45 + 56 = 156 \quad \textbf{Find the sum of the numbers.}$$
$$156 \div 4 = 39 \quad \textbf{Divide by the number of numbers.}$$

The average is 39, which is not a member of the set.

▶ *To find the average of a set of whole numbers*

1. Add the numbers.
2. Divide the sum by the number of numbers in the set.

Examples H–M	**Warm Ups H–M**

Directions: Find the average.

Strategy: Add the numbers in the set. Divide the sum by the number of numbers in the set.

H. 103, 98, and 123	H. 313, 129, and 500

$103 + 98 + 123 = 324$ **Add the numbers in the group.**

$324 \div 3 = 108$ **Divide the sum by the number of numbers.**

The average is 108.

I. Find the average of 7, 40, 122, and 211.	I. Find the average of 9, 27, 46, 58, and 65.

$7 + 40 + 122 + 211 = 380$ **Add the numbers in the group.**

$380 \div 4 = 95$ **Divide the sum by the number of numbers.**

The average is 95.

Calculator Example:

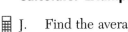

J.   Find the average of 345, 567, 824, and 960.

Enter the sum divided by 4.

$(345 + 567 + 824 + 960) \div 4$

The average is 674.

J.   Find the average of 917, 855, 1014, and 622.

K.   The average of 12, 8, 20, and ? is 12. Find the missing number.

Strategy: Since the average of the four numbers is 12 we know that the sum of the four numbers is 4(12) or 48. To find the missing number, subtract the sum of the three given numbers from four times the average.

$$4(12) - (12 + 8 + 20) = 4(12) - (40)$$
$$= 48 - 40$$
$$= 8$$

So, the missing number is 8.

K.   The average of 13, 15, 6, 7, and ? is 9. Find the missing number.

L.   In order to help Pete lose weight the dietician has him record his caloric intake for a week. He records the following: Monday, 3165; Tuesday, 1795; Wednesday, 1500; Thursday, 2615; Friday, 1407; Saturday, 1850; and Sunday, 1913. What is Pete's average caloric intake per day?

L.   The local dog food company ships the following cans of dog food: Monday, 3059; Tuesday, 2175; Wednesday, 3755; Thursday, 1851; and Friday, 2875. What is the average number of cases shipped each day?

Strategy: Add the calories for each day and then divide by 7, the number of days.

```
3165          2035
1795      7)14245
1500         14
2615          2
1407          0
1850         24
1913         21
14245        35
             35
              0
```

Pete's average caloric intake per day is 2035.

M. During the annual Salmon Fishing Derby, 35 fish are entered. The weights of the fish are recorded as shown:

Number of Fish	Weight per Fish (lb)
2	6
4	7
8	10
10	12
5	15
4	21
1	22
1	34

What is the average weight of a fish entered in the Derby?

Strategy: First find the total weight of all 35 fish.

$$2(6) = 12$$ Multiply 2 times 6 since there are 2
$$4(7) = 28$$ fish that weigh 6 pounds, for a total
$$8(10) = 80$$ of 12 pounds and so on.
$$10(12) = 120$$
$$5(15) = 75$$
$$4(21) = 84$$
$$1(22) = 22$$
$$1(34) = \underline{34}$$
$$455$$

Now divide the total weight by the number of salmon, 35.

$$455 \div 35 = 13$$

The average weight per fish is 13 pounds.

M. In a class of 30 seniors the following weights are recorded on Health Day:

Number of Students	Weight per Student (lb)
1	120
3	128
7	153
4	175
5	182
5	195
3	200
2	215

What is the average weight of a student in the class?

Answers to Warm Ups   L. 2743 cases   M. 173 pounds

Exercises 1.6

OBJECTIVE 1: *Perform any combination of operations on whole numbers.*

A.

Simplify.

1. $5 \cdot 8 + 13$

2. $17 + 5 \cdot 6$

3. $15 - 5 \cdot 3$

4. $28 \cdot 4 - 7$

5. $24 + 6 \div 2$

6. $36 \div 6 - 3$

7. $30 - (13 + 2)$

8. $(19 - 3) - 8$

9. $30 \div 6 \times 5$

10. $45 \div 5 \times 3$

11. $21 + 5 \cdot 4 - 2$

12. $25 - 3 \cdot 7 + 4$

13. $3^4 + 4^3$

14. $3^5 - 2^4 + 7^2$

B.

Simplify.

15. $4 \cdot 7 + 3 \cdot 5$

16. $6 \cdot 7 - 5 \cdot 4$

17. $3^2 - 4 \cdot 2 + 5 \cdot 6$

18. $5^2 + 12 \div 3 + 3 \cdot 3$

19. $36 \div 9 + 8 - 5$

20. $56 \cdot 3 \div 14 + 4 - 6$

21. $(14 + 28) - (34 - 27)$

22. $(56 - 8) - (17 + 7)$

23. $49 \div 7 \cdot 3^3 + 7 \cdot 4$

24. $75 \div 15 \cdot 2^4 + 3 \cdot 8$

25. $96 \div 12 \cdot 3$

26. $100 \div 4 \cdot 5$

27. $72 - 4(19 - 10) + 11 - 19$

28. $45(18 - 12) \div 3 - 8 + 12$

OBJECTIVE 2: *Find the average of a set of whole numbers.*

A.

Find the average.

29. 3, 7

30. 7, 11

31. 8, 10

32. 9, 13

33. 8, 14, 17

34. 5, 9, 13

35. 3, 5, 7, 9

36. 5, 5, 9, 9

Find the missing number to make the average correct.

37. The average of 4, 7, 9, and ? is 8.

38. The average of 6, 13, 11, and ? is 9.

B.

Find the average.

39. 14, 18, 30, 42

40. 11, 34, 41, 62

41. 25, 35, 45, 55

42. 18, 36, 41, 25

43. 7, 14, 16, 23, 30

44. 15, 8, 27, 51, 39

45. 8, 11, 19, 28, 44, 76

46. 29, 40, 48, 61, 86, 102

47. 31, 130, 238, 277

48. 88, 139, 216, 133

49. 101, 105, 108, 126

50. 281, 781, 513, 413

Find the missing number.

51. The average of 34, 81, 52, 74, and ? is 59.

52. The average of 21, 29, 46, 95, 33, and ? is 47.

C.

Simplify.

53. $50 - 12 \div 6 - 36 \div 6 + 3$

54. $80 - 24 \div 4 + 30 \div 6 + 4$

55. Find the average: 183, 526, 682, 589, 720

56. Find the average: 364, 384, 196, 736, 685

Use the following graph to answer Exercises 57–61. A chapter of Ducks Unlimited counts the following species of ducks at a lake in northern Idaho.

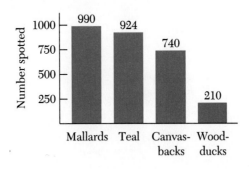

57. How many more mallards and canvasbacks were counted than teal and woodducks?

58. If twice as many canvasbacks had been counted, how many more canvasbacks would there have been than teal?

59. If four times the number of woodducks had been counted, how many more woodducks and mallards would there have been than teal and canvasbacks?

60. If twice the number of teal had been counted, how many more teal and woodducks would there have been as compared to mallards and canvasbacks?

61. Find the average number of the species of ducks.

Simplify.

62. $7(3^2 \cdot 2 - 8) \div 5 + 4$

63. $11(3^3 \cdot 4 - 98) \div 5 - 21$

64. $4(8 - 3)^3 - 4^3$

65. $3(7 - 3)^3 - 8(3 - 1)^2$

Find the average.

66. 1156, 2347, 5587, 354, 2355, 825

67. 3232, 4343, 5454, 6565, 7676, 8790

68. 23,458, 45,891, 34,652, 17,305, 15,984

69. 112,315, 236,700, 156,865, 103,674, 300,071

70. The Sing-Along Music Company last week advertised guitars for $635 each and pianos for $5125 each. They sold 15 guitars and 6 pianos. What was the total sales from the two items?

71. During a year-end sale the Neat-n-Trim clothing store hired two extra clerks. One was paid $8 per hour and the other one $10 per hour. What was the additional payroll if the first clerk worked 35 hours and the second clerk work 62 hours?

72. A golfer shoots the following scores for eleven rounds of golf: 84, 90, 103, 78, 91, 87, 75, 80, 78, 81, 77. What is the average score per round?

73. A bowler has the following scores for 9 games: 255, 198, 210, 300, 193, 213, 278, 200, 205. What is the average score per game?

74. Pete agrees to reward his son for grades earned. He pays $18 for an A, $12 for a B, and $5 for a C or a D. How much does his son earn for two A's, two B's, four C's, and one D?

75. Nanette agrees to reward her daughter for grades earned. She decides to give her $25 for each A she earns in excess of the number of B's, and $18 for each B she earns in excess of the number of C's. How much does her daughter earn if she has four A's, three B's, and one C?

76. Mr. Adams counts his caloric intake for one week before starting a diet. He reports the following intake:

Monday   4910   Tuesday   3780   Wednesday   3575   Thursday   4200
Friday   3400   Saturday   4350   Sunday   3960

What is his average caloric intake?

77. The figure shows some top speeds of running hoofed animals. What is the average top speed of these animals?

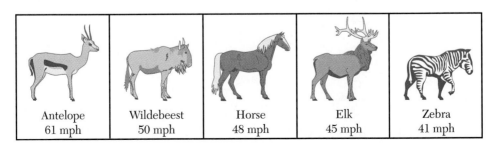

| Antelope | Wildebeest | Horse | Elk | Zebra |
| 61 mph | 50 mph | 48 mph | 45 mph | 41 mph |

78. To begin a month, Mario's Hair Barn has an inventory of 76 bottles of shampoo at $5 each, 65 bottles of conditioner at $4 each, and 35 perms at $22 each. At the end of the month they have left 18 bottles of shampoo, 9 bottles of conditioner, and 7 perms. What is the cost of the supplies used for the month?

79. Juanita's Camera Shop ordered 40 cameras that cost $346 each and 32 lenses at $212 each. She sells the cameras for $575 each and the lenses for $316 each. If they sell 27 cameras and 19 lenses, how much net income is realized on the sale of the cameras and lenses?

Exercises 80 and 81 refer to the Chapter 1 application problem. See page 1.

80. Calculate the average estimated number of HIV infections for all regions in Western Europe, North Africa/Middle East, Eastern Europe/Central Asia, and East Asia/Pacific.

81. Calculate the average estimated number of HIV infections for all regions in Africa. Do you think that using the figure you calculated gives an accurate picture of the HIV infections in Africa? Explain.

82. A consumer magazine tests 18 makes of cars for gas mileage. The results are shown in the table:

Number of Makes	Gas Mileage Based on 200 Miles (mpg)
2	14
3	20
2	25
4	31
3	36
2	40
2	45

What is the average gas mileage of the cars?

83. A home economist lists the costs of 15 brands of canned fruit drinks. The results are shown in the table:

Number of Brands	Price per Can (¢)
3	84
1	82
4	78
3	74
2	71
2	65

What is the average price per can?

84. Twenty-five football players are weighed in on the first day of practice. The weights are recorded in the table:

Number of Players	Weight (lb)
1	167
4	172
3	187
5	195
4	206
3	215
3	225
2	245

What is the average weight of the players?

85. At the Rock Creek Country Club ladies championship tournament the following scores are recorded for the 82 participants:

Number of Golfers	Score
1	66
3	68
7	69
8	70
12	71
16	72
15	74
10	76
6	78
3	82
1	85

What is the average score for the tournament?

86. A west coast city is expanding its mass transit system. It is building a 15-mile east-west light rail line for $780 million and an 11-mile north-south light rail line for $850 million. What is the average cost per mile, to the nearest million dollars, of the new lines?

STATE YOUR UNDERSTANDING

87. How would you explain which of the following is correct?

$$3 \cdot 5 + 4 = 15 + 4 \qquad 3 \cdot 5 + 4 = 3 \cdot 9$$
$$= 19 \qquad\qquad\qquad = 27$$

88. Explain what is meant by the average of two or more numbers?

89. Explain how to simplify
$2(36 \div 3^2 + 1) - 2 + 12 \div 4(3)$ using order of operations.

90. Explain how to find the average (mean) of 2, 4, 5, 5, and 9. What does the average of a set of numbers tell you about the set?

CHALLENGE

91. Simplify: $(6 \cdot 3 - 8)^2 - 50 + 2 \cdot 3^2 + 2(9 - 5)^3$

92. USA Video buys three first-run movies: #1 for $185, #2 for $143, and #3 for $198. During the first two months #1 is rented 10 times at the weekend rate of $5 and 26 times at the weekday rate of $3; #2 is rented 12 times at the weekend rate and 30 times at the weekday rate; and #3 is rented 8 times at the weekend rate and 18 times at the weekday rate. How much money must still be raised to pay for the cost of the three videos?

93. A patron of the arts estimates that the average donation to the fund drive will be $72. She will donate $150 for each dollar by which she misses the average. The 150 donors made the following contributions:

Number of Donors	Donation
5	$153
13	$125
24	$110
30	$100
30	$ 75
24	$ 50
14	$ 25
10	$ 17

How much does the patron donate to the fund drive?

GROUP ACTIVITY

94. Divide 35, 68, 120, 44, 56, 75, 82, 170, and 92 by 2 and 5. Which ones are divisible by 2 (the division has no remainder)? Which ones are divisible by 5? See if your group can find simple rules for looking at a number and telling whether or not it is divisible by 2 and/or 5.

95. Using the new car ads in the newspaper, find four advertised prices for the same model of a car. What is the average price, to the nearest $10?

1.7 Reading and Interpreting Tables

OBJECTIVE   Read and interpret information given in a table.

VOCABULARY   A **table** is a method of displaying data in an array using a horizontal and vertical arrangement to distinguish the type of data.

A **row** of a table is a horizontal line of a table and reads left to right across the page.

A **column** of a table is a vertical line of a table and reads up or down the page.

For example, in the table

Column 2

134	**56**	89	102
14	**116**	7	98
65	**45**	**12**	**67**
23	**32**	7	213

Row 3 (indicating the third row: 65, 45, 12, 67)

the number "45" is in row 3 and column 2.

HOW AND WHY
Objective   **Read and interpret information given in a table.**

Data are often displayed in the form of a *table*. We see tables in the print media, in advertisements, and in business presentations. Reading a table involves finding the correct *column* and *row* that describes the needed information and then reading the data at the intersection of that column and that row. For example,

STUDENT COURSE ENROLLMENT

Class	Mathematics	English	Science	Humanities
Freshman	950	1500	500	1200
Sophomore	600	700	650	1000
Junior	450	200	950	1550
Senior	400	250	700	950

To find the number of sophomores who take English, find the column headed English and the row headed Sophomore and read the number at the intersection. The number of sophomores taking English is 700.

We can use the table to find the difference in enrollments by class. To find how many more freshman take mathematics than juniors, we subtract the entries in the corresponding columns and rows. There are 950 freshman taking math as compared with 450 juniors. Since $950 - 450 = 500$, 500 more freshmen take mathematics than juniors.

The total enrollment in science for all four classes can be found by adding all entries in the column marked Science. Since $500 + 650 + 950 + 700 = 2800$, there are 2800 enrollments in science classes.

The table can also be used for predicting by scaling the values in the table upward or downward. If the number of seniors doubles next year, we can assume that the number of seniors taking humanities will also double. So the number of seniors taking humanities next year will be $2(950)$ or 1900.

Other ways to interpret data from a table are shown in the examples.

Examples A–C   **Warm Ups A–C**

Directions: Answer the questions associated with the table.

Strategy: Examine the rows and columns of the table to determine the values that are related.

A.   This table shows the decline in the number of railroad workers in four Western States.

RAILROAD WORKERS

State	1980	1988
Oregon	2991	1338
Idaho	3368	1748
Wyoming	3416	1486
Utah	3046	1717

1. Which state had the most railroad workers in 1980?
2. Which state had the least number of railroad workers in 1988?
3. Which state suffered the greatest loss in the number of railroad workers?
4. How many more railroad jobs did Utah have than Oregon in 1988?

Solution:

1. Wyoming   **Read down the column headed "1980" to locate the largest number of workers, 3416. Now read across the row to find the state, Wyoming.**

2. Oregon   **Read down the column headed "1988" to find the least number of workers, 1338. Then read across the row and find Oregon.**

3.   Oregon: $2991 - 1338 = 1653$ **Find the difference in**
      Idaho: $3368 - 1748 = 1620$ **the number of workers**
   Wyoming: $3416 - 1486 = 1930$ **for each state.**
      Utah: $3046 - 1717 = 1329$

Wyoming suffered the greatest loss, 1930.

A.   Use the table in Example A to answer the questions.

1. What was the total number of railroad workers in 1980 in the four states?
2. How many more railroad workers did Idaho have than Oregon in 1988?
3. What was the total number of railroad workers in Wyoming and Utah in 1988?
4. What was the combined loss in railroad workers in Idaho and Utah?

Answers to Warm Ups   A. 1. 12,821   2. 410   3. 3203   4. 2949

4. $1717 - 1338 = 379$ **Find the difference between Utah's number of jobs in 1988 and Oregon's in 1988.**

Utah has 379 more railroad jobs.

B.   This table shows the value of homes sold in the Portland metropolitan area for a given month.

VALUES OF HOUSES SOLD

Location	Lowest ($)	Highest ($)	Average ($)
N. Portland	16,000	58,500	34,833
N.E. Portland	18,000	120,000	47,091
S.E. Portland	18,000	114,000	51,490
Lake Oswego	40,000	339,000	121,080
West Portland	29,500	399,000	112,994
Beaverton	20,940	165,000	78,737

1. In which location was the highest-priced home sold?

2. What was the price difference between the average cost of a house and the lowest cost of a house in Lake Oswego?

3. What is the average lowest cost for the houses in the region?

4. If 31 houses were sold in Beaverton during the month, what was the total sales in the area?

B.   Use the table in Example B to answer.

1. Which area has the highest average sale price?

2. What is the difference between the highest- and lowest-priced house in Beaverton?

3. What is the difference in the lowest priced house in Lake Oswego and N. Portland?

4. If 67 houses were sold in Lake Oswego, what was the total sales for the month in the area?

Solution:

1. West Portland   **Read down the "highest" column and find the largest price, $399,000.**

2. $121,080
  $-$ 40,000
   $81,080   **Subtract the lowest cost from the average cost for Lake Oswego.**

The difference is $81,080.

3. $16,000   **Find the sum of the lowest prices and divide**
   $18,000   **by 6, the the number of areas.**
   $18,000
   $40,000
   $29,500
  $+$ $20,940
   $142,440

$142,440 \div 6 = \$23,740$

The average of the low-priced houses is $23,740.

4. $78,737   **Multiply the average sale price times the**
   $\times$ 31   **number of houses sold.**
   78737
   236211
   $2,440,847

The total sales in Beaverton for the month is $2,440,847.

The following table displays nutritional information about four breakfast cereals:

NUTRITIONAL VALUE PER 1-OZ SERVING

Ingredient	Oat Bran	Rice Puffs	Raisin Bran	Wheat Flakes
Calories	90	110	120	110
Protein	6 g	2 g	3 g	3 g
Carbohydrate	17 g	25 g	31 g	23 g
Fat	0 g	0 g	1 g	1 g
Sodium	5 mg	290 mg	230 mg	270 mg
Potassium	180 mg	35 mg	260 mg	4 mg

C. 1. Which cereal has the most calories per serving?

2. How many grams (g) of carbohydrates are there in 5 ounces of Rice Puffs?

3. How many more milligrams (mg) of sodium are there in a 1-ounce serving of Wheat Flakes as compared to Raisin Bran?

4. How many ounces of Oat Bran can one eat before consuming the same amount of sodium as in 1 ounce of Rice Puffs?

C. 1. Which Cereal has the most sodium per 1-oz serving?

2. How many milligrams (mg) of potassium are there in 6 ounces of Wheat Flakes?

3. Mary's doctor has counseled her to eat 18 grams (g) of protein for breakfast. How many servings of Raisin Bran will she need to eat to meet the recommendation?

4. How many ounces of Wheat Flakes can one eat before consuming the same amount of potassium as in one ounce of Oat Bran?

Solution:

1. Raisin Bran   Find the largest value in the calorie row, 120, then read the cereal heading at the top of that column.

2. 5(25 g) = 125 g   Multiply the grams of carbohydrates in Rice Puffs by 5.

3. (270 − 230) mg = 40 mg   Subtract the mg of sodium in Raisin Bran from that of Wheat Flakes.

4. $\dfrac{290 \text{ mg}}{5 \text{ mg}} = 58$ Divide the number of mg of sodium in Rice Puffs by the number in Oat Bran.

Answers to Warm Ups   C. Rice Puffs   2. 24 mg   3. 6 servings   4. 45 ounces

Exercises 1.7

Use the table for Exercises 1–8.

COST OF U.S. TELEVISION RIGHTS FOR THE OLYMPICS IN MILLIONS

	1984	1988	1992	1994	1996	1998	2000	2002	2004	2006	2008
Winter	ABC $91	ABC $309	CBS $243	CBS $295		CBS $375		NBC $545		NBC $613	
Summer	ABC $25	NBC $300	NBC $401		NBC $456		NBC $705		NBC $793		NBC $894

1. What is the least expensive year for broadcast rights for the Summer Olympics in the 1990s?

2. What is the difference in the cost for the winter games and the summer games in 1988?

3. What is the total cost of the broadcast rights for the summer Olympics in the 1990s?

4. What is the predicted increase in costs for broadcasting the summer games in 2008 as compared with the 1996 games?

5. What is the difference between the projected costs for broadcast rights to the 2004 and the 2002 games?

6. What is the total cost of broadcasting the winter games from 1984 to 2006?

7. Estimate how much will NBC pay for the rights to broadcast the games from 1984 to 2008. Find the actual cost.

8. Estimate how much will CBS pay for the rights to broadcast the games from 1984 to 2008? Find the actual cost.

Use the table for Exercises 9–18.

NUTRITIONAL INFORMATION PER SERVING OF ENTRÉE

Ingredient	Fish Cakes	Veal Chops	Chicken Dijon	Pepper Steak
Calories	259	421	247	240
Protein	28 g	34 g	31 g	28 g
Fat	12 g	24 g	12 g	10 g
Carbohydrate	7 g	15 g	2 g	9 g
Sodium	783 mg	687 mg	649 mg	820 mg
Cholesterol	147 mg	115 mg	99 mg	76 mg

9. Which entreé has the highest level of cholesterol per serving?

10. Which entreé has the least amount of calories per serving?

11. How much less cholesterol does one consume when ordering chicken Dijon as opposed to fish cakes?

12. How much more sodium is consumed when eating pepper steak as opposed to veal chops?

13. How much fat is contained in three servings of veal chops?

14. How many grams of carbohydrate are there in four servings of chicken Dijon and two servings of fish cakes?

15. At a buffet Dan eats one serving each of veal chops, chicken Dijon, and pepper steak. How many calories does he consume?

16. At a buffet Susan eats two servings of fish cakes and one serving of veal chops. How many milligrams (mg) of cholesterol does she consume?

17. Jerry's doctor puts him on a 900-calorie diet. What three entreés can he eat and stay within the 900-calorie limit?

18. Jessica is restricted to 30 grams of fat per day. Is there any combination of 3 entreés that she can eat and remain within the restriction of 30 grams of fat?

Use the table for Exercises 19–26.

FLEXIBLE LIFE INSURANCE POLICY

Age	Death Benefit	Account Value	Cash Surrender Value
57	$400,000	$144,276	$128,677
59	400,000	165,219	150,928
61	400,000	189,522	178,494
63	400,000	217,865	217,865
65	400,000	251,206	251,206

19. What is the gain in the account value from age 57 to age 65?

20. What is the gain in the cash surrender value from age 57 to 63?

21. If the policy is cashed in at age 59, what is the loss from the account value?

22. If the policy is cashed in at age 65, what is the loss from the account value?

23. What is the difference between the death benefit and the account value at age 61?

24. What is the difference between the death benefit and the surrender value at age 63?

25. Between what two consecutive ages in the table did the largest increase in the account value occur?

26. Between what two consecutive ages in the table did the least increase in the cash surrender value take place?

Use the table for Exercises 27–34.

VISITORS AT LIZARD LAKE STATE PARK

	May	June	July	August	September
Overnight camping	231	378	1104	1219	861
Picnics	57	265	2371	2873	1329
Boat rental	29	45	147	183	109
Hiking/climbing	48	72	178	192	56
Horse rental	22	29	43	58	27

27. During which month are the most picnics held?

28. Which month has the fewest horse rentals?

29. How many overnight campers use these facilities during these months?

30. How many boat rentals are there during these months?

31. If it costs $5 to hold a picnic in the park, how much income is realized from picnics in July?

32. If a horse rental costs $8, how much income is realized from horse rental in June?

33. How many more hikers/climbers are there in August than in May?

34. How many more boat rentals than horse rentals are there in August?

Use the table for Exercises 35–42.

ZOO ATTENDANCE

	1992	**1993**	**1994**	**1995**	**1996**
Fisher Zoo	2,367,246	2,356,890	2,713,455	2,745,111	2,720,567
Delaney Zoo	1,067,893	1,119,875	1,317,992	1,350,675	1,398,745
Shefford Garden	2,198,560	2,250,700	2,277,300	2,278,345	2,311,321
Utaki Park	359,541	390,876	476,200	527,893	654,345

35. Which zoo had the greatest increase in attendance from 1993 to 1996?

36. Which zoo had the smallest increase in attendance from 1993 to 1994?

37. Estimate the attendance for the 5 years at the Fisher Zoo.

38. Estimate the attendance at Utaki Park for the 5 years.

39. Find the average attendance for the 5 years at the Delaney Zoo.

40. Find the average attendance for the 5 years at Utaki Park.

41. If the average attendee, at the Fisher Zoo, spends $25 during a day at the zoo, including admission, find the revenue for 1995. Round to the nearest thousand dollars.

42. If the average attendee, at Utaki Park, spends $32 during a day at the zoo, including admission, find the revenue for 1996. Round to the nearest thousand dollars.

STATE YOUR UNDERSTANDING

43. Cite three or four examples of tables that you have used. Include at least two examples from your own experience outside of your class. Are there tables on your list that do not include numbers?

CHALLENGE

Use the table for Exercises 35–42 to solve Exercises 44–46.

44. Find the estimated total attendance at the four parks for the 5 years listed in the table.

45. What must the attendance be in 1997 for Utaki Park to average 501,200 in attendance for the years 1992 to 1997?

46. Estimate the income for 1992 to 1996 at Shefford Garden if the average attendee in 1992 spent $21, in 1993 spent $24, in 1994 spent $28, in 1995 spent $31, and in 1996 spent $35.

Exercises 1.8

OBJECTIVE 1: *Read data from bar, pictorial, and line graphs.*

The graph shows the variation in the number of phone calls during normal business hours:

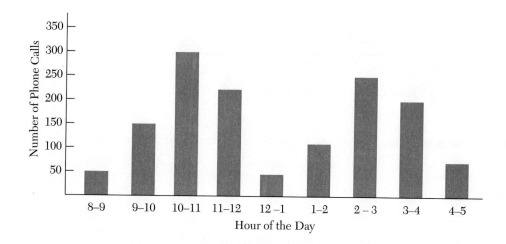

1. At what hour of the day is the number of phone calls greatest?

    10-11

2. At what hour of the day is the number of phone calls least?

    12-1

3. What is the number of phone calls made between 2 and 3?

    250

4. What is the number of phone calls made between 8 and 12?

    725

5. What is the total number of phone calls made during the times listed?

    1425

6. Are there more phone calls in the morning, 8–12, or the afternoon, 12–5?

The graph shows the number of cars in the shop for repair during a given year:

Type of car	1988 Repair Intake Record 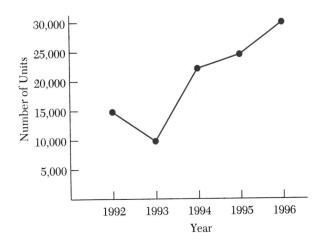 = 20 cars
Compact	🚗 🚗 🚗 🚗
Full size	🚗 🚗 🚗 🚗 🚗 🚗 🚗 🚗
Van	🚗 🚗
Subcompact	🚗 🚗 🚗 🚗 🚗 🚗

7. How many vans are in the shop for repair during the year?

8. How many compacts and subcompacts are in the shop for repair during the year?

9. What type of car has the most cars in for repair?

10. Are more subcompacts or compacts in for repair during the year?

11. How many cars are in for repair during the year?

12. If the average repair cost for compacts is $210, what is the gross income on compact repairs for the year?

The graph shows the number of production units at NERCO during the period 1992–1996

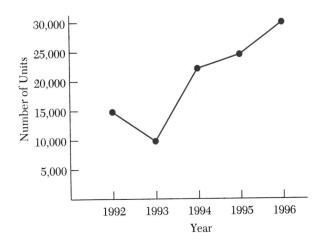

13. What is the greatest production year?

14. What is the year of least production?

15. What is the increase in production between 1993 and 1994?

16. What is the decrease in production between 1992 and 1993?

17. What is the average production per year?

18. If the cost of producing a unit in 1995 is $2750 and the unit is sold for $4560, what is the net income for the year?

The graph shows the amounts paid for raw materials at Southern Corporation during a production period.

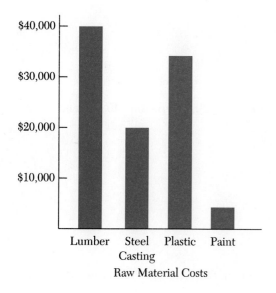

19. What is the total paid for paint and lumber?

20. What is the total paid for raw materials?

21. How much less is paid for steel castings than for plastics?

22. How much more is paid for plastic than for paint?

23. If Southern Corporation decides to double its production during the next period, what will it pay for steel casting?

24. If Southern Corporation decides to double its production during the next period, how much more will it pay for lumber and steel castings than plastic and paint?

For Exercises 25–28, draw bar graphs to display the data. Be sure to title the graph and label the parts.

25. Distribution of grades in an algebra class: A's—8; B's—6; C's—15; D's—8; F's—4.

26. The distribution of monthly income: Rent, $450; Automobile, $300; Taxes, $250; Clothes, $50; Food, $300; Miscellaneous, $100.

27. Career preference as expressed by a senior class: Business, 120; Law, 20; Medicine, 40; Science, 100; Engineering, 50; Public Service, 80; Armed Service, 10.

28. Dinner choices at the LaPlane restaurant in one week: Steak, 45; Salmon, 80; Chicken, 60; Lamb, 10; Others, 25.

For Exercises 29–32, draw line graphs to display the data. Title the graphs and label the parts.

29. Daily sales at the local men's store: Monday, $1500; Tuesday, $2500; Wednesday, $1500; Thursday, $3500; Friday, $4000; Saturday, $6000; Sunday, $4500.

30. The gallons of water used each quarter of the year by a small city in New Mexico:

Jan.–Mar.:   20,000,000 gallons
Apr.–June: 30,000,000 gallons
July–Aug.:   45,000,000 gallons
Sept.–Dec.:   25,000,000 gallons

31. Income from various sources for a given year for the Smith family: Wages, $36,000; Interest, $2000; Dividends, $4000; Sale of Property, $24,000.

32. Jobs in the electronics industry in a western state:

Year	1992	1993	1994	1995	1996	1997
Jobs	15,000	21,000	18,000	21,000	27,000	30,000

For Exercises 33–36, draw pictorial graphs to display the data. Tittle the graphs and label the parts.

33. The cost of an average three-bedroom house in a rural city:

Year	Cost
1970	$60,000
1975	$70,000
1980	$75,000
1985	$70,000
1990	$65,000
1995	$80,000

34. The population of Wilsonville over 20 years of age:

Year	1975	1980	1985	1990	1995
Population	22,500	30,000	32,500	37,500	40,000

35. The oil production from a local well over a 5-year period:

Year	Barrels Produced
1991	15,000
1992	22,500
1993	35,000
1994	32,500
1995	40,000

36. The estimated per capita income in the Great Lakes region in 1995:

State	Income
Illinois	$25,000
Michigan	$22,500
Ohio	$22,500
Wisconsin	$20,000
Indiana	$20,000

STATE YOUR UNDERSTANDING

37. Explain the advantages of each type of graph. Which do you prefer? Why?

CHALLENGE

38. The figures for United States casualties in four declared wars of the 20th century are World War I, 321,000; World War II, 1,076,000; Korean War, 158,000; Vietnam War, 211,000. Draw a bar graph, line graph, and a pictorial graph to illustrate the information. Which of your graphs do you think does the best job of displaying the data? (For the pictorial graph, you may want to round to the nearest hundred thousand.)

GROUP ACTIVITY

39. Have each member select a country and find the most recent population statistics for that country. Put the numbers together and have each member draw a different kind of graph of the populations.

CHAPTER 1

OPTIONAL Group Project *(1–2 weeks)*

All tables, graphs, and charts should be clearly labeled and computer generated if possible. Written responses should be typed and checked for spelling and grammar.

a. Go to the library and find the population and area for each state in the United States. Organize your information by geographic region. Record your information in a table.

b. Calculate the total population and the total area for each region. Calculate the population density (number of people per square mile, rounded to the nearest whole person) for each region, and put this and the other regional totals in a regional summary table. Then make three separate graphs, one for regional population, one for regional area, and the third for regional population density.

c. Calculate the average population per state for reach region, rounding as necessary. Put this information in a bar graph. What does this information tell you about the regions? How is it different from the population density of the region?

d. How did your group decide on the makeup of the regions? Explain your reasoning.

e. Are your results what you expected? Explain. What surprised you?

CHAPTER 1

True-False Concept Review

ANSWERS

Check your understanding of the language of algebra and arithmetic. Tell whether each of the following statements is True (always true) or False (not always true). For each statement that you judge to be False, revise it to make a statement that is True.

1. _____

1. Every digit can be used as a place holder in the place value notation of a number.

2. _____

2. In the number 6875, the digit "7" represents 70.

3. _____

3. The word name for 750 is seven hundred and fifty.

4. _____

4. $500 < 23$.

5. _____

5. $76 > 75$.

6. _____

6. To the nearest ten, 7449 rounds to 7500.

7. _____

7. The rounded value of a number is always smaller or larger than the number.

8. _____

8. Rounding off is one of the mathematical operations used in everyday life.

9. _____

9. The commutative and associative properties of addition are used by most persons when doing an addition problem with pencil and paper.

10. _____

10. The sum of 8 and 5 is 85.

11. _____

11. The commutative property does not apply to subtraction.

12. _____

12. "Borrowing" in a subtraction problem is based on the place values of the digits in the problem.

13. _____

13. All subtraction problems can be done without "borrowing."

14. _____   **14.** The product of 8 and 5 is 40.

15. _____   **15.** The number 9 is a factor of 36.

16. _____   **16.** The only factors of 15 are 3 and 5.

17. _____   **17.** The number 17 is a divisor of 51.

18. _____   **18.** To find the missing factor in the multiplication problem $8 \times ? = 792$, we divide 792 by 8.

19. _____   **19.** $2^4 + 2^4 = 2^5$

20. _____   **20.** $15^3 \times 15^4 = 15^7$

21. _____   **21.** $\dfrac{22,000,000}{100,000} = 22$

22. _____   **22.** In the order of operations, multiplication always takes precedence over division.

23. _____   **23.** In the order of operations, multiplication is always done before addition.

24. _____   **24.** In algebra, division problems are usually written in fraction form.

25. _____   **25.** The estimated product of 345 and 1756 is 600,000.

CHAPTER 1

Review

Section 1.1 *Objective 1*

Write the word name for

1. 892
2. 8745
3. 680,057

Write in place value notation.

4. forty-five thousand, eighty
5. two hundred eight million, twenty-five thousand, six hundred eight

Section 1.1 *Objective 2*

Insert < or > to make a true statement.

6. 82   78
7. 48   62
8. 809   908
9. 65,007   60,005
10. 435,098   435,100

Section 1.1 *Objective 3*

Round to the indicated place value.

11. 4769 (ten)
12. 4769 (hundred)
13. 67,349 (hundred)
14. 5,125,821 (thousand)
15. 3,044,999 (ten thousand)

Section 1.2 *Objective 1*

16. Add: $89 + 56 + 6 + 243$
17. Find the sum of 843, 629, 1208, 77, and 45.
18. Find the sum of 2345, 678, 1234, 45, and 99.
19. Add: $1,305,202 + 126,433 + 805 + 65,577 + 43$
20. Add: $28 + 83 + 687 + 5008 + 67,507 + 343,657$

Section 1.2 *Objective 2*

Estimate the sum.

21. $607 + 859 + 536 + 492$
22. $3900 + 4300 + 5679 + 399$
23. $972 + 851 + 736 + 689 + 432$
24. $12,560 + 693 + 59,360 + 8965$
25.
$$\begin{array}{r} 49,356 \\ 61,099 \\ 58,903 \\ 72,444 \\ + \underline{83,111} \end{array}$$

Section 1.2 *Objective 3*

26. Subtract: $7560 - 6566$
27. Subtract: $91,211 - 3368$
28. Subtract: $60,349 - 9075$
29. Find the difference of 50,008 and 30,684.
30. Find the difference between 421,066 and 65,767.

Section 1.2 *Objective 4*

Estimate the difference.

31. $895 - 733$
32. $19,755 - 14,599$
33.
$$\begin{array}{r} 28,463 \\ - \underline{21,155} \end{array}$$
34.
$$\begin{array}{r} 496,382 \\ - \underline{216,999} \end{array}$$
35.
$$\begin{array}{r} 82,399 \\ - \underline{41,932} \end{array}$$

Section 1.3 *Objective 1*

36. Multiply: 9(896)
37. Find the product of 54 and 189.
38. Find the product of 732 and 864.
39. Multiply: (6)(21)(394)
40. Multiply: 5(294)(831)

Section 1.3 *Objective 2*

Estimate the product.

41. 37(126)
42. 78(895)
43. 675(432)
44. 945(2478)
45. 23(679)(245)

Section 1.3 *Objective 3*

46. Find the quotient of 378,126 and 42.
47. Find the quotient of 3,293,988 and 482.
48. Divide: $162(?) = 1{,}006{,}344$
49. Divide: $225\overline{)15{,}440}$
50. Divide: $722{,}699 \div 901$

Section 1.3 *Objective 4*

Estimate the quotient.

51. $34{,}899 \div 532$
52. $68{,}945 \div 405$
53. $56{,}723 \div 3467$
54. $367{,}845 \div 8956$
55. $895{,}844 \div 673$

Section 1.4 *Objective 1*

Find the value.

56. 24^2
57. 12^3
58. 6^4
59. 5^4
60. 49^2

Section 1.4 *Objective 2*

Simplify.

61. 340×10^5
62. 5846×10^4
63. $2{,}700{,}000 \div 10^3$
64. $1{,}500{,}600{,}000 \div 10^5$
65. 39×10^{12}

Section 1.5 *Objective 1*

Use the property for multiplying like bases to write in exponential form.

66. $9^3 \cdot 9^8$
67. $11^4 \cdot 11^5$
68. $5^3 \cdot 5^3 \cdot 5^6$
69. $31^2 \cdot 31^2 \cdot 31^4$
70. $98^{13} \cdot 98^7$

Section 1.5 *Objective 2*

Use the property for raising a power to a power to write in exponential form.

71. $(7^5)^2$
72. $(9^3)^2$
73. $(12^2)^6$
74. $(18^4)^{10}$
75. $(22^6)^7$

Section 1.5 *Objective 3*

Use the property for raising a product to a power to write in exponential form.

76. $(7 \cdot 8)^5$
77. $(9 \cdot 5)^6$
78. $(6 \cdot 4)^7$
79. $(4 \cdot 3)^{10}$
80. $(2 \cdot 8)^{12}$

Section 1.6 *Objective 1*

Simplify.

81. $25 - 60 \div 12 \times 3$
82. $50 + 6 - 8 \cdot 2 \div 4$
83. $95 - 2 \cdot 4^2 \div 8 + 9$
84. $(3 \cdot 4)^2 - (2^2)^2 + 48 \div 3$
85. $144 \div 8 + 2^3$

Section 1.6 *Objective 2*

Find the average.

86. 24, 56, 40, and 72
87. 78, 92, 103, 129, and 143
88. 50, 62, 83, 91, 74, 99, and 101
89. 8, 37, 125, 48, 117, and 193
90. 48, 128, 168, 198, and 208

Section 1.7

Use the table to answer Exercises 91–95.

**PRODUCTIVITY CHART
AT SUNSHINE MANUFACTURING
MONTH OF SEPTEMBER**

Shift	Units Produced	Number of Employees	Days Worked
Day	480	80	20
Swing	250	50	20
Graveyard	270	50	18

91. How many units were produced at Sunshine during September?

92. How many more units does the day shift produce than the swing shift?

93. Which shift had the highest production per day?

94. Between the day and swing shifts, which has the highest production per employee?

95. If each unit wholesales for $1495, what is the value of the graveyard shift's production? Round to the nearest thousand dollars.

Section 1.8 *Objective 1*

The following graph shows the type of automobile sold last month by Acme Motors.

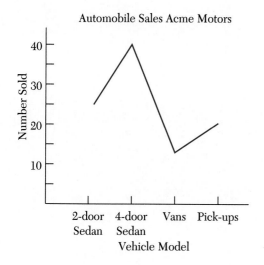

Automobile Sales Acme Motors

96. What were the total sales of these models for the month?

97. Which model had the least sales?

98. How many more 4-door sedans were sold than vans?

99. If twice as many vehicles are sold in October, what would be the expected number of 2-door sedans sold?

100. What were the total sales of vans and pick-ups?

Section 1.8 *Objective 2*

101. Draw a bar graph to display the daily gross sales at the local dairy: Monday, $1750; Tuesday, $1500; Wednesday, $2750; Thursday, $2250; Friday, $1000; Saturday, $3250; Sunday, $750.

102. Draw a bar graph to display the distribution of grades in a chemistry class: A, 12; B, 24; C, 36; D, 6; F, 2.

103. Draw a line graph to display the disposition of family income: Rent, $450; Food, $400; Clothes, $75; Car, $100; Entertainment, $50; Miscellaneous, $150.

104. Draw a pictograph to display the various sources of family income: Salary, $40,000; Interest, $4000; Dividends, $4000; Sale of Property, $16,000.

CHAPTER 1

Test

ANSWERS

1. _____

2. _____

3. _____

4. _____

5. _____

6. _____

7. _____

8. _____

9. _____

10. _____

11. _____

12. _____

13. _____

14. _____

1. Divide: $54\overline{)5886}$

2. Subtract: $9123 - 6844$

3. Simplify: $36 \div 9 + 4 \cdot 5 - 5$

4. Multiply: $53(768)$

5. Insert $<$ or $>$ to make the statement true: $278 \qquad 201$

6. Multiply: 76×10^4

7. Multiply: $709(386)$

8. Write the place value notation for four hundred fifty thousand, eighty-two.

9. Find the average of 1294, 361, 1924, 274, and 682.

10. Multiply: $35(2095)$ (Round the product to the nearest hundred.)

11. Round 17,852 to the nearest hundred.

12. Estimate the sum of 83,914, 17,348, 47,699, and 10,341.

13. Find the value of 9^3.

14. Add: $39 + 953 + 4 + 4886$

15. _____

15. Use the property of exponents for multiplying like bases to write $43^8 \cdot 43^5$ in exponential form.

16. _____

16. Subtract: $\begin{array}{r} 7040 \\ -\ 587 \\ \hline \end{array}$

17. _____

17. Write the word name for 6007.

18. _____

18. Simplify: $25 + 2^3 - 24 \div 8$

19. _____

19. Add: $\begin{array}{r} 45{,}974 \\ 31{,}900 \\ 78{,}211 \\ 12{,}099 \\ +\ 67{,}863 \\ \hline \end{array}$

20. _____

20. Divide: $6{,}050{,}000{,}000 \div 10^5$

21. _____

21. Round 524,942,664 to the nearest ten thousand.

22. _____

22. Estimate the quotient and then divide: $47{,}125 \div 76$

23. _____

23. Simplify: $72 - 8^2 + 27 \div 3$

24. _____

24. Simplify: $(4 \cdot 2)^3 + (3^2)^2 + 9 \cdot 2$

25. _____

25. Find the average of 582, 678, 425, and 979.

26. _____

26. A secretary can type an average of 80 words per minute. If there are approximately 700 words per page, how long will it take the secretary to type 12 pages?

27. _____

27. Twelve people share in the Nationwide lottery jackpot. If the jackpot is worth $9,456,000, how much will each person receive? If each person's share is to be distributed evenly over a 20-year period, how much will each person receive per year?

28. _____

28. Refer to the graph showing auto sales distribution for a local dealer to answer the following questions.

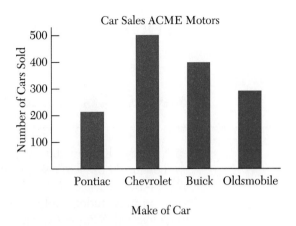

a. What make of auto had the greatest sales?

b. What was the total number of Pontiacs and Chevrolets sold?

c. How many more Buicks were sold than Oldsmobiles?

29. _____

29. Refer to the table to answer the following questions.

**EMPLOYEES BY DIVISION
EXACTO ELECTRONICS**

Division	Day Shift	Swing Shift
A	350	175
B	400	125
C	125	25

a. Which division has the greatest number of employees?
b. How many more employees are in the day shift in Division A as compared with the day shift in Division C?
c. How many employees are in the three divisions?

30. _____

30. Construct a bar graph to display the number of lunches purchased at the local fast food bar over one week: Hamburger, 1100; Fishburger, 300; Chef Salad, 500; Roll and Soup, 300; Omelet, 400.

Good Advice for Studying

New Habits From Old

If you are in the habit of studying math by only reading the examples to learn how to do the exercises, stop now! Instead, read the assigned section—all of it—before class. It is important that you read more than the examples so that you fully understand the concepts. How to do a problem isn't all that needs to be learned. Where and when to use specific skills are also essential.

When you read, read interactively. This means that you should be both writing and thinking about what you are reading. Write down new vocabulary, perhaps start a list of new terms paraphrased in words that are clear to you. Take notes on the How and Why segments, jotting down questions you may have, for example. As you read examples, work the Warm Up problems in the margin. Begin the exercise set only when you understand what you have read in the section. This process should make your study sessions go much faster and be more effective.

If you have written down questions during your study session, be sure to ask them at the next class session, seek help from a tutor, or discuss them with a classmate. Don't leave these questions unanswered.

Pay particular attention to the objectives at the beginning of each section. Read these at least twice; first, when you do your reading before class and again, after attending class. Ask yourself, "Do I understand what the purpose of this section is?" Read the objectives again before test time to see if you feel that you have met these objectives.

During your study session, if you notice yourself becoming tense and your breathing shallow (light and from your throat or upper part of your lungs), follow this simple coping strategy. Say to yourself: "I'm in control. Relax and take a deep breath." Breathe deeply and properly by relaxing your stomach muscle (that's right, you have permission to let your stomach protrude!) and inhaling so that the air reaches the bottom of your lungs. Hold the air in for a few seconds, then slowly exhale, pulling your stomach muscle in as you exhale. This easy exercise not only strengthens your stomach muscle, but gives your body and brain the oxygen you need to perform free from physical stress and anxiety. This deep breathing relaxation method can be done in one to five minutes. You may want to use it several times a day, especially during an exam.

These techniques can help you to start studying math more effectively and to begin managing your anxiety. Begin today.

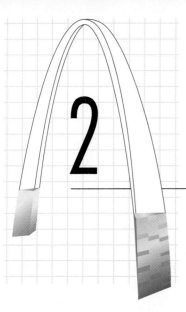

2 Measurement

APPLICATION

A landscape architect has a client who wants an English cottage garden on one side of her house. In addition to planning the garden itself, the architect designs a brick patio with a fountain in the middle and brick paths through the garden. The plans for the garden are in the drawing below. All dimensions are given in feet.

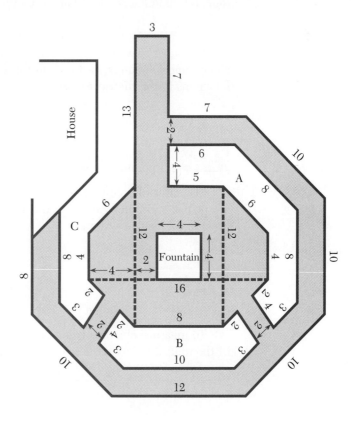

2.1   English and Metric Measurement

OBJECTIVES

1. Multiply and divide a measurement by a number.
2. Add and subtract measurements.

VOCABULARY

A **unit of measure** is the name of a fixed quantity that is used as a standard.

A **measurement** is a number together with a unit of measure.

Equivalent measurements are measures of the same amount but using different units.

The **English system** is the measurement system commonly used in the United States.

The **metric system** is the measurement system used by most of the world.

HOW AND WHY
Objective 1

Multiply and divide a measurement by a number.

One of the primary ways of describing an object is to give its *measurements*. We measure how long an object is, how much it weighs, how much space it occupies, how long it has existed, how hot it is, and so forth. The *units of measure* must be universally defined so that we all mean the same thing when we use a measurement. There are two major systems of measurement in use in the United States. One is the *English system,* so named because we adopted what was used in England at the time. The other is the *metric system,* which is currently used by almost the entire world.

Measures of length answer questions like "how long?" or "how tall?" or "how deep?" Measures of length include inches, feet, yards, and miles in the English system and millimeters, centimeters, meters, and kilometers in the metric system.

└───────────────┘   1 inch   └─────┘   1 centimeter

Measures of weight answer the question "how heavy?" Measures of weight include ounces, pounds, and tons in the English system and milligrams, grams, and kilograms in the metric system.

 1 pound   1 gram

Measures of volume answer "how much space?" Measures of volume include teaspoons, cups, and gallons in the English system and milliliters, liters, and kiloliters in the metric system.

 1 gallon   1 liter

To measure objects bigger than a single unit of measure, we count how many of the units are needed. For example, to measure the length of this line, we count how many inch units are needed for the entire length.

There are four 1-inch units in this length, so we say that it is $4 \cdot (1 \text{ inch}) = 4$ inches.

Similarly, we write

8 centimeters $= 8 \cdot (1 \text{ centimeter})$,
45 pounds $= 45 \cdot (1 \text{ pound})$, and
327 liters $= 327 \cdot (1 \text{ liter})$.

This way of interpreting measurements makes it easy to find multiples of measurements. Consider 3 boards, each 5 feet long. The total length of the boards is

$$3 \cdot (5 \text{ feet}) = 3 \cdot 5 \cdot (1 \text{ foot})$$
$$= 15 \cdot (1 \text{ foot})$$
$$= 15 \text{ feet}$$

Similarly, a case of soda holds a total of 12 liters of soda. How much does each bottle hold if there are six bottles per case?

$$\frac{12 \text{ liters}}{6} = \frac{12 \, (1 \text{ liter})}{6}$$
$$= \frac{12}{6} \, (1 \text{ liter})$$
$$= 2 \, (1 \text{ liter})$$
$$= 2 \text{ liters}$$

▶ **To multiply or divide a measurement by a number**

Multiply or divide the two numbers and write the unit of measure.

Examples A–B	**Warm Ups A–B**

Directions: Measure the length of the following lines.

Strategy: Use a ruler and count the number of units.

A.
(use centimeters)

Mark off units of centimeters and count.

The length is 6 centimeters.

A.
(use centimeters)

B.
(use inches)

Mark off units of inches and count.

The length is 3 inches.

B.
(use inches)

Answers to Warm Ups A. 3 centimeters B. 2 inches

Examples C–D	**Warm Ups C–D**

Directions: Describe each situation with a statement involving measurements and simplify.

C. What is the total weight of four pieces of cheese that each weigh 30 grams?

Strategy: To find the weight of four pieces, multiply the weight of one piece by 4.

Total weight = 4 · (30 grams)   **Multiply.**

= 4 · 30 · (1 gram)

= 120 · (1 gram)

= 120 grams

The four pieces weigh 120 grams.

C. A package of microwave popcorn weighs 101 grams. What is the weight of five packages?

D. If 140 ounces of peanut brittle are divided equally among five sacks, how much goes into each sack?

Strategy: To find how much goes in each sack, divide the total weight by 5.

1 sack = 140 ounces ÷ 5   **Divide.**

= 140 · (1 ounce) ÷ 5

= (140 ÷ 5) · (1 ounce)

= 28 · (1 ounce)

= 28 ounces

Each sack contains 28 ounces.

D. A carpenter has a 16-foot board that he must cut into four equal pieces. How long is each piece?

HOW AND WHY

Objective 2   **Add and subtract measurements.**

The expression "You can't add apples and oranges" applies to adding (and subtracting) measurements. Only measurements with the same unit of measure may be added or subtracted.

10 gallons + 3 gallons = (10 + 3) gallons

= 13 gallons

C A U T I O N   **3 gallons + 4 pints ≠ 7 gallons**

 To add or subtract measurements with the same units of measure

Add or subtract the numbers and write the unit of measure.

If the units of measure do not match, the measurements must first be converted to equivalent measures that do match.

Table 2.1 lists common English measurements, their abbreviations, and their equivalents.

TABLE 2.1   **ENGLISH MEASURES AND EQUIVALENTS**

Length	Time
12 inches (in.) = 1 foot (ft)	60 seconds (sec) = 1 minute (min)
3 feet (ft) = 1 yard (yd)	60 minutes (min) = 1 hour (hr)
5280 feet (ft) = 1 mile (mi)	24 hours (hr) = 1 day
	7 days = 1 week
Liquid Volume	**Weight**
3 teaspoons (tsp) = 1 tablespoon (tbs)	16 ounces (oz) = 1 pound (lb)
2 cups (c) = 1 pint (pt)	2000 pounds (lb) = 1 ton
2 pints (pt) = 1 quart (qt)	
4 quarts (qt) = 1 gallon (gal)	

Use the table to convert units before adding or subtracting. For instance, if a can weighs 1 lb and 3 oz and another can weighs 14 oz, what is the total weight of the two cans?

Total weight = (1 lb 3 oz) + (14 oz)

= (16 oz + 3 oz) + (14 oz)   **Convert 1 pound to 16 oz.**

= (19 oz) + (14 oz)   **Add.**

= 33 oz

= 2 lb 1 oz   **Convert 33 ounces to pounds.**
$33 \div 16 = 2$, remainder 1.

The two cans weigh 33 oz or 2 lb 1 oz.

The metric system was invented by French scientists in 1799. Their goal was to make a system that is easy to learn and use and that would be used worldwide. They based the system for length on the meter and related it to the earth by defining it as 1/10,000,000 of the distance between the north pole and the equator. (A meter is currently defined by international treaty in terms of the wavelength of the orange-red radiation of the element krypton 86.) Units of measure of volume and weight are related to water.

To make the system easy to use, the scientists based all conversions on powers of ten and gave the same suffix to all units of measure for the same characteristic. So, all measures of length end in "-meter," all measures of volume end in "-liter," and all measures of weight end in "-gram." A kilometer is 1000 meters, and a kilogram is 1000 grams. Table 2.2 shows the basic units, abbreviations, and conversions for the metric system.

TABLE 2.2 METRIC MEASURES AND EQUIVALENTS

Length (Basic Unit is 1 Meter)	Weight (Basic Unit is 1 Gram)
1000 millimeters (mm) = 1 meter (m) 100 centimeters (cm) = 1 meter (m) 1000 meters (m) = 1 kilometer (km)	1000 milligrams (mg) = 1 gram (g) 100 centigrams (cg) = 1 gram (g) 1000 grams (g) = 1 kilogram (kg)
Liquid and Dry Measure (Basic Unit is 1 Liter)	
1000 milliliters (mℓ) = 1 liter (ℓ) 100 centiliters (cℓ) = 1 liter (ℓ) 1000 liters (ℓ) = 1 kiloliter (kℓ)	

Examples E–H

Directions: Describe each situation with a statement involving measurements and simplify.

E. Ben weighs 86 kg, Chris weighs 75 kg, and Scott weighs 91 kg. What is the total weight of the three boys?

Strategy: To find the total weight, add the weight of the three boys.

Total weight = 86 kg + 75 kg + 91 kg   **Add.**

= (86 + 75 + 91) kg

= 252 kg

So the total weight of the three boys is 252 kg.

F. Change 154 sec to minutes and seconds.

Strategy: Divide by the number of seconds in a minute.

154 ÷ 60 = 2, R 34   **Since 60 sec = 1 min, divide by 60.**

So, 154 sec = 2 min 34 sec.

G. If a carpenter cuts a piece of board that is 2 ft 5 in. from a board that is 8 ft 3 in. long, how much board will be left?

Strategy: Subtract the length cut off from the length of the board.

8 ft 3 in.
− 2 ft 5 in.
remaining board

7 ft 1 ft 3 in.   **Borrow 1 ft from the 8 ft (1 ft = 12 in.).**
− 2 ft   5 in.

7 ft 15 in.   **1 ft 3 in. = 15 in.**
− 2 ft   5 in.   **Subtract.**
5 ft 10 in.

There will be 5 ft 10 in. of board remaining.

Warm Ups E–H

E. A set of mixing bowls have capacity of 2 ℓ, 5 ℓ, and 8 ℓ. What is the total capacity of the set?

F. Change 22 qt into gallons and quarts.

G. A wine maker draws off 5 gal 3 qt of wine from a 15-gal keg. How much wine is left in the keg?

Answers to Warm Ups   E. 15 ℓ F. 5 gal 2 qt   G. 9 gal 1 qt

H. To run a mile race on the indoor track at the YMCA, the runners have to go around the track eight times. If Abbey runs around the track once, how many feet has she traveled?

Strategy: Divide 1 mile by 8 to find the length of a lap.

$$\text{Distance} = 1 \text{ mile} \div 8 \qquad \textbf{Convert to feet.}$$
$$= 5280 \text{ ft} \div 8 \qquad \textbf{Divide.}$$
$$= 660 \text{ ft}$$

Abbey has run 660 ft.

H. A package of Kool-Aid makes 2 quarts of drink. How many cups of drink is this?

Answers to Warm Ups   H. 8 cups

Exercises 2.1

OBJECTIVE 1: *Multiply and divide measurements by numbers.*

Multiply or divide the following.

A.

1. $(4 \text{ ft}) \cdot 6$

2. $(4 \text{ cups}) \cdot (5)$

3. $(200 \text{ ml}) \div 25$

4. $(28 \text{ days}) \div 4$

5. $(80 \text{ gal}) \div 20$

6. $(55 \text{ mg}) \div 5$

B.

7. $3 \cdot (317 \text{ oz})$

8. $23 \cdot (18 \text{ m}\ell)$

9. $(400 \text{ hours}) \div 8$

10. $(357 \text{ cm}) \div 3$

11. $(2912 \text{ lbs}) \div 14$

12. $(9105 \text{ gal}) \div 15$

13. $(56 \text{ seconds}) \cdot (20)$

14. $(23 \text{ in.}) \cdot 174$

OBJECTIVE 2: *Add and subtract measurements.*

Add or subtract the following.

A.

15. $6 \text{ lb} + 14 \text{ lb}$

16. $5 \text{ m} + 24 \text{ m}$

17. $5 \text{ yd} + 8 \text{ yd} + 4 \text{ yd}$

18. $6 \text{ hr} + 7 \text{ hr} + 8 \text{ hr}$

19. $32 \text{ g} - 12 \text{ g}$

20. $20 \text{ mi} - 13 \text{ mi}$

B.

21. $360 \text{ k}\ell - 155 \text{ k}\ell$

22. $121 \text{ min} - 72 \text{ min}$

23. $48 \text{ mm} + 32 \text{ mm} + 10 \text{ mm}$

24. $35 \text{ m}\ell + 14 \text{ m}\ell + 23 \text{ m}\ell + 4 \text{ m}\ell$

25. 321 yd − 217 yd

26. 170 kg − 89 kg

27. 624 gal − 209 gal + 138 gal

28. 35 qt − 27 qt − 8 qt

29. 210 cm − 45 cm + 24 cm − 165 cm

30. 190 mi − 78 mi + 25 mi − 64 mi

C.

31. Estimate the length of your shoe in both inches and centimeters. Measure your shoe in both units.

32. Estimate the length of your middle finger in both inches and centimeters. Measure your middle finger in both units.

Do the indicated operations and simplify.

33. (6 ft 5 in.) + (2 ft 7 in.) + (10 ft 7 in.)

34. (7 lb 8 oz) + (2 lb 13 oz) + (11 lb 1 oz)

35. (21 min 39 sec) − (14 min 47 sec)

36. (4 yd 1 ft 5 in.) − (2 yd 2 ft 8 in.)

37. (35 min 12 sec) · 6

38. (2 yd 2 ft 1 in.) · 3

39. 2 yd 2 ft 6 in.
  + 3 yd 1 ft 8 in.

40. 2 gal 3 qt 1 pt
  + 4 gal 2 qt 1 pt

41. During one round of golf, Rick made birdie putts of 3 ft 8 in., 12 ft 10 in., 20 ft 8 in., and 7 ft 4 in. What was the total length of all the birdie putts?

42. The Corner Grocery sold 20 lb 6 oz of hamburger on Wednesday, 13 lb 8 oz on Thursday, and 21 lb 9 oz on Friday. How much hamburger was sold during the three days?

43. If a bag contains 298 grams of potato chips, how many grams are contained in seven bags?

44. Lewis, who is a lab assistant, has 312 ml of acid that is to be divided equally among 24 students. How many milliliters will each student receive?

45. The local newspaper in Green Bay, Wisconsin, charted the overnight low temperature for the last five nights. What was the average low temperature for the five nights?

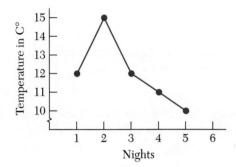

46. If a can of vegetable soup containing 48 oz is split evenly among six people, how large a serving will each person receive?

47. A doctor prescribes allergy medication of two tablets, 20 mg each, to be taken three times per day for a full week. How many milligrams of medication will the patient get in a week?

48. An elevator has a maximum capacity of 2500 lbs. A singing group of eight men and eight women get on. The average weight of the women is 125 lb, and the average weight of the men is 190 lb. Can they ride safely together?

49. The swimming pool at Tualatin Hills is 50 m long. How many meters of lane dividers should be purchased in order to separate the pool into nine lanes?

50. A decorator is wallpapering. If each length of wallpaper is 7 ft 4 in., eight lengths are needed to cover a wall, and three walls are to be covered, how much total wallpaper is needed for the project?

▼ For Exercises 51–53, refer to Chapter 2 application, see page 125.

51. What are the dimensions of the base of the fountain?

52. How wide are the walkways?

53. How long is the patio? How wide is the patio?

STATE YOUR UNDERSTANDING

54. Give two examples of equivalent measures.

55. Is it possible to add 4 g to 5 in.? Explain how or explain why it is not possible.

56. If 8 in. + 10 in. = 1 ft 6 in., why isn't it true that 8 oz + 10 oz = 1 lb 6 oz?

GROUP ACTIVITY

57. Measure, as accurately as you can, at least five parts of a typical desk in the classroom. Give measurements in both the English system and the metric system. Compare your results with the other groups. Do you all agree? Give some possible reasons for the variations in measurements.

MAINTAIN YOUR SKILLS (Sections 1.1, 1.2, 1.6)

58. Round 4566 to the nearest ten.

59. Find the sum of 8, 56, 129, 35, and 604.

60. Replace the ? with < or > in the following expressions.

    a. 14 ? 89 **b.** 4,287,984 ? 4,287,884

61. The auditorium at Lake Community College holds 850 people. If all but 68 tickets for a concert have been sold, and each ticket costs $5.00, what is the total amount of money taken in?

2.2   Perimeter

OBJECTIVE   Find the perimeter of a polygon.

VOCABULARY   A **polygon** is any closed figure whose sides are line segments.

Polygons are named according to the number of sides they have. Table 2.3 lists some common polygons.

TABLE 2.3   COMMON POLYGONS

Number of Sides	Name	Picture
3	Triangle *ABC*	
4	Quadrilateral *ABCD*	
5	Pentagon *ABCDE*	
6	Hexagon *ABCDEF*	
8	Octagon *ABCDEFGH*	
10	Decagon *ABCDEFGHIJ*	

Quadrilaterals are polygons with four sides. Table 2.4 lists the characteristics of common quadrilaterals.

TABLE 2.4   **COMMON QUADRILATERALS**

Trapezoid		One pair of parallel sides
Parallelogram		Two pairs of equal parallel sides
Rectangle		A parallelogram with four right angles
Square		A rectangle with all sides equal

The **perimeter** of a polygon is the distance around the outside of the polygon.

HOW AND WHY
Objective   **Find the perimeter of a polygon.**

The *perimeter* of a figure can be thought of in terms of the distance travelled by walking around the outside of it or by the length of a fence around the figure. The units of measure used for perimeters are length measures (such as inches, feet, meters). Perimeter is calculated by adding the length of all the individual sides.

For example, to calculate the perimeter of this figure, we add the lengths of the sides.

1 ft 6 in.

11 in.   1 ft 2 in.

1 ft 9 in.

$$\begin{array}{r} 1\text{ ft }\ 9\text{ in.} \\ 1\text{ ft }\ 2\text{ in.} \\ 1\text{ ft }\ 6\text{ in.} \\ + \qquad 11\text{ in.} \\ \hline 3\text{ ft }28\text{ in.} \end{array}$$

28 in. = 2 ft 4 in., so

3 ft 28 in. = 3 ft + 2 ft 4 in.

= 5 ft 4 in.

The perimeter is 5 ft 4 in.

 To find the perimeter of a polygon

Add the lengths of the sides.

To find the perimeter of a square

Multiply the length of one side of the square by 4.

$P = 4s$

  To find the perimeter of a rectangle

Add twice the length and twice the width.

$P = 2\ell + 2w$

Examples A–D	**Warm Ups A–D**

Directions: Find the perimeters of the given polygons.

Strategy: Add the lengths of the sides.

A.   Find the perimeter of the trapezoid.

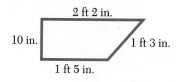

    2 ft   2 in.
    1 ft   3 in.   **Add the lengths.**
    1 ft   5 in.
 +   10 in.
    4 ft 20 in. = 5 ft 8 in.   **20 in. = 1 ft 8 in.**

The perimeter is 5 ft 8 in.

A.   Find the perimeter of the polygon.

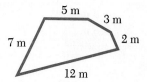

B.   Find the perimeter of the rectangle.

$P = 2\ell + 2w$ **Perimeter formula for rectangles.**

$= 2(14 \text{ cm}) + 2(6 \text{ cm})$ **Substitute.**

$= 28 \text{ cm} + 12 \text{ cm}$ **Multiply.**

$= 40 \text{ cm}$ **Add.**

The perimeter of the rectangle is 40 cm.

B.   Find the perimeter of the square.

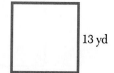

Answers to Warm Ups   A. 29 m   B. 52 yd

C. Find the perimeter of the polygon.

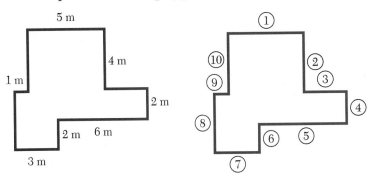

C. Find the perimeter of the polygon.

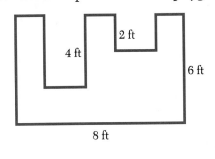

Solution:

Side	Length
1	5 m
2	4 m
3	?
4	2 m
5	6 m
6	2 m
7	3 m
8	?
9	1 m
10	?
3	3 m
8 + 10	8 m

To find the perimeter, number the sides (there are ten) and write down their lengths.

To find the length of side 3, we determine that
side 3 = side 5 + side 7 − side 9 − side 1
= 6 m + 3 m − 1 m − 5 m = 3 m

The lengths of sides 8 and 10 are not given but their sum can be found because
side 8 + side 10 = side 2 + side 4 + side 6 = 4 m + 2 m + 2 m
= 8 m

So,

$P = 5\text{ m} + 4\text{ m} + 3\text{ m} + 2\text{ m} + 6\text{ m} + 2\text{ m} + 3\text{ m} + 1\text{ m} + 8\text{ m}$

$= 34\text{ m}$

The perimeter of the polygon is 34 meters.

D. A carpenter is replacing the baseboards in a room. The floor of the room is pictured. How many feet of baseboard are needed?

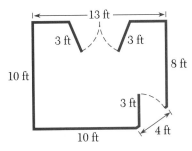

$P = 13\text{ ft} + 10\text{ ft} + 10\text{ ft} + 4\text{ ft} + 8\text{ ft}$

$P = 45\text{ ft}$

Find the perimeter of the room including all the doors.

Baseboard = 45 ft − (6 ft + 3 ft)

= 45 ft − 9 ft = 36 ft

Subtract the combined width of the doors. Simplify.

The carpenter needs 36 ft of baseboard.

D. How much baseboard lumber is needed for the room pictured?

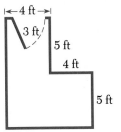

Answers to Warm Ups C. 40 ft   D. 33 ft

Exercises 2.2

OBJECTIVE: *Find the perimeter of a polygon.*

A.

Find the perimeter of the following polygons:

1.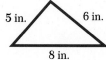
   5 in.   6 in.
   8 in.

2.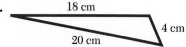
   18 cm
   4 cm
   20 cm

3.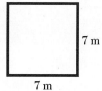
   7 m
   7 m

4.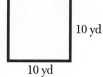
   10 yd
   10 yd

5.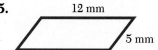
   12 mm
   5 mm

6.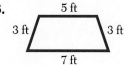
   5 ft
   3 ft   3 ft
   7 ft

7.
   2 km
   11 km

8.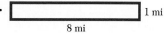
   1 mi
   8 mi

B.

9. Find the perimeter of a triangle with sides 16 mm, 27 mm, and 40 mm.

10. Find the perimeter of a square with sides 230 ft.

11. Find the distance around a rectangular field with length 45 m and width 35 m.

12. Find the distance around a rectangular swimming pool with width 20 yds and length 25 yds.

Find the perimeter of the following polygons:

13.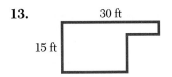

30 ft

15 ft

14.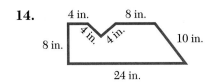

4 in.   8 in.

8 in.   4 in.   4 in.   10 in.

24 in.

15.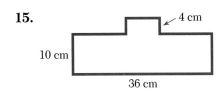

4 cm

10 cm

36 cm

16.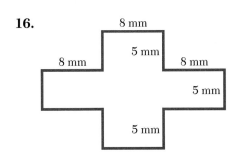

8 mm

5 mm

8 mm   8 mm

5 mm

5 mm

17.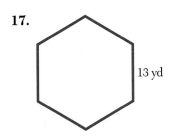

13 yd

18.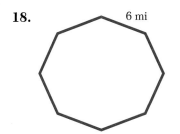

6 mi

19.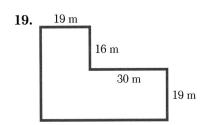

19 m

16 m

30 m

19 m

20.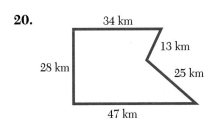

34 km

13 km

28 km   25 km

47 km

C.

21.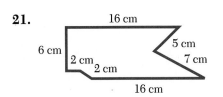

16 cm

6 cm   5 cm

2 cm   7 cm

2 cm

16 cm

22.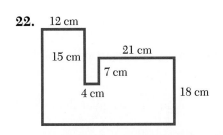

12 cm

15 cm   21 cm

7 cm

4 cm   18 cm

23.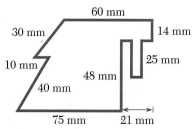

24.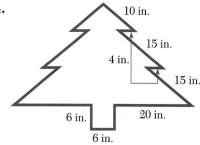

25. How many feet of picture molding are needed to frame four pictures, each measuring 8 in. by 10 in.? The molding is wide enough to require an extra inch added to each length to allow for the corners to be mitred.

26. If fencing costs $12 per meter, what will be the cost of fencing a rectangular lot that is 120 km long and 24 km wide?

27. If Hazel needs 2 minutes to put one foot of binding on a rug, how long will it take her to put the binding on a rug that is 15 ft by 12 ft?

28. How many feet of fencing is needed to fence a rectangular lot that is 92 ft wide and 35 yds long?

29. Holli has a watercolor picture that is to be matted to 14 in. by 20 in. She puts it in a mat that is 3 in. wide on all sides. What is the inside dimension of the frame she needs to buy?

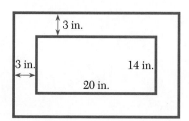

30. Jorge is lining the windows in his living room with Christmas lights. He has one picture window that is 5 ft 8 in. by 4 ft. On each side there is a smaller window that is 2 ft 6 in. by 4 ft. What length of Christmas lights does Jorge need for the three windows?

31. Jenna and Scott just bought a puppy and need to fence their backyard. How much fence should they order?

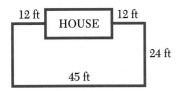

32. As a conditioning exercise, a soccer coach has his team run around the outside of the field three times. If the field measures 60 yds by 100 yds, how far did the team run?

33. A high school football player charted the number of laps he ran around the football field during the first 10 days of practice.

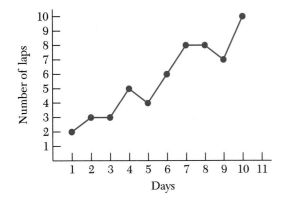

If the field measures 120 yards by 53 yards, how far did he run during the ten days?

34. A carpenter is putting baseboards in the family room/dining room pictured below. How many feet of baseboard molding are needed?

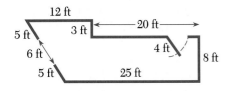

For Exercises 35–39, refer to the Chapter 2 application, see page 125.

35. What is the shape of the patio?

144

36. What is the perimeter of the patio?

37. What is the perimeter of the fountain base?

38. What is the perimeter of flower bed A?

39. What is the perimeter of flower bed B?

STATE YOUR UNDERSTANDING

40. Explain what perimeter is and how to find the perimeter of the figure below.

What possible units (both English and metric) would the perimeter be likely to be measured in if the figure is a national park? If the figure is a room in a house? If the figure is a scrap of paper?

41. Is a rectangle a parallelogram? Why or why not?

42. Explain the difference between a square and a rectangle.

CHALLENGE

43. A farmer wants to build the goat pens pictured below. Each pen will have a gate 2 ft 6 in. wide in the end. What is the total cost of the pens if the fencing is $3.00 per linear foot and each gate is $15.00?

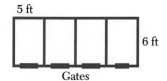

5 ft

6 ft

Gates

GROUP ACTIVITY

44. Federal Express will accept packages according to the formula:

$$L_1 + 2L_2 + 2L_3 \leq 160 \text{ in.}$$

where L_1 is the longest side, L_2 is the next longest side, and L_3 is the shortest side.

Make a table of dimensions of boxes that can be shipped by Federal Express. Are certain sizes more useful than others? Explain.

Longest Side	Next Longest Side	Shortest Side	Total

MAINTAIN YOUR SKILLS (Sections 1.1, 1.2, 1.3)

45. Write the place value notation for nine hundred thousand, fifty.

46. Write the place value notation for nine hundred fifty thousand.

47. Round 32,571,600 to the nearest ten thousand.

48. Find the difference between 733 and 348.

49. Find the product of 733 and 348.

2.3   Area

OBJECTIVE   Find the area of common polygons.

VOCABULARY   **Area** is a measure of surface. It is measured in square units.

The **base** of a geometric figure is a side, parallel to the horizon.

The **altitude** or **height** of a geometric figure is the perpendicular distance from the base to the highest point of the figure.

Table 2.5 shows the base and altitude of some common geometric figures.

TABLE 2.5   **BASE AND HEIGHT OF COMMON GEOMETRIC FIGURES**

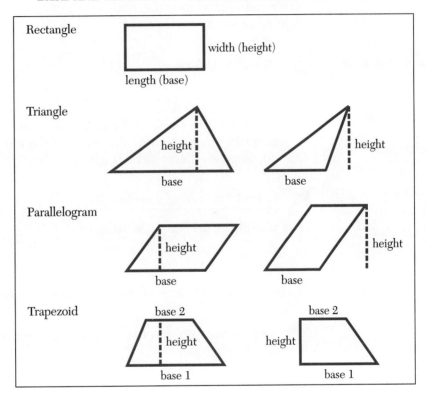

**HOW AND WHY
Objective**

Find the area of common polygons.

Suppose you wish to tile a rectangular bathroom floor that measures 5 ft by 6 ft. The tiles are 1-ft by 1-ft squares. How many tiles do you need?

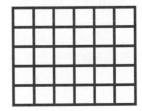

Using the picture as a model of the tiled floor, you can count that 30 tiles are necessary.

Area is a measure of the surface of a figure. It is measured in square units. Square units are literally squares that measure one unit of length on each side. For example,

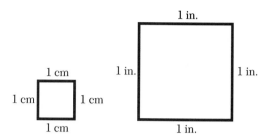

The measure of the square area on the left is 1 square centimeter, abbreviated 1 cm^2. The measure of the square area on the right is 1 square inch, abbreviated 1 in^2. The superscript in the abbreviation is simply part of the unit's name. It does not indicate an exponential operation.

C A U T I O N ⚠ **10 cm^2 ≠ 100 cm.**

In the bathroom floor example, each tile has an area of 1 ft^2. So, the number of tiles needed is the same as the area of the room, 30 ft^2. We could have arrived at this number by multiplying the length of the room by the width. This is not a coincidence. It works for all rectangles.

▶ ***To find the area of a rectangle***

Multiply the length by the width.

$A = \ell w$

To find the area of other geometric shapes, we use the area of a rectangle as a reference.

Since a square is a special case of a rectangle with all sides equal, the formula for the area of a square is

$A = \ell w$
$ = s(s)$
$ = s^2$

▶ ***To find the area of a square***

Square the length of one of the sides.

$A = s^2$

Now let's consider the area of a triangle. We start with a right triangle, that is, a triangle with one 90° angle.

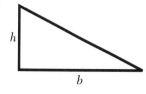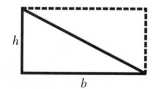

The triangle on the left has base, b, and height, h. The figure on the right is a rectangle with length, b, and width, h. According to the formula for rectangles, the area is $A = \ell w$ or $A = bh$. But the rectangle is made up of two triangles, both of which have a base of b and a height of h. Consequently, it stands to reason that the area of the rectangle (bh) is exactly twice the area of the triangle. So, we conclude that the area of the triangle is (bh) ÷ 2.

Now let's consider a more general triangle.

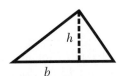 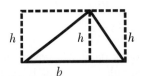

Again, the figure on the left is a triangle with base, b, and height, h. And the figure on the right is a rectangle with length, b, and width, h. Can you see that the rectangle must be exactly twice the area of the original triangle? So again we conclude that the area of the triangle is $A = (bh) \div 2$. Because it is possible to use this technique with any triangle, the general formula for the area of a triangle is the same.

 To find the area of a triangle

Multiply the base times the height and divide by 2.

$$A = (bh) \div 2 \quad \text{or} \quad A = \frac{bh}{2} \quad \text{or} \quad A = \frac{1}{2}bh$$

It is possible to use rectangles to find the formulas for the areas of parallelograms and trapezoids. This is left as an exercise.

 To find the area of a parallelogram

Multiply the base times the height.

$$A = bh$$

 To find the area of a trapezoid

Add the two bases together, multiply by the height, then divide by 2.

$$A = (b_1 + b_2)h \div 2 \quad \text{or} \quad A = \frac{(b_1 + b_2)h}{2}$$

or

$$A = \frac{1}{2}(b_1 + b_2)h$$

The lengths, widths, bases, sides, and heights must all be measured using the same units before the area formulas may be applied. If the units are different, convert to a common unit before calculating area.

Examples A–F	Warm Ups A–F

Directions: Find the area.

Strategy: Use the area formulas.

A. Find the area of a square that is 5 in. on each side.

$A = s^2$ **Formula.**

$\quad = 5^2 = 25$ **Substitute.**

The area is 25 in.^{2}

A. Find the area of a square that is 11 cm on each side.

B. A gallon of deck paint will cover 400 ft^2. A contractor needs to paint a rectangular deck that is 26 ft long and 15 ft wide. Will one gallon of paint be enough?

$A = \ell w$ **Formula.**

$\quad = 26 \text{ ft } (15 \text{ ft})$ **Substitute.**

$\quad = 390 \text{ ft}^2$

Since 390 ft^2 < 400 ft^2, one gallon of paint is enough to paint the deck.

B. A decorator found a 15 yd^2 remnant of carpet. Will it be enough to carpet a 5-yd by 4-yd playroom?

C. Find the area of this triangle.

8 cm

15 cm

$A = \dfrac{bh}{2}$ **Formula.**

$\quad = \dfrac{(15 \text{ cm})(8 \text{ cm})}{2}$ **Substitute.**

$\quad = \dfrac{120 \text{ cm}^2}{2}$ **Simplify.**

$\quad = 60 \text{ cm}^2$

The area is 60 cm^2.

C. Find the area of this triangle.

15 m

10 m

D. Find the area of the parallelogram with a base of 1 ft and a height of 4 in.

$A = bh$ **Formula.**

$\quad = (1 \text{ ft})(4 \text{ in.})$ **Substitute.**

$\quad = (12 \text{ in.})(4 \text{ in.})$ **Convert so that units match.**

$\quad = 48 \text{ in.}^2$ **Simplify.**

The area of the parallelogram is 48 in.^{2}

D. Find the area of the parallelogram with a base of 6 yd and a height of 5 ft.

Answers to Warm Ups   A. 121 cm^2 B. not enough carpet   C. 75 m^2 D. 90 ft^2

E.   Find the area of the trapezoid pictured.

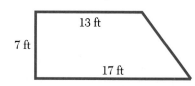

$A = (b_1 + b_2)h \div 2$ **Formula.**

$\quad = (17 \text{ ft} + 13 \text{ ft})(7 \text{ ft}) \div 2$ **Substitute.**

$\quad = (30 \text{ ft})(7 \text{ ft}) \div 2$ **Simplify.**

$\quad = (210 \text{ ft}^2) \div 2$

$\quad = 105 \text{ ft}^2$

The area is 105 ft^2.

E.   Find the area of the trapezoid with bases of 51 m and 36 m and a height of 12 m.

F.   Find the area of the polygon.

Strategy: To find the area of a polygon that is a combination of two or more common figures, first divide it into the common figure components.

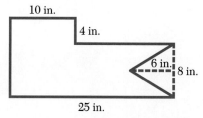

Divide into component figures.

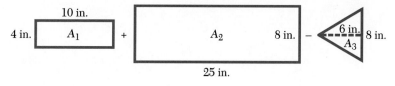

Total area $= A_1 + A_2 - A_3$

$A_1 = (10 \text{ in.})\,(4 \text{ in.}) = 40 \text{ in.}^2$ **Compute the areas of each component.**

$A_2 = (8 \text{ in.})\,(25 \text{ in.}) = 200 \text{ in.}^2$

$A_3 = (6 \text{ in.})\,(8 \text{ in.}) \div 2$

$\quad = (48 \text{ in.}^2) \div 2 = 24 \text{ in.}^2$

Total area $= 40 \text{ in.}^2 + 200 \text{ in.}^2 - 24 \text{ in.}^2$ **Combine the areas.**

$\quad\quad\quad = 216 \text{ in.}^2$

The area of the figure is 216 in.^2

F.   Find the area of the polygon.

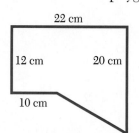

Exercises 2.3

OBJECTIVE: *Find the area of common polygons.*

A.

Find the area of the polygons:

1.

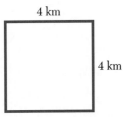

4 km

4 km

2.

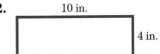

10 in.

4 in.

3.

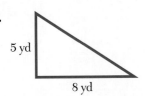

5 yd

8 yd

4.

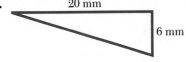

20 mm

6 mm

5.

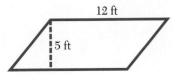

12 ft

5 ft

6.

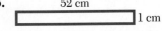

52 cm

1 cm

7.

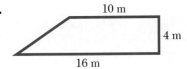

10 m

4 m

16 m

8.

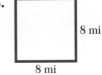

8 mi

8 mi

B.

9. Find the area of a rectangle that has a length of 12 km and a width of 11 km.

10. Find the area of a square with sides of 35 cm.

11. Find the area of a triangle with base 28 yd and height 17 yd.

12. Find the area of a parallelogram with base 56 in. and height 32 in.

Find the areas of the polygons:

13.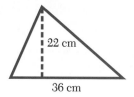

22 cm

36 cm

14.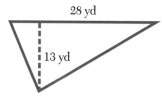

28 yd

13 yd

15.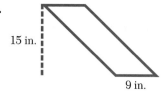

15 in.

9 in.

16.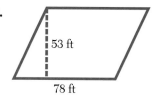

53 ft

78 ft

17.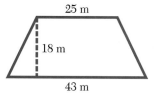

25 m

18 m

43 m

18.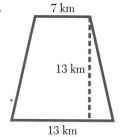

7 km

13 km

13 km

C.

Find the areas of the polygons:

19.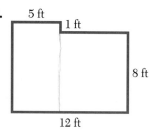

5 ft

1 ft

8 ft

12 ft

20.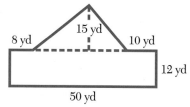

8 yd   15 yd   10 yd

12 yd

50 yd

21.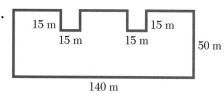

15 m   15 m

15 m   15 m

50 m

140 m

22.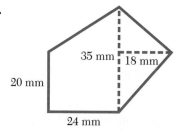

35 mm   18 mm

20 mm

24 mm

23.

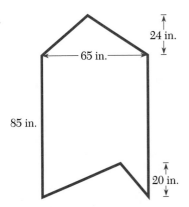

65 in.

24 in.

85 in.

20 in.

24.

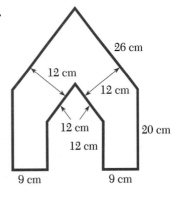

26 cm

12 cm

12 cm

12 cm

12 cm

20 cm

9 cm   9 cm

25. Use a picture to decide how many square feet there are in one square yard.

26. Use a picture to decide how many square centimeters there are in one square meter.

27. The side of Jane's house that has no windows measures 35 ft by 22 ft. If one gallon of stain will cover 250 ft^2, will 2 gallons of stain be enough to stain this side?

28. The south side of Jane's house measures 85 ft by 22 ft and has two windows, each 4 ft by 6 ft. Will 4 gallons of the stain be enough for this side?

29. If one ounce of weed killer treats 1 square meter of lawn, how many ounces of weed killer will Debbie need to treat a rectangular lawn that measures 30 m by 8 m?

30. To the nearest acre, how many acres are contained in a rectangular plot of ground if the length is 1850 ft and the width is 682 ft? (43,560 ft^2 = 1 acre)

31. How much glass is needed to replace a set of two sliding glass doors that each measures 3 ft by 6 ft?

32. How many square yards of carpet are needed to cover a rectangular room that is 9 ft by 12 ft?

33. How many square feet of sheathing are needed for the gable end of a house that has a rise of 9 ft and a span of 36 ft? (See the drawing.)

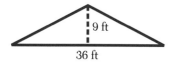

34. A farmer wants to construct a small shed that is 6 ft by 9 ft around the base and 8 ft high. How many gallons of paint will be needed to cover the outside of all four walls of the shed? (Assume 1 gallon covers 250 ft².)

35. How much padding is needed to make a pad for the hexagonal table pictured?

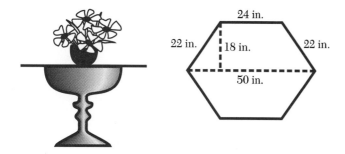

36. A window manufacturer is reviewing the plans of a home to determine the amount of glass needed to fill the order. The number and size of the windows and sliding glass doors are listed in the table.

	Dimensions	**Number Needed**
Windows	3 ft × 3 ft	4
	3 ft × 4 ft	7
	3 ft × 5 ft	2
	4 ft × 4 ft	2
	5 ft × 6 ft	1
Sliding Doors	7 ft × 3 ft	2

How much glass does he need to fill the order?

37. One 2-lb bag of wildflower seed will cover 70 ft². How many bags of seed are needed to cover the region pictured below? (Do not seed the shaded area.)

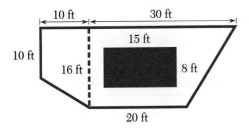

For Exercises 38 and 39, refer to Chapter 2 application, see page 125.

38. The contractor decides to build the patio by laying a concrete slab and then putting bricks on the slab. The bricklayer estimates the number of bricks and the amount of mortar needed based on the area to be covered. Subdivide the patio and the walkways into geometric figures, then calculate the total area to be covered in bricks.

39. The number of bricks and amount of mortar needed also depend on the thickness of mortar between the bricks. The bricklayer decides on a joint thickness of $\frac{1}{4}$ in. According to industry standards, a $\frac{1}{4}$-in. joint thickness will require seven bricks per square foot. Find the total number of bricks required for the patio and walkways.

STATE YOUR UNDERSTANDING

40. What kinds of units measure area? Give examples from both systems.

41. Explain how to calculate the area of the figure below. Do not include the shaded portion.

42. Describe how you could approximate the area of a geometric figure using 1-inch squares.

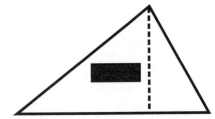

CHALLENGE

43. Joe is going to tile his kitchen floor. Along the outside he will put black squares that are 6 in. on each side. The next (inside) row will be white squares that are 6 in. on each side. The remaining inside rows will alternate between black and white squares that are 1 ft on each side. How many squares of each color and size will he need for the kitchen floor that measures 9 ft by 10 ft?

44. A rectangular plot of ground measuring 120 ft by 200 ft is to have a cement walk 5 ft wide placed around the inside of the perimeter. How much of the area of the plot will be used by the walk and how much of the area will remain for the lawn?

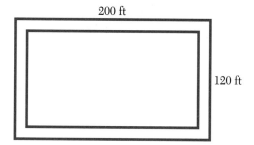

45. Ingrid is going to carpet two rooms in her house. The floor in one room measures 30 ft by 24 ft and the floor in the other room measures 22 ft by 18 ft. If the carpet costs $27.00 per square yard installed, what will it cost Ingrid to have the carpet installed?

GROUP ACTIVITY

46. Use the formula for area of a rectangle to find the area of a parallelogram and a trapezoid. Draw pictures to illustrate your argument.

47. Determine the coverage of one gallon of semi-gloss paint. How much of this paint is needed to paint your classroom, excluding chalkboards, windows, and doors? What would it cost? Compare your results with the other groups in the class. Did all the groups get the same results? Give possible explanations for the differences.

MAINTAIN YOUR SKILLS (Sections 1.1, 1.3, 1.6)

48. Find the product of 47 and 962. Round your answer to the nearest 100.

49. Find the quotient of 295,850 and 97.

50. Simplify $2(46 - 28) + 50 \div 5$.

51. A family of six attends a weaving exhibition. Parking was $4.00, adult admission to the exhibition $6.00, senior admission $4.00, and child admission $3.00. How much does it cost for two parents, one grandmother, and three children to attend the exhibition?

2.4   Volume

OBJECTIVE   Find the volume of common geometric shapes.

VOCABULARY   A **cube** is a three-dimensional geometric solid that has six sides (called **faces**), each of which is a square.

Volume is the name given to the amount of space that is contained inside a three dimensional object.

HOW AND WHY
Objective   **Find the volume of common geometric shapes.**

Suppose you have a shoe box that measures 12 in. long by 4 in. wide by 5 in. high that you want to use to store toy blocks that are 1 in. by 1 in. by 1 in. How many blocks will fit in the box?

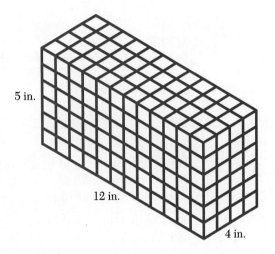

In each layer of blocks there are 12(4) = 48 blocks and there are five layers. Therefore, the box holds 48(5) = 240 blocks.

   Volume is a measure of the amount of space that is contained in a three-dimensional object. Often, volume is measured in cubic units. These units are literally *cubes* that measure one unit on each side. For example, pictured below is a cubic inch (1 in.^{3}) and a cubic centimeter (1 cm^3).

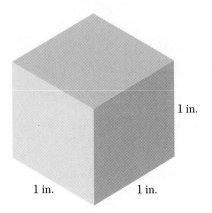

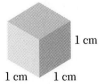

The shoebox discussed previously has a volume of 240 in.^{3} because exactly 240 blocks, which have volume 1 in.^{3}, can fit inside it and totally fill it up.

In general, volume can be thought of as the number of cubes that fill up a space. If the space is a rectangular solid, like the shoebox, it is a relatively easy matter to determine the volume by making a layer of cubes that covers the bottom, and then deciding how many layers are necessary to fill the box. Note that the number of cubes needed for the bottom layer is the same as the area of the base of the box, ℓw. The number of layers needed is the same as the height of the box, h. So, we come to the following volume formula.

 __To find the volume of a rectangular solid__

Multiply the length by the width by the height.

$V = \ell w h$

 __To find the volume of a cube__

Cube one of the sides.

$V = s^3$

The length, width, and height must all be measured using the same units before the volume formulas may be applied. If the units are different, convert to a common unit before calculating the volume.

The principle used for finding the volume of a box can be extended to any solid with sides that are perpendicular to the base. The area of the base gives the number of cubes necessary to make the bottom layer, and the height gives the number of layers necessary to fill the solid.

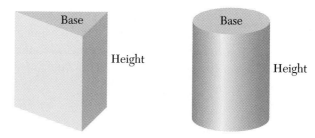

 __To find the volume of a solid with sides perpendicular to the base__

Multiply the area of the base by the height

$V = Bh$

where B is the area of the base.

When measuring the capacity of a solid to hold liquid, sometimes special units are used. Recall that in the English system, liquid capacity is measured in ounces, quarts, and gallons. In the metric system, milliliters, liters, and kiloliters are used.

One cubic centimeter measures the same volume as one milliliter. That is, $1 \text{ cm}^3 = 1 \text{ m}\ell$. So, a can whose base has area 10 cm² with a height of 5 cm has a volume of

$$V = (10 \text{ cm}^2)(5 \text{ cm})$$
$$= 50 \text{ cm}^3$$
$$= 50 \text{ m}\ell$$

The can holds 50 mℓ of liquid.

Examples A–D	**Warm Ups A–D**

Directions: Find the volume.

Strategy: Use the volume formulas.

A.   Find the volume of a cube that is 7 meters on each edge.
 	
  $V = s^3$ **Formula.**
    $= (7 \text{ m})^3$ **Substitute.**
    $= 343 \text{ m}^3$ **Simplify.**
 	
  The volume is 343 m³.
	A. Find the volume of a cube that has an edge of length 10 m.

B.   How much concrete is needed to pour a step that is 4 ft long, 3 ft wide, and 6 inches deep?
 	
  $V = \ell wh$ **Formula.**
    $= (4 \text{ ft})(3 \text{ ft})(6 \text{ in.})$ **Substitute.**
    $= (48 \text{ in.})(36 \text{ in.})(6 \text{ in.})$ **Convert to common units.**
    $= 10{,}368 \text{ in.}^3$ **Simplify.**
 	
  The step requires 10,368 in.³ of concrete.
	B. How much concrete is needed for a rectangular stepping stone that is 1 ft long, 8 in. wide, and 3 in. deep?

C.   What is the volume of a can that is 5 in. tall and has a base with area 7 in.²?
 	
  $V = Bh$ **Use the formula for volume.**
    $= (7 \text{ in.}^2)(5 \text{ in.})$ **Substitute.**
    $= 35 \text{ in.}^3$ **Simplify.**
 	
  The can holds 35 in.³.
	C. What is the volume of a garbage can that is 3 ft tall and has a base with area 4 ft²?

Answers to Warm Ups   A. 1000 m³   B. 288 in.³   C. 12 ft³

D. How many milliliters of water does this container hold?

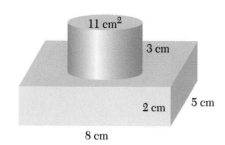

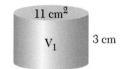

First separate the figure into common components.

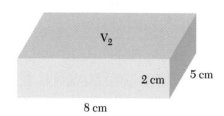

Total volume $= V_1 + V_2$

$V_1 = (11 \text{ cm}^2)(3 \text{ cm})$
   $= 33 \text{ cm}^3$

Compute the volume of each component.

$V_2 = (8 \text{ cm})(5 \text{ cm})(2 \text{ cm})$
   $= 80 \text{ cm}^3$

Total volume $= 33 \text{ cm}^3 + 80 \text{ cm}^3$
            $= 113 \text{ cm}^3$
            $= 113 \text{ m}\ell$

Combine the individual volumes.

Convert to mℓ.

The figure holds 113 mℓ of water.

D. How many milliliters of water does this container hold?

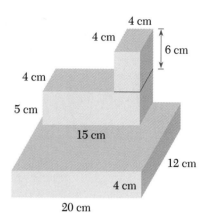

Exercises 2.4

OBJECTIVE: *Find the volume of common geometric shapes.*

A.

Find the volume of the figures.

1.

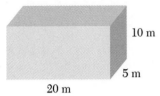

10 m

5 m

20 m

2.

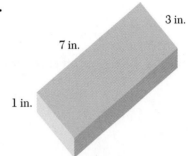

3 in.

7 in.

1 in.

3.

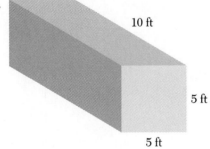

10 ft

5 ft

5 ft

4.

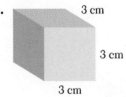

3 cm

3 cm

3 cm

5.

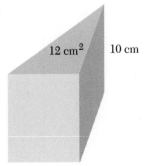

12 cm^2

10 cm

6.

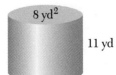

8 yd^2

11 yd

B.

7. How many milliliters of water will fill up a box that measures 52 cm long, 35 cm wide, and 12 cm high?

8. Find the volume of a cube that measures 24 mi on each side.

9. Find the volume of a garbage can that is 4 ft tall and has a base with area 12 ft^2.

10. Find the volume of two identical fuzzy dice tied to the mirror of a `57 Chevy if one edge measures 6 in.

Find the volume of the figures.

11.

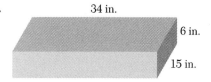

34 in.

6 in.

15 in.

12.

175 mm

45 mm^2

13.

355 cm^2

122 cm

14.

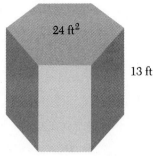

24 ft^2

13 ft

C.

15. Find the volume of a can that has a base with area 245 in.^{2} and a height of 2 ft.

16. How many cubic inches are there in a cubic foot? How many cubic inches are there in 5 ft^3?

17. How many cubic feet are there in 5184 in.^{3}?

18. How many cubic feet are there in a cubic yard? How many cubic feet are there in 4 yd^3?

19. How many cubic yards are there in 270 ft^3?

20. How many cubic centimeters are there in a cubic meter? How many cubic centimeters are there in 3 m^3?

21. How many cubic inches of concrete are needed to pour a sidewalk that is 3 ft wide, 4 in. deep, and 54 ft long? Concrete is commonly measured in cubic yards, so convert your answer to cubic yards. (*Hint:* Convert to cubic feet first, then convert that to cubic yards.)

Find the volume of the figures.

22.

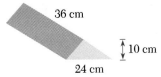

36 cm

10 cm

24 cm

23.

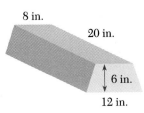

8 in.

20 in.

6 in.

12 in.

24.

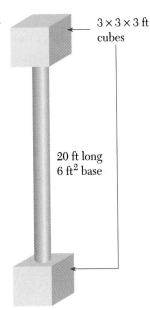

3 × 3 × 3 ft cubes

20 ft long
6 ft² base

25.

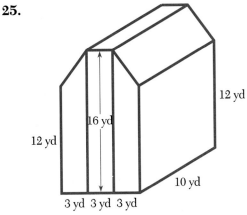

12 yd

16 yd

12 yd

10 yd

3 yd 3 yd 3 yd

26.

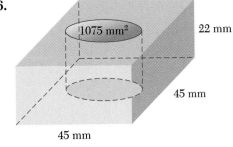

1075 mm²

22 mm

45 mm

45 mm

27.

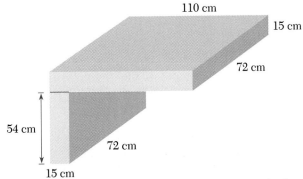

110 cm

15 cm

72 cm

54 cm

72 cm

15 cm

28. A bag of potting soil contains 3500 in.³. How many bags are needed to fill five flower boxes, each of which measures 4 ft long, 8 in. high, and 6 in. deep? Remember that you can only buy whole bags of potting soil.

29. An excavation is being made for a basement. The hole is 24 ft wide, 36 ft long, and 7 ft deep. If the bed of a truck holds 378 ft³, how many truckloads of dirt will need to be hauled away?

166

30. A farming corporation is building 4 new grain silos. The inside dimensions of the silos are given in the table.

	Area of base	**Height**
Silo A	1800 ft^2	60 ft
Silo B	1200 ft^2	75 ft
Silo C	900 ft^2	80 ft
Silo D	600 ft^2	100 ft

Find the total volume available in these silos.

For Exercises 31–33, refer to Chapter 2 application, see pages 125 and 157, Exercises 38–39.

31. According to industry standards, a joint thickness of $\frac{1}{4}$ inch means the bricklayer will need 1 ft^3 of mortar per 112 bricks. Find the total amount of mortar needed for the patio and walkways. Round to the nearest whole cubic foot.

32. The cement subcontractor orders materials based on the total volume of the slab. The industry standard for patios and walkways is 4–5 inches of thickness. Since the slab will be topped with bricks, the contractor has ordered a thickness of 4 inches. Find the volume of the slab in cubic inches. Convert this to cubic feet, rounding to the nearest whole cubic foot. Convert this to cubic yards, again rounding if necessary.

33. The cement contractor must first build a wood form that completely outlines the slab. How many linear feet of wood are needed to build the form?

STATE YOUR UNDERSTANDING

34. Explain what is meant by volume. Name three occasions in the last week when the volume of an object was relevant.

35. Explain how to find the volume of the figure below.

36. Explain why the formula for the volume of a box is a special case of the formula $V = Bh$.

CHALLENGE

37. The Bakers are constructing an in-ground pool in their backyard. The pool will be 15 ft wide and 30 ft long. It will be 3 ft deep for a distance of 10 ft on one end. It will then drop to a depth of 10 ft at the other end. How many cubic feet of water are needed to fill the pool?
If the trucks hauling away the dirt dug to make the pool have a capacity of 14 yd^3, how many loads of dirt were hauled away?

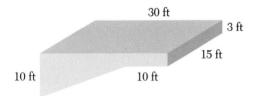

38. Norma is buying mushroom compost to mulch her garden. The garden is pictured below. How many cubic yards of compost does she need to mulch the entire garden 4 in. deep? (She cannot buy fractional parts of a cubic yard.)

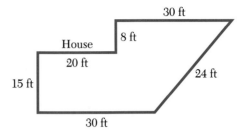

GROUP ACTIVITY

39. The insulating ability of construction materials is measured in R-values. Industry standards for exterior walls are currently R-19. An 8-in. thickness of loose fiberglass is necessary to achieve an R-19 value. Calculate the amount of cubic feet of loose fiberglass needed to insulate the exterior walls of the mountain cabin pictured. All four side windows measure 2 by 3 ft. Both doors measure 4 by 7 ft. The front window measures 5 by 3 ft.

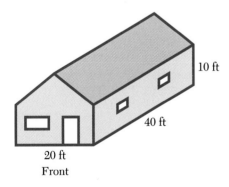

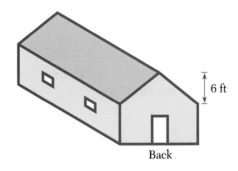

MAINTAIN YOUR SKILLS (Sections 1.4, 1.5, 1.6)

Evaluate the following.

40. 8^3

41. $4^2 + 5^2$

42. $2^3 + 3^3 + 4^3$

43. $9^2 - 4^2$

44. $(2^2 + 3^2)^2$

CHAPTER 2

OPTIONAL Group Project *(2–3 weeks)*

You are working for a kitchen design firm that has been hired to design a kitchen for the 10-ft by 12-ft room pictured below.

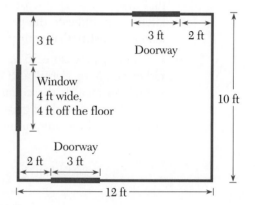

The following table lists appliances and dimensions. Some of the appliances are required and others are optional. All dimensions are in inches.

Appliance	High	Wide	Deep	Required
Refrigerator	68	30 or 33	30	Yes
Range/Oven	30	30	26	Yes
Sink	12	36	22	Yes
Dishwasher	30	24	24	No
Trash compactor	30	15	24	No
Built in microwave	24	24	24	No

The base cabinets are all 30 in. high and 24 in. deep. The widths can be any multiple of 3 from 12 in. to 36 in. Corner units are 36 in. along the wall in each direction. The base cabinets (and the range, dishwasher, and compactor) will all be installed on 4 in. bases that are 20 in. deep.

The wall (upper) cabinets are all 30 in. high and 12 in. deep. Here too, the widths can be any multiple of 3 from 12 in. to 36 in. Corner units are 24 in. along the wall in each direction.

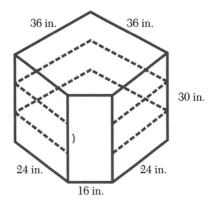

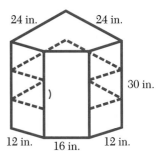

a. The first step is to place the cabinets and appliances. Your client has specified that there must be at least 80 ft³ of cabinet space. Make a scale drawing of the kitchen and indicate the placement of the cabinets and appliances. Show calculations to justify that your plan satisfies the 80-ft³ requirement.

b. Countertops measure 25 in. deep with a 4-in. backsplash. The countertops can either be tile or Formica. If the counters are Formica, there will be a 2-in. facing of Formica. If the counters are tile, the facing will be wood that matches the cabinets. (See the figure.) Calculate the amount of Formica needed for the counters, and the amount of tile and wood needed for the counters.

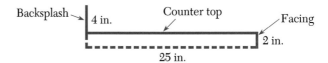

c. The bases under the base cabinets will be covered with a rubber kickplate that is 4 in. high and comes in 8 ft lengths. Calculate the total length of kickplate material needed, and the number of lengths of kickplate material necessary to complete the kitchen.

d. Take your plan to a store that sells kitchen cabinets and counters. Your goal is to get the best quality materials for the least amount of money. Prepare at least two cost estimates for the client. Do not include labor in your estimates, but do include the appliances. Include a rationale with each estimate, explaining the choices you made. Which plan will you recommend to the client and why?

CHAPTER 2

True-False Concept Review

ANSWERS

Check your understanding of the language of algebra and geometry. Tell whether each of the following statements is True (always true) or False (not always true). For each statement that you judge to be false, revise it to make a statement that is true.

1. _____

   1. English measurements are the most commonly used in the world.

2. _____

   2. Equivalent measures have different units of measurement.

3. _____

   3. A liter is a measure of weight.

4. _____

   4. The perimeter of a square can be found in inches, feet, centimeters, or meters.

5. _____

   5. Area is the measure of the inside of a solid such as a box or a can.

6. _____

   6. The volume of a square is $V = s^3$.

7. _____

   7. The formula for the area of a trapezoid is $A = \dfrac{(b_1 + b_2)h}{2}$.

8. _____

   8. It is possible to find equivalent measures without remeasuring the original object.

9. _____

   9. The metric system utilizes the base-ten place-value system.

10. _____

   10. Volume is the measure of how much a container will hold.

11. _____

   11. Weight can be measured in pounds, grams, or kilograms.

12. _____

   12. A parallelogram has three sides.

13. _____

   13. Volume can be thought of as the number of squares in an object.

14. _____

14. One mℓ is equivalent to 1 cm³.

15. _____

15. Measurements can only be added or subtracted when they are expressed with the same unit of measure.

16. _____

16. The distance around a geometric figure is called the perimeter.

17. _____

17. Volume is always measured in cubic units.

18. _____

18. A trapezoid is a quadrilateral.

19. _____

19. $1 \text{ ft}^2 = 12 \text{ in.}^2$.

20. _____

20. The prefix "kilo" means 100.

CHAPTER 2

Review

Section 2.1 *Objective 1*

1. Multiply: (6 ft)5
2. Divide: 1722 mg ÷ 7
3. If a carpenter has six boards each of length 4 ft, what is the total length?
4. If $120 is divided equally among 15 people, what will be each one's share?
5. Each of 25 oil drums contains 190 liters. What is the total number of liters contained in the 25 barrels?

Section 2.1 *Objective 2*

6. Add: 5 days 6 hours
          6 days 9 hours
        + 7 days 6 hours

7. Subtract: 2 g − 350 mg
8. Add: 4m + 67 cm
9. Subtract: 45 gal 2 qt
             − 18 gal 3 qt
10. Find the sum of 3 ft 5 in. and 7 ft 7 in. expressed in feet.

Section 2.2

11. Find the perimeter of a triangle whose sides are 3 ft, 4 ft, and 5 ft.
12. Find the perimeter of a rectangle with length 15 cm and width 10 cm.
13. Find the perimeter of an octagon with each side 4 m.
14. Find the perimeter of the figure.

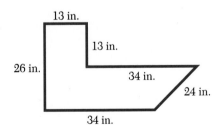

15. If fencing costs $5 per foot, what will be the cost of fencing a rectangular lot that is 1500 ft long and 250 ft wide?

Section 2.3

16. Find the area of a triangle whose base is 18 m and height is 9 m.
17. Find the area of a parallelogram that has a base of 21 ft and a height of 9 ft.
18. Find the area of a square postage stamp that measures 3 cm on each side.
19. How many square inches of glass are in a rectangular mirror that is 3 ft wide and 18 in. high?
20. Find the area of the figure.

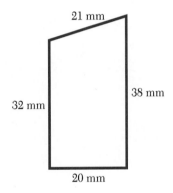

Section 2.4

21. Find the volume of a box with length of 12 in., width of 8 in., and a height of 3 in.
22. Find the volume of a garbage can that is 150 cm tall and has a base of 3800 cm².
23. A swimming pool that is 40 ft long and 8 ft wide is filled to a depth of 6 ft. How many cubic feet of water are in the pool?
24. Find the volume of the figure.

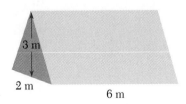

25. Find the volume of a cubic jack-in-the-box with side of 7 in.

CHAPTER 2

Test

ANSWERS

1. _____

2. _____

3. _____

4. _____

5. _____

6. _____

7. _____

1. 7 m + 454 mm = ? mm.

2. Find the perimeter of a square that is 34 cm on a side.

3. Find the volume of a drawer that is 4 in. high, 18 in. wide, and 24 in. deep.

4. How much vinyl flooring is needed to cover the room pictured.

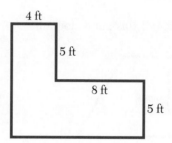

5. Anna has 135 lb of strawberries to divide equally among her five children. How many pounds of berries will each one receive?

6. Find the perimeter of the figure.

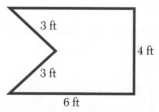

7. Find the area of this triangle.

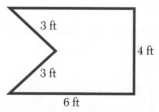

8. _____

8. Subtract: 5 gal 2 qt
                  − 3 gal 2 qt 1 pt

9. _____

9. Name two units of measure in the English system for weight. Name two units of measure in the metric system for weight.

10. _____

10. Change 2 ft² to square inches.

11. _____

11. How much weather stripping is needed to line the inside of a picture window 4 ft wide and 5 ft tall and two side windows each measuring 2 ft wide by 5 ft tall?

12. _____

12. Find the volume of the figure.

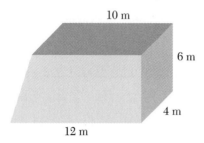

13. _____

13. Find the area of this parallelogram.

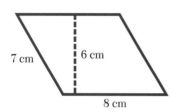

14. _____

14. Find the volume of a hot water tank that has a circular base with area of 4 ft² and a height of 5 ft.

15. _____

15. The Golden Silver Company has a bar of silver weighing 684 g. If it is melted down to form six bars of equal weight, what will each bar weigh?

16. _____

16. The cost of heavy duty steel wire fencing is $2 per linear foot. How much will it cost to build the dog runs pictured?

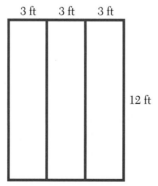

17. _____

17. Name two units of measure in the English system for volume. Name two units of measure in the metric system for volume.

18. _____

18. Find the area of the figure.

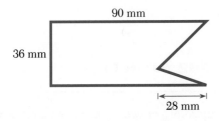

19. _____

19. A football coach is conditioning his team by having them run around the edge of the field three times each hour. A football field measures 100 yd long by 60 yd wide. How far does each player run in a 2-hour practice?

20. _____

20. Both area and volume describe interior space. Explain how they are different.

Good Advice for Studying

Managing Anxiety

For those of you who become anxious as you begin to study, we recommend that you devote a section in your math notebook to record your thoughts and feelings each time you study. Record your thoughts in the first person, as "I-statements." For example, "Nobody ever uses this stuff in real life." Rephrased as an "I" statement, it would be: "I would like to know where I would use this in my life." Research shows that a positive attitude is the single most important key to your success in mathematics. By recognizing negative thoughts and replacing them with positive thoughts, you are beginning to work on changing your attitude. The first step is to become aware of your self-talk.

Negative self-talk falls into three categories of irrational beliefs that you have about yourself and how you view the world. You think that (1) *Worrying helps.* Wrong. Worrying leads to excessive anxiety, which is distracting and hinders performance. (2) *Your worth as a person is determined by your performance.* Wrong. Not being able to solve math problems doesn't mean you won't amount to anything. If you think it does, you are a victim of "catastrophizing." (3) *You are the only one who feels anxious.* Wrong. By thinking that other students have some magical coping skills that allows them to avoid anxiety, you are comparing yourself to some irrational mythical norm.

As you begin your first section, ask yourself such questions as "What am I saying to myself now?" "What is triggering these thoughts?" "How do I feel physically?" and "What emotions am I feeling now?" Your answers will likely reveal a pattern to your thoughts and feelings. You need to analyze your statements and change them into more positive and rational self-talk. You can use a simple technique called "rational emotive therapy."[1] This method, if practiced regularly, can quickly and effectively change the way you think and feel about math. When you find yourself getting upset, watch for words such as *should, must, ought to, never, always,* and *I can't.* They are clues to negative self-talk and signals for you to direct your attention back to math. Use the following ABCD model.

A. *Triggering event:* You start to do your math homework and your mind goes blank.

B. *Negative self-talk* in response to the trigger: "I can't do math. I'll never pass. My life is ruined!"

C. *Anxiety* caused by the negative self-talk: panic, anger, tight neck, etc.

D. *Positive self-talk* to cope with anxiety: "This negative self-talk is distracting. It doesn't help me solve these problems. Focus on the problems"; or say, "I may be uncomfortable, but I'm working on it."

Recognize that negative self-talk (B) is the culprit. It causes the anxiety (C) and must be restructured to positive rational statements (D). Practice with the model using your own self-talk statements in step (B) and then complete the remaining steps.

[1]R.E.T. was created by Albert Ellis.

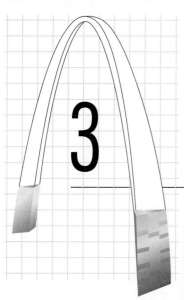

3 Introduction to Algebra

APPLICATION

A certified meeting planner is arranging for a convention of plastic surgeons in a downtown hotel in Washington, D.C. The convention is expecting somewhere between 800 and 1100 registrants for three days of meetings and social functions. The planner will need to decide on the space and personnel requirements for each of the various functions so that the hotel staff can assign rooms, equipment, and personnel as necessary. The convention is six months away so the exact number of registrants is unknown. To cover all possibilities, the planner decides to draw up a table that details the requirements for 800, 900, 1000, and 1100 registrants. As early registrations come in, the planner will be able to predict more accurately which plan to use.

The planner begins by thinking about the number of clerks needed to register and to check in the convention participants. The manual recommends one clerk for every 100 registrants. Use this information to fill in the table.

Number of registrants	800	900	1000	1100
Number of clerks				

Describe in words the process you used to fill in the table. Did it seem repetitious?

Very often, in the real world, the same process is repeated using different inputs, much like the process you went through to determine the required number of clerks for different numbers of registrants. In algebra, we use numbers for the parts of the process that do not change, and we use letters or variables for the parts of the process that might change with different situations. To use algebra to describe how to find the required number of clerks, begin with a word description of the process.

The number of registration clerks needed is calculated by dividing the total number of registrants by 100.

Next, assign variable names to the quantities that might change. Let the number of registration clerks needed be called C, and let the total number of registrants be called R. Rewrite the word description of the process using mathematical symbols.

$$C = R \div 100 \quad \text{or} \quad C = \frac{R}{100}$$

Clearly this algebraic statement is shorter that the written version, and that is precisely why many people use it. Algebra can be thought of as a mathematical shorthand for word descriptions of different processes. This shorthand uses simple names (variables) for various quantities and clearly states the relationship between the quantities. From this point on, using algebra to describe a process will become more common in your mathematical studies.

3.1 The Language of Algebra

OBJECTIVES

1. Translate an English phrase to an algebraic expression.

2. Evaluate an algebraic expression.

VOCABULARY

A **variable** is a placeholder for a number. We use letters of the alphabet for variables.

Algebraic expressions may contain numbers; operation signs for addition, subtraction, multiplication, and division; powers; and variables.

Substitute means to replace a variable by a number.

Evaluate means to substitute and simplify.

HOW AND WHY
Objective 1

Translate an English phrase to an algebraic expression.

Algebraic expressions are a combination of numbers, variables, and operation signs. For example,

$$(15 - 7) + 18, \quad 4x + 89, \quad 2p + 2q, \quad \text{and} \quad 4x^2 - ab$$

are algebraic expressions. A number and a variable $(4x)$ or two variables together (ab) indicate multiplication. So, $4x$ means "4 times x" and ab means "a times b." One of the uses of algebra is a kind of shorthand for writing general statements about relationships. For instance, to show the sums of 12 and the first ten counting numbers we need to write ten sums:

$$12 + 1, \quad 12 + 2, \quad 12 + 3, \quad 12 + 4, \quad 12 + 5, \quad 12 + 6,$$
$$12 + 7, \quad 12 + 8, \quad 12 + 9, \quad 12 + 10$$

In algebra, letters used as place holders for numbers are called *variables* or *unknowns*. For example, the ten sums can be represented by $12 + n$, where n can be replaced with 1, 2, 3, 4, 5, 6, 7, 8, 9, and 10. By substituting the ten whole numbers for n we have all of the sums.

Often, *algebra* is a generalization of arithmetic showing structure and relationships among quantities using variables, numbers, and operation symbols. In algebraic form, the order in which the numbers, variables, and operation signs appear is critical. Study Table 3.1, which shows the three forms. Carefully think about the order in each case.

TABLE 3.1

Arithmetic Form	Algebraic Form	Possible English Form
Addition		
$11 + 7$	$x + 7$	The sum of x and 7
	$2 + y$	y more than 2
	$x + y$	x increased by y
Subtraction		
$19 - 9$	$x - 9$	9 less than x or x minus 9
	$11 - y$	Subtract y from 11 or 11 minus y
	$x - y$	The difference of x and y.
	$58 - w$	The value after w is subtracted from 58 or the difference of 58 and w
Multiplication		
3×6	$3(y)$	3 times y
$3 \cdot 6$	$(3)y$	y multiplied by 3 or the product of 3 and y
$3(6)$	$3 \cdot y$	(In $3y$ the parentheses and the dot have been
$(3)(6)$	$3y$	dropped. This is the usual way to write the product of a variable and a constant.)
	xy	The product of two numbers represented by x and y
$5 \cdot 5 \cdot 5$	x^3	x cubed or a number cubed or the cube of a number
Division		
$3 \div 4$ $4\overline{)3}$	$x \div 4$ $\dfrac{x}{4}$	x divided by 4. $\left(\text{In algebra this is most often written } \dfrac{x}{4}.\right)$
$\dfrac{3}{4}$	$\dfrac{x}{y}$	The quotient of x and y, x over y, x divided by y, x per y, or y into x.
$\dfrac{5}{1000}$	$\dfrac{t}{1000}$	t per thousand

Some translations from English phrases to algebraic expressions are shown in Table 3.2.

TABLE 3.2

English Form	Variable	Operation or Relation	Algebraic Form
A number increased by 7	Let x represent the number.	"Increased" indicates addition.	$x + 7$
7 more than a given number	Let x represent the given number.	"More than" indicates addition.	$x + 7$
A number decreased by 11	Let y represent the number.	"Decreased" indicates subtraction.	$y - 11$
The difference of a number and 11	Let y represent the number.	"Difference" indicates subtraction.	$y - 11$
The product of 8 and a number	Let a represent the number.	"Product" indicates multiplication.	$8 \cdot a$ or $8a$
The quotient of a number and 18	Let b represent the number.	"Quotient" indicates division.	$\dfrac{b}{18}$
The square of a number	Let t represent the number.	"Square" indicates that the base is used as a factor twice.	$t \cdot t$ or t^2
The sum of two numbers	Let m represent one number and n the other.	"Sum" indicates addition.	$m + n$
Twice a number is equal to 21.	Let w represent the number.	"Equal" indicates the relation of having the same value.	$2w = 21$
A number plus 7 is less than 13.	Let x represent the number.	"Less than" indicates the relation of having a smaller value.	$x + 7 < 13$
50 is greater than 8 divided by a number.	Let y represent the number.	"Greater than" indicates the relation of having a larger value.	$50 > \dfrac{8}{y}$
The ratio of a number to 5	Let w represent the number.	"Ratio" indicates a division.	$\dfrac{w}{5}$

 To translate from English phrases to algebraic expressions

Represent each unknown number with a different variable and indicate the operations.

Examples A–B   **Warm Ups A–B**

Directions: Translate to algebra.

Strategy: Represent each unknown number with a variable and use operations symbols.

A.   Six less than the product of four and a number.

Let y represent the number.

$4y$ **This represents the product of 4 and a number. The term, $4y$, is written first because "six less" means "subtract six from what follows."**

$4y - 6$ **The words "less than" indicate a subtraction problem.**

C A U T I O N **One more word can change the entire meaning. If we say "six *is* less than the product of four and a number" the meaning is changed to the inequality $6 < 4y$.**

A.   Two hundred decreased by the product of three and a number.

B.   Five times the difference of a number and nine.

Let x represent the number.

$x - 9$ **This shows the difference of a number and 9.**

$5(x - 9)$ **This shows 5 times the difference.**

C A U T I O N **The parentheses are vital because they indicate the order of operations. If we write $5x - 9$, 5 is multiplied only times x, not times the difference.**

B.   Eight times the sum of a number and fourteen.

HOW AND WHY
Objective 2   **Evaluate an algebraic expression.**

Evaluating an expression means to substitute numbers for the variables and simplify. The rules for the order of operations in Section 1.6 are used for evaluating expressions. For instance:

Find the value of $x + 34$ if $x = 208$.

$x + 34 = 208 + 34$ **Substitute 208 for x.**

$ = 242$ **Add.**

Evaluate $200 - b^2$ if $b = 13$.

$200 - b^2 = 200 - 13^2$ **Substitute 13 for b.**

$ = 200 - 169$ **Do exponents first.**

$ = 31$ **Subtract.**

Answers to Warm Ups   A. $200 - 3x$ B. $8(n + 14)$

More complicated expressions may require more than one substitution.

Evaluate $3x + y - 2z$ if $x = 24$, $y = 19$, and $z = 17$.

$3x + y - 2z = 3(24) + 19 - 2(17)$	Substitute. Use parentheses to show multiplication.
$= 72 + 19 - 34$	Multiply first.
$= 57$	Add and subtract from left to right.

 To evaluate an expression

Substitute the given values and simplify.

Examples C–E	**Warm Ups C–E**

Directions: Evaluate the expression.

Strategy: Substitute the given values and simplify.

C. $5m - 19$ if $m = 16$

$5m - 19 = 5(16) - 19$	Substitute 16 for m. Use parentheses to show the product.
$= 61$	Multiply first, then subtract.

C. $58 - 4n$ if $n = 8$

C A U T I O N   **Multiplication symbols are omitted with variables but not with numerals. So $5m = 5(16)$, not 516.**

D. $\dfrac{x + 26}{y}$ if $x = 14$ and $y = 5$

$\dfrac{x + 26}{y} = \dfrac{14 + 26}{5}$	Substitute 14 for x and 5 for y.
$= 8$	Add first. The fraction bar acts as a grouping symbol. Divide.

D. $a(8 + b)$ if $a = 17$ and $b = 11$

E. Tickets to the annual homecoming game are priced at $8. Use the letter n to represent the number of tickets each person buys. Write an expression to represent the cost to each person. Then find the cost for Jim, 8 tickets; Maria, 12 tickets; Jose, 5 tickets; and Aneetra, 10 tickets.

Strategy: Multiply the cost of a ticket times the number of tickets.

Jim:	$8n = 8(8) = 64$	Substitute for the number of tickets and multiply.
Maria:	$8n = 8(12) = 96$	
Jose:	$8n = 8(5) = 40$	
Aneetra:	$8n = 8(10) = 80$	

The cost of n tickets is $8n$ dollars. Jim spent $64, Maria spent $96, Jose spent $40, and Aneetra spent $80.

E. The cost of a whale watching cruise is $6.00 per person. Use t to represent the number of tickets purchased. Write an expression to represent the cost to each person. Then find the cost for Vera, 5 tickets; Myke, 4 tickets; Karl, 12 tickets; and Peter, 15 tickets.

Answers to Warm Ups   C. 26   D. 323   E. $6t$; Vera $30; Myke $24; Karl $72; Peter $90

Exercises 3.1

OBJECTIVE 1: *Translate to algebraic expressions.*

A.

1. Twelve more than a number

2. A number increased by seventeen

3. The product of seven and a number

4. A number multiplied by nineteen

5. The difference of twenty and a number

6. Seven less than a number

7. The quotient of a number and five

8. A number divided into forty-five

B.

9. Eighteen more than twice a number

10. Twenty less than the product of a number and five

11. Fifteen less than the quotient of a number and two

12. Twenty-three less than eighteen divided by a number

13. The difference of eight times a number and twice the same number

14. The sum of twelve times a number and four times the same number

15. The product of two numbers decreased by fifteen

16. Twenty-four less than the product of two numbers

17. The quotient of two numbers increased by twenty

18. The difference of two numbers increased by twenty

19. The product of fifteen and two less than a number

20. The product of sixteen and nine more than a number

Evaluate the following expressions when $a = 6$.

A.

21. $32 + a$ **22.** $a + 43$ **23.** $5a$

24. $12a$ **25.** $\dfrac{72}{a}$ **26.** $\dfrac{a}{6}$

27. $a + 98$ **28.** $67 + a$ **29.** a^2

30. a^3

B.

31. $17a - 3$ **32.** $18a + 4$ **33.** $a^2 + 6$

34. $a^2 - 7$

Evaluate the following expressions if $x = 56$ and $y = 14$.

35. xy **36.** $\dfrac{x}{y}$ **37.** $2x - 5y$

38. $5x - 2y$ **39.** $\dfrac{x}{y + 14}$ **40.** $\dfrac{y}{x - 49}$

C.

Fill in the blanks to make a true sentence.

41. Replacing x with 17 in $x + 4$ is called _____.

42. To evaluate $2x - 7$ when $x = 4$, the first step is to _____.

43. When $x = 92$, the value of "100 less than twice x" is _____.

44. When $t = 5$, the value of "the quotient of 825 and t^2" is _____.

45. The cost of a ticket to the local production of *Swan Lake* is $18. Using n to represent the number of tickets purchased, write an expression to represent the total cost of tickets purchased. Find the cost of 4, 7, and 12 tickets.

46. The cost of a case of vegetables is $27. Write an expression to show the cost of buying n cases. Find the cost for 3, 8, and 12 cases.

47. State University advertises tickets to a football game at $16 each plus a $3 service charge for mail orders. Write an expression for the cost of buying tickets by mail. Find the cost of buying 2, 5, and 8 tickets by mail.

48. One side of a triangle is five inches longer than another. Using x to represent the measure of the shorter side, write an expression to represent the longer side. Find the length of the longer side when the shorter side is 6 inches, 8 inches, or 9 inches.

49. A taxi service estimates that it costs $11,250 to replace each old cab with a new one. Let x represent the number of cabs to be replaced. Write an expression to represent the total cost of replacing the old cabs. Find the cost to replace 5, 11, and 14 cabs.

50. A sprinkler head has a flow of 6 gallons per hour. Write an expression for the amount of water that flows through the head in h hours.

51. One sprinkler head has a flow of a gallons per hour, a second has a flow of b gallons per hour, and a third has a flow of c gallons per hour. Let h represent the number of hours the sprinklers are on. If all three heads are on the same line, write an algebraic expression for the total flow of water through the line.

52. The following graph gives the total number of airline passengers for a medium sized airport.

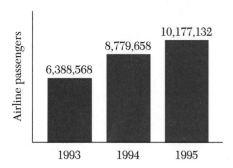

a) Calculate the average number of airline passengers over the three years (round to the nearest whole passenger).

b) Write an algebraic expression which gives the average number of airline passengers over the three years.

53. Following are data regarding accidental drownings from a Florida county over a 10-year period.

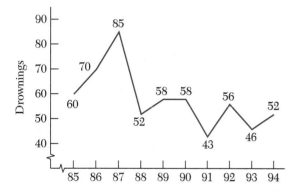

a) What was the average number of accidental drownings per year for this 10-year period?

b) Write a formula which calculates the average number of accidental drownings over a 10-year period.

Exercises 54–56 refer to the Chapter 3 Application. See page 179.

The plastic surgeon's convention begins on Thursday night with a reception. There will be hors d'oeuvres and an open bar from 5–8 PM. Let x represent the number of guests attending the reception.

54. The food and beverage industry estimates that each person attending the reception will consume 6–8 pieces of hors d'oeuvres per hour. Let H represent the total number of pieces of hors d'oeuvres needed for the reception. Write an expression for H in terms of x, using the highest estimate. Write an expression for H in terms of x, using the lowest estimate. If you were the meeting planner, which expression for the number of pieces would you use? Why?

55. The food and beverage industry estimates that one bartender is necessary for each 100 people attending the reception. Let B represent the number of bartenders needed for the reception. Write an expression for B in terms of x.

56. The industry also estimates that one cocktail server is required for every 50 people. Let C represent the number of cocktail servers needed for the reception. Write an expression for C in terms of x.

SUMMARIZE YOUR ANSWERS TO EXERCISES 54–56 IN THE TABLE.

Number of Guests	Hors d'Oeuvres	Bartenders	Cocktail Servers
x	$H =$	$B =$	$C =$

STATE YOUR UNDERSTANDING

Translate these algebraic expressions to English phrases in two different ways.

57. $2x + 3$ **58.** $3x - 7$ **59.** $6 - \dfrac{48}{x}$

60. Evaluate $(5x)^2$ and $5x^2$ when $x = 3$. Explain in your own words why the results are different.

CHALLENGE

Evaluate each expression, if $a = 20$, $b = 13$, $c = 5$, and $x = 7$.

61. $a^2 + 5a - 3b - c$ **62.** $b^2 - a + 3c - b$ **63.** $\dfrac{6a + 10b}{25}$

64. $\dfrac{19c + 27x}{71}$ **65.** $a^3 - b^2 - c^2 - x^2$ **66.** $x^3 + 3c^2$

GROUP ACTIVITY

67. Individually, go to the library and search for a formula from a field of interest to you such as business, science, and health related areas. Look for examples of how these formulas are used. Back with your group, exchange these formulas and write a sample exercise for each formula. Share these exercises with another group for review of this section.

MAINTAIN YOUR SKILLS (Sections 1.1, 1.2)

68. Round 4566 to the nearest ten.

69. Add: $34 + 86 + 38 + 65$

70. Add: $348 + 862 + 83 + 621$

71. Find the sum of 8, 56, 129, 35, and 604.

72. Find the sum of 1941, 28, 476, 8, and 4973.

3.2   Equations and Formulas

OBJECTIVES

1. Determine whether a number is a solution of an equation.

2. Evaluate a formula using whole numbers.

3. Solve an equation with the variable on one side using a table.

VOCABULARY

An **equation** states that two expressions are equal, with an equal sign, " = ", separating the two **sides** or **members** of the equation.

A **solution** of an equation is a number that makes the equation true, when it is substituted for the variable.

A **formula** is an equation that describes a relationship between two or more quantities.

HOW AND WHY
Objective 1

Determine whether a number is a solution of an equation.

An *equation* is an assertion that states two expressions name the same number. An equation can be true or false. The statements $5 + 6 = 11$, $x + 5 = 9$, and $3(5 + 8) = 300$ are equations.

To determine whether an equation is true or false, simplify each side to see if the left and right members name the same number. If the equation contains a variable, first replace the variable with its value.

Is the equation $97 + 43 + 188 = 14(23)$ true?

Does $97 + 43 + 188 = 14(23)$?

Does $\qquad 140 + 188 = 322$?   **Simplify both sides.**

Does $\qquad\qquad 328 = 322$?   **No.**

The equation is false.

In an equation with a variable, a value of the variable for which the equation is true is called a *solution of the equation.*

Is 16 a solution of the equation $2x - 17 = 15$?

Does $2(16) - 17 = 15$?   **Substitute 16 for x.**

Does $\quad 32 - 17 = 15$?   **Simplify the left side.**

Does $\qquad\quad 15 = 15$?   **Yes.**

The number 16 is a solution of $2x - 17 = 15$.

Deciding whether a number is a solution is called *checking the equation.*

▶ *To determine whether a number is a solution of an equation*

1. Replace the variable with the number.

2. Simplify both sides using the order of operations.

3. If both sides name the same number, the replacement is a solution.

| **Examples A–B** | **Warm Ups A–B** |

Directions: Determine whether the given number is a solution of the equation.

Strategy: Replace the variable with the number and simply to see if the left side is equal to the right side.

A.   Is 13 a solution of $4x - 12 = 2x + 14$?

$4(13) - 12 = 2(13) + 14$	**Substitute 13 for x on each side.**
$52 - 12 = 26 + 14$	**Simplify each side.**
$40 = 40$	**Both sides have value 40 when $x = 13$.**

The number 13 is a solution.

A.   Is 8 a solution of $4x - 20 = 3x - 12$?

B.   Is 43 a solution of $90 - 2x = x - 36$?

Does $90 - 2(43) = 43 - 36$?	**Substitute 43 for x.**
Does $90 - 86 = 7$?	**Simplify each side.**
Does $4 = 7$?	**No.**

The number 43 is not a solution.

B.   Is 36 a solution of $128 - 3a = 10$?

HOW AND WHY
Objective 2   Evaluate a formula using whole numbers.

A *formula* has two algebraic expressions separated by an equal sign. Every formula is also an equation. Some examples of formulas are $A = \ell w$ (area of a rectangle), $V = \pi r^2 h$ (volume of a cylinder), and $D = rt$ (distance in terms of speed and time). Other common formulas may be found in the appendix. Formulas are evaluated using the order of operations just as in evaluating expressions.

The formula for the perimeter of a triangle (distance around the triangle) is $P = a + b + c$, where P is the perimeter and a, b, and c represent the lengths of the sides. Find the perimeter of the following triangle.

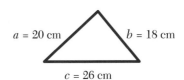

$P = a + b + c$	**Formula for perimeter.**
$P = 20 + 18 + 26$	**Substitute 20 for a, 18 for b, and 26 for c.**
$= 64$	**Add.**

The perimeter of the triangle is 64 centimeters.

 To evaluate a formula

Substitute the given values and simplify.

Examples C–E | **Warm Ups C–E**

Directions: Evaluate the formula.

Strategy: Substitute the given values and simplify.

C.   Evaluate $A = \dfrac{bh}{2}$ if $h = 40$ cm and $b = 22$ cm.

$A = \dfrac{(22)(40)}{2}$ **Substitute 40 for h and 22 for b.**

$A = \dfrac{880}{2}$ **Multiply first.**

$A = 440$ **Divide.**

The area of the triangle is 440 square centimeters (440 cm^2).

C.   Evaluate $A = \dfrac{bh}{2}$ if $h = 33$ m and $b = 44$ m.

D.   Evaluate $T = a + (n - 1)d$ if $a = 27$, $n = 13$, and $d = 5$.

$T = 27 + (13 - 1)5$ **Substitute.**

$= 27 + (12)5$ **Subtract in parentheses first.**

$= 27 + 60$ **Multiply.**

$= 87$ **Add.**

The value of T is 87.

D.   Evaluate $T = a + (n - 1)d$ if $a = 31$, $n = 13$, and $d = 7$.

E.   A rectangular section of mall area will be laid with decorative brick on a concrete base. If the rectangle has length 200 feet and width 60 feet, how many square feet of the base will the rectangular section cover?

Solution:

$A = \ell w$ **Area formula for a rectangle.**

$A = 200(60)$ **Substitute 200 for ℓ and 60 for w.**

$A = 12{,}000$

The brick will cover 12,000 square feet of the concrete base.

E.   A rectangular section of a supermarket floor is laid with tile over a concrete base. If the rectangle has length 170 feet and width 45 feet, how many square feet of tile were used?

HOW AND WHY
Objective 3   **Solve an equation with the variable on one side using a table.**

It is possible to find a solution of an equation using the "guess and check" method. A calculator is an excellent aid in speeding up the process. To keep track of the guesses and the results, a table is recommended. For example, to solve $34 - 3x = 7$, begin by making a table of possible replacements for x and for the resulting values of $34 - 3x$.

x	$34 - 3x$
1	31
2	28
3	25

Once you have made a few guesses, use the results to guide you in making your next guess. In this case, we need a value of x that will make $34 - 3x$ equal to 7. Since 31, 28, and 25 are not very close to 7, let's skip ahead and try $x = 10$.

Answers to Warm Ups   C. 726 m^2 D. 115   E. 7650 ft^2 of tile

x	$34 - 3x$
1	31
2	28
3	25
10	4

When $x = 10$, the value of $34 - 3x$ is closer to 7, but it is *smaller* than desired. We skipped too far. Let's back up and try $x = 9$.

x	$34 - 3x$
1	31
2	28
3	25
10	4
9	7

This is the desired value of x.

We conclude that $x = 9$ is a solution of the equation $34 - 3x = 7$.

Some calculators have a TABLE routine that can speed up the process considerably. Consult your calculator manual to learn how to use this routine.

> ▶ *To solve an equation, with the variable on one side using a table*
>
> **1.** Make a table with two columns. The first column is for the possible replacements for the variable. The second column will hold the values of the expression containing the variable.
>
> **2.** Make a few guesses for the value of the variable and calculate the resulting values for the second column.
>
> **3.** Based on the results of your first set of guesses, choose a larger (or smaller) number as appropriate.
>
> **4.** Continue until the desired result is located.

Example F   **Warm Up F**

Directions: Solve by making a table.

Strategy: Make a two column table. Use estimated values of the variable until you find a solution of the equation.

F.   Solve: $55 - 3a = 31$ F.   Solve: $36 - 3b = 12$

a	$55 - 3a$
1	52
2	49

Make a table and substitute a few values for a.

a	$55 - 3a$
1	52
2	49
5	40
10	25

Based on the results for $a = 1$ and $a = 2$, try a larger value for a.

The right side of the equation is 31, which is between 40 and 25, so the desired value of a is between 5 and 10.

a	$55 - 3a$
1	52
2	49
5	40
10	25
7	34
8	31

When $a = 8$, $55 - 3a = 31$.

$a = 8$ is a solution.

Exercises 3.2

OBJECTIVE 1: *Determine whether a number is a solution of an equation.*

A.

1. Is 5 a solution of $x + 16 = 21$?

2. Is 5 a solution of $16 - x = 11$?

3. Is 12 a solution of $22 - y = 10$?

4. Is 12 a solution of $2y + 1 = 25$?

5. Is 9 a solution of $w - 10 = 1$?

6. Is 9 a solution of $20 - 2w = 2$?

7. Is 7 a solution of $10 + 2a = 24$?

8. Is 7 a solution of $3a + 5 = 19$?

9. Is 3 a solution of $\dfrac{45}{x} = 15$?

10. Is 3 a solution of $\dfrac{x}{6} = 2$?

B.

11. Is 23 a solution of $3w - 15 = 54$?

12. Is 14 a solution of $3w + 22 = 64$?

13. Is 13 a solution of $55 - 2t = 33$?

14. Is 11 a solution of $47 - 3t = 15$?

15. Is 42 a solution of $764 - 2m = 670$?

16. Is 37 a solution of $671 - 4m = 520$?

17. Is 22 a solution of $\dfrac{286}{x} = 13$?

18. Is 24 a solution of $\dfrac{5x}{6} + 2 = 22$?

19. Is 18 a solution of $\dfrac{5y}{6} + 4 = 18$?

20. Is 18 a solution of $\dfrac{90}{y} - 1 = 4$?

OBJECTIVE 2: *Evaluate a formula using whole numbers.*

A.

Evaluate.

21. $A = \dfrac{bh}{2}$ if $b = 2$ ft and $h = 3$ ft (area of a triangle)   **22.** $A = \dfrac{bh}{2}$ if $b = 10$ m and $h = 4$ m

23. $D = rt$ if $r = 55$ mph and $t = 2$ hours (distance)   **24.** $D = rt$ if $r = 40$ mph and $t = 6$ hours

25. $P = a + b + c$ if $a = 12$ in., $b = 4$ in., and $c = 9$ in. (perimeter of a triangle)

26. $P = a + b + c$ if $a = 8$ yd, $b = 7$ yd, and $c = 11$ yd

27. $h = \dfrac{2A}{b}$ if $A = 14$ cm^2 and $b = 14$ cm (height of a triangle)

28. $h = \dfrac{2A}{b}$ if $A = 32$ m^2 and $b = 8$ m

B.

29. $V = s^3$ if $s = 16$ meters (volume of a cube)   **30.** $V = s^3$ if $s = 23$ inches

31. $s = \dfrac{P}{4}$ if $P = 388$ feet (side of a square)   **32.** $S = \dfrac{P}{4}$ if $P = 496$ meters

33. $V = \ell w h$ if $\ell = 12$ in., $w = 12$ in., and $h = 5$ in. (volume of a rectangular solid)

34. $V = \ell w h$ if $\ell = 11$ cm, $w = 15$ cm, and $h = 6$ cm

35. $P = 2\ell + 2w$ if $\ell = 203$ ft and $w = 76$ ft (perimeter of a rectangle)

36. $P = 2\ell + 2w$ if $\ell = 247$ m and $w = 89$ m

OBJECTIVE 3: *Solve an equation with the variable on one side using a table.*

A.

Solve.

37. $x + 8 = 15$

38. $x + 15 = 20$

39. $y - 3 = 12$

40. $y - 4 = 20$

41. $\dfrac{w}{5} = 7$

42. $\dfrac{w}{7} = 6$

43. $5m = 55$

44. $8m = 72$

B.

45. $2x + 3 = 17$

46. $2x + 5 = 27$

47. $2y - 5 = 19$

48. $2y - 8 = 6$

49. $54 + 2w = 70$

50. $37 + 2w = 65$

51. $85 - 5a = 30$

52. $92 - 4a = 48$

C.

Fill in the blanks to make a true sentence.

53. The statement "$x + 5 = x$" is a(n) _____ equation.

54. Every formula is also a(n) _____.

55. Every equation has _____ members.

56. A solution of an equation is a(n) _____ for which the equation is true.

57. Evaluate $S = \dfrac{n(a + \ell)}{2}$ if $n = 25$, $a = 1$, and $\ell = 25$ (sum of first 25 whole numbers).

58. Evaluate $S = \dfrac{n(a + \ell)}{2}$ if $n = 100$, $a = 1$, and $\ell = 100$ (sum of first 100 whole numbers).

59. Evaluate $S = \dfrac{n(a + \ell)}{2}$ if $n = 100$, $a = 2$, and $\ell = 200$ (sum of first 100 even numbers).

60. Evaluate $S = \dfrac{n(a + \ell)}{2}$ if $n = 120$, $a = 2$, and $\ell = 240$ (sum of first 120 even numbers).

61. Heather's living room is in the shape of a rectangle with a width of 18 feet and a length of 25 feet. How many square feet of carpet must she buy to cover the floor?

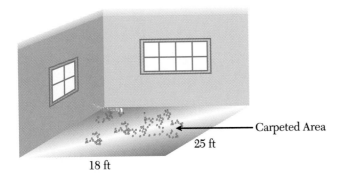

Carpeted Area

25 ft

18 ft

62. The following is part of a product list for European-made bookshelves. Use the formula for the volume of a rectangular solid to calculate the volume of each model.

Model #	Height (cm)	Width (cm)	Depth (cm)	Volume
W2524	152	78	38	
W2508	183	30	30	
W2534	152	38	38	
W2522	76	76	38	

63. Jane buys a stereo for $85 down and $41 per month for 12 months. What is the total price she pays? (*Hint:* With a deferred payment plan the total cost, T, to the consumer is the down payment, D, plus the monthly payment p, times the number of monthly payments, m. The formula is $T = D + pm$.)

64. A local auto import dealer offers a compact model for $410 down and $312 per month for 48 months. What is the total cost of the car? $T = D + pm$.

65. Knowledge of the specific gravity of a material helps to identify the mineral in a rock. Find the specific gravity of pyrite ("fool's gold") if a sample weighs 400 grams in the air and 320 grams in water. The formula is

$$\text{Specific gravity} = \frac{\text{weight in air}}{\text{weight in air} - \text{weight in water}}$$

66. Find the specific gravity of a mineral sample that weighs 24 kilograms in air and 18 kilograms in water. Use the formula in Exercise 65.

67. The formula for measuring IQ is $IQ = \dfrac{100(MA)}{CA}$ where MA is a person's mental age measured by a test and CA is the person's age in years. What is Jayne's IQ if her MA is 32 years and she is 25 years old?

68. What is Alfredo's IQ if his MA is 12 years and he is 15 years old? Use the formula in Exercise 67.

69. Financial experts recommend that everyone should have an emergency fund equal to 3–6 months income invested in savings accounts and CDs. Let I represent the amount of monthly income, and E represent the total amount of the emergency fund. Write a formula that describes the smallest recommended emergency fund. Write a formula that describes the largest recommended emergency fund. Use your formulas to calculate the range of emergency funds Tran needs if his monthly income is $2300.

70. Find the range of emergency funds that is recommended for Elisa, if her monthly income is $2575. Use the formulas from Exercise 69.

Exercise 71 refers to the Chapter 3 Application. See pages 179 and 189.

The plastic surgeon's convention begins on Thursday night with a reception. There will be hors d'oeuvres and an open bar from 5–8 PM. Let x represent the number of guests attending the reception.

71. Enlarge the table you made in Section 3.1, Exercises 54–56, to display the number of guests possible.

Number of Guests	Hors d'oeuvres	Bartenders	Cocktail Servers
x	$H =$	$B =$	$C =$
800			
900			
1000			
1100			

STATE YOUR UNDERSTANDING

72. What does it mean when we say "5 is a solution to this equation"? Write an example of an equation that has 5 as a solution. Write an example of an equation for which 5 is not a solution.

73. Explain the formula for finding the area of a rectangle, $A = \ell w$. Describe the connection between the letters in the formula and the numbers used to find the area. Also, describe the meaning of the formula as related to a rectangle. A drawing might be helpful.

CHALLENGE

74. Is 12 a solution of $x^2 + 2x = 168$?

75. Is 16 a solution of $x^2 - 7x = 144$?

76. Is 13 a solution of $2x^2 - 9x - 223 = 0$?

77. Is 13 a solution of $2x^2 - 10x = 221$?

GROUP ACTIVITY

78. Together, write five difficult exercises involving order of operations. Take the list back to class and see if other class groups can work them and get the correct results.

79. A rancher wishes to make a rectangular goat corral. He can use an existing fence for one side of the corral. He has 900 feet of fencing to construct the other three sides. Use a table to determine the dimensions of the corral which will result in the largest area. Use the formula $A = \ell w$. Explain why the area of the corral can have so many different values with the same amount of fence.

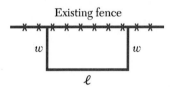

Existing fence

MAINTAIN YOUR SKILLS (Sections 1.1, 1.2, 2.1, 2.2, 2.3)

80. Round 34,601 to the nearest thousand.

81. Find the sum of 7844, 609, 2608, and 87.

82. Change 5 yards to inches.

83. Find the perimeter of a triangle with sides of 28 cm, 24 cm, and 37 cm.

84. Find the area of a triangle with base 28 cm and height 37 cm.

3.3 Adding and Subtracting Polynomials (Whole Numbers)

OBJECTIVES
1. Identify coefficients, like terms, and unlike terms.
2. Add and subtract monomials.
3. Add and subtract polynomials with whole number coefficients.

VOCABULARY

A **term** is a single number or a number multiplied by one or more variables. The parts of a term are bonded together with multiplication, division, or exponentiation.

A **monomial** is a term that contains a number, a variable, or combined numbers and variables with no variable occurring in a denominator.

Polynomials are monomials or sums or differences of monomials.

The **coefficient** of a term is its numerical factor.

Like terms have precisely the same variable factors.

To **combine like terms** is to add or subtract the terms.

HOW AND WHY
Objective 1

Identify coefficients, like terms, and unlike terms.

The *coefficient* of $17x$ is its numerical factor, 17. The coefficient of $6pst^2$ is 6. The coefficient of x is 1, that is $x = 1x$.

Monomials such as $55x$, $19x$, $7x$, and x are called *like terms*. Like terms have exactly the same variable factors.

CAUTION The variables in like terms must have the same exponents. The monomials $3x$ and $3x^2$ are not like terms.

Example A | **Warm Up A**

Directions: Identify the coefficients.

Strategy: Look for the numerical factors.

A. Identify the coefficients of $17xy$, $29x^2$, and $22pq$.

The coefficient of $17xy$ is 17.
The coefficient of $29x^2$ is 29.
The coefficient of $22pq$ is 22.

A. Identify the coefficients of $3x^2y^2$, w^2, and $15abc^2$.

Examples B–C | **Warm Ups B–C**

Directions: Identify the like terms.

Strategy: Look for common variable factors.

B. Which are like terms? $13a$, $2ab$, $5b$, $47ab$

The like terms are $2ab$ and $47ab$. **The common variables are a and b.**

B. Which are like terms? $12x$, $21xy$, $17y$, and $3x$

Answers to Warm Ups   A. 3, 1, and 15   B. $12x$ and $3x$

C. Which are like terms? $13p^2$, $5p$, $6pq$, and $7q$

There are no like terms. **The term $13p^2$ has two factors of p, whereas $5p$ has only one.**

C. Which are like terms? $8a^2$, $5ab$, $7a^2b$, a^2

HOW AND WHY
Objective 2

Add and subtract monomials.

A *monomial* is a term that contains a number or a number and one or more variables written as an indicated product. These are monomial terms:

$$17x \qquad 4 \qquad 17q^2 \qquad \frac{22y}{11} \qquad mn \qquad \text{and} \qquad 6\,pst$$

The terms $\dfrac{55x}{11y}$ and $\dfrac{18w}{a}$ are not monomials since each contains an indicated division by a variable.

To add or subtract monomials we use the distributive property from Section 1.3. Here, again, is the distributive property stated in different ways.

$$a(b + c) = ab + ac \qquad ba + ca = (b + c)a$$
$$a(b - c) = ab - ac \qquad ba - ca = (b - c)a$$

Here is how we can use the distributive property to add $55x$ and $19x$.

$55x + 19x = (55 + 19)x$ **Rewrite as a product using the distributive property.**

$\qquad\qquad = 74x$ **Add inside parentheses. We have combined the two monomials $55x$ and $19x$.**

To subtract the like terms $34xy$ and $18xy$:

$34xy - 18xy = (34 - 18)xy$ **Distributive property.**

$\qquad\qquad = 16xy$ **Subtract.**

In both of these examples, we can do the first step mentally. This is to say, we can add or subtract the coefficients and multiply the result by the common variables. So,

$5m + 12m + 19m = 36m$ **Add the coefficients $5 + 12 + 19$ and multiply the sum by the variable, m.**

 To combine like terms

Add or subtract the coefficients and multiply by the common variable factor(s).

CAUTION **Only like terms can be combined.**

For instance,

$13a + 7b \neq 20ab$ The symbol "$\neq$" is read "not equal to."

To show this, let $a = 2$ and $b = 5$.

Does $13a + 7b = 20ab$ if $a = 2$ and $b = 5$?

Does $13(2) + 7(5) = 20(2)(5)$?   Substitute.

Does $26 + 35 = 200$?   Simplify both sides.

Does $61 = 200$?   No.

The expression $13a + 7b$ cannot be simplified to $20ab$. We say that $13a + 7b$ is already in simplest form.

Examples D–G	Warm Ups D–G

Directions: Simplify by combining like terms.

Strategy: Add or subtract the coefficients of the like terms and multiply by the common variable factors.

D. $5d + 9d$	D. $4p + 7p$
$5d + 9d = (5 + 9)d$ **This step, showing the distributive**	
$= 14d$ **property, may be done mentally.**	
E. $32a + a$	E. $w + 53w$
$32a + a = 33a$ Since $a = 1a$, add 32 and 1.	
F. $49w - 35w$	F. $81y - 69y$
$49w - 35w = (49 - 35)w$	
$= 14w$	
G. $34q^2 - 12q^2 + 18q^2$	G. $145y^2 + 337y^2 - 428y^2$
$34q^2 - 12q^2 + 18q^2 = (34 - 12 + 18)q^2$	
$= 40q^2$	

HOW AND WHY
Objective 3   **Add and subtract polynomials with whole number coefficients.**

When an algebraic expression contains one or more terms that are monomials, the expression is called a *polynomial*. These expressions are polynomials:

$5ab$, $2x + 5y$, $5a - 3b + 14$, $32x + 13y + 14x$

This last polynomial contains the like terms $32x$ and $14x$. These terms can be combined.

$32x + 13y + 14x = (32x + 14x) + 13y$ Use the associative and commutative properties to group the like terms.

$= 46x + 13y$ Combine the like terms.

Answers to Warm Ups   D. $11p$ E. $54w$ F. $12y$ G. $54y^2$

To simplify, we group like terms as often as necessary. For example,

$$23a + 55b + 36c + b + 8c + a$$

$= (23a + a) + (55b + b) + (36c + 8c)$	**Group the like terms.**
$= 24a + 56b + 44c$	**Combine the like terms.**

 ### To add or subtract polynomials

Use the associative and commutative properties of addition to group the like terms. Combine like terms.

Writing the polynomials in columns provides a natural grouping of the like terms.

Sum	**Difference**
$\begin{array}{r} 47a + 23b + 85 \\ + \underline{39a + 8b + 57} \\ 86a + 31b + 142 \end{array}$	$\begin{array}{r} 47a + 23b + 85 \\ - \underline{(39a + 8b + 57)} \\ 8a + 15b + 28 \end{array}$

Examples H–L | **Warm Ups H–L**

Directions: Simplify.

Strategy: Group the like terms and combine them.

H. $26x + 45y + 19x - 13y$

$26x + 45y + 19x - 13y$

$= (26x + 19x) + (45y - 13y)$	**Group the like terms.**
$= 45x + 32y$	**Simplify.**

H. $155a + 126b + 89a - 58b$

I. $5b + 32c + 19b^2$

$5b + 32c + 19b^2$ The terms $5b$ and $19b^2$ are *not* like terms since b and b^2 are not exactly the same.

I. $99x + y - 15z$

C A U T I O N ⚠ **Only like terms can be combined.**

J. $(25a + 7b^2 + 9c^3) + (28a + 17b^2 + 13c^3)$

$\begin{array}{r} 25a + 7b^2 + 9c^3 \\ + \underline{28a + 17b^2 + 13c^3} \\ 53a + 24b^2 + 22c^3 \end{array}$ **Like terms can be grouped in columns.** **Add the like terms.**

J. $(34x + 8y) + (18x + 6y)$

K. $(46m + 19n) - (27m + 12n)$

$\begin{array}{r} 46m + 19n \\ - \underline{(27m + 12n)} \\ 19m + 7n \end{array}$ **Subtract the like terms.**

K. $(57s + 49t) - (26s + 17t)$

Answers to Warm Ups   H. $244a + 68b$ I. $99x + y - 15z$ J. $52x + 14y$ K. $31s + 32t$

L.   Find the perimeter of the rectangle having the following measurements.

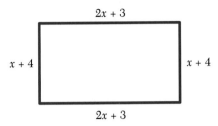

$P = (2x + 3) + (x + 4) + (2x + 3) + (x + 4)$

$P = (2x + x + 2x + x) + (3 + 4 + 3 + 4)$

$P = 6x + 14$

The perimeter of the rectangle is $6x + 14$ units.

L.   Find the perimeter of the triangle with the given dimensions.

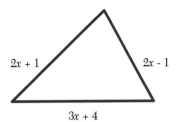

Exercises 3.3

OBJECTIVE 1: *Identify coefficients, like terms, and unlike terms.*

A.

1. What is the coefficient of $75t$?
2. What is the coefficient of $5w^2$?
3. Identify the like terms in the list: $5x^2$, $2x$, $3xy$, $8xy$
4. Identify the like terms in the list: $9ab$, $5a^2b$, $2ab^2$, $15ab^2$
5. Identify the like terms in the list: $23mn$, $14m^2n$, $39m^2n$, $8mn^2$

B.

6. What is the coefficient of $65xyz$?
7. What is the coefficient of ab^2?
8. Identify the like terms in the list: $13x$, $6xy$, $2x^2$, x
9. Identify the like terms in the list: $17pq$, $21p^2q$, pq, $13pq$
10. Identify the like terms in the list: $5y^2$, $2y$, 7, y

OBJECTIVE 2: *Add and subtract monomials.*

A.

Simplify.

11. $10a + 11a$
12. $8b + 15b$
13. $14c + 13c$

14. $18d + 9d$
15. $31x + 8x + x$
16. $42x + 2x + x$

17. $15y - 9y$
18. $20z - 17z$

19. Find the sum of $36w$ and $14w$.
20. Find the difference of $36w$ and $14w$.

B.

21. $14x + 25x + 15x + 22x$
22. $19a + 13a + 7a + 10a$

23. $48y + 33y - 13y - 9y$
24. $49z - 13z - 6z + 18z$

25. $103s + 219s - 78s$

26. $97t - 48t + 129t$

27. $38w - 19w + 56w - w$

28. $27m - 9m + 68m - m$

OBJECTIVE 3: *Add and subtract polynomials with whole number coefficients.*

A.

Simplify.

29. $(2x + 5y) + (6x + 2y)$

30. $(4a + 9) + (3a + 9)$

31. $7r + 2t + 19r + 7t$

32. $8m + 15n + 18m + 8n$

33. $17x^2 + 5x + 2x^2 + 18x$

34. $5a^2 + 20a + 13a^2 + 7a$

35. $m^2 + s + 2s + 3m^2$

36. $p^2 + q + 9q + 15p^2$

37. $10x - 7x + 10y - 2y$

38. $13a - 6a + 13b - 5b$

39. $\begin{aligned} 49xy + 39yz \\ - \underline{(31xy + 17yz)} \end{aligned}$

40. $\begin{aligned} 27z^2 + 49z \\ - \underline{(17z^2 + 39z)} \end{aligned}$

B.

41. $\begin{aligned} 55x^2 + 16x \\ + \underline{39x^2 + 6x} \end{aligned}$

42. $\begin{aligned} 28a^2 + 17a \\ + \underline{17a^2 + 19a} \end{aligned}$

43. $\begin{aligned} 43x^2 + 17x + 75 \\ - \underline{(15x^2 + 9x + 19)} \end{aligned}$

44. $\begin{aligned} 58p^2 + 16p + 50 \\ - \underline{(19p^2 + 16p + 16)} \end{aligned}$

45. $(12xy + 17yz + 19wz) + (11xy + 26yz + 18wz)$

46. $(32mn + 45nr + 19rs) + (39mn + 44nr + 17rs)$

47. $(15x^2 + 19 + 22x) + (2x^2 + 12x + 18)$

48. $(23a^2 + 21 + 19a) + (13a^2 + 16a + 9)$

49. $(15pq + 25qs) - (7pq + 12\ qs)$

50. $(48xy + 32y^2) - (30xy + 17y^2)$

51. $(72abc + 83ab + 95c) - (68abc + 77ab + 89c)$

52. $(45a + 59b + 47c) - (42b + 12c + 19a)$

C.

Fill in the blanks to make a true sentence.

53. Only _____ terms can be combined.

54. Every monomial is also a _____.

55. Find the error(s) in the statement: $45x^2 + 2x = 47x^3$. Correct the statement. Explain how you would avoid this error.

56. Find the error(s) in the statement: $(17x + 2y) - (3x + y) = 14x + 2y$. Correct the statement. Explain how you would avoid this error.

57. Write a simplified expression for the perimeter of the rectangle in the figure.

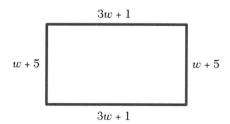

$3w + 1$

$w + 5$　　　　$w + 5$

$3w + 1$

58. Write a simplified expression for the perimeter of the triangle in the figure.

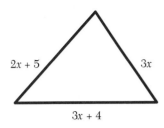

$2x + 5$

$3x$

$3x + 4$

59. To help pay for José's education, when he was born his parents invested $12,000. The money was invested so that when José is 20 years of age the interest will be two times as much as the amount they invested. The formula for the total value of the investment is $T = i + 2i$, where i is the amount of the investment. Simplify the formula and find the value of the investment when José is 20.

60. If Belinda buys each of her grandchildren a $5 toy, a $15 shirt, and a $4 box of candy, the formula for the cost is $C = 5n + 15n + 4n$, where n is the number of grandchildren. Simplify the formula and find the cost for eight grandchildren.

61. A gasoline automobile costs $15 per 100 miles to operate while a diesel automobile costs $11 per 100 miles to operate. The formula for the difference, d, in cost of operating the two cars is $d = 15m - 11m$, where m is the number of hundreds of miles driven. Simplify the formula and find the difference in cost if each is driven 800 miles.

62. Last year a bookstore purchased a math book for $37. This year the same book cost $39. The formula for the total difference, d, in the cost of the same number of books is $d = 39n - 37n$, where n is the numbers of books purchased. Simplify the formula and find the total difference in cost if the store purchased 872 books each year.

63. A gardener wishes to fence in a five-sided flower bed. The two shorter sides are the same length and each of the three remaining sides is twice as long as a shorter side. Write a formula to show the length, L, of the fence needed in terms of the length, y, of one of the shorter sides. Simplify the formula and find the length of fence needed if the shorter sides are 10 feet long.

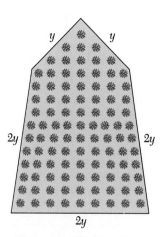

64. Banta Electronix has plants in three states. The plant in Utah can produce twice as many parts as the plant in Idaho. The plant in Montana can produce four times as many parts as the plant in Idaho. Write a formula for the weekly production, W, in all three plants. Let n represent the weekly production of the Idaho plant. Simplify the formula and find the total production if the Idaho plant can produce 310 parts in a week.

65. Nellie is planning to install a drip irrigation system in her garden. Write a simplified algebraic expression for the total number of feet of line she needs.

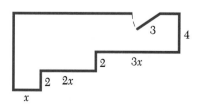

66. Write a polynomial that represents the total length of baseboard needed for the room pictured below. Do not put a baseboard on the door. Assume that the measurements are in feet.

67. Size $2d$ finishing nails are 1 inch in length and there are approximately 1351 of them per pound. Size $6d$ finishing nails are 2 inches in length and there are approximately 309 of them per pound. Write a polynomial that describes the total number of nails in P pounds of $2d$ nails and Q pounds of $6d$ nails.

▼ Exercises 68–72 refer to the Chapter 3 Application. See pages 179 and 180.

The meeting planner for the plastic surgeon's convention has turned her attention towards the total cost of the Thursday night reception. Recall that x represents the total number of guests at the reception, H represents the total number of hors d'oeuvres consumed at the reception, B represents the number of bartenders needed, and C represents the number of cocktail servers needed.

68. Write an expression, in terms of x, that represents the total number of service employees needed for the reception.

69. The hotel pays bartenders \$14 per hour and cocktail servers \$6 per hour. Write an expression in terms of B and C that represents the total cost per hour of service employees needed for the reception. Write the total cost again, this time in terms of x. Simplify your expression as much as possible.

70. The hotel charges \$1 per four hors d'oeuvres. Using the highest estimate write an expression in terms of x that represents the total cost of hors d'oeuvres for the reception.

71. Let R represent the total cost of the food plus the total hourly cost for service employees for the reception. Write an expression for R in terms of x. Simplify your expression as much as possible.

72. Add a column to the reception planning table (from Section 3.2) for the total cost, R, and fill in the column.

Number of Guests	Hors d'oeuvres	Bartenders	Cocktail Servers	Total Cost
x	$H =$	$B =$	$C =$	$R =$
800				
900				
1000				
1100				

STATE YOUR UNDERSTANDING

73. Explain how to find the sum of $(24xy + 4x^2)$ and $(2x^2 + 17xy)$.

74. How many terms does $6abc + 14ac^2 - 3ac - 2cba$ have? How many of them are like terms? What is a term?

CHALLENGE

75. Find the sum of $75a + 52b + 111c$ and $38a - 47b$.

76. Find the difference of $55x + 29y + 44z$ and $24x + y + 43z$.

77. Find the sum of the sum of $28z$ and $12z$ and the difference of $8z$ and $6z$.

78. Find the sum of the difference of $28z$ and $12z$ and the sum of $8z$ and $6z$.

79. Find the difference of the sum of $28z$ and $12z$ and the difference of $8z$ and $6z$.

80. Find the difference of the sum of $28z$ and $12z$ and the sum of $8z$ and $6z$.

GROUP ACTIVITY

81. Individually translate the expression, $5x^2 + 2x$, into an English phrase. Together, make a copy of the different ways your group members translated the expression. Compare your list with another group. Did they have any different translations? If so, make up a composite list for both groups.

MAINTAIN YOUR SKILLS (Sections 1.1, 2.3, 2.4)

82. True or false? $51 < 49$

83. True or false? $107 > 99$

84. Find the volume of the box.

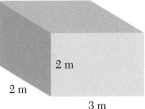

85. Find the total area of all sides of the box in Exercise 84.

86. Find the volume of a barrel which is 5 feet high and has a circular base with area 3 square feet.

3.4 Multiplying and Dividing Polynomials (Whole Numbers)

OBJECTIVES

1. Multiply and divide monomials.

2. Multiply and divide polynomials.

HOW AND WHY
Objective 1

Multiply and divide monomials.

To multiply two monomials we use the associative property of multiplication that permits us to rearrange parentheses, that is, change the grouping.

$23(15x) = (23 \cdot 15)x$ **Group the factors 23 and 15 using the associative property.**

$= 345x$ **Multiply.**

We can also use the commutative and associative properties of multiplication to change the position of the factors. When appropriate, we use the law of exponents for multiplying like bases.

$(30w^2)(5w) = (30 \cdot 5)(w^2 \cdot w)$ **Group the numerical factors and the variable factors.**

$= 150w^3$ **Multiply $30 \cdot 5$ and $w^2 \cdot w^1$.**

To multiply monomials (terms)

1. Multiply the coefficients.

2. Multiply the variable factors.

To understand the division of monomials, recall the missing factor interpretation of division. The exercise $48x^2 \div 4x$ can be restated $4x(?) = 48x^2$. Since we know that $4x(12x) = 48x^2$, we can write

$$48x^2 \div 4x = \frac{48x^2}{4x}$$

$$= 12x$$

Observe that $48 \div 4 = 12$ and that $x^2 \div x = x$. The shortcut is to divide the coefficients and divide the variables.

To divide monomials (terms)

1. Divide the coefficients.

2. Divide the variable factors.

Examples A–D	**Warm Ups A–D**

Directions: Multiply or divide the monomials.

Strategy: Multiply the coefficients and the variable factors or divide the coefficients and the variable factors.

A. Multiply: $35(4x)$	A. Multiply: $11(27y)$
$35(4x) = (35 \cdot 4)x$ **Multiply the coefficients.** $\quad = 140x$	
B. Multiply: $26x(x)$	B. Multiply: $57b(b)$
$26x(x) = 26x(1x)$ **Since $x = 1x$.** $\quad\quad\ = (26 \cdot 1)(xx)$ $\quad\quad\ = 26x^2$ **Multiply coefficients and variables.**	
C. Multiply: $(14rs)(4rt)(18rs)$	C. Multiply: $(23xy)(24xyz)(7)$
$(14rs)(4rt)(18rs) = (14 \cdot 4 \cdot 18)(r \cdot s \cdot r \cdot t \cdot r \cdot s)$ $\quad\quad\quad\quad\quad\quad\ = 1008(r^3s^2t)$ **The exponent of each shows the number of times the variable is used as a factor.**	
D. Divide: $\dfrac{81w^2}{27w}$	D. Divide: $\dfrac{75y^2}{15y}$
$\dfrac{81w^2}{27w} = \dfrac{81}{27} \cdot \dfrac{w^2}{w} = 3w$ **Divide coefficients and divide variables.**	

HOW AND WHY
Objective 2   **Multiply and divide polynomials.**

To multiply a monomial times a polynomial of more than one term, we use the distributive property.

$$2x(13x + 24y + 7) = 2x(13x) + 2x(24y) + 2x(7) \quad \text{Multiply each term in parentheses by } 2x.$$
$$= 26x^2 + 48xy + 14x \quad \text{Simplify.}$$

 To multiply a polynomial by a monomial

Multiply each term of the polynomial by the monomial using the distributive property.

Examples E–G	**Warm Ups E–G**

Directions: Multiply.

Strategy: Apply the distributive property to multiply the coefficients and the variable factors.

E. Multiply: $15x(5x + 6)$	E. Multiply: $12w(8w + 3)$

$$15x(5x + 6) = 15x(5x) + 15x(6) \quad \textbf{Distributive property.}$$
$$= 75x^2 + 90x \quad \textbf{Multiply.}$$

F. Multiply: $9ab(ac - 8ab + 12bc)$	F. Multiply: $11xy(8xy + 11xz - 14zy)$

$9ab(ac - 8ab + 12bc)$

$$= 9ab(ac) - 9ab(8ab) + 9ab(12bc) \quad \textbf{Distributive property.}$$

$$= 9a^2bc - 72a^2b^2 + 108ab^2c \quad \textbf{Multiply.}$$

G.   A rectangular plot of ground has a length that is ten more than twice its width. Let w represent the width. Express the length and the area in terms of w. Find the number of square meters in the area if the width is 23 meters.

The length is represented by $2w + 10$.

The length is represented by $2w + 10$.	The length is ten more than twice the width.
$A = \ell w$	Formula for the area of a rectangle.
$A = (2w + 10)w$	Substitute $2w + 10$ for ℓ.
$A = 2w^2 + 10w$	This is the formula for the area in terms of the width.
$A = 2(23)^2 + 10(23)$	Evaluate given $w = 23$ meters.
$A = 1288$	

The length and area in terms of w are $\ell = 2w + 10$ and $A = 2w^2 + 10w$. The area of the rectangular plot of ground is 1288 m^2.

G.   The length of a rectangular sheet of metal is three more than three times its width. Let w represent the width. Express the length and the area in terms of w. Find the number of square inches in the area when the width is 21 inches.

The product of two polynomials, each with more than one term, is found by repeated use of the distributive property. To multiply $(2x + 3)(5x + 8)$, think of $(2x + 3)$ as a single term. One way to do this is to replace $(2x + 3)$ with the variable a.

Answers to Warm Ups   E. $96w^2 + 36w$ F. $88x^2y^2 + 121x^2yz - 154xy^2z$ G. $\ell = 3w + 3$; $A = 3w^2 + 3w$; 1386 in.^{2}

$$(2x + 3)(5x + 8) = a(5x + 8) \qquad \text{Replace } (2x + 3) \text{ by } a.$$
$$= a(5x) + a(8) \qquad \text{Distributive property.}$$
$$= (2x + 3)(5x) + (2x + 3)(8) \qquad \text{Replace } a \text{ with } (2x + 3).$$
$$= 2x(5x) + 3(5x) + 2x(8) + 3(8) \qquad \text{Distributive property again.}$$
$$= 10x^2 + 15x + 16x + 24 \qquad \text{Simplify.}$$
$$= 10x^2 + 31x + 24 \qquad \text{Combine } 15x \text{ and } 16x.$$

Polynomial multiplication can be done pictorially using a box. Make a rectangle and divide it into four parts. Label the left side with the terms of one of the polynomials to be multiplied. Label the top with the terms of the other polynomial.

Now fill the box with the products of the terms at the head of each row and column.

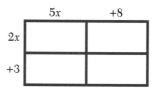

Add the contents of the boxes together to find the product.

$$(2x + 3)(5x + 8) = 10x^2 + 16x + 15x + 24 = 10x^2 + 31x + 24$$

Examples H–I	**Warm Ups H–I**

Directions: Multiply.

Strategy: Use the distributive property or the box method.

H.   Multiply: $(2a + 7)(3a + 5)$

$(2a + 7)(3a + 5)$

$= (2a + 7)(3a) + (2a + 7)(5)$ **Distributive property.**

$= 2a(3a) + 7(3a) + 2a(5) + 7(5)$ **Distributive property again.**

$= 6a^2 + 21a + 10a + 35$ **Multiply.**

$= 6a^2 + 31a + 35$ **Combine like terms.**

H.   Multiply: $(3x + 6)(2x + 4)$

Answers to Warm Ups   H. $6x^2 + 24x + 24$

I. Multiply: $(7x + 11)(x + 3)$

	x	$+3$
$7x$	$7x^2$	$21x$
$+11$	$11x$	33

Fill the box with the products of the terms.

$$(7x + 11)(x + 3) = 7x^2 + 21x + 11x + 33$$
$$= 7x^2 + 32x + 33$$

I. Multiply: $(3x + 5)(x + 6)$

To divide a polynomial with more than one term, divide each term by the divisor. This is actually the distributive property in a different form.

$$\frac{15x^2 + 18xy - 45x}{3x} = \frac{15x^2}{3x} + \frac{18xy}{3x} - \frac{45x}{3x}$$
$$= 5x + 6y - 15$$

 To divide a polynomial by a monomial

Divide each term of the polynomial by the monomial.

Example J

Directions: Divide the polynomials.

Strategy: Divide the coefficients and the variable factors.

J. $\dfrac{48abc + 32ab - 64ac}{16a}$

$$\frac{48abc + 32ab - 64ac}{16a} = \frac{48abc}{16a} + \frac{32ab}{16a} - \frac{64ac}{16a} \quad \text{Divide each term of the numerator.}$$

$$= 3bc + 2b - 4c \qquad\qquad \text{Divide.}$$

Warm Up J

J. $\dfrac{81xyz - 54xy - 72xz}{9x}$

Answers to Warm Ups   I. $3x^2 + 23x + 30$ J. $9yz - 6y - 8z$

Exercises 3.4

OBJECTIVE 1: *Multiply and divide monomials.*

Multiply or divide.

A.

1. $4(12w)$

2. $5(12t)$

3. $7(9a^2b)$

4. $11(9xy^2)$

5. $\dfrac{48y}{6}$

6. $\dfrac{56t}{4}$

7. $\dfrac{72y}{6}$

8. $\dfrac{120m}{12}$

9. Find the product of 10 and $32bc$.

10. Find the product of $12st$ and 7.

11. Find the quotient of $39x$ and 3.

12. Find the quotient of $52y$ and 4.

B.

13. $(14xy)(12y)$

14. $(22b)(9bc)$

15. $(21st)(19st)$

16. $(15xy)(23xy)$

17. $\dfrac{312abc}{24b}$

18. $\dfrac{175xyz}{35y}$

19. $\dfrac{152d^2ef}{19ef}$

20. $\dfrac{228qrst^2}{19qt^2}$

21. $\dfrac{16ax}{4x} + \dfrac{12bx}{4x} + \dfrac{20cx}{4x}$

22. $\dfrac{40ab}{8a} + \dfrac{56ac}{8a} - \dfrac{64a}{8a}$

OBJECTIVE 2: *Multiply and divide polynomials.*

Multiply or divide.

A.

23. $5z(11z + 2)$

24. $12b(6b + 1)$

25. $8xy(7yz - 2xz)$

26. $13pq(5qr - 2pr)$

27. $2xy(15xz - 22yz + 25xy)$

28. $3abc(9a + 22bc - 32c)$

29. $\dfrac{16by + 12cy + 20dy}{4y}$

30. $\dfrac{40xy + 56xz - 64x}{8x}$

31. $\dfrac{44xyz^2 + 55x^2yz + 88xy^2z}{11xy}$

32. $\dfrac{15r^2s - 30r^2t + 45rs^2}{15r}$

B.

33. $21abc(4a + 3b - 5c)$

34. $8xyz(12x + 13y - 14z)$

35. $(y + 6)(y + 3)$

36. $(a + 8)(a + 1)$

37. $(x + 15)(x + 2)$

38. $(b + 11)(b + 8)$

39. $\dfrac{336x^2 - 352xw}{16x}$

40. $\dfrac{207rst - 345stu}{23st}$

41. $\dfrac{324a^2bc - 234ab^2c + 162abc^2}{18abc}$

42. $\dfrac{168x^2y^2z - 264x^2yz^2 - 360xyz^2}{24xz}$

43. Find the product of $19a$, $4a$, and $23a$.

44. Find the product of $28x$, $9x$, and $13x$.

45. $(3x + 7)(2x + 5)$

46. $(2x + 7)(3x + 5)$

47. $(6x + 1)(4x + 9)$

48. $(4x + 1)(6x + 9)$

C.

Fill in the boxes so the statement is true. Explain your answer.

49. $17x(5x) = (17 \cdot 5)(x\square) = 85x^2$

50. $17x(5x + \square) = 85x^2 + 17x$

51. Find the error(s) in the statement: $17x(5y + 2) = 85xy + 2$. Correct the statement. Explain how you would avoid this error.

52. Find the error(s) in this statement: $\dfrac{\overset{15}{\cancel{75}}xy + 20\overset{y}{\cancel{y}^2}}{\underset{1}{\cancel{5}\cancel{y}}} = 15xy + 20y$. Correct the statement. Explain how you would avoid this error.

53. A rectangle has a length that is twice its width. Let w represent the width of the rectangle and express the area, A, of the rectangle in terms of the width. Find the area when the width is 11 feet. The formula is $A = \ell w$.

54. A decorative wall is five times longer than it is high. Let h represent the height of the wall and express the area, A, of the wall in terms of the height. Find the area of the wall if the height is 4 feet. The formula for the area of the side of the wall is $A = \ell h$.

55. The formula for the volume of a box is $V = \ell w h$. A special box has a length that is twice the width and a height that is three times the width. Express the volume of the box in terms of its width and simplify. Find the volume when the width is 2 feet.

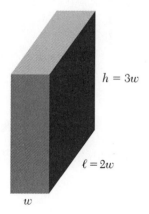

$h = 3w$

$\ell = 2w$

w

56. A bin has a length that is twice as long as its width and a height that is twice as high as its length. Let w represent the width of the bin. Write an expression that shows the volume, V, of the bin in terms of the width and simplify. Find the volume of the bin if the width is 6 feet.

57. The formula for the area of a triangle is $A = \dfrac{bh}{2}$. A special triangle has a height that is four times the base. Express the area, A, of the triangle in terms of the base and simplify. Find the area when the base is 16 inches.

58. A rectangular plot of ground has a length that is eight more than three times its width. Let w represent the width of the rectangle. Express the area of the plot in terms of the width. Find the number of square feet in the area when the width is 28 feet.

▼ Exercises 59–64 refer to the Chapter 3 Application. See page 179.

The plastic surgeon's convention will require a lecture hall that seats at least 250 people. The hall must have a stage 8 feet long and 6 feet deep at the front. The hall will have a theater setup with standard chairs that are 18 inches wide and 18 inches deep. The rows of chairs must be 8 feet from the speakers platform. There are 24 inches between rows. Because of safety restrictions, each row may be no longer than 15 chairs. Aisles between chairs must be 5 feet to 6 feet, as must be the space between the chairs and the walls. See the diagram below.

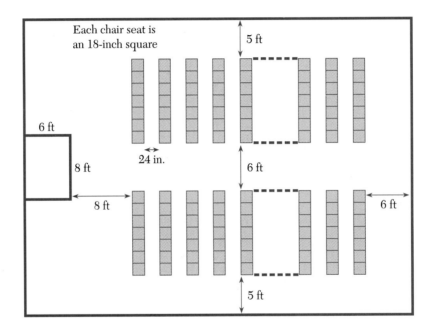

59. Let C represent the number of chairs per row. Write an expression for W, the total width in feet of the hall, in terms of C. Explain.

60. Let R be the number of rows of chairs. Write an expression for L, the total length of the hall, in terms of R. Explain.

61. What does RC represent?

62. What does LW represent? Write an expression for LW in terms of R and C and simplify.

63. Fill in the table.

C = Chairs per Row	R = Number of Rows	Total Seating Capacity	Total Area of Hall
25			
20			
16			
10			

64. The hotel industry estimates that the space requirements for a hall arranged in a theater setup is 10 square feet per person. How much space is needed for a hall that seats 250 people? How do your calculations in the table above compare with the industry estimate? How do you account for the differences?

STATE YOUR UNDERSTANDING

65. Explain how to find the product of $(2g^2 + 3)$ and $(5g^2 + 5)$.

66. Explain the procedure for finding the product of $6x^2$ and $12xy$.

67. Explain the procedure for finding the quotient of $48x^2y^2$ and $4xy$.

68. a. Simplify: $(5x)(2x)(7x)$.

b. Simplify: $5x + 2x + 7x$.

c. State in words how the two processes are different.

CHALLENGE

69. Find the volume of the box whose measurements are as shown.

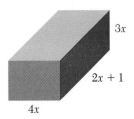

3x

2x + 1

4x

70. Find the volume of the box whose measurements are as shown.

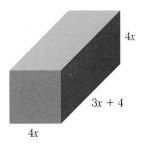

4x

3x + 4

4x

71. Find the volume of the cylinder whose measurements are shown. The formula for the volume of a cylinder is $V = \pi r^2 h$. Let $\pi \approx 3$.

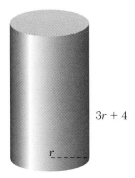

3r + 4

r

GROUP ACTIVITY

72. As a group, make up 20 exercises. Write ten squares of the form $(x + a)^2$ using different numbers for a. Write ten products of the form $(x + a)(x - a)$ using different numbers or expressions in place of a. Divide up the 20 exercises and find the products. Back together as a group, determine shortcuts for these types of exercises. Write up your results as a rule that might appear in an algebra text.

MAINTAIN YOUR SKILLS (Sections 1.2, 1.3, 2.4, 3.1)

73. Find the difference between 3412 and 1987.

74. Find the quotient of 67,854 and 43.

75. Translate to an algebraic expression: Let n represent the number: Forty-two increased by three times a number.

76. Translate to an algebraic expression. Let n represent the number: The product of a number and twelve more than the number.

77. Find the volume of a box with length 1 foot 3 inches, width of 1 foot, and height of 7 inches.

3.5 Solving Equations of the Form $x + a = b$ or $x - a = b$

OBJECTIVE

Solve equations of the form $x + a = b$ or $x - a = b$ where a and b are whole numbers.

VOCABULARY

Equivalent equations are equations with the same solutions.

To **solve an equation** means to find the replacements for the variable that makes the equation true.

The **symmetric property of equality** states that if $x = a$ then $a = x$, that is, the sides of an equation may be interchanged.

HOW AND WHY
Objective

Solve equations of the form $x + a = b$ or $x - a = b$ where a and b are whole numbers.

The equations $x + 3 = 7$, $x = 4$, $12 - x = 8$, and $5x = 20$ are *equivalent equations* since they all have the solution $x = 4$. To find the solution of an equation, we write a list of equivalent equations until the variable is isolated on one side of the equation. The final equation will have the form:

$x =$ some number   or   some number $= x$

For example,

$$x + 33 + 72 = 245$$
$$x + 105 = 245$$
$$x = 140$$

is a list of equivalent equations, since each is true when x is replaced by 140. The process of writing such a list of equivalent equations is called "solving the equation." There are four fundamental properties of equations that can be used to write such a list. We will study two of these in this section and two more in Section 3.6.

Properties of
Equality

Addition and Subtraction

Adding the same number to both sides of an equation yields an equivalent equation: If $a = b$, then $a + c = b + c$.

Subtracting the same number from both sides of an equation yields an equivalent equation. If $a = b$, then $a - c = b - c$.

We can use the subtraction property to solve $x + 89 = 127$.

$x + 89 - 89 = 127 - 89$ **Subtraction property of equality. To "undo" the addition of 89, subtract 89 from both sides.**

$x = 38$ **Simplify.**

Check by substituting 38 for x in the *original equation*.

$$x + 89 = 127$$

Does $38 + 89 = 127$?   **Substitute.**

Does $127 = 127$?   **Yes.**

The solution is 38. It is common to say that the solution is "$x = 38$" so that both the variable and the solution are identified.

To solve an equation of the form x + a = b

Subtract *a* from both sides of the equation.

In the previous example we subtracted 89 from both sides to "undo" the addition *x* + 89. To reverse or "undo" subtraction, we use the addition property. Solve 165 = *t* − 478.

$165 + 478 = t - 478 + 478$ **Addition property of equality. Add 478 to both sides to "undo" the subtraction of 478 from *t*.**

$643 = t$ or $t = 643$ **Check is left for the student.**

Many people prefer to keep the variable on the left side so they change the original equation from 165 = *t* − 478 to *t* − 478 = 165. Interchanging the sides of an equation depends on the *symmetric property of equality.*

Property of Equality	Symmetric
	If $a = x$ then $x = a$.

 To solve an equation of the form x − a = b

Add *a* to both sides of the equation.

Examples A–E   **Warm Ups A–E**

Directions: Solve.

Strategy: Add or subtract the same number from both sides of the equation so that the variable is isolated on one side.

A. $c + 476 = 704$	A. $d + 217 = 492$

$c + 476 - 476 = 704 - 476$ **Subtract 476 from each side so that *c* is isolated.**

$c = 228$ **Simplify each side.**

Check: $c + 476 = 704$

Does $228 + 476 = 704$?   **Replace *c* with 228 and add.**

Does $704 = 704$?   **Yes.**

The solution is $c = 228$.

B. $329 = 89 + m$	B. $562 = 298 + n$

$329 - 89 = 89 + m - 89$ **Subtract 89 from each side.**

$329 - 89 = m + 89 - 89$ **Commutative property of addition.**

$240 = m$ **Simplify.**

Answers to Warm Ups   A. $d = 275$ B. $n = 264$

Check: $329 = 89 + m.$

Does $329 = 89 + 240$?   **Replace m with 240 and add.**

Does $329 = 329$?   **Yes.**

The solution is $240 = m$ (or $m = 240$).

C. **Calculator Example:**

$$5000 = w + 3898$$

$$5000 - 3898 = w + 3898 - 3898$$

Use a calculator to do the subtraction.

$1102 = w$ **Check left for the student.**

The solution is $w = 1102$.

C. $3787 = x - 687$

D. The difference of a number and 129 is 88. Find the number.

Let n represent the number.   **Choose a variable.**

$n - 129$ **Translate the difference to algebra.**

$n - 129 = 88$ **The difference is 88. Write the equation.**

$n - 129 + 129 = 88 + 129$ **Solve the equation.**

$$n = 217$$

The number is 217.

D.   The sum of a number and 649 is 847. Find the number.

E. The formula for Hanh's federal income tax is $T = w - c + s$, where T is the total tax, w is the tax on his earned income, c is his total tax credit, and s the tax he owes for Medicare and Social Security. If his total tax bill is \$8842, the tax on his income is \$6547, and his credits total \$829, what is the total of his Medicare and Social Security tax?

$T = w - c + s$ **Formula.**

$8842 = 6547 - 829 + s$ **Substitute 8842 for T, 6547 for w, and 829 for c.**

$8842 = 5718 + s$ **Simplify the right side.**

$8842 - 5718 = 5718 - 5718 + s$ **Subtract 5718 from each side.**

$3124 = s$ **Simplify.**

Check: $T = w - c + s.$

Does $8842 = 6547 - 829 + 3124$?   **Substitute.**

Does $8842 = 8842$?   **Yes.**

The Medicare and Social Security taxes are \$3124.

E.   Wendee's total tax bill is \$9133, the tax on her income is \$6803, and her credits total \$879. What is the total tax she owes for Medicare and Social Security? Use the formula in Example E.

Exercises 3.5

OBJECTIVE: *Solve equations of the form $x + a = b$ or $x - a = b$.*

Solve.

A.

1. $a + 11 = 38$

2. $b + 11 = 58$

3. $c - 5 = 27$

4. $d - 7 = 32$

5. $r + 12 = 20$

6. $s + 12 = 32$

7. $t - 20 = 6$

8. $s - 20 = 17$

9. $y + 16 = 16$

10. $z + 36 = 36$

11. $m - 13 = 13$

12. $p - 34 = 34$

13. $a + 22 = 36$

14. $b + 40 = 75$

15. $c - 42 = 60$

16. $d - 57 = 50$

B.

17. $r + 37 = 76$

18. $s + 27 = 52$

19. $t - 25 = 67$

20. $s - 37 = 76$

21. $y + 92 = 120$

22. $z + 51 = 72$

23. $m - 76 = 76$

24. $m - 98 = 98$

25. $21 + y = 53$

26. $32 + t = 45$

27. $65 = t - 15$

28. $34 = b - 10$

29. $134 = 26 + z$

30. $136 = 54 + t$

31. $98 + a = 103$

32. $78 + b = 112$

33. $37 = c - 111$

34. $5 = d - 217$

35. $902 = 15 + u$

36. $335 = 28 + v$

37. $389 + y = 620$

38. $546 + z = 621$

39. $251 = p - 769$

40. $619 = q - 421$

C.

Fill in the boxes so the statement is true. Explain your answer.

41. If $\square + w = 78$, then $w = 19$.   **42.** If $w - \square = 78$, then $w = 144$.

43. Find the error(s) in the statement: If $98 + x = 100$, then $x = 198$. Correct the statement. Explain how you would avoid this error.

44. Find the error(s) in the statement: If $x - 98 = 100$, then $x = 2$. Correct the statement. Explain how you would avoid this error.

45. A triangle has a perimeter of 675 inches. If two of the sides of the triangle measure 165 inches and 273 inches, what is the length of the third side of the triangle?

46. A triangle has a perimeter of 500 yards. If two sides of the triangle measure 207 yards and 98 yards, what is the length of the third side of the triangle?

47. A truck farmer's orchard plot is in the shape of a quadrilateral. The perimeter of this plot of ground is 922 meters. If three sides of the plot measure 343 meters, 133 meters, and 209 meters, what is the length of the fourth side?

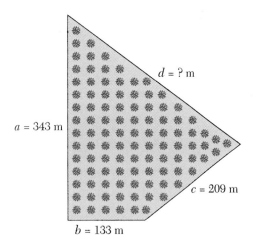

48. A corner lot has the shape of a quadrilateral. The perimeter of the lot is 3025 feet. If three sides of the lot measure 1125 feet, 436 feet, and 686 feet, what is the length of the fourth side?

49. A city treasurer made the following report to the city council regarding monies allotted and dispersed out of a City Parks bond.

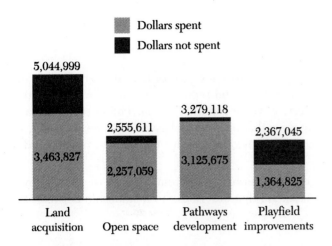

Dollars spent

Dollars not spent

5,044,999

3,279,118

2,555,611

2,367,045

3,463,827

3,125,675

2,257,059

1,364,825

Land acquisition

Open space

Pathways development

Playfield improvements

a) Write an equation which relates the total money budgeted per category to the amount of money spent and the amount of money not yet spent.

b) Use your equation to calculate the amount of money not yet spent in each of the four categories.

50. Thirty-six increased by a number results in one hundred nineteen. Translate the statement to an equation and find the number.

51. Forty-four less than a number is fifty-six. Translate the statement to an equation and find the number.

52. A number decreased by thirty-nine results in twelve. Translate the statement to an equation and find the number.

53. A Saturn with manual transmission has an EPA City rating of 4 miles per gallon more than the EPA City rating of a Subaru Impreza. Write an equation that describes this relationship. Be sure to define all variables in your equation. If the Saturn has an EPA City rating of 28 mpg, find the rating of the Impreza.

54. In 1993 in the United States, the number of deaths by drowning was 1700 less than the number of deaths by fire. Write an equation that describes this relationship. Be sure to define all variables in your equation. If there were approximately 4800 deaths by drowning that year, how many deaths by fire were there?

55. The average daily high temperature in Santiago, Chile, in January, is 26°F higher than the average daily high temperature in July. Write an equation that describes this relationship. Be sure to define all variables in your equation. Find the average high temperature in July if the average high temperature in January is 85°F.

Exercises 56–59 refer to the Chapter 3 Application. See page 179.

The hotel convention manager needs to provide the meeting planner with a report of the number of single rooms and double rooms that were occupied by convention registrants. The hotel reporting system only provides the total number of rooms occupied and the total number of guests in those rooms. The convention services manager decides that the number of double rooms is the key. She reasons that each room has at least one occupant, while the double rooms have two occupants. So, the total number of occupied rooms plus the number of occupied double rooms must be equal to the total number of guests.

56. Let d represent the total number of occupied double rooms, r represent the total number of occupied rooms, and g represent the total number of guests. Write an equation that describes the relationship between rooms and guests.

57. The first day, the hotel convention services manager gets a report that the convention has 876 guests registered in 568 rooms. How many double rooms are occupied?

58. The second day of the convention, the report is that there are 1046 guests registered in 679 rooms. How many double rooms are occupied?

59. How many more guests are there in single rooms on the second day as compared with the first day?

STATE YOUR UNDERSTANDING

60. What inverse operations are used in this section? What part do they play in solving equations symbolically?

61. Explain why $x + 3 = 7$ and $x + 113 = 117$ are equivalent equations while $x + 3 = 7$ and $x + 3 = 10$ are not equivalent.

CHALLENGE

62. In the table, find a value for n so that the sum of each row, column, and diagonal is 30.

n	14	$3n$
$4n + 2$	10	$n - 2$
$2n$	6	$4n$

GROUP ACTIVITY

63. As group members, write five word problems that can be solved using equations like those in this section. Each group is to share their problems with the class. Use these as a review for a test.

MAINTAIN YOUR SKILLS (Sections 1.3, 3.1)

64. Round the product of 47 and 962 to the nearest hundred.

65. Round the quotient of 295,850 and 97 to the nearest thousand.

66. Round the quotient of 6,472,942 and 802 to the nearest hundred.

67. Translate to an algebraic expression. Let n represent the number: The quotient of three times a number and 71.

68. Translate to an algebraic expression. Let n represent the number: Eighteen more than the quotient of twice a number and 25.

3.6   Solving Equations of the Form $ax = b$ or $\dfrac{x}{a} = b$

OBJECTIVE   Solve equations of the form $ax = b$ or $\dfrac{x}{a} = b$ where a and b are whole numbers.

HOW AND WHY
Objective   **Solve equations of the form $ax = b$ or $\dfrac{x}{a} = b$ where a and b are whole numbers.**

Employing the same idea of "undoing" we saw in Section 3.5, we can use division to reverse multiplication. Likewise, we use multiplication to reverse division. Thus, we divide $7x$ by 7 to eliminate the "multiplication by 7" and isolate the x. And we multiply $\dfrac{x}{11}$ by 11 to nullify the "division by 11" and isolate the y.

$$\frac{7x}{7} = x \qquad \text{and} \qquad 11 \cdot \frac{y}{11} = y$$

Properties of Equality

Multiplication and Division

Dividing the same number into both sides of an equation yields an equivalent equation. If $a = b$, then $\dfrac{a}{c} = \dfrac{b}{c}$.

Multiplying the same number times both sides of an equation yields an equivalent equation. If $a = b$, then $ac = bc$.

We can use the division property to solve $42n = 630$. We isolate the variable n by dividing both sides by 42.

$$\frac{42n}{42} = \frac{630}{42} \qquad \text{Divide both sides by 42.}$$

$$n = 15 \qquad \text{Simplify.}$$

Check by substituting 15 for n in the *original equation*.

$$42n = 630$$

Does $42(15) = 630$?   Substitute.

Does $630 = 630$?   Yes.

The solution is $n = 15$.

▶ **To solve an equation of the form $ax = b$, $a \neq 0$**

Divide both sides of the equation by a.

We can solve the equation $\dfrac{y}{13} = 23$ by using the multiplication property to isolate the variable y on the left side.

$$13\left(\frac{y}{13}\right) = 13(23) \qquad \text{Multiply both sides by 13 to "undo" the division of y by 13.}$$

$$y = 299 \qquad \text{Simplify. Check is left for the student.}$$

The solution is $y = 299$.

> To solve an equation of the form $\dfrac{x}{a} = b,\ a \neq 0$
>
> _____
>
> Multiply both sides of the equation by a.

Examples A–E	**Warm Ups A–E**

Directions: Solve.

Strategy: Multiply or divide both sides of the equation using the same number so that the variable is isolated on one side.

A. $12y = 6012$

$\dfrac{12y}{12} = \dfrac{6012}{12}$ **Divide each side by 12 to "undo" the multiplication of 12 times y.**

$y = 501$ **Simplify each side.**

Check: $12y = 6102$

Does $12(501) = 6012$?   **Substitute 501 for y and multiply.**

Does $6012 = 6012$?   **Yes.**

The solution is $y = 501$.

A. $14b = 686$

B. $1118 = 13x$

$\dfrac{1118}{13} = \dfrac{13x}{13}$ **Divide both sides by 13 to isolate x.**

$86 = x$ **Simplify. Check left for the student.**

The solution is $86 = x$ (or $x = 86$).

B. $1118 = 26y$

 C.   **Calculator Example:**

$347w = 21167$

$\dfrac{347w}{347} = \dfrac{21167}{347}$

Use a calculator to do the division.

$w = 61$ **Check left for the student.**

The solution is $w = 61$.

C. $3787 = 541x$

D. $16 = \dfrac{x}{21}$

$21(16) = 21\left(\dfrac{x}{21}\right)$ **Multiply both sides by 21 to "undo" the division of x by 21.**

$336 = x$ **Simplify by multiplication.**

D. $11 = \dfrac{y}{11}$

Check: $16 = \dfrac{x}{21}$

Does $16 = \dfrac{336}{21}$?   **Substitute 356 for x.**

Does $16 = 16$?   **Yes.**

The solution is $x = 336$.

E.   The wholesale cost of 25 TV sets is $4800. What is the wholesale cost of one set? Use the formula $C = np$, where C is the total cost, n is the number of sets purchased, and p is the price per set.

$C = \$4800$ **Total cost.**

$n = 25$ **Number of sets purchased.**

$C = np$ **Formula.**

$4800 = 25p$ **Substitute for C and n.**

$\dfrac{4800}{25} = \dfrac{25p}{25}$ **Divide both sides by 25 to isolate p.**

$192 = p$ **Simplify.**

Check: $C = np$

Does $4800 = 25(192)$?   **Substitute in the original formula.**

Does $4800 = 4800$?   **Yes.**

A single television set costs $192 wholesale.

E.   The wholesale cost of 24 stereo systems is $5064. What is the wholesale cost of one set? Use the formula $C = np$.

Exercises 3.6

OBJECTIVE: *Solve equations of the form* $ax = b$ *or* $\dfrac{x}{a} = b$.

Solve.

A.

1. $5x = 20$

2. $7x = 35$

3. $7u = 49$

4. $6w = 30$

5. $8v = 72$

6. $6x = 72$

7. $2y = 68$

8. $4y = 48$

9. $\dfrac{x}{2} = 10$

10. $\dfrac{x}{5} = 4$

11. $\dfrac{y}{7} = 9$

12. $\dfrac{a}{8} = 7$

13. $\dfrac{y}{12} = 3$

14. $\dfrac{b}{12} = 5$

15. $\dfrac{c}{38} = 10$

16. $\dfrac{x}{38} = 100$

B.

17. $5t = 65$

18. $6v = 114$

19. $9w = 198$

20. $8x = 136$

21. $12s = 132$

22. $6z = 264$

23. $\dfrac{s}{12} = 132$

24. $\dfrac{z}{6} = 264$

25. $13u = 910$

26. $16v = 336$

27. $\dfrac{x}{21} = 34$

28. $\dfrac{s}{17} = 35$

29. $19w = 665$

30. $18x = 702$

31. $\dfrac{r}{56} = 51$

32. $\dfrac{z}{71} = 34$

33. $31a = 775$

34. $33w = 396$

35. $\dfrac{y}{48} = 9$

36. $\dfrac{w}{76} = 10$

37. $168 = 24x$

38. $336 = 12y$

39. $168 = \dfrac{x}{24}$

40. $336 = \dfrac{y}{12}$

C.

Fill in the boxes so the statement is true. Explain your answer.

41. If $\square \cdot w = 76$, then $w = 19$.

42. If $\dfrac{w}{\square} = 78$, then $w = 312$.

43. Find the error(s) in this statement: If $5x = 100$, then $x = 95$. Correct the statement. Explain how you would avoid this error.

44. Find the error(s) in the statement: If $\dfrac{x}{6} = 36$, then $x = 6$. Correct the statement. Explain how you would avoid this error.

45. If the wholesale cost of 18 stereo sets is \$5580, what is the wholesale cost of one set? Use the formula $C = np$, where C is the total cost, n is the number of units purchased, and p is the price per unit.

46. If the wholesale cost of 24 personal computers is \$18,864, what is the wholesale cost of one computer?

47. Hank's Super Food Market held a canned-food sale by the case. If each case of food contains 24 cans, write a formula for the number, N, of cans sold where c represents the number of cases sold. Find the number of cans sold if 723 cases of food were sold.

48. Each student in Chris Raible's math classes hands in 18 homework assignments. During the semester she has corrected 2790 homework assignments. Write an equation that describes the number of students, s, in all of Chris' classes. Solve the equation to find the number of students.

49. It takes 32 bricks per linear foot to build a brick retaining wall. Write a formula for the number, N, of bricks needed to build a wall. Let d represent the length of the wall in feet. Find the number of bricks it would take to build a wall that is 212 feet long.

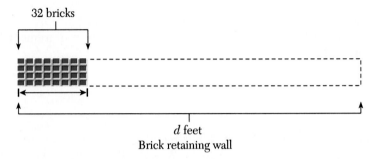

32 bricks

d feet
Brick retaining wall

50. What is the width, w, of a rectangular plot of ground on which a house is to be built if the length is 135 feet and the area is 12,420 square feet? Use the formula for the area of a rectangle.

51. Find the length of the side, s, of a square whose perimeter, P, is 100 meters. Use the formula for the perimeter of a square.

52. Find the rate of speed, r, if a distance, D, of 265 miles is traveled in the time, t, of 5 hours. Use the formula $D = rt$.

53. The average daily low temperature in Toronto in July is twice the average high temperature in January. Write an equation that describes this relationship. Be sure to define all variables in your equation. If the average daily low temperature in July is 60°F, what is the average daily high temperature in January?

54. Car manufacturers recommend that the fuel filter in a car be replaced when the mileage is ten times the recommended mileage for an oil change. Write an equation that describes this relationship. Be sure to define all variables in your equation. If a fuel filter should be replaced every 30,000 miles, how often should the oil be changed?

Exercises 55–57 refer to the Chapter 3 Application. See page 179.

When setting up a lecture hall with a projection screen, visibility is a primary concern. The meeting planner knows that audio-visual experts recommend that (1) the lower edge of the screen be at least 4 feet off the floor, (2) that the minimum distance from the screen to the first row be twice the height of the screen, and (3) that the distance from the screen to the last row be no more than eight times the height of the screen.

55. A lecture hall with theater seating is set up so that the first row is 14 feet from the screen. Let h represent the height of the screen. Write an equation in terms of h that describes the minimum distance requirement for the first row of chairs. What should the height of the screen be?

56. A lecture hall with theater seating is set up so that the last row is 48 feet from the screen. Let h represent the height of the screen. Write an equation in terms of h that describes the maximum distance requirement for the last row of chairs. What should the height of the screen be?

57. The hotel staff is planning to place a screen 8 feet wide and 6 feet high in a meeting room where the first row is 16 feet from the screen and the last row is 72 feet away from the screen. Why did the meeting planner object? What size screen did he recommend for the room?

STATE YOUR UNDERSTANDING

58. What inverse operations are used in this section? What part do they play in solving equations symbolically?

59. Explain why $\dfrac{x}{2} = 50$ and $7x = 700$ are equivalent equations while $\dfrac{x}{2} = 50$ and $0\left(\dfrac{x}{2}\right) = 0(50)$ are not equivalent.

CHALLENGE

Solve.

60. $33x = 20{,}031$

61. $44x = 17{,}996$

62. $\dfrac{y}{505} = 27$

63. $\dfrac{y}{707} = 25$

64. $\dfrac{5t}{7} = 410$

65. $\dfrac{7t}{5} = 406$

GROUP ACTIVITY

66. The adjusted temperature is calculated using a wind chill factor and the air temperature. As a group, research temperatures with and without wind chill factor. Use your findings to make a table of these temperatures including the wind speed. See if you can make up (or if you can find) a formula to show the relationship.

MAINTAIN YOUR SKILLS (Section 3.1)

Translate the following expressions to algebra. Let x represent the first number and y represent the second number.

67. Seven more than the product of two numbers.

68. The product of a number and the sum of a second number and 37.

69. The quotient of the second number and 17 diminished by the first number.

70. The product of the second number and the sum of the two numbers.

71. The product of two numbers less their quotient.

3.7 Solving Equations of the Form $ax + b = c$ or $ax - b = c$

OBJECTIVE

Solve equations of the form $ax + b = c$ or $ax - b = c$, where a, b, and c are whole numbers.

HOW AND WHY
Objective

Solve equations of the form $ax + b = c$ or $ax - b = c$, where a, b, and c are whole numbers.

Some equations require more than one step to arrive at the solution. As before, our goal is to isolate the variable. We accomplish this by "undoing" the operations. We use the properties of equality in the reverse order of operations. That is, we "undo" addition or subtraction first and then "undo" multiplication or division. The maneuver is like unwrapping a package. The ribbon is the last thing put on and the first thing taken off. So too, addition/subtraction is the last operation in formulating an expression so it is the first to go in "unformulating" the expression. Here is an example that requires both addition and division to solve.

$$13x - 25 = 66$$
$$13x - 25 + 25 = 66 + 25 \qquad \textbf{Add 25 to both sides. This isolates the term, } 13x.$$
$$13x = 91 \qquad \textbf{Simplify by combining terms on each side.}$$
$$\frac{13x}{13} = \frac{91}{13} \qquad \textbf{Divide both sides by 13 to isolate } x.$$
$$x = 7 \qquad \textbf{Simplify.}$$

Check: $13x - 25 = 66$

Does $13(7) - 25 = 66$?   **Substitute 7 for x in the original equation.**

Does $\quad 91 - 25 = 66$?

Does $\qquad 66 = 66$?   **Yes.**

The solution is $x = 7$.

 To find the solution of an equation of the form $ax + b = c$ or $ax - b = c$

1. Add or subtract on each side of the equation to isolate the term containing the variable.
2. Divide both sides of the equation by the coefficient of the variable.

Examples A–E   **Warm Ups A–E**

Directions: Solve.

Strategy: Use the properties of equality in reverse order of operations to isolate the variable.

A. $82 = 19x - 13$

$$19x - 13 = 82 \qquad \textbf{Symmetric property of equality.}$$
$$19x - 13 + 13 = 82 + 13 \qquad \textbf{Add 13 to both sides.}$$
$$19x = 95 \qquad \textbf{Simplify.}$$
$$\frac{19x}{19} = \frac{95}{19}$$
$$\qquad\qquad\qquad \textbf{Divide both sides by 19.}$$
$$x = 5 \qquad \textbf{Simplify.}$$

A. $186 = 4 + 7x$

Check: $82 = 19x - 13$

Does $82 = 19(5) - 13$?   Replace x with 5 and multiply.

Does $82 = 95 - 13$?   Simplify.

Does $82 = 82$?   Yes.

The solution is $x = 5$.

B. $7p + 173 = 264$

$7p + 173 - 173 = 264 - 173$ Subtract 173 from both sides.

$7p = 91$ Simplify.

$\dfrac{7p}{7} = \dfrac{91}{7}$ Divide both sides by 7.

$p = 13$ Simplify.

Check: $7p + 173 = 264$

Does $7(13) + 173 = 264$?   Replace p with 13 and multiply.

Does $\quad 91 + 173 = 264$?   Simplify.

Does $\qquad 264 = 264$?   Yes.

The solution is $p = 13$.

B. $16w + 137 = 345$

C. **Calculator Example:**

$448t - 524 = 5748$

$448t - 524 + 524 = 5748 + 524$ Add 524 to both sides.

$\dfrac{448t}{448} = \dfrac{5748 + 524}{448}$ Divide each side by 448.

Use a calculator to do the addition and division. Be sure to do the addition first as indicated by the fraction bar.

$t = 14$ Check left for the student.

The solution is $t = 14$.

C. $231v - 325 = 4526$

D. $432 = 13x + 25x + 52$

$432 = 38x + 52$ Combine the like terms.

$432 - 52 = 38x + 52 - 52$ Subtract 52 from both sides.

$380 = 38x$ Simplify.

$\dfrac{380}{38} = \dfrac{38x}{38}$ Divide both sides by 38.

$10 = x$ Simplify.

Check: $432 = 13x + 25x + 52$

Does $432 = 13(10) + 25(10) + 52$?   Substitute 10 for x.

Does $432 = 130 + 250 + 52$?

Does $432 = 432$?   Yes.

The solution is $x = 10$.

D. $18x + 25x + 33 = 119$

Answers to Warm Ups   B. $w = 13$ C. $v = 21$ D. $x = 2$

E. A manufacturer of personal computer accessories requires a box-shaped (rectangular solid) package that is 17 inches wide and 22 inches long.

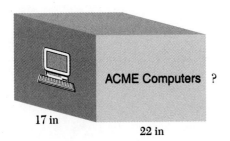

The designer plans a surface area of 1684 square inches. What is the required height of the box? We can use the formula $S = 2\ell w + 2\ell h + 2wh$, where S is the surface area, w is the width, ℓ is the length, and h is the height.

Strategy: Substitute into the formula and solve.

$S = 2\ell w + 2\ell h + 2wh$	Formula.
$1684 = 2(22)(17) + 2(22)h + 2(17)h$	Substitute.
$1684 = 748 + 44h + 34h$	Simplify.
$1684 = 748 + 78h$	Combine like terms.
$1684 - 748 = 748 - 748 + 78h$	Subtract 748 from each side.
$936 = 78h$	Simplify.
$\dfrac{936}{78} = \dfrac{78h}{78}$	Divide both sides by 78.
$12 = h$	Simplify.
The height of the box is 12 inches.	Check is left for the student.

E. A manufacturer of personal computer accessories requires a box-shaped (rectangular solid) package that is 18 inches wide and 20 inches long. The designer plans a surface area of 1860 square inches. What is the required height of the box?

Exercises 3.7

OBJECTIVE: *Solve equations of the form ax + b = c or ax − b = c.*

Solve.

A.

1. $4x + 1 = 9$

2. $3x + 1 = 10$

3. $6y - 3 = 27$

4. $5y - 4 = 21$

5. $10w + 8 = 88$

6. $10w + 1 = 11$

7. $9t + 5 = 5$

8. $7t + 11 = 11$

9. $4z - 8 = 20$

10. $8z - 4 = 20$

11. $8a - 7 = 57$

12. $6a - 6 = 48$

13. $5n + 6 = 41$

14. $6n + 5 = 71$

15. $34 = 9p - 11$

16. $33 = 8p - 15$

B.

17. $6a - 24 = 90$

18. $4a - 36 = 72$

19. $12c + 29 = 53$

20. $4c + 18 = 94$

21. $18 + 13x = 96$

22. $24 + 14x = 150$

23. $12y - 56 = 28$ **24.** $16y - 29 = 51$ **25.** $31t - 45 = 730$

26. $25t - 65 = 460$ **27.** $407 = 24a + 215$ **28.** $541 = 54a + 217$

29. $42 = 35b - 98$ **30.** $210 = 42b - 84$ **31.** $974 = 232 + 53c$

32. $909 = 450 + 51c$ **33.** $2y + 3y - 65 = 210$ **34.** $2y + y - 23 = 145$

35. $7w + 14 + 14w = 140$ **36.** $25w + 17 + 6w = 141$ **37.** $234 + 50x + 7x = 1545$

38. $118 + 35x + 8x = 935$ **39.** $1005 = 104z - 23z + 114$ **40.** $1735 = 120z - 42z + 253$

C.

Fill in the boxes so the statement is true. Explain your answer.

41. If $\square \cdot x + 9 = 21$, then $x = 6$. **42.** If $\square \cdot y - 7 = 29$, then $y = 12$.

43. Find the error(s) in the statement: If $7x + 35 = 126$, then $x + 35 = 126$. Correct the statement. Explain how you would avoid this error.

44. Find the error(s) in the statement: If $5x - 95 = 25$, then $x - 95 = 5$. Correct the statement. Explain how you would avoid this error.

45. When six is added to five times a number, the result is forty-one. Write an equation that describes this relationship. Find the number.

46. When four times a number is decreased by fourteen, the result is seventy. Write an equation which describes this relationship. Find the number.

47. Ninety-two is twelve less than four times a number. Write an equation that describes this relationship. Find the number.

48. The difference of sixteen times a number and seventy-four is fifty-four. Write an equation that describes this relationship. Find the number.

49. An electronics firm added a new automated assembly line that produces three times as many units in a week as does the old manual assembly line. If the two lines have a total output of 1024 units per week, find the number of units produced by each line. (*Hint:* Let n represent the number of units produced by the manual line. The number of units produced by the automated line is then $3n$. Write an equation that describes the relationship. Solve the equation and write the answer.)

50. On a recent vacation trip the Moriartys drove twice as far on the second day as they did on the first day. The third day they drove 165 miles. During the three days they drove 1095 miles. Find the number of miles the Moriartys drove each of the first two days.

51. A manufacturer of personal computer accessories requires a box-shaped (rectangular solid) package that is 15 inches wide and 20 inches long. The designer plans a surface area of 1510 square inches. What is the required height of the box?

52. A manufacturer of personal computer accessories requires a box-shaped (rectangular solid) package that is 2 inches wide and 18 inches long. The designer plans a surface area of 792 square inches. What is the required height of the box?

53. The following table summarizes several different long distance calling plans.

Company	Monthly Fee	Charge per Minute
ATT	none	15 ¢
Tone	$4.90	10 ¢
Pace	$6.96	9 ¢

Jessica only has $30 budgeted to spend on long distance calls per month. Let m be the number of minutes of long distance calls per month. Let C represent the monthly bill in cents. Write an equation for each long distance company which relates the monthly bill to m. Which company will give Jessica the most minutes for her $30?

54. The difference of ten times a number and three times the same number is 33 less than 250. Write an equation that describes this relationship. Find the number.

 Exercises 55–57 refer to the Chapter 3 Application. See page 179.

The hotel has four ballrooms that can be divided into lecture halls that are 100 feet by 100 feet. The meeting planner must determine how many chairs per row will fit in the room. The chairs are 18 inches wide by 18 inches deep (as before) and the aisles are 6 feet wide. For safety and comfort, there can be no more than 15 chairs in a row. Let x represent the number of chairs in a row.

55. Assume there are aisles along both walls and one center aisle. Write an equation that describes the width of the room in terms of x.

56. Assume there are aisles along both walls and two aisles in the middle of the room. Write an equation that describes the width of the room in terms of x. How many chairs can be in a row according to your equation? Is this a viable set up for the room? Explain.

57. Draw a setup for the room that meets all the specifications. Calculate how many chairs can be in one row. Tell why your solution provides the maximum number of chairs possible.

STATE YOUR UNDERSTANDING

58. Explain the steps used in solving the equation $2x - 5 = 11$.

59. Explain how to solve $4w - 5 = 22$ using a table. Explain how to solve $4w - 54 = 22$ symbolically.

CHALLENGE

Solve.

60. $17x + 554 = 10{,}805$

61. $19x - 554 = 10{,}979$

62. $\dfrac{z}{8} + 226 = 249$

63. $\dfrac{z}{9} - 626 = 7$

64. Solve $ax - b = c$ for x.

65. Solve $ax + b = c$ for x.

GROUP ACTIVITY

66. Develop a formula for calculating the average of five numbers. Test your formula to verify that it is correct.

67. The Merry Muffin Shop took in $523 on Monday, $397 on Tuesday, $468 on Wednesday, and $504 on Thursday. What does the Friday take have to be in order for the daily average to be $500?

68. A basketball player has rebounds of 9, 12, 10, 12, 4, 13, 8, 9, and 11 over nine games. What does she need in the tenth game in order to have an average of 10 rebounds per game?

MAINTAIN YOUR SKILLS (Sections 1.6, 2.2, 2.3, 2.4)

69. Evaluate $800 - 8^2(2^3)$.

70. Evaluate $800 - 2^4 + 3^2$.

71. Find the perimeter of the following figure.

72. Find the area of the figure in Exercise 71.

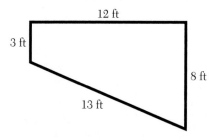

12 ft

3 ft

8 ft

13 ft

73. Find the volume of a solid whose sides are perpendicular to a base shaped like the figure in Exercise 71, if it has height 8 feet.

CHAPTER 3

OPTIONAL Group Project *(2–4 weeks)*

A meeting planner is working on a convention of psychologists to be held in Salt Lake City. The convention is expected to draw between 400 and 700 registrants. The planner is going to make up a table of convention needs for 400, 500, 600, or 700 registrants. A large lecture hall, using theater seating, must be available for the keynote speaker. The hall must seat all the convention participants. The planner also needs three lecture halls with schoolroom seating that will hold 200 people each. Saturday night there is a banquet for all participants.

a. The hotel has a ballroom that is 300 feet long and 200 feet wide. It can be divided into smaller rooms that are 100 feet by 50 feet. The speaker's platform is 20 feet long and 12 feet deep. The chairs are 18 inches wide and 18 inches deep. There must be 24 inches between the rows, the aisles must be 6 feet wide, and no more than 15 chairs can be placed in a row because of safety regulations. For each possible number of registrants, decide how many chairs are needed for the lecture hall, and draw a floor plan showing the seating arrangement. Include a written narrative of your decision process, and support your conclusion with algebraic calculations.

b. The smaller lecture halls will have a schoolroom setup that consists of tables that are 18 inches deep and 6 feet long. Each table will have chairs on one side, facing the front of the room. There must be 46 inches between the rows of tables to allow for chairs and passage. Each person is furnished with 30 inches of table width. The speaker's platform is 8 feet long and 6 feet deep, and the first table must be at least 9 feet from the platform. Each hall will contain a projection screen. Decide how many tables and chairs are needed for these lecture halls and draw a floor plan showing the seating arrangement. Give the size specifications for the projection screen. Include a written account of your decision process and support your conclusion with algebraic calculations.

c. The banquet will have seating at round tables. Tables with 60-inch diameters will seat eight and tables with 72-inch diameters will seat ten people each. There will be a platform for the head table that is 30 feet long and 6 feet deep. There must be a minimum of 60 inches between tables. Furthermore, approximately 1000 square feet in the center of the room will be left open for a dance floor. For each possible number of registrants, decide how many tables and chairs are needed for the banquet and draw a floor plan showing the seating arrangement. Include a written narrative of your decision process and support your conclusion with algebraic calculations.

CHAPTER 3

True-False Concept Review

ANSWERS

Check your understanding of the language of algebra and arithmetic. Tell whether each of the following statements is True (always true) or False (not always true). For each statement you judge to be false, revise it to make a statement that is True.

1. _____

   1. A variable is a changing number or letter.

2. _____

   2. A variable, by itself, is an algebraic expression.

3. _____

   3. A number, by itself, is an algebraic expression.

4. _____

   4. The phrase "greater than" has more than one algebraic translation depending on the way it is used.

5. _____

   5. Every equation contains an equal sign.

6. _____

   6. The statement "$34 = x + 435$" is an equation.

7. _____

   7. A formula is a special kind of equation.

8. _____

   8. The equation $A = \dfrac{bh}{2}$ is a formula.

9. _____

   9. The algebraic expressions $5ab$ and $5xy$ are like terms.

10. _____

   10. The algebraic expressions pz and $8pz$ are like terms.

11. _____

   11. Combining like terms is based on the distributive property.

12. _____

   12. Unlike terms can be combined.

13. _____

   13. Unlike terms can be multiplied.

14. _____

   14. The exponents of terms are added whenever like terms are added.

15. _____

15. When like terms are subtracted the exponents of the variables are subtracted.

16. _____

16. Equivalent equations have the same left and right members (sides).

17. _____

17. To evaluate an expression the replacement values of the variables in the expression must be given.

18. _____

18. To solve an equation we must substitute a value for the variable in the equation.

19. _____

19. The equations $x + 17 = 21$ and $5x = 20$ are equivalent equations.

20. _____

20. An algebraic expression contains an equal sign.

21. _____

21. The product of 6 and 4 is 10.

22. _____

22. The square of 6 is 12.

23. _____

23. A quotient is the answer to a division problem for both numbers and algebraic expressions.

24. _____

24. For all values of x and y, $6x + 3y = 9xy$.

25. _____

25. For all values of x, $6x(3x) = 18x^2$.

CHAPTER 3

Review

Section 3.1 *Objective 1*

Translate to algebraic expressions. Let x represent the number.

1. A number increased by thirty.
2. A number decreased by one hundred ten.
3. Fifty-six more than four times a number
4. Seventy-two less than the square of a number.
5. Ninety-five divided by the product of seven and a number.

Section 3.1 *Objective 2*

6. Evaluate $3a - 11$ if $a = 15$.
7. Evaluate $45 - b$ if $b = 23$.
8. Evaluate $56 + 23x$ if $x = 11$.
9. Evaluate $x^2 - 16$ if $x = 8$.
10. Evaluate $3x - 2y + xy$ if $x = 44$ and $y = 13$.

Section 3.2 *Objective 1*

11. Is 4 a solution of $19y - 29 = 47$?
12. Is 8 a solution of $15y + 37 = 151$?
13. Is 19 a solution of $6x - 45 = 69$?
14. Is 7 a solution of $98 - 7x = 49$?
15. Is 23 a solution of $17y + 24 = 414$?

Section 3.2 *Objective 2*

Evaluate the formulas.
16. $D = rt$ if $r = 48$ and $t = 6$.
17. $V = s^3$ if $s = 7$.
18. $P = 2\ell + 2w$ if $\ell = 34$ and $w = 16$.
19. $S = c + m$ if $c = 510$ and $m = 23$.
20. $T = D + pt$ if $D = 1200$, $p = 55$, and $t = 36$.

Section 3.2 *Objective 3*

Solve.
21. $a - 13 = 41$
22. $55 + b = 78$
23. $2x - 6 = 22$

24. $\dfrac{y}{21} = 63$
25. $2y + 12 + 3y = 67$

Section 3.3 *Objective 1*

Simplify.
26. $8a + 32a + 7a$
27. $43t - 21t + 9t$
28. $55y - 32y - 12y - y$
29. $4x^2 + 3x + 19x^2 - x$
30. $4xy + 32y + 18y - xy$

Section 3.3 *Objective 2*

Simplify.
31. $(3x + 9y) + (4x - 5y)$
32. $(5a + 3b) + (4a - b)$
33. $(7x + 12y + 4z) + (2x - 3y)$
34. $(4b + 5c) - (3b + c)$
35. $(2a + 3b + c) - (a + b + c)$

Section 3.4 *Objective 1*

Multiply or divide.
36. $23(5y)$
37. $(24z)(9y)$
38. $(7ab)(19bc)(8ac)$
39. $\dfrac{56ab}{8}$
40. $\dfrac{123a^2bc}{3ac}$

Section 3.4 *Objective 2*

Multiply or divide.
41. $3a(4a - 7)$
42. $13a(4a + 3b + 5c)$
43. $(2x + 5)(x + 4)$
44. $\dfrac{16xy - 24wx + 36xz}{4x}$
45. $\dfrac{153x^2y + 204\,xy^2 - 289xy}{17xy}$

Section 3.5

Solve.

46. $72 + y = 103$

47. $t - 45 = 80$

48. $15 = y - 45$

49. $s + 7 = 75$

50. A quadrilateral has a perimeter, P, of 456 feet. If three of the sides of the quadrilateral measure 121 feet, 96 feet, and 155 feet, what is the length of the fourth side?

Section 3.6

Solve.

51. $9t = 63$

52. $8w = 384$

53. $195 = 13y$

54. $27z = 216$

55. $\dfrac{x}{7} = 14$

56. $\dfrac{a}{12} = 28$

57. $\dfrac{w}{12} = 17$

58. $\dfrac{n}{33} = 30$

59. The cost of 62 umbrellas is \$434. Write an equation that describes this relationship and find the cost of one umbrella.

60. When a number is divided by 30, the result is 14. Write an equation that describes this relationship and find the number.

Section 3.7

Solve.

61. $41 + 5x = 76$

62. $11x - 43 = 122$

63. $22w + 14 = 146$

64. $70 = 9y - 11$

65. Twenty-seven more than the product of six and a number is one hundred twenty-nine. Write an equation which describes this relationship and find the number.

CHAPTER 3

Test

ANSWERS

1. _____

2. _____

3. _____

4. _____

5. _____

6. _____

7. _____

8. _____

9. _____

10. _____

11. _____

12. _____

13. _____

14. _____

15. _____

1. Simplify: $32z - 19z + 48z$

2. Solve: $5x + 28 = 108$

3. Evaluate $336 - 18y$ if $y = 13$.

4. Evaluate $S = \dfrac{n(a + \ell)}{2}$ if $n = 20$, $a = 2$, and $\ell = 59$.

5. Solve: $w - 336 = 792$

6. Solve: $\dfrac{t}{21} = 6$

7. Divide: $\dfrac{180ab}{15b}$

8. Divide: $\dfrac{100c^2 + 60c - 40cd}{20c}$

9. Multiply: $3xy(14x + 23y - 15z)$

10. Combine: $215x - 17x + 29x$

11. Evaluate $3p + pq - q$ if $p = 45$ and $q = 37$.

12. Solve: $r + 79 = 107$

13. Multiply: $(16x)(4x)(x)$

14. Solve: $24y = 264$

15. Solve: $4x + 19 = 91$

16. _____

16. Subtract: $25y - y - 7y$

17. _____

17. Solve: $12x - 168 = 168$

18. _____

18. Is 0 a solution of $191x - 382 = 382$?

19. _____

19. Translate to an algebraic expression. Let n represent the number. The sum of 18 and a number, divided by 5.

20. _____

20. Evaluate $C = np$ if $n = 35$ and $p = 16$.

21. _____

21. Margaret buys a bedroom set for $76 down and $21 a month for 24 months. Write an equation that describes this relationship and find the amount that she pays for the set.

22. _____

22. The perimeter of a rectangle is 204 feet. If the width is 19 feet, what is the length?

23. _____

23. The wholesale cost of 18 lawn mowers is $6264. Let c represent the cost of one mower and write an equation that describes this relationship. Find the cost of one lawn mower.

24. _____

24. The sum of fourteen times a number and ten is three hundred seventy-four. Write an equation which describes this relationship and find the number.

CHAPTERS 1-3

Cumulative Review

1. Write the place value notation of eight thousand, five.
2. Write the word name for 28,045.
3. Insert < or > to make a true statement.

   168   129

4. Round 103,845 to the nearest hundred.
5. Add: 987 + 677 + 9656 + 28.
6. Estimate the sum and then add.

   4310 + 5712 + 2049 + 1178

7. Subtract and round to the nearest ten.

   8235 − 2348

8. Estimate the difference and then subtract.

   88,066 − 9247

9. Multiply and round to the nearest thousand.

   3456(209)

10. Estimate the product and then multiply.

    206(5280)

11. Find the quotient of 101,595 and 195.
12. Estimate the quotient and then divide.

    821,766 ÷ 5046

13. Find the value of 12^3.
14. Find the value of 2893×10^4.
15. Find the value of $28,040,000 \div 10^3$.
16. Simplify: $4^3 \cdot 4^4$
17. Simplify: $(9^3)^4$
18. Simplify: $(5 \cdot 8)^7$
19. Simplify: $11^2 + 8^2 \div 2^2 - (3^2)^2$
20. Simplify: $11^2 + 8^2$
21. Simplify: $(11 + 8)^2$
22. Anne bowled four lines at the local bowling alley. She had scores of 188, 236, 208, and 188. What was her average score for the night?
23. Find the average of 234, 486, 126, 596, 222, and 196.

The following bar graph shows the occupancy of a 500-apartment complex for a four-year period:

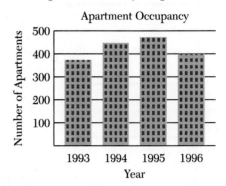

24. What year was occupancy the highest?
25. What year were the apartments all occupied?
26. What will the occupancy be in 1997 if the increase is the same as between 1993 and 1994?
27. What will the occupancy be in 1997 if the decline is the same as between 1995 and 1996?
28. How many apartments were vacant in 1996?
29. Draw a line graph to display the following data. The television models sold in one week at an appliance store: Black and White, 5; Color, 25; Color with Stereo, 21; Miniature, 7.
30. Add.   6 ft 10 in.
            11 ft 7 in.
          + 8 ft 4 in.

31. Subtract.   3 hr 25 min 41 sec
              − 1 hr 30 min 54 sec

32. A can of cherry pie filling contains 624 grams. How many grams of the pie filling are in 13 cans?
33. The In-Out Market sold 12 lb of hamburger on Monday, 22 lb on Tuesday, 35 lb on Wednesday, 28 lb on Thursday, 45 lb on Friday, and 62 lb on Saturday. What was the average number of pounds of hamburger sold during the six days?
34. LIKE International agrees to donate 1 cent for every foot that Kevin jogs during one week to the World Food Bank. Kevin jogged the

following miles during the week: Monday, 2; Tuesday, 3; Wednesday, 4; Thursday, 2; Friday, 3; Saturday, 5; Sunday, 1. How much does LIKE donate to the World Food Bank?

35. Find the perimeter of the following figure.

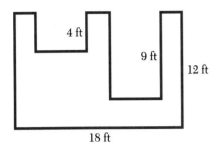

36. Find the area of the following rectangle.

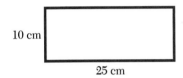

37. April's house has a large deck attached. The deck measures 16 feet by 21 feet. April plans to stain the deck. One gallon of the stain will cover 125 ft². How many gallons of stain must she purchase? (Assume the stain is sold only in 1-gallon cans.)

38. Find the volume of a barrel whose base has area 2 square feet and whose height is 20 inches.

39. Translate to an algebraic expression. Let n represent the number. The product of 11 and a number and that product decreased by 9.

40. Translate to an algebraic expression. Let n represent the number. Sixteen more than a number divided by 9.

41. Evaluate: $x^2 + 4x + 2$ if $x = 6$.

42. Evaluate: $6x - 5y + 24$ if $x = 9$ and $y = 7$.

43. Is 9 a solution of $4x - 5 = 31$?

44. Is 7 a solution of $9x + 12 = 85$?

45. Evaluate: $P = 2\ell + 2w$ if $\ell = 12$ and $w = 9$.

46. Evaluate: $A = \dfrac{bh}{2}$ when $b = 18$ and $h = 12$.

47. Add: $8x + 9x + 17x$

48. Subtract: $28x - 13x - x$

49. Simplify: $a + 9a - 6a$

50. Subtract: $58b - 47b - 11b$

51. Simplify: $21x + 17y + 9x$

52. Simplify: $12m + 19z - 12z$

53. Multiply: $2x(3x + 2y - 3z)$

54. Divide: $\dfrac{25ab + 20bc - 5b}{5b}$

55. Multiply: $12a(9a)(4a)$

56. Multiply: $9a(3b)(12c)$

57. Divide: $\dfrac{136a^2bc}{17ab}$

58. Solve: $m + 78 = 942$.

59. Solve: $942 = m - 78$

60. Solve: $38t = 11{,}742$

61. Solve: $\dfrac{t}{28} = 35$

62. Solve: $8x - 12 = 36$

63. Solve: $5x - 9 = 46$

64. Solve: $9x + 12 = 120$

65. Solve: $12a + 15 = 39$

66. The Han family pays $174 per month for a car payment, $520 per month for rent, and $297 per month for food. How much of their monthly income of $1583 is left after money for these three bills is set aside?

67. A rectangular building on an L-shaped lot has the dimensions shown in the figure. If the rest of the lot is to be landscaped, what is the total area of the landscaping?

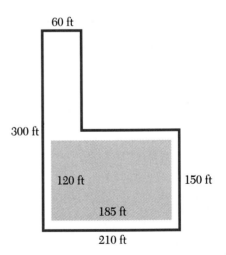

68. Galen has four test grades of 88, 72, 79, and 81. What does Galen have to get on test five to have an average of 82 for the five tests?

Good Advice for Studying

Planning Makes Perfect

Now is the time to formalize a study plan. Set aside a time of day, every day, to focus all of your attention on math. For some students, finding a quiet place in the library to study regularly for one hour is far more efficient than studying two hours at home where there are constant distractions. For others, forming a study group where you can talk about what you have learned is helpful. Decide which works best for you.

Try to schedule time as close to the class session as possible while the concepts are fresh in your mind. If you wait several hours to practice what seemed clear during class, you may find that what was clear earlier may no longer be meaningful. This may mean planning a schedule of classes that includes an hour after class to study.

If there are some days that you cannot devote one or two hours to math, find at least a few minutes and review one thing—perhaps read your notes, reread the section objectives, or if you want to do a few problems, do the section warmups. This helps to keep the concepts fresh in your memory.

If you choose to form or join a study group with other math-anxious students, don't use your time together to gripe. Instead use it to discuss and recognize the content of your negative self-talk and to write positive coping statements.

Plan, too, for your physical health. Notice how your anxious thought patterns trigger physical tension. When you wrinkle your forehead, squint your eyes, make a trip to the coffee machine, or light up a cigarette, you are looking for a way to release these tensions. Learning relaxation techniques, specifically progressive relaxation, is a healthier alternative to controlling body tension. Briefly, relaxation training involves alternately tensing and relaxing all of the major muscles in the body with the goal of locating your specific muscle tension and being able to relax it away. Use a professionally prepared progressive relaxation tape, or take a stress management class to properly learn this technique. Allow at least twenty to thirty minutes for this exercise daily. The time it takes for you to deeply relax will become briefer as you become more skilled. Soon relaxation will be as automatic as breathing and when you find yourself feeling math anxious, you can stop, take control, and relax.

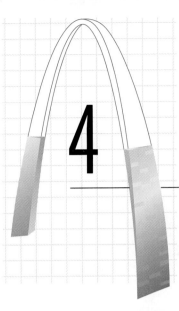

4 Integers and Equations

James Balog/Tony Stone Images ©

APPLICATION

The National Geographic Society was founded in 1888 "for the increase and diffusion of geographic knowledge." The Society has supported more than 5000 explorations and research projects with the intent of "adding to knowledge of earth, sea, and sky." Among the many facts catalogued about our planet, the Society keeps records of the elevations of various geographic features. Following is a table that lists the highest and lowest points on each continent.

Continent	Highest Point	Ft Above Sea Level	Lowest Point	Ft Below Sea Level
Africa	Kilimanjaro, Tanzania	19,340	Lake Assal	512
Antarctica	Vinson Massif	16,864	Bentley Subglacial Trench	8327
Asia	Mt. Everest, Nepal–Tibet	29,028	Dead Sea, Israel–Jordan	1312
Australia	Mt. Kosciusko, New South Wales	7310	Lake Eyre, South Australia	52
Europe	Mt. El'brus, Russia	18,510	Caspian Sea, Russia–Azerbaijan	92
North America	Mt. McKinley, Alaska	20,320	Death Valley, California	282
South America	Mt. Aconcagua, Argentina	22,834	Valdes Peninsula, Argentina	131

When measuring, one has to know where to begin, or the zero point. Notice that when measuring elevations, the zero point is chosen to be sea level. All elevations compare the high or low point with sea level. Mathematically, we represent quantities under the zero point as negative numbers.

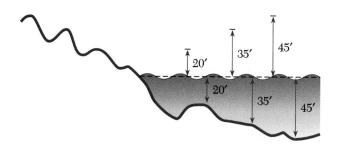

What location has the highest continental altitude on Earth? What location has the lowest continental altitude? Is there a location on Earth with a lower altitude? Explain.

4.1 Integers: Opposites, Absolute Value, and Inequalities

OBJECTIVES

1. Find the opposite of an integer.

2. Find the absolute value of an integer.

3. Write an inequality statement involving two integers.

VOCABULARY

A **number line** is a graph or picture of the placement of numbers on a line. The first number located is zero.

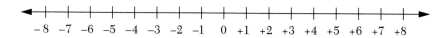

Negative numbers are located to the left of zero. **Positive numbers** are located to the right of zero.

The **opposite** of a number is the number that is the same distance from zero on the number line but on the opposite side of zero.

The **absolute value** of a number is the number of units between the number and zero on the number line regardless of direction.

The **whole numbers** are 0, 1, 2, 3, 4, 5, 6, ...

The **integers** are the whole numbers and their opposites: . . ., −5, −4, −3, −2, −1, 0, 1, 2, 3, 4, 5, . . .

HOW AND WHY
Objective 1

Find the opposite of an integer.

An *integer* is a whole number or the *opposite* of a whole number. A few integers are shown on the following *number line*.

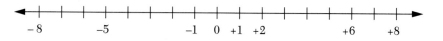

It is correct to write +7 but, more often, the positive sign is omitted and we simply write 7.

> 7 is read "seven" or "positive seven."
>
> −12 is read "negative twelve."
>
> 0 is read "zero"; it is neither positive nor negative.

Positive and negative numbers are used in many ways in the physical world to represent quantities that have opposite characteristics.

Positive	**Negative**
Temperatures above zero (82°)	Temperatures below zero (−11°)
Feet above sea level (5000 ft)	Feet below sea level (−75 ft)
Profit ($5285)	Loss (−$4065)
Right (6)	Left (−9)

Whenever quantities can be measured in opposite directions, positive and negative numbers can be used to show direction.

Opposites are pairs of numbers that are the same distance from zero on the number line but on opposite sides of zero. Opposites are indicated by a dash in front of the number.

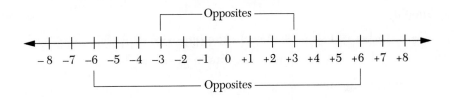

$-(6) = -6$ **The opposite of positive 6 is negative 6.**

$-(-3) = 3$ **The opposite of negative 3 is positive 3.**

$-(0) = 0$ **The opposite of 0 is 0.**

The dash in front of a number can be read in two ways:

−19   the opposite of 19   **We avoid saying "minus 19" to**

−19   negative 19   **prevent confusion with subtraction.**

The dash in front of a variable is read one way:

$-x$ **The opposite of x.**

We avoid saying "negative of x," because this implies that $-x$ represents a negative number. However, $-x$ can represent a positive number. For example, if $x = -8$, then $-x = -(-8) = 8$.

In general,

> The opposite of a positive number is negative: $-(5) = -5$
>
> The opposite of a negative number is positive: $-(-7) = 7$
>
> Zero is neither positive nor negative. It is its own opposite: $-(0) = 0$

To write the opposite of a signed number

1. Find the number that is the same number of units from zero on the number line but on the opposite side.
2. Write that number.

| **Examples A–D** | **Warm Ups A–D** |

Directions: Find the indicated opposite of the number.

Strategy: Write the number that is the same number of units from zero on the number line but on the opposite side of zero.

A. Find the opposite of 18.	A. Find the opposite of 16.

$-(18) = -18$ **Since 18 is 18 units to the right of zero on the number line, the opposite of 18 is 18 units to the left of zero.**

The opposite of 18 is -18.

B. Find the opposite: -9	B. Find the opposite: -36

$-(-9) = 9$

The opposite of negative 9 is positive 9.

C. Simplify: $-[-(-15)]$	C. Simplify: $-[-(-27)]$

$-[-(-15)] = -[15]$ **First write the opposite of -15.**

$\qquad\qquad = -15$ **Now write the opposite of 15.**

The simplest form is -15.

D. On a winter day in Buffalo, New York, the temperature goes from a low of $-12°F$ to a high that is the opposite of the low. What is the high temperature for the day?	D. On a day in December the high temperature in Boston is $32°F$. On the same day, the high temperature in Nome is the opposite of the high in Boston. What is the high temperature in Nome?

$-(-12°F) = ?$

$-(-12°F) = 12°F$ **The opposite of $-12°F$ is a positive number, $12°F$.**

The high temperature is $12°F$.

HOW AND WHY
Objective 2   **Find the absolute value of an integer.**

When the size of a number is important but its location on the number line is not, we talk about the *absolute value*. The distance between zero and the signed number on the number line is defined as the *absolute value* of a number. This distance is always positive or zero. To show the absolute value of a number we write it between two vertical bars. For instance, $|8|$ is read "the absolute value of 8." So,

---

Answers to Warm Ups   A. -16 B. 36   C. -27 D. $-32°F$

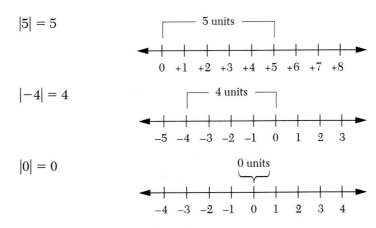

$|5| = 5$

$|-4| = 4$

$|0| = 0$

In general

The absolute value of zero is zero: $|0| = 0$

The absolute value of a positive number is the number itself: $|24| = 24$

The absolute value of a negative number is its opposite: $|-17| = -(-17) = 17$

> ### To write the absolute value of a signed number
>
> **1.** If the number is positive or zero, write the number.
> **2.** If the number is negative, write its opposite.

What does it mean when we write $|x| = 4$? The statement is read "The absolute value of some number is 4." Another way of interpreting this equation is "What numbers are 4 units away from zero on the number line?" By looking at the number line, we see that there are two different replacements for x that make the statement true, 4 and -4.

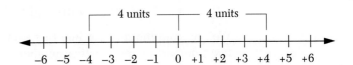

The solution to this absolute value equation is 4 or -4. Usually we write $x = 4$ or $x = -4$.

Examples E–G   **Warm Ups E–G**

Directions: Write the indicated absolute value of the number.

Strategy: Write the number, its opposite, or 0, depending upon whether the number is positive, negative, or 0.

E. The absolute value of -6.   E. The absolute value of -41.

$|-6| = 6$ **The number is negative, so write its opposite.**

The absolute value of -6 is 6.

Answers to Warm Ups   E. 41

F. Find $|x|$ if $x = 13$, -29, or -17.

If $x = 13$, we have $|x| = |13| = 13$.
If $x = -29$, we have $|x| = |-29| = 29$.
If $x = -17$, we have $|x| = |-17| = 17$.

F. Find $|x|$ if $x = 14$, -77, or 0.

G. Solve: $|x| = 35$

Strategy: Find the numbers that are the indicated distance from 0.

$|x| = 35$ **What numbers are located so that there are 35 units between them and 0?**

$x = 35$ or $x = -35$ **Thirty-five units to the right of 0 is 35, and 35 units to the left of 0 is −35.**

G. $|x| = 24$

HOW AND WHY
Objective 3   Write an inequality statement involving two integers.

As in the case of whole numbers, if two integers are not equal, then one is either *less than* or *greater than* the other. On the number line, numbers to the right of a given integer are greater (larger) than the given integer. Numbers to the left of a given integer are less (smaller) than the given integer. Look at the number line:

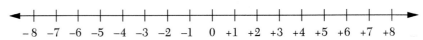

$-2 > -5$ **Since −2 is to the right of −5, −2 is greater than −5.**
$3 > -2$ **Since 3 is to the right of −2, 3 is greater than −2.**
$-6 < -1$ **Since −6 is to the left of −1, −6 is less than −1.**
$-3 < 5$ **Since −3 it to the left of 5, −3 is less than 5.**
$0 > -4$ **Since 0 is to the right of −4, 0 is greater than −4.**

We see from the number line that every negative number is less than every positive number. Every negative number is less than zero. Every positive number is greater than zero.

Examples H–I	**Warm Ups H–I**

Directions: Insert $<$ or $>$ so the statement is true

Strategy: Visualize a number line, the smaller number is the one to the left of the other on the number line.

H. -15 ? -18

$-15 > -18$ **Since −15 is to the right of −18 on a number line it is the larger number.**

H. -23 ? -21

I. 39 ? -43

$39 > -43$ **Every positive number is greater than every negative number.**

I. -89 ? 14

Answers to Warm Ups   F. 14, 77, 0   G. $x = 24$ or $x = -24$ H. $<$ I. $<$

Exercises 4.1

OBJECTIVE 1: *Find the opposite of an integer.*

A.

Find the indicated opposite of the integer.

1. $-(-3)$ 2. $-(32)$ 3. $-(14)$

4. $-(-8)$ 5. $-(-37)$ 6. $-(-43)$

7. $-(17)$ 8. $-(52)$

9. The opposite of _____ is 61.   10. The opposite of -23 is _____.

B.

Find the opposite of x for each given value of x.

11. $x = -75, 81,$ or 95 12. $x = -103, 77,$ or -131

13. $x = -65, -82,$ or -67 14. $x = 41, 129,$ or 93

Simplify.

15. $-[-(-34)]$ 16. $-[-(29)]$

17. $-[-(-75)]$ 18. $-[-(99)]$

OBJECTIVE 2: *Find the absolute value of an integer.*

A.

Find the indicated absolute value.

19. $|-11|$ 20. $|-12|$ 21. $|33|$

22. $|82|$ 23. $|-172|$ 24. $|-73|$

25. $|144|$ 26. $|0|$

27. The absolute value of _____ is 32.   28. The absolute value of _____ is 16.

B.

Find the $|x|$ for each given value of x.

29. $x = -75, 81,$ or 100 30. $x = -71, -55,$ or -150

31. $x = 99, -45,$ or -115

32. $x = -302, -76,$ or 133

Simplify.

33. $|-23| - |17|$

34. $|78| - |-56|$

35. $|88 - 45|$

36. $|114 - 102|$

OBJECTIVE 3: *Write an inequality statement involving two integers.*

A.

Insert $<$ or $>$ to make a true statement.

37. $-6 \quad 3$

38. $-18 \quad -19$

39. $-19 \quad -28$

40. $-20 \quad -14$

41. $-35 \quad -76$

42. $-66 \quad -34$

43. $19 \quad -67$

44. $-55 \quad 37$

45. -13 is _____ than -55.

46. -45 is _____ than -39.

B.

Insert $<$ or $>$ to make a true statement.

47. $-(-11) \quad 12$

48. $-(13) \quad -24$

49. $-(16 - 5) \quad -(9)$

50. $-(-19) \quad -(45 - 34)$

51. $-(-23) \quad -(-45)$

52. $-(-45) \quad -(67)$

53. $-(12 - 9) \quad -(11 - 6)$

54. $-(10 - 3) \quad -(23 + 5)$

C.

Simplify.

55. $-(-(-(-34)))$

56. $-(-(-(61)))$

57. $-[|-75| - |-45|]$

58. $-[|-56| - |-45|]$

True or false

59. $23 - 12 < 34 - 22$

60. $17 - 14 > 45 - 39$

61. At the New York Stock Exchange, positive and negative numbers are used to record stock prices on the board. What is the opposite of a gain of $(+\$2)$?

62. At the NASDAQ Exchange, a stock is shown to have taken a loss of 7 points (-7). What is the opposite of this loss?

63. On a thermometer, temperatures above zero are listed as positive and those below as negative. What is the opposite of a reading of 12°C?

64. On a thermometer such as the one in Exercise 63, what is the opposite of a reading of −23°C?

65. The modern calendar counts the years after the birth of Christ as positive numbers (1996 AD). Years before Christ are listed as BC and act as negative numbers (2045 BC means 2045 years before the birth of Christ). What is the opposite of 1875 BC or −1875?

66. If 80 miles north is represented by +80, how would you represent 80 miles south?

Solve for x.

67. $|x| = 11$

68. $|x| = 33$

69. $|x| = 101$

70. $|x| = 75$

For Exercises 71–74. A satellite records the following temperatures for a five-day period on one point on the surface of Mars.

	Day 1	**Day 2**	**Day 3**	**Day 4**	**Day 5**
5:00 AM	−92°C	−88°C	−115°C	−103°C	−74°C
9:00 AM	−57°C	−49°C	−86°C	−93°C	−64°C
1:00 PM	−52°C	−33°C	−46°C	−48°C	−10°C
6:00 PM	−45°C	−90°C	−102°C	−36°C	−42°C
11:00 PM	−107°C	−105°C	−105°C	−98°C	−90°C

71. What was the lowest temperature recorded during the five days?

72. What was the highest temperature recorded on the fourth day?

73. What was the lowest temperature recorded at 6:00 PM?

74. What was the highest temperature recorded at 5:00 AM?

 Exercises 75–77 refer to the Chapter 4 Application. See page 273.

75. Rewrite the continental altitudes in the application using integers.

76. The U.S. Department of Defense has extensive maps of the ocean floors because they are vital information for the country's submarine fleet. The table lists the deepest part and the average depth of the world's major oceans.

Ocean	Deepest Part	(ft)	Average Depth (ft)
Pacific	Mariana Trench	35,840	12,925
Atlantic	Puerto Rico Trench	28,232	11,730
Indian	Java Trench	23,376	12,598
Arctic	Eurasia Basin	17,881	3407
Mediterranean	Ionian Basin	16,896	4926

Rewrite the table using integers.

77. Is the highest point on Earth farther away from sea level than the deepest point in the ocean? Explain. What mathematical concepts allow you to answer this question?

STATE YOUR UNDERSTANDING

78. Is zero the only number that is its own opposite? Justify your answer.

79. Is there a set of numbers for which the absolute value of each number is the number itself? If yes, identify that set and tell why this is true.

80. Is there a set of numbers for which the absolute value of each number is the opposite of the number itself? If yes, identify that set and tell why this is true.

81. Explain the number −4. Draw it on a number line. On the number line, use the concepts of opposites and absolute value as they relate to −4. Give an instance in the world when −4 is useful.

CHALLENGE

Simplify.

82. $|16 - 10| - |14 - 9| + 6$

83. $8 - |12 - 8| - |10 - 8| + 2$

84. Solve: $|a - 6| = 13$

85. If n is a negative number, what kind of number is $-n$?

86. For what numbers is $|-n| = n$ always true?

GROUP ACTIVITY

87. Develop a rule (procedure) for adding a positive and a negative integer. (*Hint*: You might want to visualize that walking forward represents a positive integer and walking backwards represents a negative integer.) See if you can find answers to

$6 + (-4) = ?$ $\quad$ $7 + (-10) = ?$ $\quad$ and $\quad$ $-6 + 11 = ?$

MAINTAIN YOUR SKILLS (Sections 1.4, 1.5, 1.6, 2.3)

88. Find the product: 147×10^5

89. Write in exponent form: $37^7 \cdot 37^2 \cdot 37^{11}$

90. Find the area of the trapezoid with bases of 12 cm and 18 cm and with a height of 6 cm.

91. Find the average: 234, 176, 89, 421, 1055

92. On the first day of a sale, a grocery store has 34 boxes of macaroni and cheese on the shelf. During the week, the shelf is restocked several times. In all, the store used eight cases of 50 boxes each to restock. At the end of the sale there are 37 boxes left. How many boxes of macaroni and cheese are sold during the sale?

4.2 Adding Integers

OBJECTIVE Add integers.

HOW AND WHY
Objective **Add integers.**

Positive and negative integers are used to show quantities having opposite properties:

> +457 lb may show 457 lb loaded onto a truck.
> −322 lb may show 322 lb unloaded from a truck.
> +$112 may show $112 earned in one day.
> −$45 may show $45 spent in one day.

Using these properties, we can find the sum of two integers.

a. 53 + 41. The sum of two positive numbers is found the same way as finding the sum of two whole numbers.

53 + 41 = 94

b. 58 + (−32). If you think of $58 earned (positive) and $32 spent (negative), the end result is $26 remaining in your pocket (positive). So,

58 + (−32) = 26

c. −75 + 48. If you think of $75 spent (negative) and $48 earned (positive), the end result is that you still owe $27 (negative). So,

−75 + 48 = −27

d. −19 + (−33). If you think of $19 spent (negative) and $33 spent (negative), the end result is that you owe or spent $52 (negative). So,

−19 + (−33) = −52

The sum of two signed numbers can also be illustrated with arrows on the number line. Arrows for the positive numbers point right and arrows for the negative numbers point left. Using the number line, the starting point for adding is 0.

The following number line shows the sum of −2 and −5.

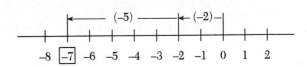

We start the arrow for −5 at the end of the arrow for −2. The sum of −2 + (−5) is −7. Note that the sum of the absolute values of −2 and −5 is the absolute value of −7, $|-2| + |-5| = |-7|$.

The following number line shows the sum of -8 and 3.

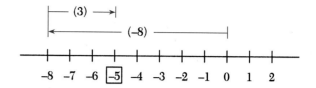

The second arrow, (3), starts at the end of the first arrow (-8). The sum of $-8 + 3$ is -5. Note that the difference between the absolute values of -8 and 3 is the absolute value of -5, $|-8| - |3| = |-5|$.

These examples lead to the following procedures:

> ### To add two integers
>
> **1.** If they have the same sign, add their absolute values and attach the common sign.
> **2.** If they have different signs, find the difference of their absolute values and attach the sign of the number with the larger absolute value.

Using the rules for adding integers, we see that

$3 + 2 = 5$	The sum of two positive integers is positive.
$-2 + (-1) = -3$	The sum of two negative numbers is negative.
$7 + (-5) = 2$	Find the difference of the absolute values. Since 7 has the larger absolute value, the sum is positive.
$-23 + (19) = -4$	Find the difference of the absolute values. Since -23 has the larger absolute value, the sum is negative.
$-75 + (75) = 0$	The sum of opposites is always zero.

Because zero is neither positive nor negative, no sign is placed in front of it.

The commutative and associative properties of addition hold when adding integers. For example,

$$-15 + 27 = 27 + (-15) \qquad \text{Commutative property.}$$
$$[24 + (-34)] + (-16) = 24 + [(-34) + (-16)] \qquad \text{Associative property.}$$

When adding three or more integers, we could follow the order of operations. Instead, we usually take advantage of the commutative and associative properties of addition to group the positive and negative integers. For example,

$$-7 + 10 + (-12) + 17 = [-7 + (-12)] + (10 + 17) \qquad \text{Group the positive and negative integers.}$$
$$= (-19) + (27) \qquad \text{Add each group.}$$
$$= 8 \qquad \text{Add.}$$

We can use the fact that the sum of opposites is zero to solve equations involving integers. In the equation

$$x + (-7) = 10$$

we want to isolate the variable. We do this by using the addition property of equality. Specifically, we isolate the variable by adding the opposite of -7 to both sides.

$$x + (-7) + 7 = 10 + 7 \qquad \text{Add the opposite of } -7 \text{ to both sides.}$$
$$x + 0 = 17 \qquad \text{Simplify.}$$
$$x = 17 \qquad \text{Addition property of zero.}$$

Check by replacing x with 17 in the original equation.

Does $17 + (-7) = 10$?

Does $\qquad 10 = 10$?   **Yes.**

The solution is $x = 17$.

Examples A–G	**Warm Ups A–G**

Directions: Add.

Strategy: Use the procedure for adding integers together with the commutative and associative properties when needed.

A. $47 + (-38)$	A. $59 + (-32)$

$\quad |47| - |-38| = 47 - 38 \qquad$ **The signs are unlike, so subtract**
$\qquad\qquad = 9 \qquad\qquad$ **their absolute values.**

$\quad 47 + (-38) = 9 \qquad\qquad$ **Since 47 is positive and has the larger absolute value, the sum is positive.**

B. Add: $-12 + 9$	B. Add: $-11 + 13$

Strategy: Use the number line.

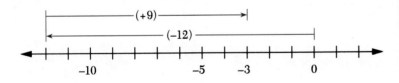

The sum is -3.

C. Simplify: $-63 + (-35)$	C. Simplify: $-46 + (-43)$

$\quad |-63| + |-35| = 63 + 35 \qquad$ **The signs are alike, so add their absolute values.**

$\qquad\qquad = 98$

$\quad -63 + (-35) = -98 \qquad$ **Use the common sign, so the sum is negative.**

D. Find the sum: $-27 + 52 + 35 + (-63) + (-18)$	D. Find the sum: $-42 + 17 + 29 + (-36) + (-43)$

Strategy: There are two ways to add. The first is to add from left to right. The second is to use the associative and commutative properties to group the positive numbers and to group the negative numbers and then add.

Answers to Warm Ups   A. 27   B. 2   C. -89 D. -75

First method:

$-27 + 52 + 35 + (-63) + (-18)$

$\quad = 25 + 35 + (-63) + (-18) \qquad -27 + 52 = 25$

$\quad = 60 + (-63) + (-18) \qquad\quad 25 + 35 = 60$

$\quad = -3 + (-18) \qquad\qquad\quad\; 60 + (-63) = -3$

$\quad = -21$

Second method:

$-27 + 52 + 35 + (-63) + (-18)$

$\quad = [-27 + (-63) + (-18)] + (52 + 35)$

Group the negative numbers and group the positive numbers.

$\quad = (-108) + 87 \qquad$ **Add in each group.**

$\quad = -21$

The sum is -21.

E.   Is -28 a solution of $x + (-51) = -79$?

Strategy: Substitute -28 for x and perform the indicated operation.

$\quad x + (-51) = -79$

$-28 + (-51) = -79 \qquad$ **Replace x with -28.**

$\quad\quad -79 = -79$

The statement is true, so -28 is a solution.

E.   Is -33 a solution of $x + 45 = 12$?

F.   Solve: $x + (-15) = -43$

Strategy: Use the properties of equations.

$x + (-15) + 15 = -43 + 15 \qquad$ **Add the opposite of -15 to each side.**

$\quad\quad x + 0 = -28 \qquad$ **Add.**

$\quad\quad\quad\; x = -28 \qquad$ **Addition property of zero.**

The solution is $x = -28$.

F.   Solve: $x + (-24) = -18$

G.   An airplane is being reloaded. A total of 2375 pounds of baggage and mail are removed (-2375 lb) and 1748 pounds of baggage and mail are loaded onto the plane ($+1748$ lb). What net change in weight should the cargo master report?

Strategy: The net weight change is the sum of the weight of the cargo loaded and unloaded.

$(-2375) + 1748 = -627$

The cargo master should report a net change of -627 lb, or 627 less lb of cargo.

G.   While a truck is being reloaded, 998 pounds of freight are removed (-998 lb) and 1243 pounds of freight are loaded on ($+1243$ lb). What net change will the driver record in his log?

Answers to Warm Ups   E. yes   F. $x = 6$ G. 245 lb

Exercises 4.2

OBJECTIVE: *Add integers.*

A.

Add.

1. $-8 + 2$

2. $-6 + 8$

3. $6 + (-7)$

4. $9 + (-4)$

5. $-8 + (-6)$

6. $-10 + (-6)$

7. $(-17) + (-17)$

8. $-17 + 17$

9. $7 + (-15)$

10. $9 + (-18)$

11. $23 + (-18)$

12. $-13 + 21$

13. Find the sum of -11 and -23.

14. Find the sum of 12 and -38.

Label the arrows and give the sum.

15.

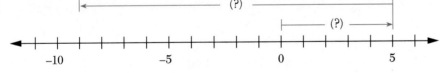

16.

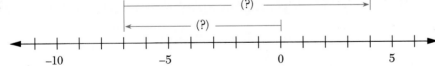

B.

Add.

17. $48 + (-39)$

18. $-56 + 63$

19. $-38 + (-57)$

20. $-26 + (-61)$

21. $(-56) + 56$

22. $100 + (-100)$

23. $-3 + (-6) + 5$ **24.** $-2 + (-5) + (-6)$ **25.** $-18 + 21 + (-3)$

26. $-22 + 25 + (-4)$ **27.** $47 + (-32) + (-14)$ **28.** $-17 + (-12) + 32$

29. Find the sum of -94 and 61.

30. Find the sum of -90 and 62.

31. Find the sum of 310 and -420.

32. Find the sum of -550 and 960.

33. If $c = 45 + (-82)$, then $c = ?$

34. If $d = -65 + 38$, then $d = ?$

35. Solve: $a - 11 = -9$

36. Solve: $b + 21 = -8$

C.

Simplify.

37. $135 + (-256)$

38. $233 + (-332)$

39. $-81 + (-32) + (-76)$

40. $-75 + (-82) + (-71)$

41. $-31 + 28 + (-63) + 36$

42. $-44 + 37 + (-59) + 45$

43. $49 + (-67) + 27 + 72$

44. $81 + (-72) + 33 + 49$

45. $356 + (-762) + (-892) + 541$

46. $-923 + 672 + (-823) + (-247)$

47. Find the sum of 542, −481, and −175.

48. Find the sum of 293, −122, and −211.

Viking II records the following temperatures for a five-day period on one point on the surface of Mars.

	Day 1	**Day 2**	**Day 3**	**Day 4**	**Day 5**
5:00 AM	−92°C	−88°C	−115°C	−103°C	−74°C
9:00 AM	−57°C	−49°C	−86°C	−93°C	−64°C
1:00 PM	−52°C	−33°C	−46°C	−48°C	−10°C
6:00 PM	−45°C	−90°C	−102°C	−36°C	−42°C
11:00 PM	−107°C	−105°C	−105°C	−98°C	−90°C

49. What is the sum of the temperatures recorded at 9:00 AM?

50. What is the sum of the temperatures recorded on Day 1?

51. Write an expression for "the sum of an integer and −67."

52. Write an expression for "the sum of −99 and an integer."

53. The sum of 22 and −12 is increased by −9. What is the result?

54. The sum of −25 and −16 is increased by 12. What is the result?

55. An airplane is being reloaded: 877 pounds of baggage and mail are removed (−877 lb) and 764 pounds of baggage and mail are loaded on (+764 lb). What net change in weight should the cargo master report?

56. At another stop, the plane in Exercise 55 unloads 1842 pounds of baggage and mail and takes on 1974 pounds. What net change should the cargo master report?

57. The change in altitude of a plane in flight is measured every 10 minutes. The figures between 3:00 PM and 4:00 PM are as follows:

3:00 PM	30,000 ft initially	(+30,000)
3:10 PM	increase of 220 ft	(+220)
3:20 PM	decrease of 200 ft	(−200)
3:30 PM	increase of 55 ft	(+55)
3:40 PM	decrease of 110 ft	(−110)
3:50 PM	decrease of 55 ft	(−55)
4:00 PM	decrease of 84 ft	(−84)

What is the altitude of the airplane at 4 PM? (*Hint*: Find the sum of the initial altitude and the six measured changes between 3 and 4 PM.)

58. What is the final altitude of an airplane if it is initially flying at 23,000 ft with the following changes in altitude?

3:00 PM	23,000 ft initially	(+23,000)
3:10 PM	increase of 315 ft	(+315)
3:20 PM	decrease of 825 ft	(−825)
3:30 PM	increase of 75 ft	(+75)
3:40 PM	decrease of 250 ft	(−250)
3:50 PM	decrease of 85 ft	(−85)
4:00 PM	decrease of 114 ft	(−114)

59. The Northwest Book Depository handles most text books for the local schools. On September 1 the inventory is 34,873 volumes. During the month the company makes the following transactions (positive numbers represent volumes received, negative numbers represent shipments): 2386, −11,705, −875, −563, +5075, −723. What is the inventory at the end of the month?

60. The Northwest Book Depository has 26,455 volumes on November 1. During the month the depository has the following transactions: shipped 3476, received 675, shipped 1185, received 345, shipped 6780, shipped 300. What is the inventory at the end of the month?

61. The Buffalo Bills made the following consecutive plays during a recent Monday night football game: 8-yard loss, 10-yard gain, 3-yard loss, and a 9-yard gain. Did they make a first down?

62. The Seattle Seahawks have these consecutive plays on Sunday: 12-yard loss, 19-yard gain, 3-yard loss, and a 7-yard gain. Do they make a first down?

63. A new retail outlet has the following weekly balances after the first month of business: a loss of $56, a gain of $18, a loss of $87, and a gain of $112. What is their net gain or loss for the month?

64. Marie plays the state lottery for one month. Drawings are held each Wednesday and Saturday, so she played nine times during the month. Her record is lost $10, lost $5, won $20, lost $8, lost $12, won $25, lost $20, won $35, and lost $10. What is the net result of her playing?

STATE YOUR UNDERSTANDING

65. When adding a positive and a negative number, explain how to determine whether the sum is positive or negative.

66. Explain how a number line can be used to add two integers.

CHALLENGE

67. What number added to (-32) equals (-17)?

68. What number added to 73 equals (-33)?

69. Simplify: $|73 + (-59)| + |-43 + (-99) + (-12)| + (-81)$

70. Simplify: $-45 + |-45 + (-23)| + |-32 + (-67) + 19| + 32$

71. Solve: $|x + 6 + (-11)| = 5 + (-45)$

GROUP ACTIVITY

72. Develop a rule for subtracting two integers. (*Hint*: Subtraction is the inverse of addition—it undoes addition. So, if adding $+6$ can be thought of as walking forward, then subtracting $+6$ may be thought of as walking backward 6 paces.) See if you can find answers to

$$4 - (+6) = ? \qquad -5 - (+6) = ? \qquad \text{and} \qquad 8 - (-4) = ?$$

MAINTAIN YOUR SKILLS (Sections 1.4, 1.5, 1.6, 3.1)

73. Multiply: 9789×10^8

74. Divide: $\dfrac{407{,}000}{10^3}$

75. Multiply: $(26^{14})(26^9)$

76. Jacqueline filled the gasoline tank of her car seven times. She paid the following prices:

39 cents per liter the first time

40 cents per liter the second time

42 cents per liter the next two times

41 cents per liter the next two times

42 cents per liter the last time

What is the average price per liter that she paid?

77. Translate to an algebraic expression: If x represents the number, sixty-five divided by the product of four and a number.

4.3   Subtracting Integers

OBJECTIVE   Subtract integers.

HOW AND WHY
Objective   **Subtract integers.**

One of the interpretations of subtraction (see Section 1.2) is "How much more" or "What is the missing addend." Thus, $11 - 8 = ?$ asks what must be added to 8 to give a sum of 11. So, $11 - 8 = 3$ since $8 + 3 = 11$.

Subtraction of integers can be approached in the same way. So, $-3 - 5 = ?$ asks what number must be added to 5 to give a sum of -3. Since $5 + (-8) = -3$, we know that $-3 - 5 = -8$. The expression $-5 - (-7) = ?$ asks what number must be added to -7 to give a sum of -5. Since $-7 + 2 = -5$, we know that $-5 - (-7) = 2$.

Compare these subtraction and addition examples.

$11 - 8$	and	$11 + (-8)$	**Both equal 3.**
$-3 - 5$	and	$-3 + (-5)$	**Both equal -8.**
$-5 - (-7)$	and	$-5 + 7$	**Both equal 2.**

Subtracting a number and adding the opposite of the number give the same result. Here are some more examples.

Subtraction	**Addition**	
$15 - 7$	$15 + (-7) = 8$	**The opposite of 7 is -7.**
$15 - (-7)$	$15 + 7 = 22$	**The opposite of -7 is 7.**
$-9 - 8$	$-9 + (-8) = -17$	**The opposite of 8 is -8.**
$-14 - (-12)$	$-14 + 12 = -2$	**The opposite of -12 is 12.**

> ### To subtract two integers
>
> 1. Change subtraction to addition of the opposite of the number subtracted.
> 2. Add.

CAUTION   **The commutative and associative properties do not hold for subtraction.**

So,

$-9 - 7 \neq 7 - (-9)$	**No commutative property for subtraction.**
$(8 - 7) - (-3) \neq 8 - [7 - (-3)]$	**No associative property for subtraction.**

Examples A–G	Warm Ups A–G

Directions: Subtract.

Strategy: Change each subtraction to addition.

A. $133 - 145$	A. $56 - 81$

$$133 - 145 = 133 + (-145) \quad \textbf{Rewrite as addition.}$$
$$= -12 \quad \textbf{Add.}$$

B. Find the difference: $-245 - (-310)$	B. Find the difference: $-96 - (-105)$

$$-245 - (-310) = -245 + 310 \quad \textbf{Rewrite as addition.}$$
$$= 65$$

C. Find the difference between 134 and -56.	C. Find the difference between 254 and -195.

$$134 - (-56) = 134 + 56 \quad \textbf{Rewrite as addition.}$$
$$= 190$$

D. Simplify: $-54 - 55 - (-77)$	D. Simplify: $-43 - (-63) - 91$

Strategy: Rewrite all subtractions as addition.

$$-54 - 55 - (-77) = -54 + (-55) + (77) \quad \textbf{Rewrite as addition.}$$
$$= -109 + 77 \quad \textbf{Add from left to right.}$$
$$= -32$$

E. Is -8 a solution of $x - 12 = -20$?	E. Is -7 a solution of $x - 9 = -16$?

Strategy: Substitute -8 for x in the equation.

$$x - 12 = -20$$

Does $\quad -8 - 12 = -20$?   **Replace x with -8.**

Does $-8 + (-12) = -20$?   **Change to addition.**

Does $\qquad -20 = -20$?   **Yes.**

A solution is -8.

F. Solve: $x + 45 = -34$	F. Solve: $x + 92 = -65$

Strategy: Use the subtraction property of equations.

$$x + 45 - 45 = -34 - 45 \quad \textbf{Subtract 45 from each side.}$$
$$x + 0 = -79 \quad \textbf{Subtract.}$$
$$x = -79 \quad \textbf{Addition property of zero.}$$

Check:

$$x + 45 = -34$$

Does $-79 + 45 = -34$?   **Replace x with -79 in the original equation.**

Does $\qquad -34 = -34$?   **Yes.**

A solution is $x = -79$.

G. Viking II records a high temperature of $-22°C$ and a low temperature of $-107°C$ for one day at a point on the surface of Mars. What is the change (difference) in temperature for the day?

Strategy: Subtract the temperatures and find the absolute value of the difference.

$$\text{Change} = |-22 - (-107)|$$
$$= |-22 + (107)| \quad \textbf{Add the opposite of } -107.$$
$$= |85|$$
$$= 85$$

The change in temperature that day was 85°C.

G. What is the difference in altitude of an airplane flying at 11,500 ft above sea level $(+11,500)$ and a submarine traveling 600 ft below the surface of the ocean (-600)?

Exercises 4.3

OBJECTIVE: *Subtract integers.*

A.

Subtract.

1. $-6 - 4$

2. $-8 - 3$

3. $-5 - (-8)$

4. $-7 - (-4)$

5. $-12 - 5$

6. $-10 - 7$

7. $17 - (-15)$

8. $21 - (-12)$

9. $-14 - 12$

10. $-22 - 19$

11. $-16 - (-17)$

12. $-27 - (-28)$

13. $-16 - (-16)$

14. $-16 - 16$

15. Find the difference between -32 and 22.

16. Find the difference between 32 and -22.

B.

Subtract.

17. $56 - (-31)$

18. $48 - (-43)$

19. $-65 - 73$

20. $-72 - 87$

21. $-65 - (-69)$

22. $-49 - (-73)$

23. $-74 - (-74)$

24. $101 - 101$

25. $145 - (-32)$

26. $136 - (-29)$

27. Find the difference between 43 and -73.

28. Find the difference between -88 and -97.

29. Subtract 328 from -349.

30. Subtract 145 from -251.

31. Is -12 a solution of $x - 12 = -24$?

32. Is 27 a solution of $a - 45 = -72$?

Simplify.

33. $-7 - (-9) - 12$

34. $-14 - (-27) - 13$

35. $-18 - (-16) - (-7)$

36. $-19 - (-14) - (-11)$

C.

37. Solve: $b + 56 = -34$

38. Solve: $c + 31 = -12$

39. Solve: $d + 19 = 7$

40. Solve: $n + 46 = 65$

Simplify.

41. $-43 - 23 - 21$

42. $-36 - 57 - 23$

43. $56 - 29 - 65$

44. $65 - 81 - 73$

45. $-47 - 71 - (-61)$

46. $-55 - 33 - (-46)$

47. $-43 - (-24) - 11 - (-28)$

48. $-56 - (-34) - 40 - (-29)$

49. Jana buys and sells real estate. She purchased a lot for $27,950. During the year she kept the piece of property, the price went down. She sold the lot for $25,250. Express her loss as a signed number.

50. In January, 100 shares of a certain stock are priced at $3950. In February, 100 shares of the same stock are priced at $3275. Express the loss in value as a signed number.

51. A satellite records high and low temperatures of $-38°C$ and $-135°C$ for one day on the surface of Mars. What is the change in temperature for that day?

52. The surface temperature of one of Jupiter's satellites is measured for one week. The highest temperature recorded is $-75°C$ and the lowest is $-139°C$. What is the difference in the extreme temperatures for the week?

53. If the difference of 12 and 23 is decreased by 32, what is the result?

54. If the difference of -15 and 19 is decreased by -34, what is the result?

▽ Exercises 55–57 refer to the Chapter 4 Application. See page 273.

55. The range of a set of numbers is defined as the difference between the largest and the smallest numbers in the set. Calculate the range of altitude for each continent. Which continent has the smallest range and what does this mean in physical terms?

56. What is the difference between the lowest point in the Mediterranean and the lowest point in the Atlantic?

57. Some people consider Mauna Kea, Hawaii, to be the tallest mountain in the world. It rises 33,476 feet from the ocean floor, but it is only 13,796 feet above sea level. What is the depth of the ocean floor at this location?

58. Marie's bank account had a balance of $235. She writes a check for $310. What is her account balance now?

59. Thomas started with $117 in his account. He writes a check for $188. What is his account balance now?

For Exercises 60–63. Viking II records the following temperatures for a five-day period on one point on the surface of Mars.

	Day 1	Day 2	Day 3	Day 4	Day 5
5:00 AM	−92°C	−88°C	−115°C	−103°C	−74°C
9:00 AM	−57°C	−49°C	−86°C	−93°C	−64°C
1:00 PM	−52°C	−33°C	−46°C	−48°C	−10°C
6:00 PM	−45°C	−90°C	−102°C	−36°C	−42°C
11:00 PM	−107°C	−105°C	−105°C	−98°C	−90°C

60. What is the difference between the high and low temperatures recorded on Day 3?

61. What is the difference between the temperatures recorded at 11:00 PM on Day 3 and Day 5?

62. What is the difference between the temperatures recorded at 5:00 AM on Day 2 and 6:00 PM on Day 4?

63. What is the difference between the highest and lowest temperatures recorded during the five days?

64. The following table gives the number of murders in various major US cities in 1995 and 1996.

City	1995	1996
Atlanta	145	156
Boston	83	58
Chicago	826	785
Dallas	276	219
Las Vegas	134	167
Los Angeles	829	688
Miami	102	120
New Orleans	362	346
New York	1128	983
Seattle	48	36
Washington, D.C.	360	396

Calculate the increase or decrease from 1995 to 1996 for each city. List the cities in order of increasing change.

STATE YOUR UNDERSTANDING

65. Explain the difference between adding and subtracting signed numbers.

66. How would you explain to a 10-year-old how to subtract -8 from 12.

67. Explain why the order in which two numbers are subtracted is important, but the order in which they are added is not.

68. Explain the difference between the problems $5 + (-8)$ and $5 - 8$.

CHALLENGE

Simplify.

69. $-34 - (-42) - |-(-32) - (-21)|$

70. $48 - |-42 - (-56)| - (-23)$

71. Solve: $|a - (-11)| = 14$

72. Solve: $|x - 31| = 29$

GROUP ACTIVITY

73. Determine the normal daily mean (average) temperature for your city for each month of the year. Find the differences from month to month. Chart this result.

MAINTAIN YOUR SKILLS (Sections 1.4, 1.6, 3.1, 3.2, 3.3)

74. Divide: $290{,}000{,}000{,}000{,}000 \div 10^{10}$

75. Simplify: $(28 + 3 \cdot 4) - 2 \cdot 3 + 4$

76. Translate to an algebraic expression: If x represents the number, two plus a number and that sum multiplied by the number.

77. Evaluate: $b^3 - 4b^2 - 10$, if $b = 5$.

78. Combine: $15y + 17y - 6y$

4.4   Adding and Subtracting Polynomials (Integers)

OBJECTIVE   Add and subtract polynomials with integer coefficients.

HOW AND WHY
Objective   **Add and subtract polynomials with integer coefficients.**

Combining like terms with integer coefficients is similar to adding and subtracting integers. In Chapter 3, we used the distributive property to combine terms with whole number coefficients. We use the same rule to combine like terms with integer coefficients, that is, we add or subtract the coefficients and multiply by the common variable factors.

$-3x + 7x = (-3 + 7)x = 4x$ Since $-3 + 7 = 4$.

$2a + (-5a) = -3a$ Since $2 + (-5) = -3$.

$-23a - 5a = -28a$ Since $-23 - 5 = -23 + (-5) = -28$.

$-7y - (-6y) = -1y = -y$ Since $-7 - (-6) = -7 + (6) = -1$.

> ► **To combine or simplify like terms with integer coefficients**
>
> **1.** Add or subtract the coefficients.
> **2.** Multiply the sum or difference times the common variable factor(s).

When adding polynomials, group the like terms using the associative and commutative properties of addition. When subtracting polynomials first change the subtraction to addition by adding the opposite. The opposite of a polynomial is found by writing the opposite of each term of the polynomial. So, to subtract:

$(3x^2 + 2x - 9) - (-x^2 - 3x + 9)$

First find the opposite of $-x^2 - 3x + 9$. To find the opposite, write the opposite of each term.

$-(-x^2 - 3x + 9) = x^2 + 3x - 9$

So,

$(3x^2 + 2x - 9) - (-x^2 - 3x + 9)$

$= (3x^2 + 2x - 9) + (x^2 + 3x - 9)$ **Rewrite as addition (add the opposite).**

$= [3x^2 + 2x + (-9)] + [x^2 + 3x + (-9)]$ **Rewrite as addition.**

$= (3x^2 + x^2) + (2x + 3x) + [-9 + (-9)]$ **Group like terms.**

$= 4x^2 + 5x + (-18)$ **Combine like terms.**

$= 4x^2 + 5x - 18$ **Write as an indicated subtraction.**

Examples A–I	**Warm Ups A–I**

Directions: Combine like terms.

Strategy: Add or subtract the coefficients of the like terms.

A. $-9a + 4a$

   $-9a + 4a = -5a$ **Add the coefficients: $-9 + 4 = -5$.**

A. $-8b + 6b$

B.   Add: $-31y + (-12y)$

   $-31y + (-12y) = -43y$ **Add the coefficients.**

B.   Add: $-23z + (-55z)$

C.   Subtract: $-43x - 55x$

   $-43x - 55x = -43x + (-55x)$ **Rewrite as addition.**

   $= -98x$ **Add the coefficients.**

C.   Subtract: $-64y - 42y$

D.   Find the difference between $-22b$ and $-27b$.

Strategy: Recall in this text we subtract the second term from the first term to find the difference.

   $-22b - (-27b) = -22b + (27b)$ **Rewrite as addition.**

   $= 5b$ **Add the coefficients.**

D.   Find the difference between $-37x$ and $-97x$.

E.   Simplify: $13x + 8x + (-17x)$

   $13x + 8x + (-17x) = 4x$ **Add and subtract the coefficients.**

E.   Simplify: $-14a + 17a + (-21a)$

F.   Combine: $12x + 19y$

   $12x + 19y = 12x + 19y$ **Since $12x$ and $19y$ are not like terms, they cannot be combined.**

F.   Combine: $26a + 19b$

G.   Combine: $(12x - 6y + 9) + (13x + 19y - 16)$

Strategy: Rewrite subtraction as addition by adding the opposite.

   $(12x - 6y + 9) + (13x + 19y - 16)$

   $= [12x + (-6y) + 9] + [13x + 19y + (-16)]$ **Rewrite as addition.**

   $= [12x + 13x] + [(-6y) + 19y] + [9 + (-16)]$ **Group like terms.**

   $= 25x + 13y + (-7)$ **Add.**

   $= 25x + 13y - 7$ **Write as an indicated subtraction.**

G.   Combine: $(10a - 15b - 32) + (17a + 19b - 21)$

An alternate method is to stack the polynomials so like terms are lined up in the same columns and then add.

$$\begin{array}{r} 12x - 6y + 9 \\ + \underline{13x + 19y - 16} \\ 25x + 13y - 7 \end{array}$$

H. Subtract: $(7x^2 - 8x - 9) - (10x^2 - 3x + 2)$

Strategy: Rewrite subtraction as addition by adding the opposite.

$(7x^2 - 8x - 9) - (10x^2 - 3x + 2)$

$= (7x^2 - 8x - 9) + (-10x^2 + 3x - 2)$ **Rewrite as addition.**

$= [7x^2 + (-8x) + (-9)] + [-10x^2 + 3x + (-2)]$ **Rewrite as addition.**

$= [7x^2 + (-10x^2)] + [(-8x) + 3x] + [(-9) + (-2)]$

$= -3x^2 + (-5x) + (-11)$

$= -3x^2 - 5x - 11$

or

$$\begin{array}{r} 7x^2 - 8x - 9 \\ - (10x^2 - 3x + 2) \end{array} \quad = \quad \begin{array}{r} 7x^2 - 8x - 9 \\ - 10x^2 + 3x - 2 \\ \hline -3x^2 - 5x - 11 \end{array} \quad \textbf{Rewrite as addition.}$$

H. Subtract: $(5x^2 - 3x - 13) - (7x^2 - 4x + 6)$

I. Two rockets are fired at the same time from ground level. The height H (in feet) of the first rocket at any time t (in seconds) is $H_1 = 429t - 16t^2$. The height of the second and slower rocket is $H_2 = 212t - 16t^2$. Write a formula, and simplify, for the vertical distance (D) between the rockets at any time t. Find the distance between the rockets after 20 seconds.

Strategy: To find the distance, subtract the height of the slower rocket from the height of the faster one.

$D = (429t + 16t^2) - (212t - 16t^2)$

$= [429t + (-16t^2)] + -[212t + (-16t^2)]$

$= [429t + (-16t^2)] + [-212t + 16t^2]$

$= [429t + (-212t)] + [(-16t^2) + 16t^2]$

$= 217t + 0$

$= 217t$

To find the distance between the rockets after 20 seconds, replace t with 20 in the formula.

$D = 217t$ **Replace t with 20.**

$= 217(20)$

$= 4340$

The vertical distance between the rockets is given by $D = 217t$. The vertical distance between the rockets in 20 seconds is 4340 ft.

I. Two rectangles share a common width. The perimeter (P) of the first rectangle, in terms of the width (w), is $P_1 = 10w + 24$. The perimeter of the second and smaller rectangle is $P_2 = 9w - 3$. Write a formula for the difference (D) in the perimeters. Find the difference if the width is 4 feet.

Answers to Warm Ups H. $-2x^2 + x - 19$ I. $D = w + 27$; 31 ft

Exercises 4.4

OBJECTIVE: *Add and subtract polynomials with integer coefficients.*

A.

Simplify.

1. $5x + 9x$

2. $6a + 9a$

3. $-11x + 8x$

4. $-15y + 7y$

5. $8a + (-6a)$

6. $9d + (-11d)$

7. $-17a - 13a$

8. $-14x - 7x$

9. $-13y + (-5y)$

10. $-15c + (-4c)$

11. $-12y - (-9y)$

12. $-8x - (-7x)$

B.

13. $4a + 6a + (-5a)$

14. $5b + 7b + (-8b)$

15. $7x + (-6x) + (-9x)$

16. $9y + (-11y) + (-9y)$

17. $3x^2 - 8x^2 + 7x^2$

18. $-7b^2 - 9b^2 + (-3b^2)$

19. $(3a - 7) + (-4a + 8)$

20. $(-6b - 9) + (2b - 10)$

21. $(x^2 - 8x) + (-5x^2 + 6x)$

22. $(-7y^2 - 18y) + (3y^2 + 4y)$

23. $(4x - 17) - (7x + 9)$

24. $(-5a - 11) - (3a - 15)$

25. $9xy + 12xy - 8xy + (-24xy) - (-15xy)$

26. $3ab - (-7ab) - 8ab + (-12ab) - (-9ab)$

C.

27. $(8y^2 - 4y - 7) + (3y^2 + 7y - 5)$

28. $(-3a^2 + 4a + 5) + (11a^2 - 6a - 9)$

Add.

29.
$$\begin{array}{r} 3a - 2b + 9 \\ + \underline{-7a - 3b + 11} \end{array}$$

30.
$$\begin{array}{r} 6x - y + 8 \\ + \underline{3x - 2y - 15} \end{array}$$

Subtract.

31.
$$\begin{array}{r} 7c - 8d - 3 \\ - \underline{(-2c - 5d - 6)} \end{array}$$

32.
$$\begin{array}{r} 9s - 13t + 12 \\ - \underline{(5s + 6t - 21)} \end{array}$$

33. Add $3a$ to the sum of $5a$ and $-7a$.

34. Find the difference of $12x$ and $-2x$.

35. Subtract $15b$ from the sum of $-11b$ and $-7b$.

36. Subtract $18c$ from the sum of $-19c$ and $-5c$.

Simplify.

37. $(5x^2 - 9x - 11) - (3x^2 + 8x - 17)$

38. $(21x^2 - 5x + 20) - (2x^2 - 10x + 21)$

39. $(3a - 2b + 7) + (2a - 8b - 9) - (4a + 6b + 3)$

40. $(3s + 9t - 12) - (3s - 12t + 6) - (-10s + 4t + 11)$

41. Subtract $(4a - 3b)$ from $(2a + 5b)$.

42. Subtract $(-6x - 18)$ from $(3x - 33)$.

43. Add $(3x + 7y - 9)$ to the difference of $(6x + 3y - 8)$ and $(4x - 7y - 2)$.

44. Add $(4a - 6b - 19)$ to the difference of $(4a - 8b - 12)$ and $(-4a - 8b + 11)$.

45. In a music store, Ben bought 3 CDs and 4 tapes. Josh bought 5 CDs and one tape. Let c represent the cost of one CD and t represent the cost of one tape. Write a polynomial that represents the amount of money Ben spent in the store. Write a polynomial that represents the amount of money Josh spent. Write a polynomial that represents how much more money Josh spent than Ben. Did Josh spend more money than Ben? Explain.

46. Alamo Express charges $15 plus $2 for every pound over 45 pounds to ship a package from Phoenix to Dallas. Thrifty Movers charges $17 plus $3 for every pound over 75 pounds. Let p represent the total weight of a package ($p > 100$). Write a polynomial that represents Alamo's total cost of shipping a package. Write a polynomial that represents Thrifty's total cost of shipping a package. Write a polynomial that represents how much more it will cost to ship by Thrifty than by Alamo. Write a polynomial that represents how much more it will cost to ship by Alamo than by Thrifty.

47. Write a polynomial that represents the distance indicated.

48. Write a polynomial that represents the distance indicated.

STATE YOUR UNDERSTANDING

49. Explain how to find the opposite of a polynomial.　**50.** Explain in words how to subtract $-15x^2$ and $5x^2$.

51. Explain how to simplify $(4p - 7q + 8) - (9p + 2q - 5)$. Be sure to mention which properties you use.

CHALLENGE

Simplify.

52. $3(3x - 4y) + 6(2x + 2y) + 5(3y - 5x)$　　　**53.** $3(2a - 7b) + 2(3a + 2b + 6) + 7(4a - 7)$

54. $6(a - 3b) + 7(4a^2 + 2b) - 4(3a^2 - 2a)$

GROUP ACTIVITY

55. Have all members of your group write down what they eat tomorrow. Have them determine how many total calories, C, they eat and the number of fat grams. Each fat gram is worth 9 calories. Each member should calculate the number of calories that are from fat and the number that are not. Letting n represent the number of fat grams, write a general formula for those calories (T) that were not from fat.

MAINTAIN YOUR SKILLS (Sections 3.2, 3.3, 3.4)

56. Evaluate $6a - 4b$ if $a = 10$ and $b = 8$.　　**57.** Is 11 a solution of $5y - 32 = 23$?

58. Simplify: $76a - 34a - 15a$　　　　　　　**59.** What is the product of 8 and $3x - 12$?

60. Find the quotient of $132p$ and 12.

4.5   **Multiplying and Dividing Integers**

OBJECTIVES   **2.** Divide integers.

HOW AND WHY   **Multiply integers.**
Objective 1

Consider the following products and differences.

$$3 \cdot 4 = 12$$
$$3 \cdot 3 = 9 \qquad \text{and} \qquad 12 - 9 = 3$$

The difference of the products $3 \cdot 4$ and $3 \cdot 3$.

$$3 \cdot 2 = 6 \qquad \text{and} \qquad 9 - 6 = 3$$

The difference of the products $3 \cdot 3$ and $3 \cdot 2$.

$$3 \cdot 1 = 3 \qquad \text{and} \qquad 6 - 3 = 3$$
$$3 \cdot 0 = 0 \qquad \text{and} \qquad 3 - 0 = 3$$
$$3 \cdot (-1) = ? \qquad \text{and} \qquad 0 - (?) = 3$$
$$3 \cdot (-2) = ? \qquad \text{and} \qquad ? - (?) = 3$$

Note that each product is 3 smaller than the one before it. If we continue that pattern then

$$3 \cdot (-1) = -3 \qquad \text{and} \qquad 0 - (-3) = 3$$
$$3 \cdot (-2) = -6 \qquad \text{and} \qquad -3 - (-6) = 3$$

Rule

The product of a positive integer and a negative integer is a negative integer.

So,

$$(-4)(2) = -8$$
$$(-3)(9) = -27$$
$$(5)(-80) = -400$$

A similar pattern shows the product of two negative integers.

$$(-3)(4) = -12$$
$$(-3)(3) = -9 \qquad \text{and} \qquad -12 - (-9) = -3$$

The difference of the products $(-3)(4)$ and $(-3)(3)$.

$$(-3)(2) = -6 \qquad \text{and} \qquad -9 - (-6) = -3$$

The difference of the products $(-3)(3)$ and $(-3)(2)$.

$$(-3)(1) = -3 \qquad \text{and} \qquad -6 - (-3) = -3$$
$$(-3)(0) = 0 \qquad \text{and} \qquad -3 - 0 = -3$$
$$(-3)(-1) = ? \qquad \text{and} \qquad 0 - (?) = -3$$
$$(-3)(-2) = ? \qquad \text{and} \qquad ? - (?) = -3$$

Each product is 3 larger than the one before it. Continuing the pattern we have

$$(-3)(-1) = 3 \quad \text{and} \quad 0 - 3 = -3$$
$$(-3)(-2) = 6 \quad \text{and} \quad 3 - 6 = -3$$
$$(-3)(-3) = 9 \quad \text{and} \quad 6 - 9 = -3$$

Rule

The product of two negative integers is a positive integer.

So,

$$(-6)(-8) = 48$$
$$(-9)(-11) = 99$$
$$(-7)(-14) = 98$$

▶ *To multiply two integers*

With like signs: Multiply their absolute values. The product is positive.

$$(+)(+) = +$$
$$(-)(-) = +$$

With unlike signs: Multiply their absolute values. The product is negative.

$$(+)(-) = -$$
$$(-)(+) = -$$

The commutative and associative properties of multiplication hold for integers the same as for whole numbers, as does the distributive property.

Examples A–F	**Warm Ups A–F**

Directions: Multiply.

Strategy: The product of integers with like signs is positive. The product of integers with unlike signs is negative.

A. $(-7)(8)$	A. $(-13)(3)$
$(-7)(8) = -56 \qquad (-)(+) = -$	
B. Find the product of -14 and -15.	B. Find the product of -21 and -16.
$(-14)(-15) = 210 \qquad (-)(-) = +$	

Answers to Warm Ups   A. -39 B. 336

Calculator Example:

C. Multiply: $(-345)(211)$

$(-345)(211) = -72{,}795$

C. Multiply: $(675)(-425)$

D. Simplify: $(-4)(7)(6)$

Strategy: Use the order of operations, multiply from left to right.

$(-4)(7)(6) = (-28)(6)$ $(-4)(7) = -28$

$= -168$

Alternatively, use the associative property to group the positive integers together.

$(-4)(7)(6) = (-4)[(7)(6)]$ **Group the positive integers.**

$= (-4)(42)$ **Multiply inside brackets first.**

$= -168$

Or, count the number of negative numbers to decide the sign of the product. An odd number of negatives means the product is negative. Then multiply the absolute values.

$(-4)(7)(6) = -|-4| \cdot |7| \cdot |6|$ **One negative sign, so the product is negative.**

$= -(4 \cdot 7 \cdot 6)$

$= -168$

D. Simplify: $(-8)(3)(9)$

E. Is -12 a solution of $-5x = 60$?

Strategy: Substitute -12 for x and multiply.

$-5x = 60$ **Original equation.**

Does $-5(-12) = 60$?   **Replace x with -12.**

Does $60 = 60$?   **Yes.**

A solution is -12.

E. Is -15 a solution of $-6x = 90$?

F. In order to attract business, the Auto Tire Shop ran a "loss leader" sale last week. The shop sold a brand of tire at a loss of $3 ($-3) per tire. If 47 tires of that brand are sold over the weekend, what is the total loss? Express the amount as a signed number.

Strategy: To find the total loss, multiply the loss per tire times the number of tires sold.

$(-3)(47) = -141$

The shop lost $141 that week. Expressed as a signed number, the amount is $-$141$.

F. Pete lost $35 on each of 15 consecutive rolls of the dice at the craps table. Express the amount as a signed number.

Answers to Warm Ups   C. $-286{,}875$ D. -216 E. yes   F. $-$525$

HOW AND WHY
   Objective 2   **Divide integers.**

To divide two integers we use the "missing factor" interpretation of division. Study the examples in the chart.

Division	Missing Factor Interpretation	Conclusion
$-8 \div 4 = ?$	$4(?) = -8$ $4(-2) = -8$	$-8 \div 4 = -2$
$-75 \div (-15) = ?$	$(-15)(?) = -75$ $(-15)(5) = -75$	$-75 \div (-15) = 5$
$66 \div (-11) = ?$	$(-11)(?) = 66$ $(-11)(-6) = 66$	$66 \div (-11) = -6$

These examples lead to rules similar to those for multiplication.

Rule

The quotient of two negative integers is positive. The quotient of a positive and a negative integer is negative.

So,

$$-20 \div 5 = -4$$
$$24 \div -6 = -4$$
$$-30 \div -10 = 3$$
$$-70 \div -5 = 14$$

▶ *To divide two integers*

With like signs: Divide their absolute values. The quotient is positive.

$$(+) \div (+) = +$$
$$(-) \div (-) = +$$

With unlike signs: Divide their absolute values. The quotient is negative.

$$(+) \div (-) = -$$
$$(-) \div (+) = -$$

C A U T I O N   **The commutative and associative properties do not hold for division.**

For example,

$8 \div (-2) \neq (-2) \div 8$ No commutative property for division.
$(16 \div 4) \div (-2) \neq 16 \div [4 \div (-2)]$ No associative property for division.

Examples G–J | **Warm Ups G–J**

Directions: Divide.

Strategy: The quotient of two integers with like signs is positive. The quotient of two integers with unlike signs is negative.

G. $98 \div (-49)$

$98 \div (-49) = -2$ $(+) \div (-) = -$

G. $84 \div (-6)$

H.   Find the quotient: $\dfrac{-85}{-5}$

$\dfrac{-85}{-5} = 17$ $(-) \div (-) = +$

H.   Find the quotient $\dfrac{-132}{-12}$

 **Calculator Example:**

I.   Divide $-158{,}627$ by 301.

$\dfrac{-158{,}627}{301} = -527$

I.   Divide $259{,}560$ by -412.

J.   Is -48 a solution of $\dfrac{x}{8} = -6$?

$\dfrac{x}{8} = -6$ **Original equation.**

Does $\dfrac{-48}{8} = -6$?   **Replace x with -48.**

Does $-6 = -6$?   **Yes.**

A solution is -48.

J.   Is -30 a solution of $\dfrac{y}{6} = -5$?

Exercises 4.5

OBJECTIVE 1: *Multiply integers.*

A.

1. The product of two negative integers is _____.
2. The product of a positive and a negative integer is _____.

Multiply.

 3. $(-3)(2)$ **4.** $(-4)(3)$ **5.** $(-5)(-6)$

 6. $(-9)(-4)$ **7.** $6(-8)$ **8.** $7(-5)$

 9. $(-12)(-3)$ **10.** $(-11)(-4)$

11. The product of -5 and _____ is 55.   **12.** The product of 6 and _____ is -42.

B.

Multiply.

 13. $(-11)(10)$ **14.** $(-13)(11)$ **15.** $(-12)(-12)$

 16. $(-15)(-15)$ **17.** $(43)(-6)$ **18.** $(-21)(5)$

 19. $(-14)(-15)$ **20.** $(-21)(-4)$ **21.** $(-7)(-2)(-3)$

 22. $(2)(5)(-6)$ **23.** $(-8)(-2)(3)$ **24.** $(9)(-3)(-4)$

 25. $(-1)(-1)(-1)(-1)(-1)(-6)$ **26.** $(-1)(-1)(-1)(-1)(-1)(-1)(-7)$

OBJECTIVE 2: *Divide integers.*

A.

27. $8 \div (-2)$

28. $9 \div (-3)$

29. $(-15) \div 3$

30. $(-24) \div 4$

31. $\dfrac{-14}{-2}$

32. $\dfrac{-21}{-3}$

33. $\dfrac{-33}{11}$

34. $\dfrac{-28}{14}$

35. $\dfrac{-66}{-11}$

36. $\dfrac{72}{-6}$

37. The quotient of -32 and _____ is 4.

38. The quotient of -72 and _____ is -8.

B.

Divide.

39. $125 \div (-25)$

40. $225 \div (-15)$

41. $(-260) \div 13$

42. $(-480) \div 24$

43. $\dfrac{-300}{-12}$

44. $\dfrac{-400}{16}$

45. $\dfrac{252}{-14}$

46. $\dfrac{182}{-13}$

47. $\dfrac{-918}{-9}$

48. $\dfrac{808}{-8}$

49. $\dfrac{-475}{25}$

50. $\dfrac{-520}{26}$

51. $\dfrac{594}{-18}$

52. $\dfrac{465}{-15}$

C.

53. Find the product of 39 and -18.

54. Find the product of 27 and -24.

55. Find the quotient of -384 and -24.

56. Find the quotient of -357 and 21.

Simplify.

57. $(13)(-2)(-10)(5)$ **58.** $(21)(-10)(2)(3)$ **59.** $(-13)(-12)(-8)(2)$

60. $(-20)(-15)(-12)(3)$ **61.** $\dfrac{-560}{-28}$ **62.** $\dfrac{-540}{-12}$

63. $\dfrac{-1071}{17}$ **64.** $\dfrac{-988}{19}$ **65.** $(-34)(-123)(65)$

66. $(-55)(-73)(-104)$ **67.** $\dfrac{-16{,}272}{36}$ **68.** $\dfrac{-34{,}083}{-63}$

69. Is -15 a solution of $8x = -120$?   **70.** Is -11 a solution of $-12y = 122$?

71. Is -18 a solution of $\dfrac{z}{9} = -162$?   **72.** Is -567 a solution of $\dfrac{a}{27} = -21$?

73. For six consecutive weeks Mr. Obese loses 4 lb every week. If each loss is represented by -4 lb, what is his total weight loss for the six weeks?

74. If Mrs. Tomlison loses 2 lb every week for nine consecutive weeks, what is her total weight loss, expressed as a signed number?

75. The membership of the Burlap Baggers Investment Club took a loss of \$522 ($-\$522$) on the sale of stock. If there are six members in the club and each shares equally in the loss, what is each member's share of the loss? Express your answer as a signed number.

76. The temperature in Fairbanks, Alaska, dropped $32°$ ($-32°$) in an eight-hour period. What was the average drop per hour, expressed as a signed number?

Exercises 77–79 refer to the Chapter 4 Application. See page 273.

77. Which continent has a low point that is approximately ten times the low point of South America?

78. Which continent has a high point that is approximately twice the absolute value of the lowest point?

79. Which continent has a high point that is approximately one-fourth of the height of Mt. Everest?

80. A gambler lost $25 on each of 11 consecutive rolls of the dice at the craps table. Express his total loss as a signed number.

81. Miss Jones lost $32 on each of seven hands of twenty-one. Express her total loss as a signed number.

82. A scientist is studying the movement, within its web, of a certain spider. Any movement up is considered to be positive, while any movement down is negative. Determine the net movement of a spider that goes up 2 cm five times and down 3 cm six times.

83. A certain junk bond trader purchased 950 shares of stock at $13 per share. She sold the shares for $11. What did she pay for the stock? How much did she receive when she sold the stock? How much did she lose or gain? Represent her loss or gain with a signed number.

84. A company bought 360 items for $3 each. They tried to sell them for $8 each and only sold 45. They lowered the price to $7 and sold 38 more. The price was lowered a second time to $4 and 142 were sold. Finally, they advertised a close-out sale at $2 and sold the rest. Did they make a profit or lose money on this item? Express the profit or loss as a signed number.

85. The Jimco Corp. loses $862,200 during one 20-month period. Determine the average monthly loss (written as a signed number). If there are 30 stockholders in this company, determine the total loss per stockholder (written as a signed number).

STATE YOUR UNDERSTANDING

86. When dividing two integers, care must be taken if one of the numbers is zero. Why?

87. Explain the difference between -3^2 and $(-3)^2$.

88. Explain the procedure for multiplying two integers.

89. Does $\dfrac{36 + 4}{-9} = -4 + 4$? If not, why not?

90. The sign rules for multiplication and division of two signed numbers may be summarized as follows:

If the numbers have the same sign, the answer is positive.
If the numbers have different signs, the answer is negative.

Explain why the rules for division are the same as the rules for multiplication.

CHALLENGE

Simplify.

91. $|-(-5)|(-9 - [-(-5)])$

92. $|-(-8)|(-8 - [-(-9)])$

93. $[-|-9|(8 - 12)] \div [(9 - 13)(8 - 7)]$

94. $[(14 - 20)(-5 - 9)] \div [-(-12)(-8 + 7)]$

95. Evaluate: $6ab - a^2$, if $a = -3$ and $b = -5$.

96. Evaluate: $-4cd - 5cd^2$, if $c = -5$ and $d = -7$.

GROUP ACTIVITY

97. Throughout the years, mathematicians have used a variety of examples to show students that the product of two negative numbers is positive. Talk to science and math instructors and record their favorite explanation. Discuss your findings with the class.

MAINTAIN YOUR SKILLS (Sections 3.3, 3.4, 3.5)

98. Combine like terms: $15z^2 - 9z^2 + 17z^2$

99. Combine like terms: $23s + 42t - 52s - 23t + 18s - 16t - 13s$

100. Find the quotient of $672bc$ and 14.

101. Solve: $y + 37 = 59$

102. Find the product of $6m$ and $13mn$.

4.6   Multiplying and Dividing Polynomials (Integers)

OBJECTIVES

1. Multiply polynomials.

2. Divide polynomials.

HOW AND WHY
Objective 1

Multiply polynomials.

In Chapter 3, you learned to multiply polynomials with whole number coefficients. Polynomials with integer coefficients are multiplied using the same properties.

$$-7(12x) = [(-7)(12)]x$$ 　　　**To multiply a term by a constant, multiply the constant times the coefficient.**

$$= -84x$$

$$(-6a)(-15a) = (-6)(-15)(aa)$$ 　　**Use the associative and commutative properties to group the coefficients and the variables.**

$$= 90a^2$$ 　　**Multiply.**

The box method of multiplication is useful when multiplying polynomials. Make a rectangle and label the length and the width with the two polynomials, then make subdivisions in the rectangle for each subproduct. Fill in the subdivisions by multiplying the headings. Then add the subproducts. To illustrate, consider $(3y - 5)(2y + 9)$.

	2y	**+9**
3y		
−5		

	2y	**+9**
3y	$6y^2$	$27y$
−5	$-10y$	-45

So,

$$(3y - 5)(2y + 9) = 6y^2 + 27y + (-10y) + (-45) = 6y^2 + 17y - 45$$

The box method is a way of using the distributive property. Each term of the first polynomial is multiplied by each term of the other. Let's demonstrate the distributive property using the same polynomials.

$$(3y - 5)(2y + 9) = (3y - 5)(2y) + (3y - 5)(9)$$ 　　**Distributive property.**

$$= (3y)(2y) - (5)(2y) + (3y)(9) - (5)(9)$$ 　　**Distributive property.**

$$= 6y^2 - 10y + 27y - 45$$ 　　**Multiply.**

$$= 6y^2 + 17y - 45$$ 　　**Combine like terms.**

When the coefficient of a term is negative, the sign is distributed with the term. For example,

$$-7t(3t - 11) = -7t(3t) - (-7t)(11)$$ 　　**Distributive property.**

$$= -21t^2 - (-77t)$$ 　　**Multiply the terms.**

$$= -21t^2 + 77t$$ 　　**Change subtract to add the opposite.**

Examples A–D   **Warm Ups A–D**

Directions: Multiply.

Strategy: Use the commutative, associative, and distributive properties as needed to multiply.

A. $-12(-7c)$ A. $-21(4d)$

$$-12(-7c) = [-12(-7)]c$$
$$= 84c$$

B.   Find the product: $(-16c)(5c)$ B.   Find the product: $(-8a)(-11a)$

$(-16c)(5c) = [(-16)(5)](cc)$ **Group the numerical and the variable factors.**

$$= -80c^2$$

C.   Multiply: $-11m(3m - 8)$ C.   Multiply: $-21c(2c - 7)$

$-11m(3m - 8) = -11m(3m) - (-11m)(8)$ **Distributive property.**

$$= -33m^2 - (-88m)$$ **Multiply.**
$$= -33m^2 + 88m$$ **Change subtract to add the opposite.**

D. $(3x - 8)(4x - 7)$ D. $(5x + 8)(3x - 7)$

Strategy: We show two methods of multiplying: the box method and the distributive property.

Method 1:

	$4x$	-7
$3x$	$12x^2$	$-21x$
-8	$-32x$	56

$$(3x - 8)(4x - 7) = 12x^2 - 32x - 21x + 56$$
$$= 12x^2 - 53x + 56$$

Method 2:

$(3x - 8)(4x - 7)$

$= [3x + (-8)][4x + (-7)]$ **Change subtraction to addition.**

$= [3x + (-8)]4x + [3x + (-8)](-7)$ **Distributive property.**

$= 3x \cdot 4x + (-8)4x + 3x(-7) + (-8)(-7)$ **Distributive property.**

$= 12x^2 + (-32x) + (-21x) + 56$ **Simplify.**

$= 12x^2 + (-53x) + 56$ **Combine like terms.**

$= 12x^2 - 53x + 56$

Answers to Warm Ups   A. $-84d$ B. $88a^2$ C. $-42c^2 + 147c$ D. $15x^2 - 11x - 56$

HOW AND WHY
Objective 2

Divide polynomials.

Division of terms with integer coefficients is done in the same manner as terms with whole numbers.

$$45x^2 \div (-9) = \frac{45}{-9} x^2 \qquad \textbf{Divide the coefficient by the integer.}$$

$$= -5x^2$$

and

$$-91b^2 \div (-7b) = \left(\frac{-91}{-7}\right)\left(\frac{b^2}{b}\right) \qquad \textbf{Divide the coefficients and the variables.}$$

$$= 13b$$

The box method can also be used to divide polynomials using the missing factor interpretation of division. Make a rectangle and label the width with the divisor and place the dividend inside the box. So, for

$$(15ab - 12ac + 24a) \div (-3a)$$

we have the box:

	?	?	?
$-3a$	$15ab$	$-12ac$	$24a$

Now determine what polynomial must be the length in order to make the box complete.

	$-5b$	$+4c$	-8
$-3a$	$15ab$	$-12ac$	$24a$

$(-3a)(-5b) = 15ab; (-3a)(4c) = -12ac;$
$(-3a)(-8) = 24a.$

So,

$$(15ab - 12ac + 24a) \div (-3a) = -5b + 4c - 8$$

The polynomial can also be divided by the monomial by dividing each term of the polynomial by the monomial.

$$(15ab - 12ac + 24a) \div (-3a) = \frac{15ab}{-3a} - \frac{12ac}{-3a} + \frac{24a}{-3a}$$

$$= -5b - (-4c) + (-8)$$

$$= -5b + 4c - 8$$

Examples E–G	**Warm Ups E–G**

Directions: Divide.

Strategy: Divide each term by the monomial divisor.

E. $-70x^2 \div (-10)$

$$-70x^2 \div (-10) = \frac{-70}{-10}x^2 \qquad \textbf{Divide the coefficients.}$$
$$= 7x^2$$

E. $115ab \div (-5)$

F. Divide $-72d^2$ by $18d$.

$$-72d^2 \div 18d = \frac{-72}{18}\left(\frac{d^2}{d}\right) \qquad \textbf{Divide the coefficients}$$
$$\textbf{and the variables.}$$
$$= -4d$$

F. Divide $-85b^2$ by $-17b$.

G. Divide: $(7x^2 - 35xy + 77xy^2) \div (-7x)$

G. Divide:
$$(8abc - 24abd - 16abe) \div (-8ab)$$

Strategy: We show two methods: The box method and dividing term by term.

	?	?	?
$-7x$	$7x^2$	$-35xy$	$77xy^2$

	$-x$	$+5y$	$-11y^2$
$-7x$	$7x^2$	$-35xy$	$77xy^2$

$$(7x^2 - 35xy + 77xy^2) \div (-7x) = -x + 5y - 11y^2$$

or

$$(7x^2 - 35xy + 77xy^2) \div (-7x) = \frac{7x^2}{-7x} - \frac{35xy}{-7x} + \frac{77xy^2}{-7x}$$
$$= -x - (-5y) + (-11y^2)$$
$$= -x + 5y - 11y^2$$

Answers to Warm Ups E. $-23ab$ F. $5b$ G. $-c + 3d + 2e$

Exercises 4.6

OBJECTIVE 1: *Multiply polynomials.*

A.

Multiply.

1. $7(-5c)$

2. $9(-11b)$

3. $-8(-5d)$

4. $-6(-11t)$

5. Find the product of -7 and $-ab$.

6. Find the product of 9 and $-xy$.

Multiply by filling in the box.

7. -3
$2x$	-5

8. 7
$5c$	-8

Multiply.

9. $6(3t - 7)$

10. $-5(2a - 7)$

11. $-2(3s - 7)$

12. $8(-2r - 9)$

B.

13. $3x(-2x)$

14. $3y(-6y)$

15. $-9b(-6b)$

16. $-7c(-13c)$

17. $-11m(-12n)$

18. $-12x(-12y)$

19. What is the product of $-10b$ and $-8c$?

20. What is the product of $-11x$ and $5z$?

Multiply.

21. $-4(4x - 5)$

22. $-5(4x - 5)$

23. $4(-7x - 2)$

24. $7(-3y - 7)$

25. $-13x(4x - 7)$

26. $-14a(5a + 4)$

Multiply by filling in the box and then adding the products.

27. $(x - 2)(x - 7)$

	x	-7
x		
-2		

28. $(x + 5)(x + 8)$

	x	$+8$
x		
5		

Multiply

29. $(x + 6)(x - 5)$

30. $(x - 7)(x + 4)$

OBJECTIVE 2: *Divide polynomials.*

A.

Divide.

31. $12x \div (-4)$

32. $-18z \div 3$

33. $\dfrac{-25y}{-5}$

34. $\dfrac{-36a}{-9}$

35. $\dfrac{-64b}{16}$

36. $\dfrac{56c}{-7}$

37. $\dfrac{-77ab}{7a}$

38. $\dfrac{-120xy}{15y}$

39. Find the quotient of $105mn$ and $-35m$.

40. Find the quotient of $-110pq$ and $-11q$.

Find the quotient by filling in the length of the box.

41. 3 | ? | ?
| $12x$ | -9 |

42. -7 | ? | ?
| $21y$ | -35 |

B.

Divide.

43. $\dfrac{-120a^2}{-15a}$

44. $\dfrac{-132b^2}{-11b}$

45. $\dfrac{160m^2}{-20m}$

46. $\dfrac{135p^2}{-15p}$

47. $\dfrac{2ab - 3ac + 5ad}{a}$

48. $\dfrac{7mn - 3ms + 2mt}{m}$

49. $\dfrac{3x^2 - 9x}{3x}$

50. $\dfrac{6a^2 - 12a}{3a}$

51. $(-18abc - 24bcd + 42bce) \div (-6bc)$

52. $(45stw - 75swy + 90swz) \div (-15sw)$

53. Find the quotient of $(-123b^2 - 82b)$ and $(-41b)$.

54. Find the quotient of $(-203z^2 + 87z)$ and $(-29z)$.

C.

Simplify.

55. $(b + 9)(b + 10)$

56. $(a - 11)(a - 7)$

57. $\dfrac{9x^2 - 15x}{3x}$

58. $\dfrac{30c^2 - 42c}{-6c}$

59. $(y - 7)(y + 6)$

60. $(a + 11)(a - 4)$

61. $(c - 8)(c + 9)$

62. $(m - 5)(m + 5)$

63. $(3x - 13)(5x + 8)$

64. $(5y - 9)(7y - 8)$

65. $(3a - 2b)(7c - 8)$

66. $(2m - 7n)(3m - 5)$

67. $(5c - 1)(2c + 8)$

68. $(7a + 3)(2a - 4)$

69. $(3x + 9)(2x - 1)$

70. $(5x + 3)(x - 2)$

71. Twentieth Century Securities has outlined three different portfolio strategies: aggressive, moderate, and conservative. The aggressive strategy puts 1 part in money market, 4 parts in bonds, and the remaining 15 parts in stocks. The moderate strategy puts 1 part in money market, 13 parts in bonds, and the remaining 6 parts in stocks. The conservative strategy puts 3 parts in money market, 9 parts in bonds, and the remaining 8 parts in stocks. Let x represent one part of a portfolio investment. Fill in the following table using polynomials in terms of x

	Aggressive	Moderate	Conservative
Money Market			
Bonds			
Stocks			
Total			

(a) What part of the aggressive fund is in stocks?

(b) What part of the moderate fund is in money market?

(c) What part of the conservative fund is in bonds?

72. Write a polynomial that represents the area of the rectangle.

$2x - 8$

$3x + 1$

73. Write a polynomial that represents the area of the figure.

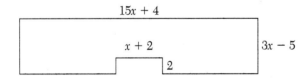

$15x + 4$

$x + 2$

2

$3x - 5$

STATE YOUR UNDERSTANDING

74. Explain how to find the quotient of $20y^2$ and $-4y$.   **75.** Explain how to simplify $(3m - 7)(2m - 5)$.

CHALLENGE

Simplify.

76. $3x(7x + 2y + 9) + 4x(2x - 5y - 6)$

77. $(x - 9)(x + 5) + (x + 8)(x - 2)$

78. $(2x + 5)(x - 2) - (x - 8)(3x + 1)$

79. $(2x + 3)(2x - 3)(x + 6)$

GROUP ACTIVITY

80. Multiply the following binomials:

$(x - 1)(x + 1)$	$(x - 3)(x + 3)$	$(x - 6)(x + 6)$
$(x - 5)(x + 5)$	$(x - 11)(x + 11)$	$(2x - 3)(2x + 3)$

Can you write a rule for a shortcut for multiplying two binomials that are the sum and difference of the same two terms? If so, use it to multiply $(97)(103) = (100 - 3)(100 + 3)$ mentally.

MAINTAIN YOUR SKILLS (Sections 3.2, 3.4, 3.5)

81. Find the value of P in $P = 2L + 2W$, if $W = 13$ and $L = 21$.

82. Find the value of S in $S = \dfrac{n(a + 1)}{2}$, if $a = 12$ and $n = 22$.

83. Find the product of $13w$ and $4wz$.

84. Solve: $96 = r + 54$.   **85.** Solve: $78 = m - 37$.

4.7   **Order of Operations and Average (Integers)**

OBJECTIVES

1. Perform any combination of operations with integers.

2. Find the average of a set of integers.

HOW AND WHY
Objective 1

Perform any combination of operations with integers.

The rules for order of operations do not depend on the type of numbers involved. Therefore, the order of operations for integers is the same as that for whole numbers. That is,

Order of Operations

In an expression with two or more operations, perform the operations in the following order.

1. Grouping symbols: Perform operations included within grouping symbols first.
2. Exponents: Perform operations indicated by exponents.
3. Multiply and divide: Perform multiplication and division from left to right as you come to them.
4. Add and subtract: Perform addition and subtraction from left to right as you come to them.

Examples A–F   **Warm Ups A–F**

Directions: Simplify.

Strategy: Perform the indicated operations according to the order of operations.

A. $-84 + (-36) \div (-3)$ A. $-90(-3) + 112$

$-84 + (-36) \div (-3) = -84 + 12$ **Divide first.**
$\qquad\qquad\qquad\qquad = -72$ **Add.**

B. $-21(6) - 45 \div (-3)$ B. $(-13)(-17) - 99 \div (-3)$

$-21(6) - 45 \div (-3) = -126 - (-15)$ **Multiply and divide as they occur.**

$\qquad\qquad\qquad\qquad = -126 + 15$ **Rewrite as addition.**
$\qquad\qquad\qquad\qquad = -111$ **Add.**

C. $(-5)^3 - 4(6^2)$ C. $(-3)^3(4) - 4(-2)^2$

$(-5)^3 - 4(6^2) = -125 - 4(36)$ **Do exponents first.**
$\qquad\qquad\qquad = -125 - 144$ **Multiply.**
$\qquad\qquad\qquad = -269$ **Subtract.**

Answers to Warm Ups   A. 382   B. 254   C. -124

D. $\dfrac{8^2 - 7(4) + 3(-21) - 8}{5(-7)}$

Strategy: The fraction bar acts as a grouping symbol. So first do the operations in the numerator and denominator.

$\dfrac{8^2 - 7(4) + 3(-21) - 8}{5(-7)}$ **Do exponents first.**

$= \dfrac{64 - 7(4) + 3(-21) - 8}{5(-7)}$ **Multiply in each group.**

$= \dfrac{64 - 28 - 63 - 8}{-35}$ **Add and subtract.**

$= \dfrac{-35}{-35}$ **Divide.**

$= 1$

D. $\dfrac{6^3 - 12(13) + 8(-15)}{3(4)}$

E. Find the difference of the product of -5 and 12 and the quotient of 16 and -4.

Strategy: To find the difference we subtract. First, determine what two numbers are to be subtracted. The first number is the product of -5 and 12 and the second number is the quotient of 16 and -4.

$-5(12)$ **The product of -5 and 12.**

$16 \div (-4)$ **The quotient of 16 and -4.**

Now subtract the quotient from the product.

$-5(12) - 16 \div (-4) = -60 - (-4)$ **Multiply and divide.**

$= -56$ **Subtract.**

E. Find the sum of the quotient of -25 and 5 and the product of -6 and -8.

F. During a special 2-hour sale a tire store sold 24 tires at a loss of $4 ($-4) per tire. During the remainder of the day the store sold 44 tires at a profit of $8 per tire. Express their profit or loss on the sale of tires as a signed number.

Strategy: To determine the amount of profit or loss, multiply the loss per tire times the number sold at a loss. Multiply the profit per tire times the number sold at a profit. Find the sum of the two numbers.

$24(-4)$ **Amount of loss.**

$44(8)$ **Amount of profit.**

$24(-4) + 44(8) = -96 + 352$ **Find the sum of the two numbers and multiply.**

$= 256$ **Add.**

Since the sum is positive, the store has a net profit of $256.

F. During a special sale, the Ace Hardware store sold 34 regular coffee makers at a loss of $6 ($-6) per maker. During the same sale the store sold 29 deluxe models of the coffee maker at a profit of $10 per maker. Express their profit or loss on the sale of the coffee makers as a signed number.

Answers to Warm Ups   D. -5 E. 43 F. $86 (profit)

HOW AND WHY
Objective 2   Find the average of a set of integers.

Finding the average of a set of numbers does not depend on the type of number. Therefore, to calculate the average of a set of integers use the same procedure that we used to find the average of whole numbers.

Examples G–I	**Warm Ups G–I**

Directions: Find the average.

Strategy: Find the sum of the numbers in the set and then divide the sum by the number of numbers in the set.

G. 24, -18, -45, 27, and -48

$24 + (-18) + (-45) + 27 + (-48) = -60$ **Add the numbers.**

$-60 \div 5 = -12$ **Divide the sum by the number of numbers.**

The average is -12.

G. -38, 46, -21, -18, and -19

H. -78, 90, -134, -37, 132, 45, and -25

$-78 + 90 + (-134) + (-37) + 132 + 45 + (-25) = -7$

$-7 \div 7 = -1$

The average is -1.

H. 89, -65, -126, -12, -67, and 301

I. During a five-day period the low temperatures recorded in Freeport, Maine, are $-13°F$, $-9°F$, $-10°F$, $5°F$, and $-3°F$. What is the average low temperature for the five days?

Strategy: To find the average, add the temperatures and divide by 5, the number of days.

$-13 + (-9) + (-10) + 5 + (-3) = -30$

$-30 \div 5 = -6$

The average low temperature is $-6°F$.

I. During a 4-hour period a submarine recorded the following depths: -457 ft, -385 ft, -267 ft, and -295 ft. What is the average depth of the submarine during the 4-hour period?

Answers to Warm Ups   G. -10 H. 20   I. -351 ft

Exercises 4.7

OBJECTIVE 1: *Perform operations on integers.*

A.

Simplify.

1. $-5(6) + (-22)$ **2.** $-7(8) + 21$ **3.** $34 - 7(-3)$

4. $45 - 9(6)$ **5.** $-4 \cdot 6 + 7(-3)$ **6.** $-8 \cdot 5 + 9(-2)$

7. $6 - 3^2$ **8.** $7 - 4^2$ **9.** $-28 - (-3)(-4)$

10. $-32 - (-6)(5)$

11. In $8(-3) + 5^2 - 7(-3)$, which operation is performed first?

12. In $7(-4) - 7^2 - (5 - 9)(-5)$, which operation is performed first?

B.

13. $5[10 + 3(-8)] - 33$ **14.** $6[11 - 4(4)] - 24$

15. $6(-10 + 4) - 33 \div (-11)$ **16.** $7(14 - 18) - 42 \div (-2)$

17. $-35 \div (-5) + (-2)(-3)$ **18.** $3(-11) - 14(-2) + 3(-2)$

19. $8(-3) + 5^2 - 7(-3)$

20. $-6(5) + 7^2 - 9(-4)$

21. $(-5)^2 + (-2)^2 - (-5)(-2)$

22. $(-7)^2 - (-6)^2 + 3(-4) - (-5)^2$

23. $\dfrac{18 - 18(2 - 3)}{-2(3)}$

24. $\dfrac{45 - 6(7 - 9) - 50}{-1(16 - 9)}$

25. $\dfrac{-5(6) - 3(8)}{6(-3) + 6(6)}$

26. $\dfrac{-8(-6) + 4(12)}{7(-8) - 6(-8)}$

OBJECTIVE 2: *Find the average of a set of integers.*

A.

Find the average.

27. $-3, -4, 6, 5, 1$

28. $-5, 7, -8, 3, -2$

29. $-4, -7, -2, -3$

30. $-9, -1, -5, -5$

31. $-1, -2, -2, -3, 5, -3$

32. $-3, 6, -4, 7, -5, 5$

33. −8, −11, −14, −7

34. −5,−16, −4, −9, −1

35. If the average of −3, −5, 7, and ? is −1, what is the missing number?

36. If the average of −5, −2, 7, 4, and ? is −2, what is the missing number?

B.

Find the average.

37. −49, −72, −28, 37

38. 39, 36, −24, −19

39. −10, −24, 45, −16, −30

40. 34, −23, −8, 19, −7

41. −8, −3, 12, −21, 27, −19

42. −11, 43, 15, −34, −54, −19

43. −23, −34, −21, −6, −15, 0, −13

44. −25, −3, 0, −17, −54, −15, −47

45. 62, −54, −81, −34, −21, 55, 82, 87   **46.** 66, −75, −53, 99, −33, 84, 43, −91

C.

47. Find the sum of the product of 12 and −4 and the product of −3 and −12.

48. Find the difference of the product of 3 and 9 and the product of −8 and 3.

For Exercises 49 to 52. A satellite records the following temperatures for a five-day period on one point on the surface of Mars.

	Day 1	Day 2	Day 3	Day 4	Day 5
5:00 AM	−92°C	−88°C	−115°C	−103°C	−74°C
9:00 AM	−57°C	−49°C	−86°C	−93°C	−64°C
1:00 PM	−52°C	−33°C	−46°C	−48°C	−10°C
6:00 PM	−45°C	−90°C	−102°C	−36°C	−42°C
11:00 PM	−107°C	−105°C	−105°C	−98°C	−90°C

49. What is the average temperature recorded during Day 5?

50. What is the average temperature recorded at 6:00 PM?

51. What was the average high temperature recorded for the five days?

52. What was the average low temperature recorded for the five days?

Simplify.

53. $[-3 + (-6)]^2 - [-8 - 2(-3)]^2$

54. $[-5(-9) - (-6)^2]^2 + [(-8)(-1)^3 + 2]^2$

55. $[46 - 3(-4)^2]^3 - [-7(1)^3 + (-5)(-8)]$

56. $[30 - (-5)^2]^2 - [-8(-2) - (-2)(-4)]^2$

57. $-15 - \dfrac{8^2 - (28)}{3^2 + 3}$

58. $-22 + \dfrac{9^2 - 6}{6^2 - 11}$

59. $\dfrac{12(8 - 24)}{5^2 - 3^2} \div (-12)$

60. $\dfrac{15(12 - 45)}{6^2 - 5^2} \div (-9)$

61. $-8|125 - 321| - 21^2 + 8(-7)$

62. $-9|482 - 632| - 17^2 + 9(-9)$

63. $-6(8^2 - 9^2)^2 - (-7)20$

64. $-5(6^2 - 7^2)^2 - (-8)19$

65. Find the difference of the quotient of 28 and -7 and the product of -4 and -3.

66. Find the sum of the product of −3 and 7 and the quotient of −15 and −5.

67. Keshia buys a TV for $95 down and $47 per month for 15 months. What is the total price she pays for the TV? (*Hint*: When a deferred payment plan is used, the total cost of the article is the down payment plus the total of the monthly payments.)

68. The E-Z Chair Company advertises recliners for $40 down and $17 per month for 24 months. What is the total cost of the recliner?

69. During a "blue light" special K-Mart sold 24 fishing poles at a loss of $3 per pole. During the remainder of the day they sold 9 poles at a profit of $7 per pole. Express the profit or loss on the sale of fishing poles as a signed number.

70. Fly America sells 40 seats on Flight 402 at a loss of $52 per seat (−$52). Fly America also sells 67 seats at a profit of $78 per seat. Express the profit or loss on the sale of the seats as a signed number.

Exercises 71–75 refer to the Chapter 4 Application. See page 273.

71. For each continent, calculate the average of the highest and lowest point. Which continent has the largest average and which has the smallest average?

The following table gives the altitudes of selected cities around the world.

City	Altitude (ft)	City	Altitude (ft)
Athens, Greece	300	Mexico City, Mexico	7347
Bangkok, Thailand	0	New Delhi, India	770
Berlin, Germany	110	Quito, Ecuador	9222
Bogota, Columbia	8660	Rome, Italy	95
Jakarta, Indonesia	26	Tehran, Iran	5937
Jerusalem, Israel	2500	Tokyo, Japan	30

72. Find five cities with an average altitude of less than 100 ft.

73. Find three cities with an average altitude of approximately 350 ft.

74. Find four cities with an average altitude of approximately 7000 ft.

75. The treasurer of a local club records the following transactions during one month:

Opening balance	$4756
Deposit	$345
Check #34	$212
Check #35	$1218
New check cost	$15
Deposit	$98
National dues paid	$450
Electric bill	$78

What is the balance at the end of the month?

76. Try this game on your friends. Have them pick a number. Tell them to double it, then add 20 to that number, divide the sum by 4, subtract 5 from that quotient, square the difference, and multiply the square by 4. They should now have the square of the original number. Write a mathematical representation of this riddle.

STATE YOUR UNDERSTANDING

Locate the error in Exercises 77 and 78. Indicate why each is not correct. Determine the correct answer.

77. $2[3 + 5(-4)] = 2[8(-4)]$
$$= 2[-32]$$
$$= -64$$

78. $3 - [5 - 2(6 - 4^2)^3] = 3 - [5 - 2(6 - 16)^3]$
$$= 3 - [5 - 2(-10)^3]$$
$$= 3 - [5 - (-20)^3]$$
$$= 3 - [5 - (-8000)]$$
$$= 3 - [5 + 8000]$$
$$= 3 - 8005$$
$$= -8002$$

79. Is there ever a case when exponents are not computed first? If so, give an example.

80. Explain the difference between -2^4 and $(-2)^4$. Explain the difference between -5^3 and $(-5)^3$.

81. Write a paragraph that explains the process of calculating the average depth of a body of water.

CHALLENGE

Simplify.

82. $\dfrac{3^2 - 5(-2)^2 + 8 + [4 - 3(-3)]}{4 - 3(-2)^3 - 18}$

83. $\dfrac{(5 - 9)^2 + (-6 + 8)^2 - (14 - 6)^2}{[3 - 4(7) + 3^3]^2}$

84. $\dfrac{3(4 - 7)^2 + 2(5 - 8)^3 - 18}{(6 - 9)^2 + 6}$

GROUP ACTIVITY

85. Engage the entire class in a game of KRYPTO. This card game consists of 41 cards numbered from -20 to 20. Each group gets four cards. A card is chosen at random from the remaining cards and the number is put on the board. Each group must find a way using addition, subtraction, multiplication, and/or division to combine their given cards to equal the number on the board. Operations may be used more than once.

MAINTAIN YOUR SKILLS (Sections 3.2, 3.3, 3.4, 3.5)

86. Evaluate $S = \dfrac{n(a + r)}{2}$ for S when $n = 100$, $a = 1$, and $r = 100$.

87. Simplify: $45y - 18y - 21y$

88. Multiply: $12(8a)$

89. Solve: $14 + w = 57$

90. Solve: $b - 124 = 43$

4.8   Evaluating Algebraic Expressions and Formulas (Integers)

OBJECTIVE   Evaluate algebraic expressions and formulas with integers.

HOW AND WHY
Objective   **Evaluate algebraic expressions and formulas with integers.**

The procedures for evaluating algebraic expressions and formulas are the same regardless of the type of numbers used.

When replacing a variable with a negative value it is helpful to enclose the number in parentheses.

Evaluate the expression $3x - 2y$ if $x = -4$ and $y = -6$.

$3x - 2y = 3(-4) - 2(-6)$	**Substitute −4 for x and −6 for y. Write these substitutions in parentheses.**
$= -12 - (-12)$	**Multiply first.**
$= -12 + 12$	**Change subtract to add the opposite.**
$= 0$	**Add.**

There are also formulas in which integers are used. A formula to find the speed of a falling object is given by

$$s = v - gt$$

where s is the speed of the falling object measured in feet per second, v is the initial velocity measured in feet per second, t is the time measured in seconds, and g is the gravitational force measured in feet per second per second. The gravitational force is a constant that is approximately 32.

To find the speed of a ball tossed upward from the top of a building at an initial velocity of 5 feet per second (v) after 6 seconds (t), we do the following.

$s = v - gt$	**Original formula.**
$s = 5 - (32)6$	**Substitute 5 for v, 32 for g, and 6 for t.**
$s = 5 - 192$	**Simplify.**
$s = -187$	

After 6 seconds the ball will be traveling downward at approximately 187 feet per second.

Examples A–F	**Warm Ups A–F**

Directions: Evaluate

Strategy: Substitute the values for the variables and then use the order of operations to simplify.

A. $34 - y; y = -17$		A. $-45 - t; t = -13$	
$34 - y = 34 - (-17)$	**Substitute −17 for y.**		
$= 34 + 17$	**Simplify.**		
$= 51$			

B. Evaluate: $-6abc$; $a = -3$, $b = 9$, $c = -5$

$-6abc = -6(-3)(9)(-5)$ **Substitute -3 for a, 9**
for b, and -5 for c.

$= 18(9)(-5)$ **Simplify.**

$= 162(-5)$

$= -810$

B. Evaluate: $3xyz$; $x = -4$, $y = -7$, $z = -11$

⊞ **Calculator Example**

C. Evaluate: $5cd - 6de$; $c = -7$, $d = -13$, $e = 23$

$5cd - 6de = 5(-7)(-13) - 6(-13)(23)$ **Substitute -7 for**
c, -13 for d, and
23 for e.

$= 455 - (-1794)$ **Simplify.**

$= 2249$

C. Evaluate: $9ab - 7bc$; $a = -4$, $b = 16$, $c = -8$

D. Find t if $t = 3s^2 - 4d$, when $s = -6$ and $d = 25$.

Strategy: Substitute the values of the variables in the formula.

$t = 3s^2 - 4d$

$= 3(-6)^2 - 4(25)$ **Substitute -6 for s and 25 for d.**

$= 3(36) - 4(25)$ **Simplify.**

$= 108 - 100$

$= 8$

So, when $s = -6$ and $d = 25$, $t = 8$.

D. Find D if $D = -14c - 4n^2$ when $c = 16$ and $n = -9$.

E. Is -12 a solution of $3x - 5 = 4x + 7$?

Strategy: Check by substituting -12 for x in the equation.

$3x - 5 = 4x + 7$ **Original equation.**

Does $3(-12) - 5 = 4(-12) + 7$?   **Substitute -12 for x.**

Does $-36 - 5 = -48 + 7$?

Does $-41 = -41$?   **Yes.**

A solution is -12.

E. Is -15 a solution of $7a - 12 = 10a + 33$?

F. Find the speed (s), in feet per second, a skydiver will reach in a time (t) of 5 seconds if $v = -3$ and $g = 32$. The formula is $s = v - gt$.

Strategy: Substitute in the formula and simplify.

$s = v - gt$ **Formula.**

$s = -3 - 32(5)$ **Substitute.**

$= -3 - 160$

$= -163$

The skydiver reaches a falling speed of 163 feet per second in 5 seconds.

F. Find the speed (s), in feet per second, a skydiver will reach in a time (t) of 11 seconds if $v = -3$ and $g = 32$. The formula is $s = v - gt$.

Answers to Warm Ups   B. -924 C. 320   D. -548 E. yes   F. 355 feet per second

Exercises 4.8

OBJECTIVE: *Evaluate expressions and formulas.*

A.

For Exercises 1–8, evaluate when $a = -5$, $b = -7$, $c = 12$, $x = -8$, $y = 14$, and $z = -9$.

1. $y - 21$

2. $z + 14$

3. $a + b$

4. $x + z$

5. $b - c$

6. $x - b$

7. $a - b - c$

8. $x - y - z$

9. Find C in $C = D - R + T$, if $D = 85$, $R = 98$, and $T = 11$.

10. Find S in $S = 3t - q$, if $t = -7$ and $q = -5$.

B.

For Exercises 11–16, evaluate if $a = -7$, $b = -4$, and $c = 3$.

11. $15a - 3b - 5c$

12. $19b - 4c + 8a$

13. $a(3b - 4c) - 5a$

14. $c(4a + 7b) - 12c$

15. $2a^2 - 4b^2 - 3c^2$

16. $6c^2 - 5b^2 - 2a^2$

17. Is 10 a solution of $3x - 6 = 24$?

18. Is 8 a solution of $4y + 8 = -24$?

19. Find S in $S = v + gt$, if $v = 45$, $g = -32$, and $t = 3$.

20. Find C in $C = P - np$, if $P = 845$, $n = 12$, and $p = 21$.

For Exercises 21–26, evaluate if $m = -6$, $n = -8$, and $p = -9$.

21. $-4mn - 9p$

22. $-7mp - 4np$

23. $mnp - 6m + 3np$

24. $4p - 8mn + 4mnp$

25. $m^2 - np^2 + 14p$

26. $m^2p^2 - (13n)^2$

C.

For Exercises 27–30. A satellite records the following temperatures for a five-day period on one point on the surface of Mars.

	Day 1	**Day 2**	**Day 3**	**Day 4**	**Day 5**
5:00 AM	$-92°C$	$-88°C$	$-115°C$	$-103°C$	$-74°C$
9:00 AM	$-57°C$	$-49°C$	$-86°C$	$-93°C$	$-64°C$
1:00 PM	$-52°C$	$-33°C$	$-46°C$	$-48°C$	$-10°C$
6:00 PM	$-45°C$	$-90°C$	$-102°C$	$-36°C$	$-42°C$
11:00 PM	$-107°C$	$-105°C$	$-105°C$	$-98°C$	$-90°C$

27. Convert the degrees Centigrade, C, recorded at 11:00 PM on Day 5 to degrees Fahrenheit (F). Use the formula, $F = \dfrac{9C + 160}{5}$.

28. Convert the degrees Centigrade recorded at 5:00 AM on Day 3 to degrees Fahrenheit.

29. Convert the degrees Centigrade recorded at 1:00 PM on Day 5 to degrees Fahrenheit.

30. Convert the degrees Centigrade recorded at 11:00 PM on Day 3 to degrees Fahrenheit.

31. Is 8 a solution of $5x - 12 = 2x + 6$?

32. Is 6 a solution of $5x - 12 = 2x + 6$?

33. Is -2 a solution of $7a + 12 = 5a + 10$?

34. Is -5 a solution of $9y - 15 = 2y - 20$?

35. Is -5 a solution of $-2b - 13 = -7b - 12$?

36. Is 6 a solution of $-8x + 15 = -4x + (-9)$?

For Exercises 37–44, evaluate when $a = -11$, $b = -12$, $c = 25$, $x = 12$, and $y = -15$.

37. $3abc - xy$

38. $5bxy - 7ac$

39. $-7abx - 12cy$

40. $-10acx - 15bx$

41. $\dfrac{-cy^2 - 3}{x}$

42. $\dfrac{ax^2 + 9}{y^2}$

43. $\dfrac{a^2 - 2b^2 - 8xy + 3}{-c - 2y - 9}$

44. $\dfrac{-7x^2 - 5y^2 + ab}{x + y}$

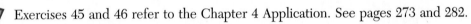

Exercises 45 and 46 refer to the Chapter 4 Application. See pages 273 and 282.

45. Three times the depth of the Eurasia Basin is 2035 ft deeper than the Java Trench and the Puerto Rico Trench combined. Let E represent the depth of the Eurasia Basin, J represent the depth of the Java Trench, and P represent the depth of the Puerto Rico Trench. Write an algebraic equation that describes this relationship. Is the statement true? Explain.

46. Let us suppose the highest point in Europe plus twenty times the lowest point in Europe is equal to the highest point in Asia. Let h_E represent the highest point in Europe, l_E represent the lowest point in Europe, and h_A represent the highest point in Asia. Write an algebraic equation that describes this relationship. Is this statement true? Explain.

STATE YOUR UNDERSTANDING

47. Describe how to evaluate $3abc$ when $a = 3$, $b = 6$, and $c = -7$.

48. Explain why x cannot be replaced by 7 in $\dfrac{2x + 3}{x - 7}$.

49. Does the expression $-5de$ always represent a negative number? Under what conditions does it represent a negative number? Does $-5t^2$ always represent a negative number? Explain.

CHALLENGE

Simplify.

50. $\dfrac{a^2 + b^2 - c^2}{abc}$ when $a = 3$, $b = -1$, and $c = -2$

51. $\dfrac{3a^2b^2 - 2b^2c^2 - (c - a)}{(a + b)^2 - (b - c)^2}$ when $a = 3$, $b = -1$, and $c = -2$

52. $\dfrac{(a - b)^2 + (a - c)^2 + (b - c)^2 + 3}{a^2 + b^2 - c^2}$ when $a = 2$, $b = 3$, and $c = -2$

GROUP ACTIVITY

53. In Exercise 31 you were asked to verify that 8 is a solution of $5x - 12 = 2x + 6$. Have each group member find an equation of the form $5x - 12 = 2x + k$, where the solutions are $x = 10, 20, 30, 40,$ or 50. That is, find values of k so that the answers will be correct.

MAINTAIN YOUR SKILLS (Sections 3.6, 3.7)

Solve.

54. $5x = 115$

55. $7t = 126$

56. $8x + 23 = 71$

57. $9y - 7 = 92$

58. $7d - 15 = 139$

4.9   Solving Equations (Integers)

OBJECTIVE   Solve equations containing integers.

HOW AND WHY

Objective   **Solve equations containing integers.**

Equations with integers are solved using the properties of equations in Chapter 3. You may want to review Sections 3.5, 3.6, and 3.7.

The addition and subtraction properties of equality together with the multiplication and division properties of equality are used to isolate the variable.

Examples A–H   **Warm Ups A–H**

Directions: Solve.

Strategy: Use the properties of equality to isolate the variable.

A. $x + 11 = -21$	A. $y + 16 = -17$

$x + 11 - 11 = -21 - 11$ Subtract 11 from both sides.

$x = -32$

Check:

$x + 11 = -21$

Does $-32 + 11 = -21$?   Substitute -32 for x in the original equation.

Does $\quad -21 = -21$?   **Yes.**

The solution is $x = -32$.

B. Solve: $-7x = 105$ B. Solve: $6b = -132$

$\dfrac{-7x}{-7} = \dfrac{105}{-7}$ **Divide both sides by -7.**

$x = -15$

Check:

$-7x = 105$

Does $-7(-15) = 105$?   Substitute -15 for x.

Does $\quad 105 = 105$?   **Yes.**

The solution is $x = -15$.

C. Solve: $-4x + 7 = -25$ C. Solve: $-7a + 12 = -65$

$-4x + 7 - 7 = -25 - 7$ **Subtract 7 from both sides.**

$-4x = -32$

$\dfrac{-4x}{-4} = \dfrac{-32}{-4}$ **Divide both sides by -4.**

$x = 8$ **The check is left for the student.**

The solution is $x = 8$.

Answers to Warm Ups   A. $y = -33$ B. $b = -22$ C. $a = 11$

D. Solve: $8a - 9 - 5a = 27$

Strategy: First combine the terms containing a.

$$8a + (-9) + (-5a) = 27$$ Change subtraction to add the opposite.

$$3a + (-9) = 27$$ Combine terms.

$$3a + (-9) + 9 = 27 + 9$$ Add the opposite of -9 to both sides.

$$3a = 36$$

$$\frac{3a}{3} = \frac{36}{3}$$ Divide both sides by 3.

$$a = 12$$

Check:

$$8a - 9 - 5a = 27$$ Use original equation to check.

Does $8(12) - 9 - 5(12) = 27$? Substitute 12 for a.

Does $96 - 9 - 60 = 27$?

Does $27 = 27$? Yes.

The solution is $a = 12$.

D. Solve: $8y - 12 - 4y = 16$

E. Solve: $-9(2b - 3) = -27$

Strategy: Use the distributive property to simplify the left side.

$$-9[2b + (-3)] = -27$$ Change subtraction to add the opposite.

$$-18b + 27 = -27$$ Distributive property.

$$-18b + 27 - 27 = -27 - 27$$ Subtract 27 from both sides.

$$-18b = -54$$

$$\frac{-18b}{-18} = \frac{-54}{-18}$$ Divide both sides by -18.

$$b = 3$$ The check is left for the student.

The solution is $b = 3$.

E. Solve: $5(7c - 8) = -180$

F. Solve: $5w + 7 = 34 + 8w$

Strategy: First gather the terms containing the variable on the left side by subtracting $8w$ from both sides.

$$5w - 8w + 7 = 34 + 8w - 8w$$ Subtract $8w$ from both sides.

$$-3w + 7 = 34$$ Combine terms.

$$-3w + 7 - 7 = 34 - 7$$ Subtract 7 from both sides.

$$-3w = 27$$

$$\frac{-3w}{-3} = \frac{27}{-3}$$ Divide both sides by -3.

$$w = -9$$

F. Solve: $11 - 5x = x - 19$

Answers to Warm Ups D. $y = 7$ E. $c = -4$ F. $x = 5$

Check:

$$5w + 7 = 34 + 8w \qquad \textbf{Original equation.}$$

Does $5(-9) + 7 = 34 + 8(-9)$?   **Substitute -9 for w.**

Does $-45 + 7 = 34 + (-72)$?

Does $-38 = -38$?   **Yes.**

The solution is $w = -9$.

G.   The product of 8 and the difference of 4 times a number and 3 is 40. What is the number?

Strategy: Eight times the difference is 40. To find a difference two numbers are needed. These numbers are four times a number and three.

Select a variable:

Let n represent the number.

$4n$ **Four times the number.**

$4n - 3$ **The difference of four times the number and 3.**

$8(4n - 3) = 40$ **Eight times the difference is 40.**

$32n - 24 = 40$ **Distributive property.**

$32n - 24 + 24 = 40 + 24$ **Add 24 to both sides.**

$32n = 64$

$n = 2$ **Divide both sides by 32.**

Check:

The difference of 4 times 2 and 3 is $8 - 3 = 5$.

Eight times that difference is $8(5) = 40$.

The number is 2.

H.   If the area (A) of a rectangular lot is 6765 square feet and the width (w) is 55 feet, find the length (ℓ).

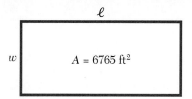

ℓ

w $A = 6765 \text{ ft}^2$

Formula:

$A = \ell w$ **The area of a rectangle.**

Solve.

$\ell w = A$ **Use the symmetric property of equality to get ℓ on the left side.**

$\ell(55) = 6765$ **Substitute 55 for w and 6765 for A.**

$\dfrac{\ell(55)}{55} = \dfrac{6765}{55}$ **Divide both sides by 55.**

$\ell = 123$

G.   The product of 6 and the difference of 3 times a number and 4 is 12. What is the number?

H.   If the area (A) of a rectangular lot is 7296 square feet and the width (w) is 48 feet, find the length (ℓ).

Answers to Warm Ups   G.   The number is 2   H.   152 feet

Check:

If the length is 123 feet and the width is 55 feet, then the area will be 123(55).

Does $123(55) = 6765$?

Does $6765 = 6765$?   **Yes.**

The rectangle is 123 feet long.

Exercises 4.9

OBJECTIVE: *Solve equations with integers.*

Solve.

A.

1. $a + 11 = -12$

2. $b - 8 = -21$

3. $x - 12 = -22$

4. $x + 21 = 15$

5. $4x = -36$

6. $5a = -65$

7. $-12c = -60$

8. $-9y = -63$

9. $-2x - 8 = 16$

10. $-3c + 4 = -11$

B.

11. $-14 = 37y - 88$

12. $19 = -8w + 3$

13. $29y + 2 = 205$

14. $74w + 3 = 225$

15. $-2x + 5 = -17$

16. $9z - 15 = -87$

17. $67w - 34 = -369$

18. $50w + 16 = -384$

19. $\dfrac{a}{9} - 8 = -2$

20. $\dfrac{b}{7} + 6 = -5$

21. $3y - 14 = 2y - 14$

22. $21 - 2t = 8t - 19$

23. $-9x - 26 = 5x + 44$

24. $11x - 23 = 15x + 53$

25. $19 - 2x - 5 = 5x + 7$

26. $9 - 5a - 5 = 3a - 12$

C.

27. $-3k + 25 = 1 - 6k$

28. $2q + 15 = 4 + q$

29. $-2(4r - 21) = -78$

30. $-3(2p + 18) = -30$

31. Find the time (t) it takes a free-falling skydiver to reach a falling speed (s) of 147 feet per second (-147 fps) if $v = -19$ and $g = 32$. The formula is $s = v - gt$.

32. Find the time (t) it takes a free-falling skydiver to reach a falling speed (s) of 211 feet per second (-211 fps) if $v = -19$ and $g = 32$. The formula is $s = v - gt$.

33. If 98 is added to six times some number, the sum is 266. What is the number?

34. If 73 is added to 11 times a number, the sum is -158. What is the number?

35. The difference of 15 times a number and 181 is -61. What is the number?

36. The difference of 24 times a number and 32 is -248. What is the number?

Solve.

37. $3x + 15 + 21 = 22 + 4x$ **38.** $2x + 14 + 16 = 23 - 5x$

39. $40p - 24 - 33p - 77 = 102$ **40.** $27p - 24 - 44p - 77 = -390$

41. $37q + 9 - 8q = 8q - 9 - 87$ **42.** $9q + 73 - q = 109 + q - 78$

43. The temperature scales for Fahrenheit and Celsius are related by the formula: $5F = 9C + 160$. Complete the following table.

°F	122°F	104°F				−13°F	−40°F
°C			25°C	10°C	−10°C		

44. (a) Water freezes at 32°F. What is the corresponding Celsius temperature?

(b) A person with a body temperature of 35°C has a fever. What is the corresponding Fahrenheit temperature?

45. The formula for the balance of a loan (D) is $D = B - NP$, where P represents the monthly payment, n represents the number of payments, and B represents the money borrowed. Find N when $D = \$575$, $B = \$925$, and $P = \$25$.

46. Use the formula in Exercise 45 to determine the monthly payment (P) if $D = \$820$, $B = \$1020$, and $N = 5$.

Exercises 47–49 refer to the Chapter 4 Application. Use negative numbers to represent feet below sea level. See pages 273 and 282.

47. The high point of Australia is 12,558 ft more than four times the lowest point of one of the continents. Write an algebraic equation that describes this relationship. Which continent's lowest point fits the description?

48. The lowest point of Antarctica is 2183 ft less than 12 times the lowest point of one of the continents. Write an algebraic equation that describes this relationship. Which continent's lowest point fits this description?

49. The Mariana Trench in the Pacific Ocean is about 2000 ft deeper than twice one of the other oceans' deepest parts. Write an algebraic equation that describes this relationship. Which ocean's deepest part fits this description?

STATE YOUR UNDERSTANDING

50. Write a process that may be used to solve equations containing:
  (a) like terms on the same side of the equation.   **(b)** like terms on each side of the equation.

51. Why is it a good idea to combine terms before using the properties of equality?

52. Explain how to solve the equation $4y - 12 = 8y + 8$ using a table. Explain how to solve the equation $4y - 12 = 8y + 8$ symbolically.

CHALLENGE

Solve.

53. $3(x - 2) + 5(x + 7) = 69$ **54.** $-2(2a - 5) + 3(a + 6) + 4(a - 5) = -1$

55. $5(x - 4) + 2(x - 3) - 3(x - 5) = 9$ **56.** $-4(a - 1) - 6(a + 3) + 2(2a - 12) = 22$

GROUP ACTIVITY

57. Consider the following equations:

$$3x - (3x + 2) = -4 \quad \text{and} \quad 5x + (7 - 5x) = 7$$

Does either equation have a solution? More than one? What makes these equations special?

MAINTAIN YOUR SKILLS (Sections 3.6, 3.7)

Solve.

58. $\dfrac{c}{15} = 13$ **59.** $\dfrac{z}{19} = 15$

60. $337 = 71 + 14x$ **61.** $45y + 17 = 1052$

62. If the wholesale cost of 34 swivel chairs is \$4216, what is the cost of one swivel chair? Use the formula $C = np$, where C is the total cost, n is the number of chairs, and p is the price per chair.

OPTIONAL Group Project *(3–4 weeks)*

For three consecutive weeks, on Monday, locate the final scores for each of the three major professional golf tours in the United States, the Professional Golf Association, PGA; the Ladies Professional Golf Association, LPGA; and the Senior Professional Golf Association, Senior PGA. These scores can usually be found on the summary page in the sports section of the daily newspaper.

a. Record the scores, against par, for the 30 top finishers and ties on each tour. Display the data using bar graphs for week one, line graphs for week two, and pictorial graphs for week three. Which type of graph best displays the data? Why?

b. Calculate the average score, against par, for each tour for each week. When finding the average, if there is a remainder and if it is half or more than the divisor, round up, otherwise round down. Now average the average scores for each tour. Which tour scored the best? Why?

c. What is the difference between the best and worst scores on each tour for each week?

d. What is the average amount of money earned by the players whose scores were recorded on each tour for each week? Which tour pays the best?

e. How much did the winner on each tour earn per stroke under par in the second week of your data? Compare the results. Is this a good way to compare the earnings on the tour? If not, why not?

CHAPTER 4

True-False Concept Review

ANSWERS

Check your understanding of the language of algebra and arithmetic. Tell whether each of the following statements is True (always true) or False (not always true). For each statement you judge to be false, revise it to make a statement that is true.

1. _____

1. The opposite of an integer is a negative integer.

2. _____

2. The absolute value of an integer is never a negative integer.

3. _____

3. The difference of two negative integers is a negative integer.

4. _____

4. The product of two negative integers is a positive integer.

5. _____

5. The cube (third power) of a negative integer is a negative integer.

6. _____

6. The sum of two negative integers is a negative integer.

7. _____

7. The quotient of a positive integer and a negative integer is a positive integer.

8. _____

8. The difference of a positive integer and a negative integer is a positive integer.

9. _____

9. Like terms with integer coefficients are combined using the distributive property together with the rules for adding integers.

10. _____

10. The product of negative integer factors is positive.

11. _____

11. Every equation containing integers (coefficients) has a solution that is an integer.

12. _____

12. $-78 < -2$

13. _____

13. $(-222)(-666) > -888$

14. _____

14. It is possible for two different integers to have the same absolute value.

15. _____

15. The order of operations for integers is the same as the order of operations for whole numbers.

16. _____

16. A subtraction problem with integers can be checked in the same way as a subtraction problem for whole numbers.

17. _____

17. The commutative property applies to the subtraction of integers even though it does not apply to the subtraction of whole numbers.

18. _____

18. The rule for determining the sign (positive or negative) of the product and the quotient of two integers is the same.

19. _____

19. The rules for determining the sign (positive or negative) of the sum and the difference of two integers is the same.

20. _____

20. $-5^2 = 25$

CHAPTER 4

Review

Section 4.1 *Objective 1*

1. What is the opposite of -22?
2. What is the opposite of 0?
3. If $x = 32$, then $-x = $?
4. If $x = -85$, then $-x = $?
5. $-(-8) = $?

Section 4.1 *Objective 2*

6. Find the value of $|56|$.
7. Find the value of $|-87|$.
8. Find the value of $|8 + 2|$.
9. Find $|x|$ if $x = -12$.
10. Find $|x|$ if $x = 43$.

Section 4.1 *Objective 3*

Insert $<$ or $>$ to make a true statement.

11. $-43 \quad\quad -25$
12. $54 \quad\quad -15$
13. $-79 \quad\quad -32$
14. $-32 \quad\quad -43$
15. $-11 \quad\quad -8$

Section 4.2

Add:

16. $-43 + (-18) + (-32)$
17. Find the sum of 87, 43, and -89.
18. $-76 + (-89) + 90$
19. If $x = -123 + (-83)$, then $x = $?
20. If $x = 200 + (-75)$, then $x = $?

Section 4.3

21. Find the difference of -103 and 29.
22. Subtract: $-75 - (-65)$

23. Subtract: $97 - (-83)$
24. Find the difference between 87 and 98.
25. If $k = -98 - 98$, then $k = $?

Section 4.4

26. Simplify: $-20a + (-12a) + 7a$
27. Combine like terms: $43x - 21x - 45x$
28. Combine like terms: $60y - (-60y)$
29. Simplify: $(2a - 5b + 6) + (4a + 3b + 9)$
30. Simplify: $(x^2 - 3x + 7) - (2x^2 + 4x + 9)$

Section 4.5 *Objective 1*

31. Find the product of -23 and -12.
32. Multiply: $(26)(-18)$
33. Find the product of -29 and 12.
34. Multiply: $(12)(-12)(-6)$
35. If $x = (-3)(-9)(-21)$, then $x = $?

Section 4.5 *Objective 2*

36. Find the quotient of -128 and -32.
37. Divide: $-189 \div 9$
38. Find the quotient of 324 and -36.
39. Divide: $\dfrac{-448}{-14}$
40. If $x = -136 \div 17$, then $x = $?

Section 4.6 *Objective 1*

41. Find the product of 12 and $11a$.
42. Multiply: $(13x)(5x)$
43. Find the product of $-18z$ and $-6z$.
44. Multiply: $2a(b + 6)$
45. Multiply: $(2x + 1)(x - 3)$

367

46. Find the quotient of $35xy$ and $7x$.

47. Divide: $-75y^2 \div 5y$

48. Find the quotient of $-121ab$ and $-11b$.

49. Divide: $144b^2 \div (-16b)$

50. Divide: $(6ab - 4ac + 10a) \div 2a$

Section 4.7 *Objective 1*

Perform the indicated operations.

51. $-9[-12 + 2(3 \cdot 4 - 15)]$

52. $12[-2(2 \cdot 5 - 13)^2] - 15$

53. $2(12 - 19)^2 + 3(-5 - 18)$

54. $5(-8)^2 + (-7)(-18 - 12)$

55. $7(3 - 8)^2 - 4(9 - 13)^2 - 3(4)$

Section 4.7 *Objective 2*

Find the average of the set of numbers.

56. $-14, -16, 12$

57. $-4, -7, 5, -13, -16$

58. $33, -43, 75, -81$

59. $47, -101, -341, -45, 153, 5$

60. $64, -82, -201, -432, 313, -28$

Section 4.8

Evaluate the following when $a = -4$, $b = -3$, and $c = 9$.

61. $2a - 3b + 4c$

62. $-3a + 5b - 2c$

63. $-8abc + a$

64. $ab + ac - bc$

65. Solve for y given $y = -8x - 5$ and $x = -2$.

Section 4.9

Solve:

66. $5x - 8 = 32$

67. $-3(a + 4) = 15$

68. $3(y + 6) = 24$

69. $2x + 5 - 3x = -2$

70. $4z - 9 - 2z = 7$

CHAPTER 4

Test

ANSWERS

1. _____

2. _____

3. _____

4. _____

5. _____

6. _____

7. _____

8. _____

9. _____

10. _____

11. _____

12. _____

13. _____

14. _____

1. Add: $-46 + 34 + (-23)$

2. Multiply: $(-11x)(-3x)(-5)$

3. Evaluate $4a + 5b - 6c$, when $a = -2$, $b = 8$, and $c = -11$.

4. Multiply: $-3x(2x - 7y - 13)$

5. What is the opposite of -88?

6. Subtract: $87 - 106$

7. Simplify: $-15b + 25b - 37b - 23b$

8. Divide: $-595 \div 35$

9. Simplify: $2[-3(-2 \cdot 12 - 34) - 4(5)]$

10. Evaluate: $3abc - 2b^2 - 4c$, if $a = -3$, $b = 7$, $c = -4$.

11. Solve: $5x + 19 = -36$

12. Insert $<$ or $>$ to make the statement true: $-7 \quad -19$

13. Add: $-68 + (-43) + (-34) + 55$

14. Find the difference between -44 and -56.

15. _____

15. Find the product of $-3x$ and $4x - 5y + 10$.

16. _____

16. Evaluate: $\dfrac{9x - xy^2 - 7y}{3x - 4y}$, when $x = -6$ and $y = -4$.

17. _____

17. Solve: $8a - 12 - 3a = -32 + 7a + 18$

18. _____

18. Combine: $(4b^2 - 6b - 11) + (-6b^2 + 3b - 45)$

19. _____

19. Combine: $(6a^2 - 7ab + 5b^2) - (9a^2 + 12ab + 21b^2)$

20. _____

20. Multiply: $-5(-7ab)(-4bc)$

21. _____

21. Is -4 a solution of $5x - 22 - 3x = -50$?

22. _____

22. Find the value of $|-135|$.

23. _____

23. Find the quotient of $-84m^2$ and $-7m$.

24. _____

24. Solve: $5d - 32 = -97$

25. _____

25. Multiply: $-18(-14)(-1)(11)$

26. _____

26. Simplify: $\dfrac{4(-5) - (-3)^3 + 5(-6) - 4}{4^2 - 5^2}$

27. _____

27. Find the average of -35, 67, -12, -43, 124, and -89.

28. _____

28. Solve: $7k - 35 - 9k = 3k - 2(k + 4)$

29. _____

29. The temperature in Tampa, Florida, hit a high of 102°F while the temperature in Fairbanks, Alaska, hit a high of -18°F. What is the difference between the temperatures?

30. _____

30. The Uptown Appliances sold 12 TV sets at a loss of $56 each during a Saturday sale. During the same sale they also sold five sets at a profit of $83 each. What is their profit or loss on the sale of the TVs, expressed as a signed number?

Good Advice for Studying

Learning to Learn Math

Learning mathematics is a building process. For example, if you have not mastered fractions, rational expressions are difficult to learn because they require an understanding of the rules for fractions. Therefore, if you are having difficulty with the current topic, you may not have mastered a previous skill that you need. It will be necessary to go back and learn/relearn this skill before you can continue.

Learning math also means learning not just skills, but how and where and when to *apply* the skills. For example: if it takes 16 gallons of gas to travel 320 miles, how many miles to the gallon are you getting? What skill would you use to solve this problem? (Answer: dividing) Reading the application problems and thinking of situations where you have used or could use these concepts helps integrate the concept into your experience.

Learning mathematics is learning something basic to daily life and to virtually every field of science and business. The examples in the book have a statement of the problem and a *strategy* for solving the problem. This strategy applies to several related problems. Learning mathematics is learning strategies to solve related problems.

The more you begin to appreciate mathematics as relevant to your life, the more you will see mathematics as worthy of your time, and the more committed you will feel to studying mathematics. Here is an activity that may give you fresh opportunities to see how much mathematics relates to your life. First, create a simple "web" or "map" with "math" at the center and spokes out from this center naming areas in life where math comes up—areas where math is useful. Capture as many areas as you can in a five-minute period.

Second, turn the page over and construct a second "web," but this time choose one of your "math areas" for the center and create spokes that capture subtopics or subheadings in this particular math area. Take another five minutes for this second web. Next, create a problem that you believe to be solvable, from your own experience, and that might be enticing for someone else to solve.

If you are working with a partner, trade problems and see if you can solve each others' problems. Talk about how you might approach the problems, and whether they are stated clearly and believably. Here are some criteria for a good problem:

- Enough data and information to solve the problem
- Clear statement of what you need to find
- Not too many questions included
- Appropriate reading level and clearly written
- Appears solvable and not too scary
- Makes the reader care about wanting to solve the problem

Going through this activity may help you become more aware of mathematics in your daily life, and give you a greater understanding of problems and problem-solving.

5

Rational Numbers: Fractions and Equations

Myron Taplin/Tony Stone Images ©

APPLICATION

One of the major applications of statistics is their value in predicting future occurrences. Before the future can be predicted, statisticians study what has happened in the past and look for patterns. If a pattern can be detected, and it is reasonable to assume that nothing will happen to interrupt the pattern, then it is a relatively easy matter to predict the future simply by continuing the pattern. Insurance companies, for instance, study the occurrences of traffic accidents among various groups of people. Once they have identified a pattern, they use this to predict future accident rates. These predictions then help to set insurance rates. When a group, such as teenage boys, is identified as having a higher incidence of accidents, their insurance rates are set higher.

Although predicting accident rates is a very complicated endeavor, there are other activities in which the patterns are relatively easy to find. Take, for instance, the act of rolling a die. The die has six sides marked from 1 to 6. Theoretically, each side has an equal chance of ending in the up position after a roll. Fill in the following table by rolling a die 120 times.

Side up	.	:·	·.	::	:·:	:::
Number of times rolled						

Theoretically, each side will be rolled the same amount as the others. Since you rolled the die 120 times and there are six possible outcomes, each side should come

373

up $\dfrac{120}{6}$ = 20 times. How close to 20 are your outcomes in the table? Compare your outcomes with others in your class. Did anyone come out exactly right? Why do you suppose there are differences in the tables?

Mathematicians like to express the relationships in this situation using the concept of *probability*, which is a measure of the likelihood of a particular event occurring. We describe the probability of an event with a fraction. The numerator of the fraction is the number of different ways the desired event can occur, and the denominator of the fraction is the total number of possible outcomes. So, the probability of rolling a two on the die is $\dfrac{1}{6}$ because there is only one way to roll a two but there are six possible outcomes when rolling a die. What is the probability of rolling a five? What is the probability of rolling a six? Nonmathematicians are more likely to express this relationship using the concept of odds. They would say that the odds of rolling a two are 1 in 6. This means that, for every six times you roll a die, you can expect one roll to show a two.

5.1   Rational Numbers (Fractions)

OBJECTIVES

1. Write a fraction to describe parts of a unit.

2. Select proper or improper fractions from a list of fractions.

3. Write the opposite of a fraction or the absolute value of a fraction.

4. Change improper fractions to mixed numbers.

5. Change mixed numbers to improper fractions.

VOCABULARY

A **numerical fraction** is a symbol of the form $\dfrac{a}{b}$, where a and b are numerals.

Rational numbers are numbers that can be written in the form $\dfrac{a}{b}$, where a and b are integers with $b \neq 0$.

The **numerator** of the fraction $\dfrac{a}{b}$ is the expression on top, a.

The **denominator** of the fraction $\dfrac{a}{b}$ is the expression on the bottom, b.

A **proper fraction** is one in which the absolute value of the numerator is less than the absolute value of the denominator.

An **improper fraction** is one in which the absolute value of the numerator is greater than or equal to the absolute value of the denominator.

A **mixed number** is the sum of a whole number and a fraction.

HOW AND WHY
Objective 1   **Write a fraction to describe parts of a unit.**

The integers we studied in earlier chapters are rational numbers with a denominator of 1. Thus $5 = \dfrac{5}{1}$, $-12 = -\dfrac{12}{1}$, and $0 = \dfrac{0}{1}$. Both integers and rational numbers are part of a larger system of numbers called "the real numbers." The real number system is a subject of algebra.

Symbols of the form $\dfrac{a}{b}$ are called fractions, which are used to describe parts of a unit. The expression, a, is called the *numerator* and the expression, b, is called the *denominator.* The denominator tells the number of parts in a unit and the numerator tells the number of the parts that are counted. In the fraction $\dfrac{6}{7}$, the unit is divided into 7 parts. Six of the 7 parts are indicated by $\dfrac{6}{7}$. The one unshaded part is indicated by $\dfrac{1}{7}$.

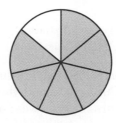

A fraction is often illustrated using a rectangle, a circle, or a number line. The number line in the figure below shows the fraction $\dfrac{6}{10}$ as the distance from 0 to the end of the arrow.

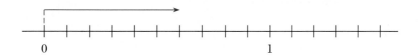

0   1

> *To write a fraction that describes the parts of a unit*
>
> **1.** Using a geometric figure:
>
> $$\frac{\text{numerator}}{\text{denominator}} = \frac{\text{number of shaded parts}}{\text{total number of parts in one unit}}$$
>
> **2.** Using a number line:
>
> $$\frac{\text{numerator}}{\text{denominator}} = \frac{\text{number of spaces between zero and end of arrow}}{\text{number of spaces between zero and one}}$$

Examples A–C   **Warm Ups A–C**

Directions: Write a fraction that describes the parts of the unit.

Strategy: For number lines, count the spaces between 0 and the arrow. This number is the numerator. Count the spaces between 0 and 1. This is the denominator. For a geometric figure, count the shaded parts and write the number in the numerator. Count the total number of parts in one unit and write the number in the denominator.

A.

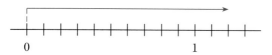

The fraction is $\dfrac{12}{10}$.   **There are 12 spaces between 0 and the arrow. There are 10 spaces between 0 and 1.**

A.

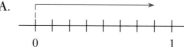

B.

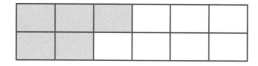

The fraction is $\dfrac{5}{12}$.   **There are 5 shaded parts out of a total of 12 parts.**

B.

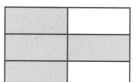

C.   At the beginning of the month, the High Pressure Tire Company had 53 used tires in stock. During the month they sold 40 of those tires. What fraction of the tires in stock were not sold?

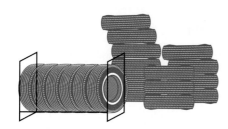

Solution: If 40 out of 53 tires were sold, 13 were not sold. So, the fraction of tires not sold is 13 out of 53 or $\dfrac{13}{53}$.

C.   In one month, the Hardly Used Car Agency sold 11 automobiles. Three of these were blue. What fraction of the cars sold were blue? What fraction of the cars sold were not blue?

HOW AND WHY
Objective 2   **Select proper or improper fractions from a list of fractions.**

In arithmetic and algebra we can take either of two points of view with respect to fractions.

1. A fraction is one of the ways to name a rational number. For example, $\dfrac{5}{8}$.

Answers to Warm Ups   A. $\dfrac{7}{8}$ B. $\dfrac{4}{6}$ C. $\dfrac{3}{11}$, $\dfrac{8}{11}$

2. A fraction is one of the ways to write a division problem. For example, $\dfrac{5}{8}$ is another way to show $5 \div 8$ or $8\overline{)5}$.

Every unit has at least one part; therefore, no fraction can have a denominator of zero. This is consistent with the rules for division that state that division by zero is undefined. This means that symbols such as $1 \div 0$, $\dfrac{3}{0}$, and $0\overline{)7}$ have no value.

In a fraction, if the absolute value of the numerator is smaller than the absolute value of the denominator, the fraction is called a *proper fraction*. Thus, $\dfrac{3}{8}$, $\dfrac{8}{16}$, and $\dfrac{-8}{14}$ are proper fractions. Otherwise the fraction is called an *improper fraction*. The fractions $\dfrac{8}{3}$, $\dfrac{6}{6}$, and $\dfrac{14}{1}$ are improper fractions. Proper fractions represent quantities less than one whole unit.

Example D	**Warm Up D**

Directions: Find which of the fractions are proper.

Strategy: Compare the absolute value of the numerator and denominator. If the absolute value of the numerator is smaller, the fraction is proper.

D. $\dfrac{5}{6}$, $\dfrac{-3}{6}$, $\dfrac{7}{-8}$, $\dfrac{24}{25}$, $\dfrac{-24}{25}$, $\dfrac{25}{25}$, $\dfrac{25}{30}$

The fractions $\dfrac{5}{6}$, $\dfrac{-3}{6}$, $\dfrac{7}{-8}$, $\dfrac{24}{25}$, $\dfrac{-24}{25}$, $\dfrac{25}{30}$ are proper. In each case the numerator has the smaller absolute value.

D. $\dfrac{9}{9}$, $\dfrac{-11}{19}$, $\dfrac{6}{9}$, $\dfrac{15}{9}$, $\dfrac{-7}{3}$, $\dfrac{2}{-7}$, $\dfrac{-1}{-1}$

HOW AND WHY
Objective 3

Write the opposite of a fraction or the absolute value of a fraction.

The opposite of $\dfrac{3}{4}$ can be written in three ways: $-\dfrac{3}{4}$, $\dfrac{-3}{4}$, or $\dfrac{3}{-4}$. The most common way is to write $-\dfrac{3}{4}$. On a measuring device such as a ruler (number line) or

---

thermometer, opposites are equal distances from zero but on opposite sides of zero, as shown in the following figure.

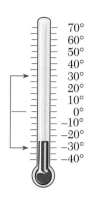

The absolute value of a fraction is determined the same way as the absolute value of an integer. (See Section 4.1.) Thus, $\left|-\dfrac{5}{6}\right| = \dfrac{5}{6}$, $\left|\dfrac{22}{19}\right| = \dfrac{22}{19}$, and $\left|-\dfrac{1}{12}\right| = \dfrac{1}{12}$.

Examples E–F   **Warm Ups E–F**

Directions: Find the opposite or absolute value.

Strategy: The opposite of a positive number is negative and vice versa. The absolute value of a positive number is positive, of zero is zero, and of a negative number is positive.

E. Find the opposites of $\dfrac{5}{9}$, $\dfrac{-8}{13}$, and $-\dfrac{17}{29}$.

$$-\left(\dfrac{5}{9}\right) = -\dfrac{5}{9} \quad\text{or}\quad \dfrac{-5}{9} \quad\text{or}\quad \dfrac{5}{-9}$$
The opposite of positive number is negative. The form $-\dfrac{5}{9}$ is preferred.

$$-\left(\dfrac{-8}{13}\right) = \dfrac{8}{13}$$

$$-\left(-\dfrac{17}{29}\right) = \dfrac{17}{29}$$

E. Find the opposites of $-\dfrac{3}{7}$, $\dfrac{-6}{13}$, and $\dfrac{15}{4}$.

F. Simplify: $\left|-\dfrac{8}{17}\right|$ and $\left|\dfrac{7}{19}\right|$

$$\left|-\dfrac{8}{17}\right| = \dfrac{8}{17}$$
The absolute value of a negative number is positive.

$$\left|\dfrac{7}{19}\right| = \dfrac{7}{19}$$
The absolute value of a positive number is positive.

F. Simplify: $\left|-\dfrac{17}{11}\right|$ and $\left|\dfrac{18}{23}\right|$

Answers to Warm Ups   E. $\dfrac{3}{7}, \dfrac{6}{13}, -\dfrac{15}{4}$ F. $\dfrac{17}{11}, \dfrac{18}{23}$

HOW AND WHY

Objective 4 **Change improper fractions to mixed numbers.**

A *mixed number* is the sum of a whole number and a fraction. The plus sign is usually omitted but understood. Hence, $8\frac{2}{3}$ $\left(\text{or } 8 + \frac{2}{3}\right)$ is a mixed number.

A mixed number, in fraction form, is always an improper fraction. This is because a mixed number is at least one entire unit (the whole number part). As a result, in fraction form, the numerator will be at least as large as the denominator.

To change an improper fraction to a mixed number, we divide the numerator by the denominator.

$$\frac{51}{3} = 51 \div 3 = 17 \qquad \text{and} \qquad \frac{51}{7} = 51 \div 7 = 7\overline{)51} = 7\frac{2}{7}$$
$$\underline{49}$$
$$2$$

> **To change an improper fraction to a mixed number**
>
> **1.** Divide the numerator by the denominator.
>
> **2.** If there is a remainder, write the whole number followed by $\dfrac{\text{remainder}}{\text{divisor}}$.

If an improper fraction is negative, such as $-\dfrac{9}{5}$, then the mixed number is also negative: $-\dfrac{9}{5} = -\left(\dfrac{9}{5}\right) = -\left(1\dfrac{4}{5}\right)$. To avoid using the parentheses, we prefer to leave such improper fractions in fraction form. In fact, improper fractions are often the more convenient and useful form in algebra.

CAUTION **Do not confuse the process of changing an improper fraction to a mixed number with "simplifying." Simplifying fractions is a totally different procedure. See Section 5.3.**

Examples G–I	**Warm Ups G–I**

Directions: Change to a mixed number.

Strategy: Divide the numerator by the denominator. The remainder is the numerator of the fraction part.

G. $\dfrac{144}{8}$	G. $\dfrac{133}{7}$

$$\frac{144}{8} = 8\overline{)144}^{\underline{18}}$$

Divide. The result can be $18\frac{0}{18}$, but the

$$\begin{array}{r} 8 \\ \hline 64 \\ 64 \\ \hline 0 \end{array}$$

fraction $\frac{0}{18}$ is seldom written.

$$\frac{144}{8} = 18$$

H. $\dfrac{147}{8}$

$$\frac{147}{8} = 8\overline{)147}^{\underline{18}}$$

Divide. The remainder, 3, is the numerator of the fraction part.

$$\begin{array}{r} 8 \\ \hline 67 \\ 64 \\ \hline 3 \end{array}$$

$$\frac{147}{8} = 18\frac{3}{8}$$

H. $\dfrac{138}{7}$

Calculator Example:

I. $\dfrac{847}{16}$

$$\frac{847}{16} = 52.9375$$

Divide, using a calculator. The whole number part is 52.

$$847 - 52(16) = 15$$

To find the remainder, multiply the whole number part, 52, by the divisor, 16, and subtract from 847.

$$\frac{847}{16} = 52\frac{15}{16}$$

Many calculators have a fraction key that will display the result in fraction form in one step. Consult your manual for instructions.

I. $\dfrac{755}{16}$

HOW AND WHY
Objective 5   **Change mixed numbers to improper fractions.**

To change a mixed number to an improper fraction, we reverse the steps we used to change the improper fraction to a mixed number.

$$\frac{38}{5} = 7 \text{ R } 3 = 7\frac{3}{5}$$

Divide 38 by 5 to get whole number 7 and remainder 3.

In reverse, for $7\dfrac{3}{5}$:

$5(7) + 3 = 38$ **Multiply 5(7) and add 3 to find the numerator.**

$\dfrac{38}{5}$ **Write the fraction with denominator 5.**

Written in one step, we have

$$7\frac{3}{5} = \frac{5(7) + 3}{5} = \frac{38}{5}$$

 To change a mixed number to an improper fraction

Multiply the denominator by the whole number, add the numerator, and write the sum over the denominator.

Every whole number can be written as a mixed number by adding a fraction with numerator 0. So $5 = 5\dfrac{0}{4}$. We use this later to aid in subtracting mixed numbers.

Example J	**Warm Up J**

Directions: Change to an improper fraction.

Strategy: Add the product of the whole number and the denominator to the numerator. Write the sum over the denominator.

J. $4\dfrac{5}{8}$	J. $10\dfrac{7}{12}$
$4\dfrac{5}{8} = \dfrac{8 \cdot 4 + 5}{8} = \dfrac{37}{8}$	

Answer to Warm Up J. $\dfrac{127}{12}$

Exercises 5.1

OBJECTIVE 1: *Write a fraction to describe parts of a unit.*

Write a fraction for the shaded part of the figure or as indicated by the arrow.

A.

1.

2.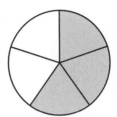

3.

4.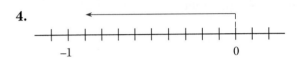

B.

5.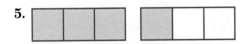

6.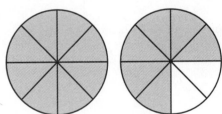

7.

8.

OBJECTIVE 2: *Select proper or improper fractions from a list of fractions.*

A.

List the proper fractions.

9. $\dfrac{4}{6}, \dfrac{5}{6}, \dfrac{6}{6}, \dfrac{7}{6}, \dfrac{8}{6}$

10. $\dfrac{3}{4}, \dfrac{6}{5}, \dfrac{12}{13}, \dfrac{14}{16}, \dfrac{17}{15}$

11. $\dfrac{8}{12}, \dfrac{2}{14}, \dfrac{10}{12}, \dfrac{8}{14}, \dfrac{11}{22}$

12. $\dfrac{6}{10}, \dfrac{9}{10}, \dfrac{10}{10}, \dfrac{9}{8}, \dfrac{10}{8}$

383

B.

13. $\dfrac{3}{3}, \dfrac{4}{5}, -\dfrac{15}{15}, \dfrac{16}{17}, -\dfrac{99}{100}, \dfrac{144}{144}, \dfrac{145}{144}, \dfrac{244}{144}$

14. $\dfrac{2}{3}, -\dfrac{3}{4}, \dfrac{4}{5}, -\dfrac{7}{7}, \dfrac{8}{9}, \dfrac{9}{10}, -\dfrac{11}{10}, \dfrac{12}{12}$

OBJECTIVE 3: *Write the opposite of a fraction or the absolute value of a fraction.*

A.

Write the opposite of each fraction.

15. $\dfrac{13}{19}, -\dfrac{17}{14}, \dfrac{-12}{17}$

16. $\dfrac{-8}{25}, \dfrac{27}{15}, -\dfrac{31}{27}$

Write the absolute value.

17. $\left|\dfrac{13}{17}\right|$

18. $\left|\dfrac{5}{-6}\right|$

B.

Write the opposite of each fraction.

19. $\dfrac{-16}{19}, -\dfrac{14}{3}, \dfrac{-16}{-13}$

20. $\dfrac{83}{127}, \dfrac{-45}{-97}, -\dfrac{41}{7}$

Write the absolute value.

21. $\left|-\dfrac{14}{27}\right|$

22. $\left|\dfrac{11}{25}\right|$

OBJECTIVE 4: *Change improper fractions to mixed numbers.*

A.

23. $\dfrac{5}{3}$

24. $\dfrac{7}{5}$

25. $\dfrac{11}{5}$

26. $\dfrac{19}{4}$

B.

27. $\dfrac{91}{25}$ **28.** $\dfrac{95}{31}$ **29.** $\dfrac{214}{41}$

30. $\dfrac{316}{11}$

OBJECTIVE 5: *Change mixed numbers to improper fractions.*

A.

31. $2\dfrac{1}{6}$ **32.** $3\dfrac{1}{7}$ **33.** 9

34. 12

B.

35. $47\dfrac{2}{3}$ **36.** $33\dfrac{1}{5}$ **37.** $106\dfrac{7}{8}$

38. $109\dfrac{4}{5}$

C.

Fill in the boxes so the statement is true. Explain your answer.

39. The fraction $\dfrac{0}{\square}$ is a proper fraction.   **40.** $55\dfrac{\square}{7} = \dfrac{387}{7}$

41. Find the error(s) in the statement: $\left|\dfrac{-7}{16}\right| = -\dfrac{7}{16}$. Correct the statement. Explain how you would avoid this error.

42. Find the error(s) in the statement: The opposite of $\dfrac{-6}{17}$ is $-\dfrac{-6}{-17}$. Correct the statement. Explain how you would avoid this error.

385

© 1997 Saunders College Publishing.

43. Draw a number line divided into parts of equal size and draw an arrow that shows $\dfrac{5}{8}$.

44. Draw a number line divided into parts of equal size and draw an arrow that shows $\dfrac{7}{6}$.

45. In a sample of bricks taken from a home construction, 25 were used and 32 were new. What fractional part were used bricks?

46. If a six-cylinder motor has one cylinder that is not firing, what fractional part of the cylinders are not firing?

47. The Adams' family monthly income is $2147. They spend $843 per month for housing. What fractional part of their income is spent for housing?

48. What fraction of a full tank of gas is indicated by the gas gauge?

49. A scale is marked with a whole number at each pound. What whole number mark is closest to the weight of $\dfrac{50}{16}$ lb?

50. A ruler is marked with a whole number at each centimeter. What whole number mark is closest to a length of $\dfrac{87}{10}$ cm?

51. Sue's construction company must place section barriers, each $\dfrac{1}{352}$ mile long, between two sides of a freeway. How many such sections are needed for $7\dfrac{31}{352}$ miles of freeway?

52. How many sections from Exercise 51 are needed for $37\dfrac{197}{352}$ miles of freeway?

 For Exercises 53–58, refer to the Chapter 5 Application. See page 373.

For Exercises 53–56. According to the Almanac of American People, the odds that a U.S. male will have significant hair loss at various ages are as follows:

Age	Odds
20–29	1 in 5
30–39	3 in 10
40–49	2 in 5
50–59	1 in 2
60–69	2 in 3
70–79	3 in 4

53. What is the probability that a 56-year-old man is balding?

54. What is the probability that a 37-year-old man is balding?

55. Which age group has the highest probability of being bald? Explain.

56. Make an estimate for the odds of being bald for men 80–89. Justify your estimate.

57. Suppose you and a friend each flip a coin. What are all the possible outcomes? What is the probability of getting two heads? What is the probability of getting two tails? What is the probability of getting one head and one tail?

58. What does it mean if the probability of an event is $\frac{3}{3}$? Is it possible for the probability of an event to be $\frac{5}{4}$? Explain.

STATE YOUR UNDERSTANDING

59. Tell why mixed numbers and improper fractions are both useful. Give examples of the use of each.

60. Explain how to change $\frac{34}{5}$ to a mixed number. Explain how to change $7\frac{3}{8}$ to an improper fraction.

61. Find some examples in books or newspapers of the use of opposites where positive and negative fractions could be used. Write a few sentences describing such use.

62. Explain why a proper fraction cannot be changed into a mixed number.

CHALLENGE

63. Write the whole number, 13, as an improper fraction: (a) using 117 as the numerator, (b) using 117 as the denominator.

64. Write the whole number, 14, as an improper fraction: (a) using 154 as the numerator, (b) using 154 as the denominator.

65. Write $\frac{47}{6}$ as a mixed number in five different ways. (*Hint:* The fraction part may be improper.)

66. Write $\frac{47}{8}$ as a mixed number in five different ways.

GROUP ACTIVITY

67. Many times data about a population is presented as a pie chart. Pie charts are based on fractions. If the fractions are easy to draw, a pie chart can be drawn quickly. Sometimes a drawing that is a reasonable estimate is adequate. A survey revealed that $\frac{1}{2}$ of the class preferred pepperoni pizza, $\frac{1}{4}$ of them preferred cheese-tomato pizza, $\frac{1}{8}$ preferred Canadian bacon pizza, and $\frac{1}{8}$ preferred sausage pizza. Make a pie chart that illustrates this survey. Explain your strategy.

68. According to a radio advertising survey, 77 out of 100 people in the United States say they listen to the radio daily. Make a pie chart that illustrates this survey. Explain your strategy.

69. Make a pie chart for a survey in which $\frac{2}{5}$ of the people surveyed preferred brand X, $\frac{1}{10}$ preferred brand Y, and half preferred brand Z. Explain your strategy.

MAINTAIN YOUR SKILLS (Sections 4.1 and 4.3)

70. If $x = -87$, then $-x = ?$

71. Find the value of $|-29|$.

72. Is -43 a solution of $x - 43 = 0$?

73. Fill in the box with $<$ or $>$: $72 \; \square \; -73$

74. Solve: $|x| = 57$.

5.2   Multiples, Factors, and Prime Factorization

OBJECTIVES

1. List multiples of a given whole number or determine whether a whole number is a multiple of a given number.

2. Determine whether a natural number is divisible by 2, 3, 5, or 10.

3. List all the factors of a given whole number.

4. Write the prime factorization of a given whole number.

VOCABULARY

A **multiple** of a number is the product of that number and a natural number.

A whole number is a **factor** or a **divisor** of another number if it divides the number evenly (with zero remainder).

A **prime number** is a whole number, greater than one, with exactly two different factors. These factors are 1 and the number itself.

A **composite number** is a whole number with more than two different factors.

The **prime factorization** of a number is the number written as the product of prime factors.

HOW AND WHY
Objective 1

List multiples of a given whole number or determine whether a whole number is a multiple of a given number.

A multiple of 6 is the product of 6 and a natural number. For instance, 78 is a multiple of 6 because 6(13) = 78. To list the multiples of 6, we can multiply 6 by each natural number.

Natural numbers:	1	2	3	4	5	6	...	12	...	16	...	41
Multiples of 6:	6	12	18	24	30	36	...	72	...	96	...	246

To determine if 852 is a multiple of 6 divide $\dfrac{852}{6}$. Since $\dfrac{852}{6} = 142$, then 6(142) = 852, and we conclude that 852 is a multiple of 6.

  To list the first five multiples of a given whole number

Multiply it by 1, 2, 3, 4, and 5.

  To determine whether a natural number is a multiple of a given number

Divide by the given number. If the remainder is zero, the dividend is a multiple of the number.

Example A	**Warm Up A**

Directions: List the first five multiples.

Strategy: Multiply the number by 1, 2, 3, 4, and 5.

A. 19	A. 23
The first five multiples of 19 are 1(19), 2(19), 3(19), 4(19), and 5(19); that is, 19, 38, 57, 76, and 95.	

Example B	**Warm Up B**

Directions: Determine whether the first number is a multiple of the second.

Strategy: Divide the first number by the second. If the remainder is zero, the first number is a multiple.

B. 198 and 9	B. 78 and 3
Since $198 \div 9 = 22$, we can say that **Divide 198 by 9.** 198 is a multiple of 9 **The remainder is 0.**	

HOW AND WHY
Objective 2

Determine whether a natural number is divisible by 2, 3, 5, or 10.

Divisibility tests can tell us whether one number will divide another (with a remainder of zero) before we do the division. These shortcuts are time-savers when finding the factors and divisors of a number.

Every even number is divisible by 2. That is, if the ones digit of a number is 0, 2, 4, 6, or 8, it is divisible by 2. So, 78, 126, 3002, and 56,734 are divisible by 2.

If the sum of the digits of a number is divisible by 3, then the number itself is divisible by 3. So, 78 ($7 + 8 = 15$), 126 ($1 + 2 + 6 = 9$), 3003 ($3 + 0 + 0 + 3 = 6$), and 53,001 ($5 + 3 + 0 + 0 + 1 = 9$) are divisible by 3.

If the ones digit of a number is 5 or 0, the number is divisible by 5. So, 110, 3245, and 77,770 are divisible by 5.

If the ones digit of a number is 0, the number is divisible by 10. The numbers 90, 960, 5010, and 62,220 are divisible by 10.

 To test for divisibility of a natural number by 2, 3, 5, or 10

If the ones-place digit is 0, 2, 4, 6, or 8, the number is divisible by 2.

If the sum of the digits is divisible by 3, the number is divisible by 3.

If the ones-place digit is 0 or 5, the number is divisible by 5.

If the ones-place digit is 0, the number is divisible by 10.

Examples C–D	**Warm Ups C–D**

Directions: Determine whether the number is divisible by 2, 3, 5, or 10.

Strategy: Use the divisibility tests.

C.	840		C. 720
	840 is divisible by 2.	The ones digit is 0.	
	840 is divisible by 3.	8 + 4 + 0, or 12, is divisible by 3.	
	840 is divisible by 5.	The ones digit is 0.	
	840 is divisible by 10.	The ones digit is 0.	

D.	7011		D. 1005
	7001 is *not* divisible by 2.	The ones digit is not 0, 2, 4, 6, or 8.	
	7011 is divisible by 3.	7 + 0 + 1 + 1, or 9, is divisible by 3.	
	7011 is *not* divisible by 5.	The ones digit is not 0 or 5.	
	7011 is *not* divisible by 10.	The ones digit is not 0.	

HOW AND WHY
Objective 3

List all the factors of a given whole number.

When a whole number is written as the product of two other whole numbers, each of the numbers in the product is a factor of the original whole number. The number 6 is a factor of 54 since 6(9) = 54. We can also say that 6 is a *divisor* of 54 since 54 ÷ 6 = 9. Furthermore, 54 is *divisible* by 6. Every multiple of 6 is divisible by 6 and has a divisor of 6.

It is possible to find all the factors (or divisors) of a number, say 250, by guess and check but it can take a long time. A systematic procedure will save time.

First: Chart all the natural numbers from 1 to the first number whose square is larger than 250. We stop at 16 because 16^2 is 256, which is larger than 250. Any factor of 250 that is larger than 16 will be paired with a factor that is smaller than 16. For example, 25, which is larger than 16, is paired with 10, and 10 is smaller than 16.

1	6	11	16	We stop at 16 since $16^2 = 256$ which is larger
2	7	12		than 250. Checking all the factors less than 16
3	8	13		guarantees that we will find all the factors.
4	9	14		
5	10	15		

Second: Divide each number in the chart into 250.

1 · 250	6̶	1̶1̶	1̶6̶	Cross out the numbers that do
2 · 125	7̶	1̶2̶		not divide evenly.
3̶	8̶	1̶3̶		
4̶	9̶	1̶4̶		
5 · 50	10 · 25	1̶5̶		

Third: List all the factors of 250. The factors are 1, 2, 5, 10, 25, 50, 125, and 250. These are the only whole numbers that will divide into 250 "evenly."

 To list all the factors of a number

1. List all the counting numbers from one to the first number whose square is larger than the whole number.
2. Test whether each number on the list is a divisor.
3. If not, cross the number off the list.
4. If so, write the indicated product of the two factors.
5. List all of the factors.

If there are only two factors in the list, the number one (1) and the number itself, the number is called a *prime number.* The whole numbers less than 50 that are prime numbers are 2, 3, 5, 7, 11, 13, 17, 19, 23, 29, 31, 37, 41, 43, and 47. It is helpful to memorize, at least, the prime numbers less than 20. Notice that the only even number in this short list of prime numbers is 2.

If a number has *more* than two factors, the number is called a *composite number.* The whole numbers 4, 6, and 32 are composite numbers.

The whole numbers 0 and 1 are neither prime nor composite. The first prime number is 2, since 2 and 1 are the only factors of 2 ($2 = 2 \cdot 1$). The next whole number, 3, is prime, and the one after, 4, is composite. The number 4 has three factors, 1, 2, and 4.

Example E   **Warm Up E**

Directions: List all the factors.

Strategy: Chart all whole numbers whose square is less than the number. Divide the number by each of these.

E.   List all factors of 180. | E.   List all the factors of 208.

$1 \cdot 180$	$6 \cdot 30$	$1\!\!1$
$2 \cdot 90$	7	$12 \cdot 15$
$3 \cdot 60$	8	$1\!\!3$
$4 \cdot 45$	$9 \cdot 20$	$1\!\!4$
$5 \cdot 36$	$10 \cdot 18$	

Make a list of the possible factors. Stop at 14 since $14^2 = 196$, which is larger than 180.

The factors of 180 are 1, 2, 3, 4, 5, 6, 9, 10, 12, 15, 18, 20, 30, 36, 45, 60, 90, and 180.

Answer to Warm Up   E. 1, 2, 4, 8, 13, 16, 26, 52, 104, 208

Example F   **Warm Up F**

Directions: Is the number prime or composite?

Strategy: Look for the factors. As soon as a factor other than one and the number itself is found, you know that the number is composite.

F.   Is 101 prime or composite?

   1 · 101 $\cancel{6}$ $\cancel{11}$ **Make a list of the possible factors. Stop at 11 since 11^2 is larger than 101.**
   $\cancel{2}$ $\cancel{7}$
   $\cancel{3}$ $\cancel{8}$
   $\cancel{4}$ $\cancel{9}$
   $\cancel{5}$ $\cancel{10}$

   Since 101 has only two factors it is a prime number.

F.   Is 71 prime or composite?

HOW AND WHY
Objective 4   **Write the prime factorization of a given whole number.**

The *prime factorization* of a number is the number written as the product of only prime factors. Thus, $21 = 3 \cdot 7$ and $30 = 2 \cdot 3 \cdot 5$ are prime factorizations.

  To write the prime factorization of a composite number

   1. Divide the number repeatedly by prime numbers until the quotient is one.
   2. Write the number as the product of these primes.

To write the prime factorization of 360, start by dividing 360 by 2. Then divide the quotient by 2, and so on. When the quotient is not divisible by 2, divide by the next prime, 3, and continue until the quotient is 1.

$2\overline{)360}$
$2\overline{)180}$
$2\overline{)\ 90}$
$3\overline{)\ 45}$
$3\overline{)\ 15}$
$5\overline{)\ \ 5}$
$\qquad 1$

So, $360 = 2 \cdot 2 \cdot 2 \cdot 3 \cdot 3 \cdot 5 = 2^3 \cdot 3^2 \cdot 5$. The *Fundamental Theorem of Arithmetic* states that there is only one set of prime factors for a composite number. That is to say, the prime factorization of a number is unique.

Example H shows the "tree method" for prime factorization.

Examples G–H   **Warm Ups G–H**

Directions: Write the prime factorization.

Strategy: Divide by prime numbers repeatedly until the quotient is 1. The number is the product of these prime factors.

G. Prime factor 848.

$$2)\overline{848}$$
$$2)\overline{424}$$
$$2)\overline{212}$$
$$2)\overline{106}$$
$$53)\overline{\ 53}$$
$$1$$

$848 = 2 \cdot 2 \cdot 2 \cdot 2 \cdot 53 = 2^4 \cdot 53.$

G. Prime factor 459.

H. Use the tree method to prime factor 68.

Strategy: The "tree method" uses "factor branches" with *any* two factors whose product is 68. Draw new branches using factors of numbers at the end of each branch. The branch stops splitting when it ends in a prime number. The prime factorization is the product of the primes at the ends of the branches.

H. Use the tree method to prime factor 855.

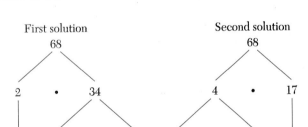

First solution   Second solution

$68 = 2 \cdot 2 \cdot 17 = 2^2 \cdot 17.$

Exercises 5.2

OBJECTIVE 1: *List multiples of a given whole number or determine whether a whole number is a multiple of a given number.*

List the first five multiples of each number.

A.

1. 2   **2.** 4   **3.** 5   **4.** 7   **5.** 11   **6.** 12

B.

7. 25   **8.** 22   **9.** 135   **10.** 142

A.

Is each number a multiple of 4?

11. 38   **12.** 54   **13.** 48   **14.** 52

B.

15. 116   **16.** 192   **17.** 290   **18.** 358

19. List all the multiples of 8 from 72 to 112.   **20.** List all the multiples of 7 from 252 to 280.

OBJECTIVE 2: *Determine whether a natural number is divisible by 2, 3, 5, or 10.*

Determine whether the number is divisible by 2, 3, 5, or 10.

A.

21. 30   **22.** 40   **23.** 135

24. 145   **25.** 370   **26.** 375

B.

27. 7880   **28.** 8100   **29.** 11,305   **30.** 22,305

OBJECTIVE 3: *List all the factors of a given whole number.*

List all the factors.

A.

31. 12   **32.** 15   **33.** 13   **34.** 29

B.

35. 36   **36.** 104   **37.** 111   **38.** 95   **39.** 405   **40.** 615

A.

State whether the number is prime or composite.

41. 29   **42.** 39   **43.** 57   **44.** 67

B.

45. 497   **46.** 597   **47.** 797   **48.** 997

OBJECTIVE 4: *Write the prime factorization of a given whole number.*

Write the prime factorization.

A.

49. 63   **50.** 66   **51.** 44   **52.** 99

B.

53. 121

54. 143

55. 190

56. 192

57. 1536

58. 1539

C.

Fill in the boxes with a single digit so the statement is true. Explain your answer.

59. 37□1 is a multiple of 3

60. 7 is a factor of 5□1

61. Find the error(s) in the statement: Six is a multiple of 66. Correct the statement. Explain how you would avoid this error.

62. Find the error(s) in the statement: The prime factorization of 221 is 1 · 221. Correct the statement. Explain how you would avoid this error.

63. A teacher assigns the problems from 1 to 52 that are multiples of 4. Which problems should the students work?

64. A teacher assigns the problems from 1 to 52 that are multiples of 6. Which problems should the students work?

65. The Office Group has a discount price on desk chairs for retail outlets that buy the chairs in multiples of 16. Does the outlet that places an order for 176 chairs get the discount price?

66. Canyon Cars will receive a bonus of $1500 for each group of 23 new cars it sells during the months of January–March. The chart shows sales through March 25. What bonus can the dealership expect at this time and how many more cars must they sell to gain the next level in the bonus system?

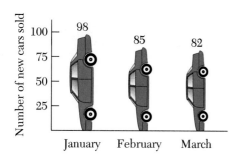

67. What is the first year in the 21st century that is a multiple of 75?

68. What was the last year in the 19th century that was a multiple of 63?

69. The years 1993 and 1997 are prime numbers. What is the next year that is a prime number?

70. What is the first year of the 21st century that will be a prime number?

71. To play the lottery, Jorge bets using the smallest four-digit prime number. What number does Jorge play?

STATE YOUR UNDERSTANDING

72. Write a short statement to explain why *every* multiple of 12 is also a multiple of 6.

73. One of the factors of 126 is 9. The number 9 is also a divisor of 126. Every factor of a number is also a divisor of that number. Explain in your own words why you think we have two different words, "factor" and "divisor" for such numbers.

74. What is a prime factor of a number? How many prime factors does a number have? Explain how to prime factor 4500.

75. Is the number 2 the only even prime number? Justify your answer. Explain why the numbers 0 and 1 are *not* prime numbers.

CHALLENGE

76. Find the largest number less than 5000 that is a multiple of 6, 9, and 17.

77. How many multiples of 3 are there between 1000 and 5000?

78. What is special about numbers that have an *odd* number of factors?

79. A number that is divisible by both 2 and 5 is also divisible by 10. Write divisibility tests for 6 and 15.

80. Find divisibility tests for 4, 8, and 9.

GROUP ACTIVITY

81. Study the answer to Exercise 76. Find the number of multiples of 6 between 1000 and 5000. Can your group discover a short procedure or formula for the number of multiples of 13 between 1000 and 5000? Explain.

82. Go to the library or resource center to find the value of the largest prime number that you can. Divide up the search so that one person checks an encyclopedia, another checks math books, another checks journals and magazines. Look through any other sources you have available. See if you can find out why people do computer searches for prime numbers. Report your finding to the class.

MAINTAIN YOUR SKILLS (Sections 4.5 and 4.6)

83. Find the product of -2, 111, and 18.

84. Find the product of $-3x$, $-44x$, and $-17x$.

85. Find the quotient of -1176 and 14.

86. Find the quotient of $-153x^3$ and $-17x$.

87. Find the quotient of $468xy^2$ and $-6xy$.

401

5.3 Multiplying and Dividing Fractions

OBJECTIVES

1. Simplify a fraction.

2. Build a fraction by finding the missing numerator.

3. Multiply numeric and algebraic fractions.

4. Divide numeric and algebraic fractions.

VOCABULARY

Equivalent fractions are fractions that are different names for the same number.

Simplifying a fraction means renaming a fraction by dividing both the numerator and denominator by a common factor.

Building a fraction means renaming a fraction by multiplying both the numerator and denominator by a common factor.

A fraction is **completely simplified** when its numerator and denominator have no common factors other than one.

When two or more fractions have the same denominator, we say they have a **common denominator**.

Two rational numbers are **reciprocals** if they have a product of 1. Informally, the reciprocal of a fraction is the result of switching the numerator and denominator.

HOW AND WHY
Objective 1

Simplify a fraction.

Fractions are *equivalent* if they represent the same quantity. The fractions, $\frac{4}{8}$ and $\frac{5}{10}$ both represent $\frac{1}{2}$ of an original quantity.

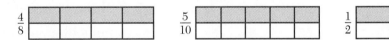

The *Basic Principle of Fractions* states that a common factor can be eliminated from or introduced into both the numerator and denominator of a fraction. The result is an equivalent fraction.

Basic Principle of Fractions

$$\frac{A}{B} = \frac{AC}{BC} \qquad \text{or} \qquad \frac{AC}{BC} = \frac{A}{B}$$

where $B \neq 0$ and $C \neq 0$.

A fraction is *simplified* after both the numerator and the denominator are divided by a common factor. A fraction is *completely simplified* after both numerator and denominator are divided by their *greatest* common factor. For instance, $\frac{5}{10} = \frac{1}{2}$. To show that $\frac{5}{10} = \frac{1}{2}$, we can factor numerator and denominator and eliminate the common factors. Or, we can divide both numerator and denominator by 5, which is the greatest common factor of 5 and 10

$$\frac{5}{10} = \frac{\cancel{5}^1 \cdot 1}{\cancel{5}_1 \cdot 2} = \frac{1}{2} \qquad \text{or} \qquad \frac{5}{10} = \frac{5 \div 5}{10 \div 5} = \frac{1}{2}$$

> ### ▶ *To simplify a fraction completely*
>
> Eliminate all common factors (other than one) in the numerator and the denominator.

Algebraic fractions are simplified in the same way. We eliminate common numerical and variable factors. Thus,

$$\frac{24xy}{40} = \frac{12xy}{20} = \frac{6xy}{10} = \frac{3xy}{5} \qquad \begin{array}{l}\textbf{Divide numerator and}\\ \textbf{denominator by 2 repeatedly.}\end{array}$$

and

$$\frac{24ab}{12a} = \frac{\cancel{12}^1 \cdot \cancel{a}^1 \cdot 2b}{\cancel{12}_1 \cdot \cancel{a}_1 \cdot 1} = 2b \qquad \textbf{Eliminate the common factors.}$$

CAUTION **Simplifying fractions is done by dividing, the inverse of multiplying. Therefore, fractions that contain plus or minus signs cannot be simplified by canceling across the plus or minus signs. For example,**

Does 4 = 7?

$$4 = \frac{8}{2} = \frac{6 + \cancel{2}^1}{\cancel{2}_1} = \frac{7}{1} = 7 \qquad \begin{array}{l}\textbf{This means that fractions like } \dfrac{x+4}{4}\\[2mm] \textbf{cannot be simplified. This is a}\\ \textbf{common error to be avoided.}\end{array}$$

If the largest common factor cannot be found easily, find the prime factorization of the numerator and denominator. See Example B.

Examples A–C	**Warm Ups A–C**

Directions: Simplify.

Strategy: Eliminate the common factors.

A. Simplify: $-\dfrac{16}{24}$

$$-\frac{16}{24} = -\frac{\cancel{8}^1 \cdot 2}{\cancel{8}_1 \cdot 3} = -\frac{2}{3}$$

$$-\frac{16}{24} = -\frac{16 \div 8}{24 \div 8} = -\frac{2}{3}$$

Eliminate the common factor of 8, or divide numerator and denominator by 8. Or, you could divide by 2 three times.

A. Simplify: $-\dfrac{18x}{45}$

Answer to Warm Up A. $-\dfrac{2x}{5}$

B. Simplify: $\dfrac{56a^2b^2}{84ab}$

B. Simplify: $\dfrac{18c^2d}{12cd^2}$

$$\frac{\cancel{2}\cdot\cancel{2}\cdot 2\cdot\cancel{7}\cdot\cancel{a}\cdot a\cdot\cancel{b}\cdot b}{\cancel{2}\cdot\cancel{2}\cdot 3\cdot\cancel{7}\cdot\cancel{a}\cdot\cancel{b}}=\frac{2ab}{3}$$

Find the prime factors and divide them out.

C. Morris washes cars on Saturday to earn extra money. One Saturday he has 12 cars to wash. After he has washed eight of them, what fraction of the total has he washed? Write the answer as a simplified fraction.

C. In Example C, if Morris had washed only four cars, what fraction of the total had he washed?

Solution:

$$\frac{\text{number of cars washed}}{\text{total number of cars}}=\frac{8}{12}$$

$$=\frac{2\cdot\cancel{4}}{3\cdot\cancel{4}}$$

Eliminate the common factor of 4.

$$=\frac{2}{3}$$

Morris has washed $\dfrac{2}{3}$ of the cars.

HOW AND WHY
Objective 2

Build a fraction by finding the missing numerator.

Building a fraction means renaming the fraction by multiplying both the numerator and denominator by a common factor. A common denominator is needed to add and subtract fractions of arithmetic and algebra. To accomplish this we "build fractions" to a common denominator. This is the opposite of "simplifying" fractions.

Simplifying a Fraction	Building a Fraction
$\dfrac{6}{10}=\dfrac{3\cdot\cancel{2}}{5\cdot\cancel{2}}=\dfrac{3}{5}$	$\dfrac{3}{5}=\dfrac{3\cdot 2}{5\cdot 2}=\dfrac{6}{10}$
$\dfrac{14ab}{21b}=\dfrac{\cancel{7}\cdot 2\cdot a\cdot\cancel{b}}{\cancel{7}\cdot 3\cdot\cancel{b}}=\dfrac{2a}{3}$	$\dfrac{2a}{3}=\dfrac{2a\cdot 7b}{3\cdot 7b}=\dfrac{14ab}{21b}$

The following chart shows five fractions built to equivalent fractions.

	Multiply Numerator and Denominator by					
	2	3	4	6	10	15
$\dfrac{3}{5}$	$\dfrac{6}{10}=$	$\dfrac{9}{15}=$	$\dfrac{12}{20}=$	$\dfrac{18}{30}=$	$\dfrac{30}{50}=$	$\dfrac{45}{75}$
$\dfrac{1}{2}$	$\dfrac{2}{4}=$	$\dfrac{3}{6}=$	$\dfrac{4}{8}=$	$\dfrac{6}{12}=$	$\dfrac{10}{20}=$	$\dfrac{15}{30}$

(Chart continues next page)

	Multiply Numerator and Denominator by					
	2	**3**	**4**	**6**	**10**	**15**
$\dfrac{5}{8}$	$\dfrac{10}{16}=$	$\dfrac{15}{24}=$	$\dfrac{20}{32}=$	$\dfrac{30}{48}=$	$\dfrac{50}{80}=$	$\dfrac{75}{120}$
$\dfrac{4}{9}$	$\dfrac{8}{18}=$	$\dfrac{12}{27}=$	$\dfrac{16}{36}=$	$\dfrac{24}{54}=$	$\dfrac{40}{90}=$	$\dfrac{60}{135}$
$\dfrac{7}{3}$	$\dfrac{14}{6}=$	$\dfrac{21}{9}=$	$\dfrac{28}{12}=$	$\dfrac{42}{18}=$	$\dfrac{70}{30}=$	$\dfrac{105}{45}$

If we want a fraction equivalent to $\dfrac{2}{3}$ with a denominator of 39, we write $\dfrac{2}{3}=\dfrac{?}{39}$. Since $39 \div 3 = 13$, we multiply the numerator and the denominator of $\dfrac{2}{3}$ by 13.

$$\frac{2}{3} = \frac{2 \cdot 13}{3 \cdot 13} = \frac{26}{39} \qquad \text{We say that } \frac{2}{3} \text{ has been "built" to } \frac{26}{39}.$$

 To build a fraction by finding a missing numerator

1. Divide the required denominator by the denominator of the given fraction.
2. Multiply the numerator and denominator by the result found in Step 1.

Examples D–E   **Warm Ups D–E**

Directions: Build each fraction by finding the missing numerator.

Strategy: Divide the required denominator by the given denominator. Multiply this quotient times the given numerator.

D. $\dfrac{3}{5} = \dfrac{?}{85}$

$85 \div 5 = 17$ **Divide 85 by 5.**

$\dfrac{3}{5} = \dfrac{3(17)}{5(17)}$ **Multiply the numerator and denominator by 17.**

$= \dfrac{51}{85}$ **The product, 51, is the missing numerator.**

D. $\dfrac{6}{7} = \dfrac{?}{84}$

E. $\dfrac{13}{11}\,st = \dfrac{?}{121}\,st$

$121 \div 11 = 11$

$\dfrac{13}{11}\,st = \dfrac{13 \cdot 11}{11 \cdot 11}\,st$ **Building is the reverse of simplifying. A built-up fraction can be simplified to the original fraction.**

E. $\dfrac{23}{16}\,cd = \dfrac{?}{128}\,cd$

Answers to Warm Ups   D. $\dfrac{72}{84}$ E. $\dfrac{184}{128}\,cd$ or $\dfrac{184cd}{128}$

$$= \frac{143}{121} st$$

or

$$= \frac{143st}{121}$$ **The term can also be written in fraction form.**

HOW AND WHY

Objective 3   **Multiply numeric and algebraic fractions.**

What is $\frac{1}{2}$ of $\frac{1}{3}$ or $\frac{1}{2} \cdot \frac{1}{3}$? See the following figure. The rectangle is divided into three parts. One part, $\frac{1}{3}$, is shaded blue. To find $\frac{1}{2}$ of the shaded third, divide each of the thirds into two parts (halves). The second part of the figure shows the rectangle divided into six parts. So $\frac{1}{2}$ of the shaded third is $\frac{1}{6}$ of the rectangle, which is shaded black.

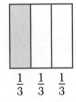

$$\frac{1}{2} \text{ of } \frac{1}{3} = \frac{1}{2} \cdot \frac{1}{3} = \frac{1}{6} = \frac{\text{number of parts shaded black}}{\text{total number of parts}}$$

The shortcut is to multiply the numerators and multiply the denominators.

$$\frac{12}{35} \cdot \frac{25}{18} = \frac{12(25)}{35(18)} = \frac{300}{630}$$

$$= \frac{300 \div 30}{630 \div 30} \qquad \textbf{Simplify by dividing both numerator and denominator by 30.}$$

$$= \frac{10}{21}$$

The product can be found more quickly by simplifying before multiplying.

$$\frac{12}{35} \cdot \frac{25}{18} = \frac{\overset{2}{\cancel{12}}}{35} \cdot \frac{25}{\underset{3}{\cancel{18}}} \qquad \textbf{Simplify by dividing 12 and 18 by 6.}$$

$$= \frac{\overset{2}{\cancel{12}}}{\underset{7}{\cancel{35}}} \cdot \frac{\overset{5}{\cancel{25}}}{\underset{3}{\cancel{18}}} \qquad \textbf{Simplify by dividing 25 and 35 by 5.}$$

$$= \frac{10}{21} \qquad \textbf{Multiply the numerators and denominators.}$$

C A U T I O N   **Simplifying before doing the operation only works for multiplication since it is based upon the multiplication property of one. It *does not* work for addition, subtraction, or division.**

We multiply algebraic fractions by the same method.

▶ *To multiply numeric or algebraic fractions*

1. Write the product of the numerators over the product of the denominators.

2. Simplify

To find the product of algebraic fractions, regroup the coefficients and the variables using the commutative and associative properties of multiplication. Here are two examples.

$$\frac{44}{5}xy \cdot \frac{3}{11} = \left(\frac{44}{5} \cdot \frac{3}{11}\right)(xy) = \frac{4 \cdot \overset{1}{\cancel{11}} \cdot 3}{5 \cdot \underset{1}{\cancel{11}}}(xy) = \frac{12}{5}xy \quad \text{or} \quad \frac{12xy}{5}$$

$$\left(\frac{4a}{21}\right)\left(\frac{35b}{2}\right) = \left(\frac{4}{21} \cdot \frac{35}{2}\right)(ab) = \frac{\overset{1}{\cancel{2}} \cdot 2 \cdot 5 \cdot \overset{1}{\cancel{7}}}{3 \cdot \underset{1}{\cancel{7}} \cdot \underset{1}{\cancel{2}}}(ab) = \frac{10}{3}ab \quad \text{or} \quad \frac{10ab}{3}$$

If any of the factors are mixed numbers, change them to improper fractions and then multiply. See Example G.

Examples F–I   **Warm Ups F–I**

Directions: Simplify by multiplying.

Strategy: Simplify if possible and write the product of numerators over the product of the denominators.

F.   Multiply: $\dfrac{1}{3}\left(-\dfrac{3}{7}\right)\left(\dfrac{5}{8}\right)$ | F.   Multiply: $\dfrac{1}{4}\left(-\dfrac{3}{7}\right)\left(-\dfrac{8}{5}\right)$

$\dfrac{1}{3}\left(-\dfrac{3}{7}\right)\left(\dfrac{5}{8}\right) = -\dfrac{1}{\underset{1}{\cancel{3}}} \cdot \dfrac{\overset{1}{\cancel{3}}}{7} \cdot \dfrac{5}{8}$ The product of two positives and one negative is negative.

$= -\dfrac{5}{56}$ Simplify.

G.   Multiply: $\left(5\dfrac{1}{3}\right)\left(4\dfrac{1}{2}\right)$ | G.   Multiply: $\left(4\dfrac{1}{3}\right)\left(1\dfrac{1}{8}\right)$

$\left(5\dfrac{1}{3}\right)\left(4\dfrac{1}{2}\right) = \dfrac{16}{3} \cdot \dfrac{9}{2}$ Write the mixed numbers as fractions.

$= \dfrac{16 \cdot 9}{3 \cdot 2}$ Write products of numerators and denominators.

Answers to Warm Ups   F. $\dfrac{6}{35}$ G. $\dfrac{39}{8}$ or $4\dfrac{7}{8}$

$$= \frac{\overset{1}{\cancel{2}} \cdot 8 \cdot \overset{1}{\cancel{3}} \cdot 3}{\underset{1}{\cancel{3}} \cdot \underset{1}{\cancel{2}}} \qquad \textbf{Simplify}$$

$$= \frac{24}{1} = 24 \qquad \textbf{Multiply.}$$

H. Multiply: $\left(-\dfrac{12cd}{25}\right)\left(\dfrac{35}{6d}\right)$

$$\left(-\frac{12cd}{25}\right)\left(\frac{35}{6d}\right) = -\frac{2 \cdot \overset{1}{\cancel{6}} \cdot c \cdot \overset{1}{\cancel{d}} \cdot \overset{1}{\cancel{5}} \cdot 7}{\underset{1}{\cancel{5}} \cdot 5 \cdot \underset{1}{\cancel{6}} \cdot \underset{1}{\cancel{d}}} \qquad \begin{array}{l}\textbf{Write products and}\\ \textbf{simplify.}\end{array}$$

$$= -\frac{14c}{5} \quad \text{or} \quad -\frac{14}{5}c \qquad \textbf{Multiply.}$$

I. In one year, $\dfrac{7}{8}$ of all cars sold by Trustis Used Cars had automatic transmissions. Of the cars sold with automatic transmissions, $\dfrac{1}{70}$ had to be repaired before they were sold. What fraction of the total number of cars sold had automatic transmissions and had to be repaired?

Solution: Multiply the fraction of cars that had automatic transmissions by the fraction that had repairs.

$$\frac{7}{8} \cdot \frac{1}{70} = \frac{\overset{1}{\cancel{7}}}{8} \cdot \frac{1}{\underset{10}{\cancel{70}}}$$

$$= \frac{1}{80}$$

Therefore, $\dfrac{1}{80}$ of the total number of cars sold had automatic transmissions and required repair.

H. Multiply: $\dfrac{15a}{28}\left(\dfrac{36b}{5a}\right)$

I. In one year, $\dfrac{2}{3}$ of all the tires sold by Tyre Factory were highway tread tires. If $\dfrac{1}{40}$ of the highway treads had be replaced, what fractional part of the tires sold were highway treads that needed to be replaced?

HOW AND WHY

Objective 4 **Divide numeric and algebraic fractions.**

Suppose we would like to divide a class of 24 students into two equal groups. Mathematically we do the operation $24 \div 2 = 12$ to determine the size of each group. Another way to approach the same problem is to find the size of half the class. Mathematically we do the operation $24\left(\dfrac{1}{2}\right) = \dfrac{24}{2} = 12$. So, dividing by 2 and multiplying by $\dfrac{1}{2}$ bring about the same result. We use this principle to describe the general procedure for dividing fractions. The quotient of two fractions $\dfrac{4}{5}$ and $\dfrac{1}{10}$ can be found by multiplying $\dfrac{4}{5}$ by the reciprocal of $\dfrac{1}{10}$.

Answers to Warm Ups   H. $\dfrac{27b}{7}$ or $\dfrac{27}{7}b$ I. $\dfrac{1}{60}$

The *reciprocal* of a fraction is the fraction inverted or "flipped." Technically, the reciprocal of a rational number is the number you multiply by to get a product of 1. For example, the reciprocal of $\frac{1}{10}$ is $\frac{10}{1}$ because $\frac{1}{10} \cdot \frac{10}{1} = 1$. So

$$\frac{4}{5} \div \frac{1}{10} = \frac{4}{5} \cdot \frac{10}{1} = \frac{40}{5} = 8$$

The result can be verified by multiplication. The product of the divisor, $\frac{1}{10}$, and the quotient, 8, should be $\frac{4}{5}$.

Check: $\frac{1}{10} \cdot 8 = \frac{8}{10} = \frac{4}{5}$

We divide algebraic fractions in the same way as we divide numeric fractions.

▶ *To divide numeric or algebraic fractions*

Multiply the first number by the reciprocal of the second number. In other words, invert the divisor and multiply.

$$\frac{2}{3}xy \div \frac{1}{4}x = \frac{2xy}{3} \div \frac{x}{4}$$ Write the expressions as fractions.

$$= \frac{2xy}{3} \cdot \frac{4}{x}$$ Invert the divisor and multiply.

$$= \frac{2\overset{1}{\cancel{x}}y}{3} \cdot \frac{4}{\underset{1}{\cancel{x}}}$$ Simplify.

$$= \frac{8y}{3} \quad \text{or} \quad \frac{8}{3}y$$ Simplify.

Examples J–K   **Warm Ups J–K**

Directions: Simplify by dividing.

Strategy: Invert the divisor and multiply.

J. Divide: $\frac{8}{25} \div \frac{16}{25}$ J. Divide: $\frac{13}{21} \div \left(-\frac{7}{21}\right)$

$\frac{8}{25} \div \frac{16}{25} = \frac{8}{25} \cdot \frac{25}{16}$ **Invert the divisor.**

$= \frac{\overset{1}{\cancel{8}}}{\underset{1}{\cancel{25}}} \cdot \frac{\overset{1}{\cancel{25}}}{\underset{2}{\cancel{16}}}$ **Simplify.**

$= \frac{1}{2}$ **Multiply.**

Answer to Warm Up J. $-\frac{13}{7}$

K.   Divide: $\dfrac{23}{4}xy \div \dfrac{117}{10}y$

$\dfrac{23}{4}xy \div \dfrac{117}{10}y = \dfrac{23xy}{4} \div \dfrac{117y}{10}$ **Write both terms as fractions.**

$= \dfrac{23xy}{4} \cdot \dfrac{10}{117y}$ **Invert the divisor.**

$= \dfrac{23 \cdot x \cdot \overset{1}{\cancel{y}}}{\underset{2}{\cancel{4}}} \cdot \dfrac{\overset{5}{\cancel{10}}}{117 \cdot \underset{1}{\cancel{y}}}$ **Simplify.**

$= \dfrac{115x}{234}$ or $\dfrac{115}{234}x$ **Multiply.**

K.   Divide: $\dfrac{21}{8}x^2y \div \dfrac{7}{5}x$

Exercises 5.3

OBJECTIVE 1: *Simplify a fraction.*

Simplify.

A.

1. $\dfrac{16}{20}$ 2. $\dfrac{18}{20}$ 3. $\dfrac{44}{88}$

4. $\dfrac{32}{64}$ 5. $\dfrac{28}{40}$ 6. $\dfrac{32}{40}$

B.

7. $\dfrac{45}{75}$ 8. $\dfrac{65}{75}$ 9. $\dfrac{56b}{4b}$ 10. $\dfrac{56b^2}{12b}$

11. $\dfrac{64w}{72}$ 12. $\dfrac{90x^2}{126x}$ 13. $\dfrac{546xy^2}{910x}$ 14. $\dfrac{630abc^2}{1050c}$

OBJECTIVE 2: *Build a fraction.*

Build each fraction by multiplying each numerator and denominator by 2, 3, 4, and 5.

A.

15. $\dfrac{2}{3}$ 16. $\dfrac{4}{5}$ 17. $\dfrac{5c}{4}$ 18. $\dfrac{9w}{8}$

B.

19. $\dfrac{7}{16}$ 20. $\dfrac{9}{16}$ 21. $\dfrac{8}{15}$ 22. $\dfrac{11}{15}$

Build each fraction by finding the missing numerator.

A.

23. $\dfrac{2}{3} = \dfrac{?}{24}$

24. $\dfrac{2}{3} = \dfrac{?}{36}$

25. $\dfrac{6}{7} = \dfrac{?}{28}$

26. $\dfrac{6}{7} = \dfrac{?}{49}$

B.

27. $\dfrac{11x}{13} = \dfrac{?}{52}$

28. $\dfrac{13y}{14} = \dfrac{?}{56}$

29. $\dfrac{6x^2}{7} = \dfrac{?}{84}$

30. $\dfrac{11p^2}{15} = \dfrac{?}{75}$

OBJECTIVE 3: *Multiply numeric and algebraic fractions.*

Multiply and simplify.

A.

31. $\dfrac{5}{8} \cdot \dfrac{4}{7}$

32. $\dfrac{5}{9} \cdot \dfrac{4}{15}$

33. $\dfrac{5}{8} \cdot \dfrac{4}{15}$

34. $\dfrac{5}{9} \cdot \dfrac{3}{20}$

35. $\dfrac{6}{7} \cdot \dfrac{14}{15}$

36. $\dfrac{6}{7} \cdot \dfrac{14}{18}$

B.

37. $\dfrac{9}{12}\left(\dfrac{10}{15}\right)$

38. $\dfrac{5}{24}\left(\dfrac{8}{10}\right)$

39. $\left(-\dfrac{3}{5}\right)\left(\dfrac{5}{2}\right)\left(-\dfrac{2}{3}\right)$

40. $\left(\dfrac{8}{10}\right)\left(-\dfrac{3}{4}\right)\left(\dfrac{2}{9}\right)$

41. $\left(2\dfrac{5}{8}\right)\left(\dfrac{4}{21}\right)$

42. $\left(5\dfrac{1}{2}\right)\left(3\dfrac{1}{3}\right)$

43. $\left(3\dfrac{1}{8}\right)\left(4\dfrac{3}{5}\right)(3)$

44. $\left(3\dfrac{3}{5}\right)\left(4\dfrac{1}{8}\right)(5)$

45. $\left(\dfrac{56a}{65}\right)\left(\dfrac{39b}{48}\right)\left(\dfrac{18b}{25}\right)$

46. $\left(\dfrac{30x}{48}\right)\left(\dfrac{5x}{14}\right)\left(\dfrac{7y}{25}\right)$

414

OBJECTIVE 4: *Divide numeric and algebraic fractions.*

Divide and simplify.

A.

47. $\dfrac{2}{5} \div \dfrac{3}{5}$

48. $\dfrac{3}{16} \div \dfrac{8}{16}$

49. $\dfrac{8}{9} \div \left(-\dfrac{7}{18}\right)$

50. $\left(-\dfrac{6}{7}\right) \div \left(\dfrac{3}{14}\right)$

51. $\dfrac{8}{15} \div \dfrac{16}{5}$

52. $\dfrac{9}{20} \div \dfrac{18}{5}$

B.

53. $\left(-\dfrac{5}{18}\right) \div \left(\dfrac{10}{27}\right)$

54. $\left(\dfrac{4}{15}\right) \div \left(-\dfrac{5}{8}\right)$

55. $\left(-\dfrac{20}{21}xy\right) \div \left(\dfrac{9}{10}\right)$

56. $\left(-\dfrac{14}{15}mn\right) \div \left(-\dfrac{5}{7}\right)$

57. $\left(\dfrac{5a}{72}\right) \div \left(\dfrac{2a}{25}\right)$

58. $\left(\dfrac{8x}{75}\right) \div \left(\dfrac{5x}{2}\right)$

C.

Fill in the boxes with a single digit so the statement is true. Explain your answer.

59. $\dfrac{2}{3} \cdot \dfrac{\square}{16} = \dfrac{5}{24}$

60. $\dfrac{3}{\square} \cdot \dfrac{7}{12} = \dfrac{7}{64}$

61. Find the error(s) in the statement: $\dfrac{3}{8} \cdot \dfrac{4}{5} = \dfrac{15}{32}$. Correct the statement. Explain how you would avoid this error.

62. Find the error(s) in the statement: $\dfrac{3}{8} \div \dfrac{4}{5} = \dfrac{32}{15}$. Correct the statement. Explain how you would avoid this error.

63. Find the product of $\dfrac{243a}{1000}$, $\dfrac{25}{81}$, and $\dfrac{8}{9}$.

64. Find the product of $\dfrac{27}{45}$, $\dfrac{144t}{100}$, and $\dfrac{10}{36}$.

65. Find the quotient of $-\dfrac{16s^2t}{81}$ and $-\dfrac{8st}{108}$.

66. Find the quotient of $-\dfrac{70mn^2}{63}$ and $-\dfrac{154n}{3}$.

67. Is 4 a solution of $\dfrac{11}{6}x = \dfrac{22}{3}$?

68. Is $\dfrac{2}{3}$ a solution of $6x + 1 = 5$?

69. Build the four fractions $\dfrac{1}{2}$, $\dfrac{2}{3}$, $\dfrac{1}{6}$, and $\dfrac{5}{8}$ so that each has a denominator of 24.

70. Build the four fractions $\dfrac{1}{4}$, $\dfrac{4}{13}$, $\dfrac{5}{26}$, and $\dfrac{12}{13}$ so that each has a denominator of 52.

71. The weight of a cubic foot of sandstone rock is approximately $2\dfrac{13}{20}$ times the weight of a cubic foot of water. If a cubic foot of water weighs approximately $62\dfrac{1}{2}$ pounds and a sandstone rock contains 5800 cubic feet, what is the approximate weight of the rock?

416

72. A plastic container holds $1\frac{3}{4}$ gallons. How many gallons does it contain when it is $\frac{3}{4}$ full of a cleaning chemical?

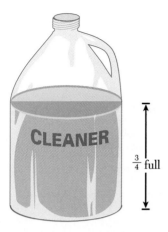

73. The float on a tank registers 12 feet. If the tank is full when it registers 28 feet, what fraction of the tank is full? Simplify.

74. One-third of Mohammed's paycheck is deducted for taxes and insurance. If $\frac{2}{5}$ of the deductions are for insurance, what fractional part of his paycheck is deducted for taxes?

75. Yoshi has $2\frac{7}{8}$ pounds of cheese. If an omelet recipe calls for $\frac{1}{4}$ pound of cheese, how many omelets can Yoshi make?

76. As part of her job at the Fyne Pet Shop, Becky feeds each gerbil $\frac{1}{8}$ cup of seed each day. If the seed comes in packages of $\frac{3}{4}$ cup, how many gerbils can be fed from one package?

77. According to U.S. postal regulations, mail that is less than $\frac{1}{4}$ inch thick must be rectangular in shape, at least $3\frac{1}{2}$ inches wide, and at least 5 inches long. What is the minimum volume of a letter $\frac{1}{8}$ inch thick?

78. The May Company offered the following discounts during its gigantic President's Day Sale:

Item	Men's suits	Polo shirts	Casual pants	Sport shoes
Regular Price	$315	$55	$32	$96
Discount	$\frac{1}{3}$ off	$\frac{1}{5}$ off	$\frac{1}{4}$ off	$\frac{1}{6}$ off

What is the sale price on each item?

79. Floors over unheated crawl spaces are generally recommended to be insulated at the R-19 level. It takes a thickness of $8\frac{3}{4}$ inches of fiberglass to produce an R-19 insulation value. How many cubic inches of fiberglass are needed to properly insulate a one story home of 1100 square feet?

 Exercises 80–83 refer to the Chapter 5 Application. See page 373.

80. Suppose you pick a card at random out of a deck of playing cards (52 cards). What is the probability that the card will be a queen? What is the probability that the card will be a heart? What is the probability that the card will be the queen of hearts? Fill out the table below and try to discover the relationship between these three probabilities.

Probability of a Queen	Probability of a Heart	Probability of the Queen of Hearts

For a card to be the queen of hearts, two conditions must hold true at the same time. The card must be a queen *and* the card must be a heart. Make a guess about the relationship of the probabilities when two conditions must occur at the same time.

81. Test your guess in Exercise 80 by considering the probability of drawing a black seven. What are the two conditions which must be true in order for the card to be a black seven? What are their individual probabilities? Was your guess correct?

82. What two conditions must be true for you to draw a red face card? What is the probability of drawing a red face card?

83. Suppose that a class of 30 students has 18 males. Twenty-five of the students have brown eyes. If a student is chosen at random, what is the probability that the student will be a brown-eyed male?

STATE YOUR UNDERSTANDING

84. Using an example, write an explanation for the statement: "Simplifying before multiplying does not change the product of two fractions, *but* simplifying before dividing gives an incorrect result."

85. Draw a picture that illustrates that $\dfrac{4}{8} = \dfrac{2}{4} = \dfrac{1}{2}$.   **86.** Explain how to simplify $\dfrac{36xy}{15x^2} \div \dfrac{56yz}{35xz}$.

CHALLENGE

87. Find four fractions equivalent to $\dfrac{15}{25}$ with denominators 10, 20, 30, and 60.

88. Find four fractions equivalent to $\dfrac{14}{21}$ with numerators 8, 12, 20, and 26.

GROUP ACTIVITY

To make accurate pie charts, it is necessary to use a measuring instrument for angles. One such instrument is called a protractor. It is also necessary to convert the fractions of the components into equivalent fractions with denominator of 360 because there are 360° in a complete circle. Starting with any radius, use the protractor to measure the correct angle for each component.

89. An aggressive investment strategy allocates $\dfrac{3}{4}$ of a portfolio to stocks, $\dfrac{1}{5}$ to bonds, and $\dfrac{1}{20}$ to money market funds. Make an accurate pie chart for this strategy.

90. A moderate investment strategy allocates $\dfrac{3}{5}$ of a portfolio to stocks, $\dfrac{3}{10}$ to bonds, and $\dfrac{1}{10}$ to money market funds. Make an accurate pie chart for this strategy.

91. A conservative investment strategy allocates $\dfrac{2}{5}$ of a portfolio to stocks, $\dfrac{9}{20}$ to bonds, and $\dfrac{3}{20}$ to money market funds. Make an accurate pie chart for this strategy.

92. Divide your group into two subgroups. Then have each group take opposite sides to discuss the following statement: "Since computers and calculators use decimals more frequently than fractions, the use of fractions will eventually disappear." Have each subgroup write down its arguments for or against the statement.

MAINTAIN YOUR SKILLS (Sections 4.8 and 5.1)

For Exercises 93 and 94, write the fraction indicated by the arrow.

93.

94.

95. Evaluate $-6abc - 5a$, when $a = -3$, $b = -3$, and $c = -3$.

96. Evaluate $\dfrac{5a + 12}{b}$, when $a = -6$ and $b = -18$.

97. Evaluate $\dfrac{4x - 12}{a}$, when $a = -18$ and $x = -6$.

5.4   Conversion of Units Within a System

OBJECTIVE   Convert units within the English system or within the metric system.

HOW AND WHY
Objective   **Convert units within the English system or within the metric system.**

Because 12 inches = 1 foot, they are equivalent measures. When we write 12 inches in place of 1 foot, or vice versa, we say that we have "converted the units." The division (12 inches) ÷ (1 foot) asks, "How many units of measure 1 foot does it take to make 12 inches?" Since they measure the same length, the answer is 1.

$$\frac{12 \text{ inches}}{1 \text{ foot}} = \frac{1 \text{ foot}}{12 \text{ inches}} = 1$$ **See Appendix II for a table of equivalent measures.**

This idea, along with the multiplication property of one, is used to convert from one measure to another. The units of measure are treated like factors and are simplified before multiplying. For example, to convert 48 inches to feet we multiply.

$$48 \text{ inches} = (48 \text{ inches}) \cdot 1 \qquad \textbf{Multiply by 1.}$$

$$48 \text{ inches} = \frac{48 \text{ inches}}{1} \cdot \frac{1 \text{ foot}}{12 \text{ inches}} \qquad \textbf{Substitute } \frac{1 \text{ foot}}{12 \text{ inches}} \textbf{ for 1.}$$

$$= \frac{48}{12} \text{ feet} \qquad \textbf{Simplify.}$$

$$= 4 \text{ feet}$$

In the second step we chose to multiply by $\dfrac{1 \text{ foot}}{12 \text{ inches}}$ deliberately so the unit "inches" would simplify or divide out. Therefore, 48 inches can be converted to 4 feet.

In some cases, it is necessary to multiply by various names for the number 1 as when we convert 7200 seconds to hours.

$$7200 \text{ seconds} = (7200 \text{ seconds}) \cdot 1 \cdot 1 \qquad \textbf{Multiply by 1 twice.}$$

$$7200 \text{ seconds} = \frac{7200 \text{ seconds}}{1} \cdot \frac{1 \text{ minute}}{60 \text{ seconds}} \cdot \frac{1 \text{ hour}}{60 \text{ minutes}} \qquad \textbf{Substitute.}$$

$$= \frac{7200(1)(1) \text{ hour}}{1(60)(60)}$$

$$= \frac{7200}{3600} \text{ hours} \qquad \textbf{Simplify.}$$

$$= 2 \text{ hours} \qquad \textbf{Simplify.}$$

Hence, 7200 seconds is equivalent to 2 hours.

▶ *To convert the units of a measurement*

1. Multiply by fractions formed by equivalent measurements, that is, names for 1.
2. Simplify.

The examples include converting measures within the metric system. See Chapter 6 for an alternate method of converting metric units.

When calculating areas we get square units of measure such as 5 square meters. The symbol "5 m^2" literally means "5 · 1 meter · 1 meter." See Example E.

Examples A–G	**Warm Ups A–G**

Directions: Convert units of measure.

Strategy: Multiply the given unit of measure by fractions formed by equivalent measures. Use a conversion chart if necessary.

A. Convert 5 gallons to pints.

Strategy: Multiply by quarts/gallon to get quarts and then by pints/quart to get pints.

5 gallons = (5 gallons) · 1 · 1   **Multiply by 1 twice.**

$$= \frac{5 \text{ gallons}}{1} \cdot \frac{4 \text{ quarts}}{1 \text{ gallon}} \cdot \frac{2 \text{ pints}}{1 \text{ quart}}$$

$$= 5 \cdot 4 \cdot 2 \text{ pints}$$

$$= 40 \text{ pints}$$

5 gallons = 40 pints.

A. Convert 5 hours to seconds.

B. Convert 91 kilograms to grams.

$$\frac{91 \text{ kg}}{1} \cdot \frac{1000 \text{ g}}{1 \text{ kg}} = 91 \cdot 1000 \text{ g}$$

91 kilograms = 91,000 grams.

B. Convert 4 meters to centimeters.

C. Five hundred meters is how many kilometers?

$$\frac{500 \text{ m}}{1} \cdot \frac{1 \text{ km}}{1000 \text{ m}} = \frac{500}{1000} \text{ km}$$

$$= \frac{1}{2} \text{ km}$$

500 meters = $\frac{1}{2}$ kilometer.

C. How many feet are in 2 miles?

D. Convert 60 miles per hour to feet per second.

$$60 \text{ mph} = \frac{\overset{1}{60 \text{ miles}}}{1 \text{ hour}} \cdot \frac{1 \text{ hr}}{\underset{1}{60 \text{ min}}} \cdot \frac{1 \text{ min}}{60 \text{ sec}} \cdot \frac{5280 \text{ ft}}{1 \text{ mi}}$$

$$= \frac{1 \cdot 1 \cdot 1 \cdot 5280 \text{ ft}}{1 \cdot 1 \cdot 60 \text{ sec} \cdot 1}$$

$$= \frac{88 \text{ ft}}{1 \text{ sec}}$$

60 miles per hour = 88 feet per second.

D. Convert 66 feet per second to miles per hour.

E. How many square inches are in 3 square feet?

Strategy: When we write "in.^{2}" the exponent means to use the factor twice. So, 1 ft^2 = 1 ft · 1 ft and 1 in.^{2} = 1 in. · 1 in. Thus,

$$3 \text{ ft}^2 = 3 \cdot 1 \text{ ft} \cdot 1 \text{ ft}$$

$$= \frac{3 \cdot \cancel{1\text{ ft}} \cdot \cancel{1\text{ ft}}}{1} \cdot \frac{12 \text{ in.}}{\cancel{1\text{ ft}}} \cdot \frac{12 \text{ in.}}{\cancel{1\text{ ft}}} \qquad \textbf{Multiply by } \frac{12 \text{ in.}}{1 \text{ ft}}$$

$$= 3(12)(12)(\text{in.})(\text{in.}) \qquad \textbf{twice.}$$

$$= 432 \text{ in.}^2$$

3 ft^2 is equivalent to 432 in.^{2}.

E. How many square feet are in 6 square yards?

F. Convert 480 grams per liter to milligrams per kiloliter.

$$\frac{480 \text{ g}}{\ell} = \frac{480 \cancel{\text{ g}}}{1 \cancel{\ell}} \cdot \frac{1000 \text{ mg}}{1 \cancel{\text{ g}}} \cdot \frac{1000 \cancel{\ell}}{1 \text{ k}\ell}$$

$$= \frac{480(1000)(1000) \text{ mg}}{1(1)(1) \text{ k}\ell}$$

$$= \frac{480\,000\,000 \text{ mg}}{\text{k}\ell}$$

480 grams per liter is equivalent to 480,000,000 milligrams per kiloliter.

F. Convert 50 meters per minute to kilometers per hour.

G. Belinda's employer wants to change her hourly wage of $9.00 per hour to an equivalent piecework wage. If she makes an average of four circuit boards per hour, what will be the equivalent wage per board?

Strategy: Multiply by $\dfrac{1 \text{ hr}}{4 \text{ boards}}$, since in one hour, Belinda makes 4 boards.

$$\frac{\$9.00}{1 \text{ hr}} = \frac{\$9.00}{1 \cancel{\text{ hr}}} \cdot \frac{1 \cancel{\text{ hr}}}{4 \text{ boards}}$$

$$= \frac{\$9}{4 \text{ boards}}$$

$$= \frac{\$2\frac{1}{4}}{1 \text{ board}}$$

The equivalent piecework wage is $2.25 per circuit board.

G. Lucas can type an average of 70 words per minute. If a page averages 600 words, how many pages can he type in one hour?

Answers to Warm Ups   E. 54 ft^2 F. 3 kilometers per hour   G. 7 pages per hour

Exercises 5.4

OBJECTIVE : *Convert units within the English system or within the metric system.*

Convert.

A.

1. 2 weeks = ___ days
2. 120 seconds = ___ minutes
3. 2 years = ___ months
4. 21 days = ___ weeks
5. 2 feet = ___ inches
6. 36 inches = ___ yards
7. 1 mile = ___ feet
8. 1760 yards = ___ miles
9. 2 meters = ___ centimeters
10. 2 kilometers = ___ meters
11. 1 yd^2 = ___ ft^2
12. 1 m^2 = ___ cm^2

B.

13. 80 ounces = ___ pounds
14. 8 quarts = ___ gallons

15. 15 cm = ___ mm
16. 4 g = ___ cg

17. 40 pints = ___ gallons
18. 12 ft = ___ yd

19. 6000 g = ___ kg
20. 9 feet = ___ inches

21. 10 kg = ___ g
22. 3 pounds = ___ ounces

23. 108 inches = ___ yards
24. 9 yards = ___ feet

25. 5 m = ___ cm
26. 1080 inches = ___ yards

27. $\dfrac{144 \text{ lb}}{1 \text{ ft}} = \dfrac{\text{lb}}{1 \text{ in.}}$

28. $\dfrac{\$660}{\text{hr}} = \dfrac{\$}{\text{min}}$

29. 1000 mm = ___ m

30. 1000 cm = ___ m

31. 32 inches = ___ feet

32. 16 feet = ___ yards

33. 3 yards 1 foot = ___ inches

34. 4 feet 6 inches = ___ yards

35. 3000 pounds = ___ tons

36. 2640 feet = ___ mile

37. 10,080 minutes = ___ days

38. 371 days = ___ weeks

39. $6\dfrac{1}{2}$ tons = ___ pounds

40. $5\dfrac{1}{2}$ days = ___ hours

41. $\dfrac{45 \text{ miles}}{1 \text{ hour}} = \dfrac{\text{feet}}{1 \text{ second}}$

42. $\dfrac{\$180}{1 \text{ ton}} = \dfrac{\text{cents}}{1 \text{ pound}}$

43. $\dfrac{9 \text{ tons}}{1 \text{ ft}} = \dfrac{\text{pounds}}{1 \text{ in.}}$

44. $\dfrac{10 \text{ ounces}}{1 \text{ cup}} = \dfrac{\text{pounds}}{1 \text{ gallon}}$

45. $\dfrac{180 \text{ kilometers}}{1 \text{ hour}} = \dfrac{\text{meters}}{1 \text{ second}}$

46. $\dfrac{24,000 \text{ miles}}{1 \text{ hour}} = \dfrac{\text{feet}}{1 \text{ second}}$

47. $\dfrac{24 \text{ g}}{1 \text{ m}^2} = \dfrac{\text{kg}}{1 \text{ km}^2}$

48. $\dfrac{\$36}{1 \text{ gallon}} = \dfrac{\text{cents}}{1 \text{ pint}}$

49. $\dfrac{30 \text{ tons}}{1 \text{ day}} = \dfrac{\text{pounds}}{1 \text{ min}}$

50. $\dfrac{\$90}{1 \text{ ft}^2} = \dfrac{\text{cents}}{1 \text{ in.}^2}$

C.

51. If a secretary can type 90 words per minute $\left(\dfrac{90 \text{ words}}{1 \text{ minute}} \right)$, how many words can he type in one second?

52. Mark is on a diet that causes him to lose 8 ounces every day. At this rate, how many pounds will he lose in 6 weeks?

53. A physician orders 0.03 g of Elixir Chlor-Trimeton. The available dose has a label that reads 2 mg per cc (cubic centimeter). How many cubic centimeters are needed to fill the doctor's order?

54. Larry's family eats approximately 78 kilograms of Wheat Bran Flakes in a year. If Wheat Bran is sold in boxes containing 500 grams of wheat bran, what is the average number of boxes of Wheat Bran that Larry's family consumes in one week?

55. During a fundraiser to preserve land to save the African elephant, the Jungle Society agrees to donate 12¢ a milligram for the largest bass caught during the club's annual Bass Contest. If the largest bass caught weighs 2 kilograms, how much does the Jungle Society donate to save the African elephant?

56. During a fundraiser to save the black rhino, Greg's Sports Shop agrees to donate 5¢ for every yard Rita jogs during one week. Rita jogs 2 miles a day for 5 weekdays. How much does Greg's Sports Shop donate?

57. A snail can crawl $\frac{5}{8}$ inch in one minute. How long will it take the snail to cover three inches?

3 in.

58. If Dan averages 50 miles per hour during an 8-hour day of driving, how many days will it take him to drive 2000 miles?

59. Water weighs 64 pounds per ft³. How many pounds does 1 cubic inch weigh?

60. During a bicycle trip the Trong family averages a speed of 22 feet per second. At this rate, how many miles will they average in a 7-hour day of cycling? How many days will it take the family to cover 350 miles?

61. An insect repellent weighs 100 kilograms per cubic meter. The Non-Sting Company sells the repellent in bottles containing 300 cubic centimeters. How many grams of repellent are sold in one bottle? If the repellent is priced at 20¢ per gram, what is the price of one bottle?

STATE YOUR UNDERSTANDING

62. Explain how to change $\dfrac{x \text{ ft}^2}{\text{sec}}$ to $\dfrac{\text{mi}^2}{\text{hr}}$.

63. Explain each step in the procedure for converting 12,000 seconds to hours.

64. Explain the role that the Basic Principle of Fractions plays in converting units.

CHALLENGE

65. A pharmacy pays \$5 per gram for a new drug. They resell the drug at $7\frac{1}{2}$ ¢ per centigram. What is the profit on the sale of 1 kilogram of the drug?

66. A chain made of precious metals is priced at \$7.50 per inch. What is the cost of 4 yd 2 ft 4 in. of the chain?

GROUP ACTIVITY

67. A ft^3 of water weighs about 1000 ounces. Iron is $7\frac{4}{5}$ times heavier than water. An open rectangular tank of $\frac{1}{4}$ inch thick iron has outside dimensions of 4 ft long, 2 ft 6 in. wide, and 2 ft high.

  (a) Find the volume of the iron used to construct the tank.

  (b) Find the weight of the tank, to the nearest pound.

  (c) Suppose the tank is filled with water. Find the weight of the filled tank, to the nearest pound.

MAINTAIN YOUR SKILLS (Sections 4.3, 4.7, 5.1)

68. Find the difference of -86 and -77.

69. Find the difference of 188 and -124.

70. Is -16 a solution of $x^2 + x = 240$?

71. Is -14 a solution of $x^2 - 3x = 238$?

72. Write the opposites of the fractions $-\frac{13}{19}$, $\frac{-12}{17}$, and $\frac{13}{21}$.

5.5   Adding and Subtracting Rational Numbers (Fractions)

OBJECTIVES

1. Find the LCM, least common multiple, of two or more numbers.

2. Compare two fractions.

3. Add like and unlike fractions.

4. Subtract fractions.

5. Add or subtract mixed numbers.

VOCABULARY

Here are four equivalent meanings of **least common multiple (LCM)** of a set of natural numbers.

1. The smallest natural number that is a multiple of each number in the set.

2. The smallest natural number for which each number in the set is a factor.

3. The smallest natural number for which each number in the set is a divisor.

4. The smallest natural number that each number in the set will divide "evenly."

The least common denominator of two or more fractions is the LCM of the denominators.

Like fractions have the same denominator.

Unlike fractions do not have the same denominator.

HOW AND WHY
Objective 1

Find the LCM, least common multiple, of two or more numbers.

To add or subtract fractions we need a common denominator. The LCM of the denominators is exactly the number we will use.

The LCM of 4, 6, and 18 is 36. This means that

1. 36 is the smallest number that is a multiple of 4, 6, and 18.

2. 36 is the smallest number for which 4, 6, and 18 are factors.

3. 36 is the smallest number for which 4, 6, and 18 are divisors.

4. 36 is the smallest number that 4, 6, and 18 divide "evenly."

We can verify this by looking at lists of multiples for 4, 6, and 18.

4, 8, 12, 16, 20, 24, 28, 32, $\boxed{36}$, 40
6, 12, 18, 24, 30, $\boxed{36}$, 42, 48
18, $\boxed{36}$, 54, 72, 90

To find the LCM of larger numbers, such as 21 and 35, we write them in prime factored form:

$21 = 3 \cdot 7$ and $35 = 5 \cdot 7$

By picking out each *different* prime factor, that is, 3, 5, and 7, then multiplying $3 \cdot 5 \cdot 7$, we get 105, which is the LCM of 21 and 35.

Repeated factors of any number in the set are also repeated in the LCM. To find the LCM of 16, 15, and 24, find the prime factors.

$$16 = 2^4 \qquad 16 = 2 \cdot 2 \cdot 2 \cdot 2 \quad \text{or} \quad 16 = 2 \cdot 2 \cdot 2 \cdot 2$$
$$15 = 3^1 \cdot 5^1 \qquad 15 = 3 \cdot 5 \qquad\qquad 15 = \qquad\qquad 3 \cdot 5$$
$$24 = 2^3 \cdot 3^1 \qquad 24 = 2 \cdot 2 \cdot 2 \cdot 3 \qquad 24 = 2 \cdot 2 \cdot 2 \cdot \quad 3$$
$$\text{LCM} = 2 \cdot 2 \cdot 2 \cdot 2 \cdot 3 \cdot 5$$

We do not use all seven factors of two because we want the *least* common multiple. So we use the largest power of two, 2^4, because if we used fewer factors of two then 16 would not be a factor of the LCM. The LCM is the product of the largest power of each prime factor.

$$\text{LCM} = 2^4 \cdot 3^1 \cdot 5^1 = 240$$

▶ ***To find the least common multiple of a set of numbers***

1. Write each number in prime factored form using exponents for repeated factors.

2. Find the product of the largest power of each prime factor.

Examples A–B   **Warm Ups A–B**

Directions: Find the LCM.

Strategy: Write the prime factorization of each number. Use the largest power of each prime number to build the LCM.

A.   Find the LCM of 18, 30, and 36.

$$18 = 2 \cdot 3 \cdot 3 = 2^1 \cdot \boxed{3^2}$$
$$30 = 2 \cdot 3 \cdot 5 = 2^1 \cdot 3^1 \cdot \boxed{5^1}$$
$$36 = 2 \cdot 2 \cdot 3 \cdot 3 = \boxed{2^2} \cdot 3^2$$

Write the prime factors of each number. The prime factors of the LCM are 2, 3, and 5.

$$\text{LCM} = 2^2 \cdot 3^2 \cdot 5^1 = 180$$

The LCM is 180.

The largest exponent of 2 is 2, the largest exponent of 3 is 2, and the largest exponent of 5 is 1.

A.   Find the LCM of 20, 24, and 40.

B.   Jan and Robin have been saving coins. Jan saved dimes and Robin saved quarters. The girls went shopping together, each bought the same thing and each spent all of her money. What is the least amount the item could cost?

Strategy: The least amount the item could cost is the smallest number that 10 and 25 both divide evenly. That number is the LCM of 10 and 25.

$$10 = 2 \cdot 5$$
$$25 = 5^2$$
$$\text{LCM} = 2 \cdot 5^2 = 50$$

The least the item could cost is 50 cents.

B.   If Jan uses only nickels and Robin uses only quarters, what is the least each could pay for the same thing?

Answers to Warm Ups   A. 120   B. 25 cents

HOW AND WHY

Objective 2   **Compare two fractions.**

We can see that the fraction $\dfrac{6}{17}$ has a larger value than the fraction $\dfrac{5}{17}$, since $\dfrac{6}{17}$ has a larger numerator and the denominators are the same.

$$\frac{6}{17} > \frac{5}{17}$$ **Six-seventeenths is greater than five-seventeenths.**

It is less clear whether $\dfrac{8}{15}$ is larger than $\dfrac{7}{12}$. In this case, we build the fractions so that they have a common denominator. Then we make the comparison. The common denominator is the LCM of 15 and 12, which is 60. We know that $15 = 3 \cdot 5$ and $12 = 2^2 \cdot 3$, so the LCM is $2^2 \cdot 3 \cdot 5 = 60$. Now build the two fractions.

$$\frac{8}{15} = \frac{?}{60} \quad \text{and} \quad \frac{7}{12} = \frac{?}{60}$$

$$\frac{8}{15} = \frac{32}{60} \quad \text{and} \quad \frac{7}{12} = \frac{35}{60}$$

We can see that $\dfrac{32}{60} < \dfrac{35}{60}$, so we know that $\dfrac{8}{15} < \dfrac{7}{12}$.

If both fractions are negative, first write them in the form $\dfrac{-a}{b}$ and then build to a common denominator. See Example D.

 To compare fractions

1. Build the fractions to a common positive denominator.
2. The fraction with the largest numerator is the fraction with the largest value.

Examples C–D   **Warm Ups C–D**

Directions: Compare the fractions. Write the result as an inequality.

Strategy: Build each fraction to a common denominator and compare the numerators.

C.   Compare $\dfrac{7}{18}$ and $\dfrac{11}{24}$.

The LCM of 18 $(2 \cdot 3^2)$ and 24 $(2^3 \cdot 3)$ is 72.

$$\frac{7}{18} = \frac{7 \cdot 4}{72} = \frac{28}{72}$$ **Build each fraction to the common denominator, 72.**

$$\frac{11}{24} = \frac{11 \cdot 3}{72} = \frac{33}{72}$$

$$\frac{7}{18} < \frac{11}{24}$$ **Since 28 *is less than* 33.**

C.   Compare $\dfrac{13}{18}$ and $\dfrac{17}{24}$.

D. Compare $-\dfrac{11}{21}$ and $-\dfrac{15}{28}$.

$-\dfrac{11}{21} = \dfrac{-11}{21} = \dfrac{-44}{84}$ **Build each fraction to the common denominator, 84.**

$-\dfrac{15}{28} = \dfrac{-15}{28} = \dfrac{-45}{84}$

$-\dfrac{11}{21} > -\dfrac{15}{28}$ **Since -44 is greater than -45.**

D. Compare $-\dfrac{4}{9}$ and $-\dfrac{3}{8}$.

HOW AND WHY
Objective 3   **Add like and unlike fractions.**

Like fractions have the same denominator. The sum of the like fractions $\dfrac{7}{13}$ and $\dfrac{4}{13}$ is $\dfrac{11}{13}$. The reason for adding 7 and 4, but not the common denominators, 13, is that the denominators tell us the number of parts in the unit. The numerators tell how many of the parts are counted. Since the size of the parts, thirteenths, is the same, the sum represents a total of 11 parts each of equal size, that is, one-thirteenth.

Visually, we notice that when we add fractions, the size of the parts does not change, but the number of parts does change. For example,

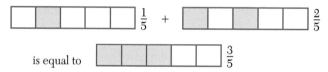

is equal to

To add like fractions

Add the numerators and keep the common denominator. Simplify if possible.

$$\dfrac{a}{c} + \dfrac{b}{c} = \dfrac{a+b}{c}$$

Unlike fractions have different denominators. To add $\dfrac{1}{2}$ and $\dfrac{1}{5}$, we need to rewrite them with a common denominator and use the procedure for adding like fractions.

To add unlike fractions

1. Write the fractions with positive denominators if necessary.
2. Build the fractions to a common denominator. Use the LCM of the denominators for the common denominator.
3. Add the numerators, keep the common denominator, and simplify if possible.

Answer to Warm Up   D. $-\dfrac{4}{9} < -\dfrac{3}{8}$

So, $\dfrac{1}{2} + \dfrac{1}{5} = \dfrac{5}{10} + \dfrac{2}{10} = \dfrac{7}{10}$.

Terms and algebraic fractions are added the same way.

$$\dfrac{3}{4}xy + \dfrac{-5}{12}xy + \dfrac{2}{3}xy = \dfrac{9}{12}xy + \dfrac{-5}{12}xy + \dfrac{8}{12}xy$$ **Build the fractions using the LCM, 12, for the common denominator.**

$$= \left(\dfrac{9 + (-5) + 8}{12}\right)xy$$ **Add the numerators and keep the common denominator.**

$$= \dfrac{12}{12}xy$$

$$= 1xy \quad \text{or} \quad xy$$ **Simplify.**

Examples E–F   **Warm Ups E–F**

Directions: Add the like fractions.

Strategy: Add the numerators and write the sum over the common denominator.

E. $\dfrac{9}{16} + \left(-\dfrac{3}{16}\right)$

$\dfrac{9}{16} + \left(-\dfrac{3}{16}\right) = \dfrac{9 + (-3)}{16}$ **Add the numerators.**

$= \dfrac{6}{16}$ **Keep the common denominator.**

$= \dfrac{3}{8}$ **Simplify.**

E. $\dfrac{13}{9} + \left(-\dfrac{16}{9}\right)$

F. $\dfrac{5}{9}x + \dfrac{7}{9}x + \dfrac{3}{9}x$

$\dfrac{5}{9}x + \dfrac{7}{9}x + \dfrac{3}{9}x = \left(\dfrac{5 + 7 + 3}{9}\right)x$ **Add the coefficients of these like terms and multiply times the variable factor. Then simplify.**

$= \dfrac{15}{9}x$

$= \dfrac{5}{3}x$

F. $\dfrac{11}{20}b + \dfrac{1}{20}b + \dfrac{3}{20}b$

Answers to Warm Ups   E. $-\dfrac{1}{3}$ F. $\dfrac{3}{4}b$

Examples G–J   **Warm Ups G–J**

Directions: Add the unlike fractions.

Strategy: Build the fractions to a common denominator and add the like fractions. Simplify if possible.

G. $\left(\dfrac{5}{9}\right) + \left(-\dfrac{7}{12}\right)$

$\left(\dfrac{5}{9}\right) + \left(-\dfrac{7}{12}\right) = \left(\dfrac{20}{36}\right) + \left(-\dfrac{21}{36}\right)$ **Use the LCM, 36, of 9 and 12 for the common denominator.**

$= \dfrac{20 + (-21)}{36}$

$= \dfrac{-1}{36}$ or $-\dfrac{1}{36}$ **Add.**

G. $-\dfrac{5}{14} + \dfrac{7}{6}$

H. $-\dfrac{9}{64}a + \dfrac{15}{56}a$

$-\dfrac{9}{64}a + \dfrac{15}{56}a = -\dfrac{63}{448}a + \dfrac{120}{448}a$ **The LCM of 64 and 56 is 448.**

$= \dfrac{-63 + 120}{448}a$

$= \dfrac{57}{448}a$

H. $-\dfrac{11}{96}y + \dfrac{35}{72}y$

I.   Sheela Williams is assembling a tricycle for her son's birthday. She needs a bolt that must reach through a $\dfrac{1}{32}$-inch washer, a $\dfrac{3}{16}$-inch plastic bushing, a $\dfrac{3}{4}$-inch piece of steel tubing, a second $\dfrac{1}{32}$-inch washer, and a $\dfrac{1}{4}$-inch thick nut. How long a bolt does she need?

Strategy: The bolt must be long enough to reach through all five pieces to hold them together. Add all the thicknesses together.

$\dfrac{1}{32} + \dfrac{3}{16} + \dfrac{3}{4} + \dfrac{1}{32} + \dfrac{1}{4}$

$= \dfrac{1}{32} + \dfrac{6}{32} + \dfrac{24}{32} + \dfrac{1}{32} + \dfrac{8}{32}$ **Since the fractions are unlike we use the common denominator, 32.**

$= \dfrac{40}{32} = \dfrac{5}{4}$ **Add and simplify.**

$= 1\dfrac{1}{4}$ **Change to a mixed number.**

The bolt must be $\dfrac{5}{4}$ inches or $1\dfrac{1}{4}$ inches long.

I.   A nail must reach through three thicknesses of wood and penetrate the fourth thickness $\dfrac{1}{4}$ inch. If the first piece of wood is $\dfrac{5}{16}$ inch, the second is $\dfrac{3}{8}$ inch, and the third is $\dfrac{9}{16}$ inch, how long must the nail be?

Answers to Warm Ups   G. $\dfrac{17}{21}$ H. $\dfrac{107}{288}y$ I. $\dfrac{3}{2}$ inches or $1\dfrac{1}{2}$ inches   J. $\dfrac{31}{54}$

Calculator Example:

J. $\dfrac{5}{18} + \dfrac{7}{27}$

$\dfrac{5}{18} + \dfrac{7}{27} = \dfrac{29}{54}$ **Add, using the fraction key on your calculator. The display may look like** $29 \rfloor 54$, $29/54$, or $\dfrac{29}{54}$ **depending on your calculator.**

J. $\dfrac{7}{18} + \dfrac{5}{27}$

HOW AND WHY
Objective 4

Subtract fractions.

The steps for subtracting fractions are similar to those for adding fractions.

 To subtract two fractions

1. Build the fractions so they have a common denominator.
2. Subtract the numerators and write the difference over the common denominator.
3. Simplify.

$\dfrac{2}{3} - \dfrac{3}{4} = \dfrac{8}{12} - \dfrac{9}{12}$ **Build the fractions to a common denominator.**

$= \dfrac{8 - 9}{12}$ **Subtract the numerators by adding** $8 + (-9)$.

$= \dfrac{-1}{12}$ or $-\dfrac{1}{12}$

Examples K–N

Warm Ups K–N

Directions: Subtract.

Strategy: Build the fractions to a common denominator. Subtract the numerators.

K. $\dfrac{4}{5} - \dfrac{4}{9}$

$\dfrac{4}{5} - \dfrac{4}{9} = \dfrac{36}{45} - \dfrac{20}{45}$ **The LCM of 5 and 9 is 45.**

$= \dfrac{36 - 20}{45}$ **Subtract numerators.**

$= \dfrac{16}{45}$

K. $\dfrac{5}{8} - \dfrac{5}{9}$

Answer to Warm Up   K. $\dfrac{5}{72}$

L. $\dfrac{5}{6}x - \dfrac{7}{8}x$

$\dfrac{5}{6}x - \dfrac{7}{8}x = \left(\dfrac{5}{6} - \dfrac{7}{8}\right)x$ **Subtract the coefficients and multiply by the variable factor.**

$= \left(\dfrac{20}{24} - \dfrac{21}{24}\right)x$

$= \left(\dfrac{20 - 21}{24}\right)x$

$= -\dfrac{1}{24}x$

L. $\dfrac{25}{48}x^2 - \dfrac{15}{64}x^2$

M. $\dfrac{8}{15} - \left(-\dfrac{1}{4}\right)$

$\dfrac{8}{15} - \left(-\dfrac{1}{4}\right) = \dfrac{32}{60} - \dfrac{-15}{60}$ **The LCM of 15 and 4 is 60.**

$= \dfrac{32 - (-15)}{60}$

$= \dfrac{32 + 15}{60}$

$= \dfrac{47}{60}$

M. $\dfrac{1}{8} - \left(-\dfrac{11}{12}\right)$

N. Lumber mill operators must plan for the shrinkage of "green" (wet) boards when they cut logs. If the shrinkage for a $\dfrac{5}{8}$-inch-thick board is expected to be $\dfrac{1}{16}$ inch, what is the thickness of the dried board?

Strategy: Subtract the shrinkage from the thickness of the "green" board to find the thickness of the dried board.

$\dfrac{5}{8} - \dfrac{1}{16} = \dfrac{10}{16} - \dfrac{1}{16}$ **The LCM is 16.**

$= \dfrac{9}{16}$

The dried board will be $\dfrac{9}{16}$ inch thick.

N. Mike must plane $\dfrac{3}{32}$ inch from the thickness of a board. If the board is now $\dfrac{3}{4}$ inch thick, how thick will it be after he has planed it?

HOW AND WHY
Objective 5   **Add or subtract mixed numbers.**

To add mixed numbers, add the whole numbers and add the fractions. Pictorially we can show the sum of $3\dfrac{1}{6}$ and $5\dfrac{1}{4}$ by drawing rectangles.

Answers to Warm Ups   L. $\dfrac{55}{192}x^2$ M. $\dfrac{25}{24}$ N. $\dfrac{21}{32}$ inch

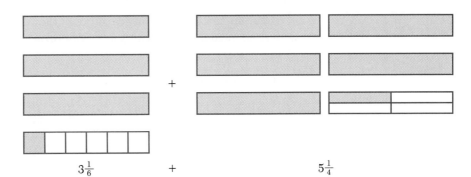

$3\frac{1}{6}$ + $5\frac{1}{4}$

It is easy to see that the sum contains eight whole units. The sum of the fraction parts requires finding a common denominator. The LCM of 6 and 4 is 12.

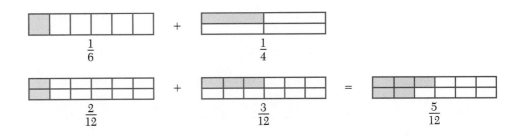

$\frac{1}{6}$ + $\frac{1}{4}$

$\frac{2}{12}$ + $\frac{3}{12}$ = $\frac{5}{12}$

So, the sum is $8\frac{5}{12}$.

Symbolically, mixed numbers can be added horizontally or in columns. The sum of $3\frac{1}{6}$ and $5\frac{1}{4}$ is shown both ways.

$$\left(3 + \frac{1}{6}\right) + \left(5 + \frac{1}{4}\right) = (3 + 5) + \left(\frac{1}{6} + \frac{1}{4}\right)$$

$$= 8 + \left(\frac{2}{12} + \frac{3}{12}\right)$$

$$= 8\frac{5}{12}$$

$$\begin{array}{r} 3\frac{1}{6} = 3\frac{2}{12} \\ +5\frac{1}{4} = 5\frac{3}{12} \\ \hline 8\frac{5}{12} \end{array}$$

The column arrangement has the advantage of aligning the whole numbers and the fractions.

Sometimes the sum of the fraction parts is greater than 1. In this case, change the fraction sum to a mixed number and add it to the whole number part. See Example O.

▶ **To add mixed numbers**

1. Add the whole number parts.
2. Add the fraction parts. If the sum of the fractions is more than 1, change the fraction to a mixed number and add again.
3. Simplify.

To subtract mixed numbers, subtract the whole numbers and subtract the fractions. As in addition, we can subtract horizontally or in columns.

$$7\frac{9}{10} - 4\frac{3}{10}$$

$$\left(7 + \frac{9}{10}\right) - \left(4 + \frac{3}{10}\right) = (7-4) + \left(\frac{9}{10} - \frac{3}{10}\right)$$

$$= 3 + \frac{6}{10}$$

$$= 3\frac{3}{5}$$

$$\begin{array}{r} 7\frac{9}{10} \\ -4\frac{3}{10} \\ \hline 3\frac{6}{10} = 3\frac{3}{5} \end{array}$$

When subtracting mixed numbers it is sometimes necessary to borrow from the whole number. In this event, "borrowing" means rewriting a mixed number using an improper fraction. For instance, to subtract $9\frac{1}{6} - 4\frac{5}{6}$, rename the fraction $9\frac{1}{6}$ by writing $9\frac{1}{6} = 9 + \frac{1}{6} = 8 + 1\frac{1}{6} = 8\frac{7}{6}$. We do this so we can subtract the fraction parts without getting a negative number.

$$\begin{array}{ll} 9\frac{1}{6} = 8\frac{7}{6} & \text{Rename } 9\frac{1}{6} \text{ by "borrowing."} \\[2mm] -4\frac{5}{6} = 4\frac{5}{6} & \\ \hline = 4\frac{2}{6} & \text{Subtract the whole numbers and fractions.} \\[2mm] = 4\frac{1}{3} & \text{Simplify.} \end{array}$$

▶ *To subtract mixed numbers*

1. Build the fractions so they have a common denominator.
2. If the second fraction part is larger than the first fraction part, rename the first mixed number by "borrowing" 1 from the whole number part to add to the fraction part.
3. Subtract the fraction parts and the whole number parts.
4. Simplify.

Examples O–P

Directions: Add.

Strategy: Build the fractions to a common denominator and add the whole numbers and the fractions. Simplify if possible.

O. $25\frac{1}{6} + 13\frac{7}{8} + 7\frac{5}{9}$

O. $13\frac{2}{3} + 21\frac{8}{9} + 4\frac{1}{4}$

$$25\frac{1}{6} = 25\frac{12}{72}$$

Write the fractions with a common denominator.

$$13\frac{7}{8} = 13\frac{63}{72}$$

$$+ \ 7\frac{5}{9} = \ 7\frac{40}{72}$$

The LCM of 6, 8, and 9 is 72.

$$\overline{\phantom{+7\frac{5}{9}} \ 45\frac{115}{72}}$$

Write $\frac{115}{72}$ as a mixed number and add.

$$= 45 + 1\frac{43}{72}$$

$$= 46\frac{43}{72}$$

P.   Raoul works part time and takes two or three classes per term. His employer allows him to choose the number of hours he works each day. He worked the following schedule one week: Monday, $6\frac{2}{3}$ hours; Tuesday, $1\frac{3}{4}$ hours; Wednesday, $10\frac{5}{6}$ hours; Thursday, $6\frac{1}{2}$ hours; Friday, $7\frac{1}{4}$ hours. How many hours did he work that week?

Strategy: Find the sum of the hours he worked. Write the fraction parts with the common denominator, 12.

$$6\frac{2}{3} = \ 6\frac{8}{12}$$

$$1\frac{3}{4} = \ 1\frac{9}{12}$$

$$10\frac{5}{6} = 10\frac{10}{12}$$

$$6\frac{1}{2} = \ 6\frac{6}{12}$$

$$+ \ 7\frac{1}{4} = \ 7\frac{3}{12}$$

$$\overline{\phantom{+7\frac{1}{4}} \ 30\frac{36}{12}}$$

$$= 30 + 3$$ **Change the improper fraction to a whole number.**

$$= \ 33$$

Raoul worked a total of 33 hours last week.

P.   The next week, Raoul worked Monday, $5\frac{3}{4}$ hours, Tuesday $6\frac{11}{12}$ hours; Wednesday, $4\frac{1}{2}$ hours; Thursday, $8\frac{1}{4}$ hours; and Friday, $9\frac{5}{6}$ hours.

How many hours did he work that week?

Answers to Warm Ups   O. $39\frac{29}{36}$ P. $35\frac{1}{4}$ hours

| **Examples Q–T** | **Warm Ups Q–T** |

Directions: Subtract.

Strategy: Write the fractions with a common denominator. If necessary, "borrow" 1 from the whole number part of the first mixed number to add to the fraction part. Subtract the fraction parts and the whole number parts.

Q. Subtract: $14\dfrac{7}{8} - 9\dfrac{2}{3}$

$$14\dfrac{7}{8} = 14\dfrac{21}{24}$$
$$-\ \ 9\dfrac{2}{3} = \ \ 9\dfrac{16}{24}$$
$$\overline{\qquad\qquad 5\dfrac{5}{24}}$$

The difference is $5\dfrac{5}{24}$.

Q. Subtract: $28\dfrac{11}{12} - 23\dfrac{3}{4}$

R. Subtract: $19 - 5\dfrac{3}{4}$

R. Subtract: $17 - 9\dfrac{5}{8}$

C A U T I O N   **We cannot subtract the whole numbers and get $14\dfrac{3}{4}$. We must borrow from 19 to get $18\dfrac{4}{4}$ to be able to subtract both whole numbers and fractions.**

$$19\ \ = 19\dfrac{0}{4} = 18 + 1\dfrac{0}{4} = 18\dfrac{4}{4}$$
$$-\ \ 5\dfrac{3}{4} = \qquad\qquad\quad = \ \ 5\dfrac{3}{4}$$
$$\qquad\qquad\qquad\qquad\qquad 13\dfrac{1}{4}$$

The difference is $13\dfrac{1}{4}$.

Calculator Example:

S. $13\dfrac{1}{8} - 9\dfrac{11}{12}$

$13\dfrac{1}{8} - 9\dfrac{11}{12} = 3\dfrac{5}{24}$ **Subtract, using a calculator. The display may look like 3⌐5⌐24, 77/24, or $3\dfrac{5}{24}$ depending on your calculator.**

S. $18\dfrac{1}{3} - 11\dfrac{7}{12}$

Answers to Warm Ups   Q. $5\dfrac{1}{6}$ R. $7\dfrac{3}{8}$ S. $6\dfrac{3}{4}$

T. Shawn McKillip bought a roast for Sunday dinner that weighed 7 pounds. He cut off some fat and took out a bone. The meat left weighed $4\frac{1}{3}$ pounds. How many pounds of bone and fat did he trim off?

Strategy: Subtract the weight of the remaining meat from the starting weight to find the amount trimmed off.

$$7 \quad = 7\frac{0}{3} = 6 + 1\frac{0}{3} = 6\frac{3}{3}$$
$$-4\frac{1}{3} = \qquad\qquad\qquad = 4\frac{1}{3}$$
$$2\frac{2}{3}$$

$2\frac{2}{3}$ pounds were trimmed off.

T. Jaime weighed $142\frac{1}{2}$ pounds and decided to lose some weight. He lost a total of $6\frac{3}{4}$ pounds last month. What is his weight after the loss?

Exercises 5.5

OBJECTIVE 1: *Find the LCM, least common multiple, of two or more numbers.*

Find the LCM.

A.

1. 7, 28

2. 8, 40

3. 2, 3, 5

4. 2, 5, 7

5. 2, 3, 6

6. 2, 5, 10

7. 2, 14, 21

8. 3, 6, 22

B.

9. 4, 12, 10, 15, 20

10. 2, 6, 8, 12, 24

11. 48, 96, 120

12. 48, 72, 80

13. $12x, 17x^2, 51, 68$

14. $14, 21, 26y^2, 91y$

OBJECTIVE 2: *Compare two fractions.*

Compare the fractions. Write the result as an inequality.

A.

15. $\dfrac{19}{21}$ and $\dfrac{20}{21}$

16. $\dfrac{7}{23}$ and $\dfrac{3}{23}$

17. $-\dfrac{3}{13}$ and $-\dfrac{5}{13}$

18. $-\dfrac{22}{5}$ and $-\dfrac{32}{5}$

19. $-\dfrac{4}{21}$ and $\dfrac{3}{7}$

20. $\dfrac{13}{18}$ and $-\dfrac{5}{6}$

B.

21. $\dfrac{13}{15}$ and $\dfrac{21}{25}$

22. $\dfrac{11}{20}$ and $\dfrac{17}{32}$

23. $\dfrac{20}{33}$ and $\dfrac{8}{11}$

24. $\dfrac{5}{42}$ and $\dfrac{4}{39}$ **25.** $-\dfrac{7}{30}$ and $-\dfrac{2}{9}$ **26.** $-\dfrac{13}{43}$ and $-\dfrac{17}{60}$

OBJECTIVE 3: *Add like and unlike fractions.*

Add.

A.

27. $\dfrac{2}{15} + \dfrac{4}{15} + \dfrac{3}{15}$ **28.** $\dfrac{5}{24} + \dfrac{2}{24} + \dfrac{1}{24}$ **29.** $\dfrac{3}{8}x + \dfrac{5}{8}x$

30. $\dfrac{7}{18}m + \dfrac{5}{18}m$ **31.** $\dfrac{5}{48} + \left(-\dfrac{7}{48}\right) + \dfrac{3}{48}$ **32.** $\dfrac{3}{16} + \left(-\dfrac{1}{16}\right) + \dfrac{5}{16}$

33. $\dfrac{1}{8} + \dfrac{7}{24}$ **34.** $\dfrac{7}{15} + \dfrac{1}{3}$ **35.** $\dfrac{1}{2}y + \dfrac{1}{3}y + \dfrac{1}{6}y$

36. $\dfrac{1}{3}pt + \dfrac{1}{4}pt + \dfrac{1}{12}pt$ **37.** $\dfrac{5}{14} + \left(-\dfrac{1}{7}\right) + \left(-\dfrac{2}{7}\right)$ **38.** $\dfrac{5}{21} + \left(-\dfrac{1}{7}\right) + \left(-\dfrac{2}{7}\right)$

B.

39. $\dfrac{7}{16} + \dfrac{3}{20} + \dfrac{1}{5}$ **40.** $\dfrac{5}{12} + \dfrac{9}{16} + \dfrac{7}{24}$ **41.** $\dfrac{3}{10}w + \dfrac{7}{20}w + \dfrac{11}{30}w$

42. $\dfrac{3}{16}z + \dfrac{7}{20}z + \dfrac{5}{12}z$ **43.** $\dfrac{5}{6}x + \left(-\dfrac{7}{8}\right) + \dfrac{3}{4} + \left(-\dfrac{1}{2}x\right)$

44. $\dfrac{9}{10}\,y + \dfrac{4}{5}\,z + \left(-\dfrac{7}{15}\,y\right) + \left(-\dfrac{11}{18}\,z\right)$

45. $\dfrac{29xy}{200} + \dfrac{7xy}{50} + \left(-\dfrac{9xy}{25}\right)$

46. $\dfrac{10a^2b}{33} + \dfrac{9a^2b}{22} + \left(-\dfrac{4a^2b}{55}\right)$

OBJECTIVE 4: *Subtract fractions.*

Subtract.

A.

47. $\dfrac{3}{8} - \dfrac{1}{8}$

48. $\dfrac{11}{12} - \dfrac{3}{12}$

49. $\dfrac{17}{30} - \dfrac{7}{30}$

50. $\dfrac{85}{90} - \dfrac{75}{90}$

51. $\dfrac{5}{18} - \dfrac{2}{9}$

52. $\dfrac{5}{6} - \dfrac{1}{3}$

53. $\dfrac{5}{6} - \left(-\dfrac{1}{2}\right)$

54. $\dfrac{7}{15} - \left(-\dfrac{2}{3}\right)$

B.

55. $\dfrac{8}{9} - \dfrac{5}{6}$

56. $\dfrac{13}{21} - \dfrac{5}{14}$

57. $-\dfrac{9}{10} - \dfrac{3}{4}$

58. $-\dfrac{7}{12} - \dfrac{11}{15}$

59. $\dfrac{33}{35}st - \dfrac{17}{28}st$

60. $\dfrac{7}{10}cd - \dfrac{19}{20}cd$

61. $\dfrac{3}{10}x^2 - \dfrac{1}{5}x^2 - \dfrac{4}{15}x^2$

62. $\dfrac{1}{2}a^2 - \dfrac{1}{16}a^2 - \dfrac{1}{6}a^2$

OBJECTIVE 5: *Add or subtract mixed numbers.*

Perform the indicated operations.

A.

63. $1\dfrac{3}{7} + 5\dfrac{2}{7}$

64. $1\dfrac{6}{13} + 8\dfrac{5}{13}$

65. $13\dfrac{6}{7} - 8\dfrac{4}{7}$

66. $27\dfrac{7}{9} - 23\dfrac{2}{9}$

67. $8\dfrac{5}{12} + 6\dfrac{1}{6}$

68. $6\dfrac{3}{8} + 2\dfrac{1}{16}$

69. $10\dfrac{2}{5} - 3\dfrac{1}{10}$

70. $11\dfrac{4}{7} - 5\dfrac{3}{14}$

71. $1\dfrac{1}{3} + 8\dfrac{1}{3} - 5\dfrac{1}{6}$

72. $4\dfrac{1}{12} + 2\dfrac{7}{12} - 3\dfrac{1}{3}$

B.

73. $2\dfrac{2}{5}$

$7\dfrac{1}{6}$

$+1\dfrac{4}{15}$

74. $7\dfrac{1}{6}$

$1\dfrac{4}{15}$

$+3\dfrac{1}{10}$

75. $18\dfrac{7}{12}$

$-9\dfrac{1}{4}$

76. $18\dfrac{8}{15}$

$-8\dfrac{1}{12}$

77. $11\dfrac{1}{5}$

$3\dfrac{7}{10}$

$+12\dfrac{1}{2}$

78. $3\dfrac{7}{10}$

$9\dfrac{1}{2}$

$+7\dfrac{16}{25}$

79. 45

$-16\dfrac{2}{3}$

80. 76

$-26\dfrac{3}{8}$

81. $5\dfrac{31}{32}$

$-1\dfrac{3}{16}$

82. $17\dfrac{13}{18}$

$-6\dfrac{5}{12}$

83. $8\dfrac{1}{2} + 17\dfrac{7}{8} - 6\dfrac{3}{4}$

84. $5\dfrac{2}{3} + 12\dfrac{8}{9} - 7\dfrac{13}{18}$

C.

Fill in the boxes with a single digit so the statement is true. Explain your answer.

85. The LCM of $3, 5, 8,$ and $\square$ is 120.

86. $4\dfrac{5}{8} + \square\dfrac{2}{3} = 10\dfrac{7}{24}$

87. Find the error(s) in the statement: $16 - 13\dfrac{1}{4} = 3\dfrac{3}{4}$. Correct the statement. Explain how you would avoid this error.

88. Find the error(s) in the statement: $5\dfrac{1}{2} - 2\dfrac{3}{4} = 3\dfrac{1}{4}$. Correct the statement. Explain how you would avoid this error.

89. Find the sum of $12\frac{11}{12}$, 22, and $8\frac{5}{8}$.

90. Find the sum of $26\frac{6}{7}$, 18, and $11\frac{15}{28}$.

91. Find the sum of $15\frac{3}{8}$, $22\frac{1}{2}$, $19\frac{5}{9}$, and $36\frac{2}{3}$.

92. Find the sum of $41\frac{17}{25}$, $34\frac{11}{15}$, 16, and $25\frac{3}{5}$.

93. Find the difference of $6\frac{23}{25}$ and $5\frac{14}{15}$.

94. Find the difference of $8\frac{19}{32}$ and $7\frac{43}{48}$.

95. Find the difference of $-\frac{121}{144}x^2$ and $-\frac{13}{36}x^2$.

96. Find the difference of $-\frac{153}{175}w^3$ and $-\frac{12}{25}w^3$.

For Exercises 97–100, add the polynomials.

97. $\left(\frac{2}{3}x - \frac{3}{4}y\right) + \left(\frac{4}{3}x - \frac{5}{4}y\right)$

98. $\left(\frac{3}{10}y - \frac{3}{5}z\right) + \left(\frac{7}{15}y - \frac{4}{15}z\right)$

99. $\left(\frac{5}{3}x + \frac{3}{8}y - \frac{1}{2}\right) + \left(\frac{4}{5} - \frac{5}{6}x\right)$

100. $\left(\frac{1}{5}x^2 - \frac{3}{5}x - \frac{3}{4}\right) + \left(\frac{1}{8} - \frac{5}{8}x + \frac{9}{10}x^2\right)$

101. The stock of the Eastern Corporation rose $\frac{5}{8}$ points on Monday, $\frac{7}{8}$ points on Tuesday, $\frac{3}{8}$ points on Wednesday, $\frac{5}{8}$ points on Thursday, and $\frac{7}{8}$ points on Friday. What was the total rise for the week?

102. The stock of Northwestern Corporation dropped $\frac{4}{8}$ points on Monday, $\frac{9}{8}$ points on Tuesday, $\frac{5}{8}$ points on Wednesday, 0 points on Thursday, and $\frac{3}{8}$ points on Friday. What was the total decline for the week?

103. If the shrinkage of a $\frac{5}{4}$-inch-thick green board is $\frac{1}{8}$ inch, what will be the thickness of the dried board?

104. Ivan has a bolt that is $\frac{3}{8}$ inch in length. He discovered that it is $\frac{3}{16}$ inch too long. What is the length of the bolt Ivan needs?

105. Jonnie Lee is assembling a rocking horse for his granddaughter. He needs a bolt to reach through a $\frac{7}{8}$-inch piece of steel tubing, a $\frac{1}{16}$-inch bushing, a $\frac{1}{2}$-inch piece of tubing, a $\frac{1}{8}$-inch-thick washer, and a $\frac{1}{4}$-inch-thick nut. How long a bolt does he need?

106. Juanita worked the following hours at her part time job during the month of October:

Week	Oct. 1–7	Oct. 8–14	Oct. 15–21	Oct. 22–28	Oct. 29–31
Hours	$25\frac{1}{2}$	$19\frac{2}{3}$	10	$16\frac{5}{6}$	$4\frac{3}{4}$

How many hours did Juanita work during October?

107. The graph displays the average yearly rainfall for five cities.

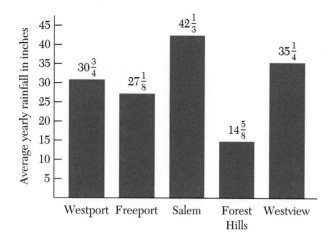

(a) How much more rain falls in Westport during a year than in Freeport?

(b) What is the total amount of rain that falls in a given year in Salem, Forest Hills, and Westview?

(c) In a ten-year period, how much more rain falls in Salem than in Forest Hills?

(d) If the average rainfall in Westview doubles, how much more rain would it receive than Salem?

108. From a tank containing $9\frac{3}{4}$ gallons of gas, Charlie filled a can that holds $2\frac{1}{2}$ gallons. How much gas is still in the tank?

109. Find the length of this pin.

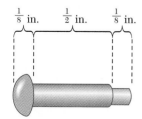

110. What is the overall length of the bolt?

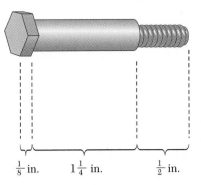

$\frac{1}{8}$ in. $1\frac{1}{4}$ in. $\frac{1}{2}$ in.

111. Is $\frac{5}{8}$ a solution of $x + \frac{2}{3} = \frac{31}{24}$?

112. Is $-\frac{4}{9}$ a solution of $a + \frac{5}{6} = \frac{1}{3}$?

113. A landscaper is building a brick border, one brick wide, around a formal rose garden. The garden is a 10-foot by 6-foot rectangle. Standard bricks are 8 inches × $3\frac{3}{4}$ inches × $2\frac{1}{4}$ inches, and the landscaper is planning on using a $\frac{3}{8}$-inch-wide mortar in the joints. How many bricks are needed for the project? Explain your reasoning.

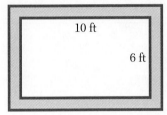

10 ft

6 ft

Exercises 114–118 refer to the Chapter 5 Application. See page 373.

114. Suppose you pick a card at random out of a deck of playing cards. What is the probability that the card will be a three? a four? What is the probability that the card will be a three or a four? Fill in the table below to try to discover the relationship between these probabilities.

Probability of a Three	Probability of a Four	Probability of a Three or a Four

A card is a three *or* four if *either* condition holds. Make a guess about the relationship of the two probabilities when either of two conditions can be true.

115. Test your guess from Exercise 114 by calculating the probability that a card will be a heart or a club. Was your guess correct?

116. Sometimes a complicated probability is easier to calculate using a back-door approach. For example, suppose you want to calculate the probability that a card drawn is an ace or a two or a three or a four or a five or a six or a seven or an eight or a nine or a ten or a jack or a queen. You can certainly add the individual probabilities. (What do you get?) However, another way to look at this situation is to ask, "What is the probability of *not* getting a king?" We reason that if you do not get a king then you do get one of the desired cards. We calculate this by subtracting the probability of getting a king from the number 1. This is because 1 is the sum of all the probabilities that are possible. In this case, it is the sum of the probability of getting a king and the probability of getting one of the other cards. Verify that you get the same probability by doing the subtraction.

117. What is the probability of drawing any card less than 10? Consider aces less than 10s.

118. A bus has 45 passengers on board. Twenty-three of them are male. If a passenger is picked at random, what is the probability that the passenger is female?

STATE YOUR UNDERSTANDING

119. Explain why $-\dfrac{4}{5} = \dfrac{-4}{5} = \dfrac{4}{-5}$.

120. Explain how to simplify $\left(7\dfrac{1}{2}x + 1\dfrac{3}{4}y\right) + \left(\dfrac{1}{6}x - \dfrac{3}{10}y\right)$.

121. Now that you understand addition of fractions, explain why the procedure for changing a mixed number into an improper fraction (Section 5.1) works. Use $4\dfrac{2}{5}$ to illustrate.

122. Explain, in your own words, why we do *not* add the denominators when finding the sum of fractions.

CHALLENGE

123. List the first five common multiples of both 6 and 15. This is also a list of the multiples of what number?

124. Find the sum of $\dfrac{107}{372}$ and $\dfrac{41}{558}$.

125. Subtract $3\dfrac{3}{4}$ from $2\dfrac{3}{8}$.

GROUP ACTIVITY

126. There are several procedures for finding the LCM of two or more numbers. Have each member of your group look for some procedures that are different from the two in your text. Other texts and journals devoted to teacher education may be a good place to look. You might also ask other instructors if they have a favorite (different) method. Write down one other method you discover and report back to the class.

127. Fractions of the form $\dfrac{1}{n}$, where n is a counting number, are called unit fractions. The ancient Egyptians used unit fractions for writing the answers to division exercises. We write $2 \div 5 = \dfrac{2}{5}$, whereas they wrote $2 \div 5 = \dfrac{1}{3} + \dfrac{1}{15}$, using hieroglyphic symbols. Working together, write $\dfrac{3}{5}, \dfrac{2}{7}, \dfrac{6}{7}$, and $\dfrac{8}{9}$ as sums of *different* unit fractions.

MAINTAIN YOUR SKILLS (Sections 4.3, 4.7, 5.2, 5.3)

128. Find the difference of -68 and -55.

129. Is -14 a solution of $x^2 - 3x = 238$?

130. Is 143 a prime number or composite number?

131. Simplify the fraction $\dfrac{374}{561}$.

132. Find the average of 45, -17, -21, 33, and -10.

5.6   Evaluating Expressions and Average (Fractions)

OBJECTIVES

1. Evaluate algebraic expressions and formulas with fractions.
2. Find the average of a set of fractions.
3. Evaluate geometric formulas that involve the number π.

VOCABULARY

The **circumference** of a circle is the distance around the circle.

The **radius** of a circle is the distance from the center to any point on the circle.

The **diameter** of a circle is twice the radius.

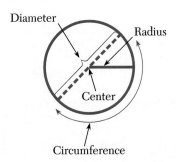

HOW AND WHY
Objective 1

Evaluate algebraic expressions and formulas with fractions.

To evaluate formulas and expressions we use the rules for order of operations. These rules apply to all types of numbers including fractions.

Order of Operations

1. Grouping symbols—perform operations included within the grouping symbols first.
2. Exponents—perform operations indicated by exponents.
3. Multiply and divide—perform multiplication and division from left to right as you come to them.
4. Add and subtract—perform addition and subtraction from left to right as you come to them.

Evaluate $ab + ac + bc$ when $a = \dfrac{2}{3}$, $b = \dfrac{4}{5}$, and $c = \dfrac{1}{4}$.

$$ab + ac + bc = \frac{2}{3}\left(\frac{4}{5}\right) + \frac{2}{3}\left(\frac{1}{4}\right) + \frac{4}{5}\left(\frac{1}{4}\right)$$ **Substitute.**

$$= \frac{8}{15} + \frac{1}{6} + \frac{1}{5}$$ **Multiply from left to right and simplify.**

$$= \frac{16}{30} + \frac{5}{30} + \frac{6}{30}$$ **Build to the common denominator, 30.**

$$= \frac{27}{30}$$ **Add.**

$$= \frac{9}{10}$$ **Simplify.**

The fraction bar used in this chapter is also a grouping symbol. Operations in the numerator and the denominator take precedence over the division shown by the fraction bar. This means that the fractions

$$\frac{16 - 4}{18 + 3} \quad \text{and} \quad \frac{(16 - 4)}{(18 + 3)}$$

have the same value. So,

$$\frac{16 - 4}{18 + 3} + 2 = \frac{12}{21} + 2 = \frac{4}{7} + 2 = 2\frac{4}{7}$$

The following chart summarizes the operations as they relate to rational numbers and algebraic fractions.

Operation	Do We Need a Common Denominator?	Should We Change Mixed Numbers to Improper Fractions?	Should We Multiply by the Reciprocal?	Should We Simplify the Result?
Add	Yes	No	No	Yes
Subtract	Yes	No	No	Yes
Multiply	No	Yes	No	Yes
Divide	No	Yes	Yes	Yes

Examples A–B | **Warm Ups A–B**

Directions: Evaluate the expression for the given values.

Strategy: Substitute the values into the expression and use the order of operations.

A. Evaluate: $xy^2 - z$, if $x = \dfrac{16}{15}$, $y = \dfrac{7}{8}$, and $z = \dfrac{3}{4}$

$xy^2 - z = \dfrac{16}{15}\left(\dfrac{7}{8}\right)^2 - \dfrac{3}{4}$ **Substitute.**

$\quad = \dfrac{16}{15}\left(\dfrac{49}{64}\right) - \dfrac{3}{4}$ **Exponents are done first.**

$\quad = \dfrac{49}{60} - \dfrac{3}{4}$ **Multiply and simplify.**

$\quad = \dfrac{49}{60} - \dfrac{45}{60}$ **Build using the LCM, 60.**

$\quad = \dfrac{4}{60}$ **Subtract.**

$\quad = \dfrac{1}{15}$ **Simplify.**

A. Evaluate: $a^2 - b^2$, if $a = \dfrac{4}{5}$ and $b = \dfrac{2}{3}$

Answer to Warm Up   A. $\dfrac{44}{225}$

B. Evaluate: $P = 2w + 2\ell$, if $w = \dfrac{4}{5}$ and $\ell = \dfrac{3}{8}$

$P = 2\left(\dfrac{4}{5}\right) + 2\left(\dfrac{3}{8}\right)$ **Substitute.**

$= \dfrac{8}{5} + \dfrac{3}{4}$ **Multiply** $\left(2 = \dfrac{2}{1}\right)$ **and simplify.**

$= \dfrac{32}{20} + \dfrac{15}{20}$ **Build. The LCM is 20.**

$= \dfrac{47}{20}$ or $2\dfrac{7}{20}$ **Add.**

B. Evaluate: $P = 2w + 2\ell$, if $w = 1\dfrac{4}{5}$

and $\ell = 2\dfrac{1}{8}$

HOW AND WHY

Objective 2   **Find the average of a set of fractions.**

To find the average of a set of numbers, divide the sum by the number of numbers. The procedure is the same for all types of numbers.

Examples C–D

Directions: Find the average.

Strategy: Divide the sum by the number of fractions.

C. Find the average: $1\dfrac{1}{2}$, $\dfrac{2}{3}$, and $2\dfrac{3}{4}$

$1\dfrac{1}{2} + \dfrac{2}{3} + 2\dfrac{3}{4}$ **Add the three fractions.**

$= \dfrac{3}{2} + \dfrac{2}{3} + \dfrac{11}{4}$ **Change mixed numbers to fractions.**

$= \dfrac{18}{12} + \dfrac{8}{12} + \dfrac{33}{12}$ **The common denominator is 12.**

$= \dfrac{59}{12}$ **Add.**

Now divide the sum by 3.

$\dfrac{59}{12} \div 3 = \dfrac{59}{12} \cdot \dfrac{1}{3}$ **Invert and multiply.**

$= \dfrac{59}{36}$ or $1\dfrac{23}{36}$

The average is $1\dfrac{23}{36}$.

Warm Ups C–D

C. Find the average: $2\dfrac{5}{6}$, $\dfrac{3}{8}$, and $1\dfrac{1}{4}$

Answers to Warm Ups   B. $7\dfrac{17}{20}$ C. $1\dfrac{35}{72}$

D. A class of ten students took a 12-problem test. Their results are listed in the table. What is the class average?

Number of Students	Fraction of Problems Correct
1	$\dfrac{12}{12}$
2	$\dfrac{11}{12}$
3	$\dfrac{10}{12}$
4	$\dfrac{9}{12}$

D. What is the average of the top four scores of Example D?

Strategy: To find the class average, add all the grades together and divide by 10. Because there were two scores of $\dfrac{11}{12}$, three scores of $\dfrac{10}{12}$, and four scores of $\dfrac{9}{12}$ we add $\dfrac{12}{12}, \dfrac{22}{12}, \dfrac{30}{12}$, and $\dfrac{36}{12}$.

$$\frac{12}{12} + \frac{22}{12} + \frac{30}{12} + \frac{36}{12} = \frac{100}{12} \qquad \textbf{Add.}$$

The sum is $\dfrac{100}{12}$.

$$\frac{100}{12} \div 10 = \frac{100}{12} \cdot \frac{1}{10} \qquad\qquad \textbf{Divide by 10.}$$
$$= \frac{10}{12}$$

The class average was $\dfrac{10}{12}$ correct.

C A U T I O N **Do not simplify the answer since the test score is based on 12 problems.**

HOW AND WHY

Objective 3 **Evaluate geometric formulas that involve the number π.**

Formulas for geometric figures that involve circles contain the number called pi (π). The number, π, is the quotient of the *circumference* of (distance around) the circle and its diameter. This number is the same for *every* circle no matter how large or small. This remarkable fact was discovered over a long period of time historically and during that time a large number of approximations have been used. Here are some of the approximations:

$$3 \qquad 3\frac{1}{8} \qquad 3\frac{1}{7}\left(\text{or } \frac{22}{7}\right) \qquad \frac{355}{113}$$

Answer to Warm Up   D. $\dfrac{11}{12}$

Because $\dfrac{22}{7}$ is an approximate value, when we use this fraction we write the symbol for approximately equal to, $\pi \approx \dfrac{22}{7}$. For decimal approximations, see Chapter 6.

Formulas

Circle

If C is the circumference, d is the diameter, and r is the radius of a circle, then

$$C = \pi d \quad \text{or} \quad C = 2\pi r$$

If A is the area, then

$$A = \pi r^2$$

Semicircle

If L is half the circumference of a circle, then

$$L = \pi r \quad \text{or} \quad L = \frac{1}{2}\pi d$$

Since cylinders, spheres, and cones contain circles, their formulas also contain the number π.

FORMULAS

Name	Picture	Formula
Cylinder (right circular cylinder)	Cylinder	$V = \pi r^2 h$ r = radius of the base h = height of the cylinder
Sphere	Sphere	$V = \dfrac{4}{3}\pi r^3$ r = radius of the sphere
Cone (right circular cone)	Cone	$V = \dfrac{1}{3}\pi r^2 h$ r = radius of the base h = height of the cone

Examples E–H	**Warm Ups E–H**

Directions: Evaluate the formula.

Strategy: Substitute the values of the measurements into the appropriate formula and evaluate.

E. Find the circumference of the circle.

$C = 2\pi r$ Formula for circumference.

$C \approx 2\left(\dfrac{22}{7}\right)(7)$ in.   Substitute; let $\pi \approx \dfrac{22}{7}$.

$C \approx 44$ in.   Multiply.

The circumference is approximately 44 inches.

E. Find the circumference of the circle.

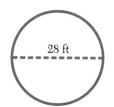

F. Find the area of a circle with a radius of $\dfrac{3}{4}$ inch.

$A = \pi r^2$ Formula for area.

$A \approx \left(\dfrac{22}{7}\right)\left(\dfrac{3}{4}\ \text{in.}\right)^2$ Substitute; let $\pi \approx \dfrac{22}{7}$.

$A \approx \left(\dfrac{22}{7}\right)\left(\dfrac{9}{16}\ \text{in.}^2\right)$ Exponents are done first.

$A \approx \dfrac{99}{56}\ \text{in.}^2$ Multiply.

$A \approx 1\dfrac{43}{56}\ \text{in.}^2$ Change to a mixed number.

The area is approximately $1\dfrac{43}{56}$ in.^{2}.

F. Find the area of a circle with radius 14 centimeters.

G. Find the volume of the cylinder.

G. Find the volume of a cylinder that has a circular base with radius 12 cm and a height of 13 cm. Write the approximate answer as a mixed number.

Answers to Warm Ups   E. 88 ft   F. 616 cm^2 G. $5883\dfrac{3}{7}$ cm^3

$V = \pi r^2 h$ **Formula for cylinder volume.**

$V \approx \dfrac{22}{7}\,(2\text{ ft})^2(6\text{ ft})$ **Substitute. Let $\pi \approx \dfrac{22}{7}$.**

$V \approx \dfrac{528}{7}\text{ ft}^3$

$V \approx 75\dfrac{3}{7}\text{ ft}^3$

The volume is approximately $75\dfrac{3}{7}$ ft^3.

H. Find the volume of the cone.

5 m

3 m

$V = \dfrac{1}{3}\pi r^2 h$ **Formula for cone volume.**

$V \approx \dfrac{1}{3}\left(\dfrac{22}{7}\right)(3\text{ m})^2(5\text{ m})$ **Substitute. Let $\pi \approx \dfrac{22}{7}$.**

$V \approx \dfrac{330}{7}\text{ m}^3$

$V \approx 47\dfrac{1}{7}\text{ m}^3$

The volume is approximately $47\dfrac{1}{7}$ m^3.

H. Find the volume of a right circular cone that has a base diameter of 3 ft and a height of 7 ft. Write the approximate answer as a mixed number.

Answer to Warm Up   H. $16\dfrac{1}{2}$ ft^3

Exercises 5.6

OBJECTIVE 1: *Evaluate algebraic expressions and formulas with fractions.*

A.

1. Evaluate $x + y$, if $x = \frac{2}{3}$ and $y = \frac{1}{6}$.

2. Evaluate $x - y$, if $x = \frac{2}{3}$ and $y = \frac{1}{6}$.

3. Evaluate $-x + y$, if $x = \frac{2}{3}$ and $y = \frac{1}{6}$.

4. Evaluate $-x - y$, if $x = \frac{2}{3}$ and $y = \frac{1}{6}$.

5. Evaluate $\frac{y}{x}$, if $x = \frac{2}{3}$ and $y = \frac{1}{6}$.

6. Evaluate $\frac{x}{y}$, if $x = \frac{2}{3}$ and $y = \frac{1}{6}$.

7. Evaluate $x^2 + y^2$, if $x = \frac{2}{3}$ and $y = \frac{1}{6}$.

8. Evaluate $x^2 - y^2$, if $x = \frac{2}{3}$ and $y = \frac{1}{6}$.

B.

Evaluate each of the expressions if $x = \frac{5}{12}$, $y = \frac{3}{16}$, and $z = \frac{3}{20}$.

9. $x - y + z$

10. $x^2 z^2$

11. $(x + y) \div z$

12. $y \div (x - z)$

Evaluate each of the expressions if $a = -\dfrac{4}{5}$, $b = \dfrac{7}{15}$, and $c = -\dfrac{5}{9}$.

13. $a + b + c$

14. abc

15. $(b - c) \div (a + c)$

16. $(a - c) \div (b + c)$

17. $b^2 - a$

18. $a^2 - b$

OBJECTIVE 2: *Find the average of a set of fractions.*

Find the average.

A.

19. $\dfrac{1}{6}$ and $\dfrac{7}{12}$

20. $\dfrac{2}{3}$ and $\dfrac{11}{15}$

21. $\dfrac{1}{2}, \dfrac{1}{4}$, and $\dfrac{3}{4}$

22. $\dfrac{1}{3}, \dfrac{2}{3}$, and $\dfrac{5}{6}$

B.

23. $\dfrac{1}{2}, \dfrac{3}{4}, \dfrac{5}{8}$, and $\dfrac{13}{16}$

24. $\dfrac{2}{3}, \dfrac{2}{9}, \dfrac{4}{27}$, and $\dfrac{8}{81}$

25. $3\dfrac{1}{3}, 4\dfrac{1}{6}$, and $2\dfrac{2}{9}$

26. $3\frac{5}{8}$, $2\frac{3}{4}$, and $4\frac{1}{2}$

OBJECTIVE 3: *Evaluate geometric formulas that involve the number* π.

A.

27. Find the circumference of a circle with radius 7 cm. **28.** Find the area of a circle with radius 7 cm.

29. Find the circumference.

14 ft

30. Find the area of the circle in Exercise 29.

31. Find the circumference.

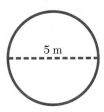

5 m

32. Find the area of the circle in Exercise 31.

33. Find the volume.

2 in.

7 in.

34. Find the volume.

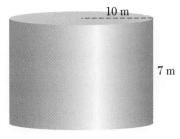

10 m

7 m

B.

35. Find the perimeter.

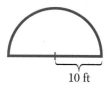

10 ft

36. Find the perimeter.

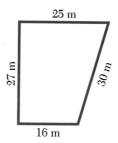

25 m

27 m

30 m

16 m

37. Find the area.

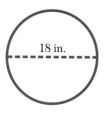

18 in.

38. Find the area.

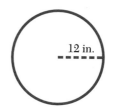

12 in.

39. Find the area.

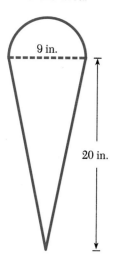

9 in.

20 in.

40. Find the area.

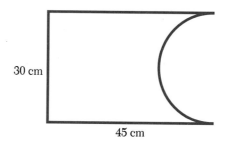

30 cm

45 cm

41. Find the volume of a cylinder with radius $6\frac{1}{2}$ inches and height 4 feet.

42. Find the volume of a sphere with diameter 4 cm.

C.

43. Find the area of a circle if the radius is $\frac{7}{11}$ inch. Let $\pi \approx \frac{22}{7}$.

44. Find the area of a circle if the radius is $\frac{7}{8}$ cm. Let $\pi \approx \frac{22}{7}$.

45. The simple interest, i, earned on an investment of p dollars at an interest rate, r, of t years is given by the formula $i = prt$. Find the interest earned on $50 at 6% $\left(\frac{6}{100}\right)$ for 9 months $\left(\frac{3}{4} \text{ year}\right)$.

46. Use the formula in Exercise 45 to find the interest earned on an investment of $500 at 8% $\left(\frac{8}{100}\right)$ for $1\frac{1}{2}$ years.

47. Five students took a 15-problem makeup test. Three scored $\frac{11}{15}$ correct and the other two scored $\frac{8}{15}$ correct. What was their average score?

48. What is the average of the four highest scores in Exercise 47?

49. A class of ten students took a 20-problem test. Their results are listed in the table. What is the class average?

Number of Students	Fraction of Problems Correct
2	$\dfrac{20}{20}$
1	$\dfrac{19}{20}$
3	$\dfrac{17}{20}$
2	$\dfrac{16}{20}$
2	$\dfrac{15}{20}$

50. What is the average of the five highest scores in Exercise 49?

51. Louise caught six salmon. The salmon measured $23\frac{1}{4}$ inches, $31\frac{5}{8}$ inches, $42\frac{3}{4}$ inches, $28\frac{5}{8}$ inches, $35\frac{3}{4}$ inches, and 40 inches in length. What is the average length of the salmon?

52. Nurse Wayne weighs five newborns at General Hospital. They weigh $6\frac{1}{2}$ lb, $7\frac{3}{4}$ lb, $9\frac{3}{8}$ lb, $7\frac{1}{2}$ lb, and $8\frac{7}{8}$ lb. What is the average weight of the infants?

53. Kohough Inc. packs a variety carton of canned seafood. Each carton contains three $3\frac{1}{2}$ oz cans of smoked sturgeon, five $6\frac{3}{4}$ oz cans of tuna, four $5\frac{1}{2}$ oz cans of salmon, and four $10\frac{1}{2}$ oz cans of sardines. How many ounces of seafood are in the carton? If the carton sells for $52, to the nearest cent, what is the average cost per ounce?

54. In a walk for charity seven people walk $2\frac{7}{8}$ miles, six people walk $3\frac{4}{5}$ miles, nine people walk $4\frac{1}{4}$ miles, and five people walk $5\frac{3}{4}$ miles. What is the total number of miles walked? If the charity raises $2355, what is the average amount raised per mile rounded to the nearest dollar?

55. Find the volume. Let $\pi \approx \frac{22}{7}$.

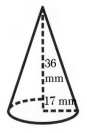

56. Find the volume of a sphere with a diameter of 2 feet 4 inches in cubic inches. Let $\pi \approx \frac{22}{7}$.

57. Find the volume of the solid formed by a cone with a diameter of 5 inches and a height of 8 inches that is topped with a hemisphere of the same diameter. Let $\pi \approx \dfrac{22}{7}$.

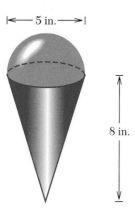

|← 5 in. →|

8 in.

58. Wind and rain erosion caused this arch formation that is nearly circular in shape. Geologists estimate the weight of rock that has been removed by viewing the hole in the formation as approximately a cylinder. If the diameter is 30 feet, the thickness is 8 feet, and the rock weighs 165 pounds per cubic foot, find the estimated weight of the rock that was removed. Round to the nearest ton. Let $\pi \approx \dfrac{22}{7}$.

Exercises 59–60 refer to the Chapter 5 Application. See pages 373 and 418.

59. Let A represent a certain condition, and let B represent a different condition. $P(A)$ represents the probability that condition A is true, and $P(B)$ represents the probability that condition B is true. In Section 5.3, Exercise 80, you discovered the relationship

$$P(A \text{ and } B) = P(A) \cdot P(B)$$

Verify the formula by calculating the probability of drawing the jack of spades.

60. Use the formula in Exercise 59 to answer the following questions. A survey shows that 49 out of 100 people think the President is doing a good job. The same survey showed that only 32 out of 100 people approve of the President's health plan. What is the probability that a person chosen a random both approves how the President is doing his job and approves of his health plan? Can you assume that 51 out of 100 people disapprove of the job the President is doing? Explain.

STATE YOUR UNDERSTANDING

61. Explain how to evaluate $\dfrac{a-b}{5+c}$, if $a = 4\frac{1}{2}$, $b = \frac{1}{8}$, and $c = \frac{1}{4}$.

62. Explain how the formula for the volume of a cylinder, $V = \pi r^2 h$, is consistent with the volume formula, $V = Bh$, used in Chapter 2.

63. Must the average of a set of fractions be larger than the smallest fraction and smaller than the largest fraction? Why?

64. If the average of five fractions is multiplied by 5, the product is *always* the same as the sum of the five fractions. Explain why this is true.

CHALLENGE

65. Find the value of $x^3 + y^2 + z$, if $x = 1\frac{1}{2}$, $y = 2\frac{2}{3}$, and $z = 3\frac{3}{4}$.

66. Find the value of $x + y^2 + z^3$, if $x = 1\frac{1}{2}$, $y = 2\frac{2}{3}$, and $z = 3\frac{3}{4}$.

67. The Acme Fish Company pays \$4500 per ton for crab. Jerry catches $3\frac{2}{5}$ tons, his brother Joshua catches $1\frac{1}{2}$ times as much as Jerry. Their sister, Salicita, catches $\frac{7}{8}$ the amount that Joshua does. What is the total amount paid to the three people by Acme, to the nearest dollar?

GROUP ACTIVITY

68. A satellite travels in a circular orbit around the Earth once every $2\frac{1}{2}$ hours. The satellite is orbiting 2900 km above the surface of the Earth. The diameter of the Earth is approximately 12,800 km. Calculate the speed of the satellite and explain your strategy.

MAINTAIN YOUR SKILLS (Sections 4.9, 5.2)

69. Find the prime factorization of 299.

70. Find the prime factorization of 622.

71. Solve: $2x - 8 = 10 - x$

72. Solve: $3a + 12 = 4 - a$

73. Solve: $-2x + 12 = 3x + 32$

5.7 Solving Equations Involving Rational Numbers (Fractions)

OBJECTIVE   Solve equations involving fractions.

HOW AND WHY
Objective   **Solve equations involving fractions.**

The four properties of equations from Chapter 3 are used to solve equations with all types of numbers including fractions. For instance:

$$\frac{3}{4}x + \frac{2}{5} = \frac{7}{10}$$

$$\frac{3}{4}x + \frac{2}{5} + \left(-\frac{2}{5}\right) = \frac{7}{10} + \left(-\frac{2}{5}\right) \qquad \text{Add } -\frac{2}{5} \text{ to both sides.}$$

$$\frac{3}{4}x = \frac{3}{10} \qquad \text{Add.}$$

$$\left(\frac{4}{3}\right)\left(\frac{3}{4}x\right) = \left(\frac{4}{3}\right)\left(\frac{3}{10}\right) \qquad \text{Multiply both sides by } \frac{4}{3}, \text{ the reciprocal of the coefficient of } x, \text{ to get } 1x.$$

$$x = \frac{2}{5} \qquad \text{Multiply.}$$

Using these steps requires a lot of work with fractions. It is possible to eliminate the fractions early by forming a new, equivalent equation. We use the LCM of the denominators of the fractions and the multiplication law of equality. Here is the same example done this way. There are no fewer steps, but there are fewer fractions to deal with.

$$\frac{3}{4}x + \frac{2}{5} = \frac{7}{10} \qquad \text{The LCM of the denominators 4, 5, and 10 is 20.}$$

$$20\left(\frac{3}{4}x + \frac{2}{5}\right) = 20\left(\frac{7}{10}\right) \qquad \text{Multiply both sides by the LCM, 20.}$$

$$20\left(\frac{3}{4}x\right) + 20\left(\frac{2}{5}\right) = 20\left(\frac{7}{10}\right) \qquad \text{Multiply, using the distributive property.}$$

$$15x + 8 = 14 \qquad \text{Simplify. The fractions have been "cleared."}$$

$$15x = 6 \qquad \text{Subtract 8 from both sides.}$$

$$x = \frac{6}{15} \qquad \text{Divide both sides by 15.}$$

$$x = \frac{2}{5} \qquad \text{Simplify.}$$

▶ *To solve an equation involving fractions*

1. Find the LCM of all the denominators.
2. Multiply both sides of the equation by the LCM.
3. Solve the resulting equation.

Examples A–E	Warm Ups A–E

Directions: Solve.

Strategy: Clear the fractions by multiplying both sides by the LCM of the denominators. Then, use the properties of equations to isolate the variable.

A. Solve: $\dfrac{2}{3}a - \dfrac{4}{5} = \dfrac{8}{15} + \dfrac{1}{3}a$

$$15\left(\dfrac{2}{3}a - \dfrac{4}{5}\right) = 15\left(\dfrac{8}{15} + \dfrac{1}{3}a\right)$$ Multiply both sides by the LCM, 15.

$$15\left(\dfrac{2}{3}a\right) - 15\left(\dfrac{4}{5}\right) = 15\left(\dfrac{8}{15}\right) + 15\left(\dfrac{1}{3}a\right)$$ Simplify using the distributive property.

$$10a - 12 = 8 + 5a$$ Simplify. The fractions are cleared.

$$5a - 12 = 8$$ Subtract $5a$ from both sides.

$$5a = 20$$ Add 12 to both sides.

$$a = 4$$ Divide both sides by 5.

The solution is $a = 4$.   See check in Example B.

A. Solve: $b - \dfrac{3}{8} = \dfrac{1}{8} + \dfrac{1}{2}b$

Calculator Example:

B. Check Example A.

$\dfrac{2}{3}(4) - \dfrac{4}{5} \approx 1.866667$ Substitute 4 for a on the left side of the equation and evaluate using a calculator.

$\dfrac{8}{15} + \dfrac{1}{3}(4) \approx 1.866667$ Substitute 4 for a on the right side of the equation and evaluate using a calculator.

The value, $a = 4$, checks.   The left and right sides have the same value when $a = 4$.

B. Check Warm Up A.

C. Solve: $\dfrac{1}{2}\left(\dfrac{y}{2} + \dfrac{10}{3}\right) = \dfrac{5}{6}$

$$\dfrac{1}{2}\left(\dfrac{y}{2}\right) + \dfrac{1}{2}\left(\dfrac{10}{3}\right) = \dfrac{5}{6}$$ Multiply using the distributive property.

$$\dfrac{y}{4} + \dfrac{5}{3} = \dfrac{5}{6}$$ Simplify.

$$12\left(\dfrac{y}{4} + \dfrac{5}{3}\right) = 12\left(\dfrac{5}{6}\right)$$ Multiply both sides by the LCM, 12.

$$\dfrac{12y}{4} + \dfrac{60}{3} = \dfrac{60}{6}$$

$$3y + 20 = 10$$ Simplify.

$$3y = -10$$ Subtract 20 from each side.

C. Solve: $\dfrac{2}{3}\left(\dfrac{21t}{20} + \dfrac{9}{4}\right) = \dfrac{2}{5}$

Answers to Warm Ups   A. $b = 1$ B. Both sides are equal to 0.625 when $b = 1$.   C. $t = -\dfrac{11}{7}$

$$y = -\frac{10}{3}$$ **Divide both sides by 3. Check is left for the student.**

The solution is $y = -\frac{10}{3}$.

D. Solve: $5c - \frac{14}{15} - 2c = \frac{2}{5}$

$15\left(5c - \frac{14}{15} - 2c\right) = 15\left(\frac{2}{5}\right)$ **Multiply each side by the LCM.**

$75c - 14 - 30c = 6$ **Simplify.**

$45c - 14 = 6$

$45c = 20$

$c = \frac{20}{45}$

$c = \frac{4}{9}$ **Check is left for the student.**

The solution is $c = \frac{4}{9}$.

D. Solve: $6d - \frac{7}{20} - 3d = \frac{13}{4}$

E. The perimeter, P, of a rectangle is $8\frac{1}{3}$ inches. If the width, w, is $\frac{5}{6}$ inch, what is the length, ℓ, of the rectangle?

$P = 2w + 2\ell$ **Perimeter formula.**

$8\frac{1}{3} = 2\left(\frac{5}{6}\right) + 2\ell$ **Substitute $8\frac{1}{3}$ for P and $\frac{5}{6}$ for w.**

$\frac{25}{3} = \frac{5}{3} + 2\ell$ **Write $8\frac{1}{3}$ as an improper fraction.**

$3\left(\frac{25}{3}\right) = 3\left(\frac{5}{3} + 2\ell\right)$ **Multiply both sides by 3.**

$3\left(\frac{25}{3}\right) = 3\left(\frac{5}{3}\right) + 3(2\ell)$ **Multiply using the distributive property.**

$25 = 5 + 6\ell$ **Simplify.**

$20 = 6\ell$ **Subtract 5 from each side.**

$\frac{20}{6} = \ell$ **Divide each side by 6.**

$\frac{10}{3} = \ell$

$3\frac{1}{3} = \ell$ **Change to a mixed number.**

The length of the rectangle is $3\frac{1}{3}$ inches.

E. The perimeter, P, of a rectangle is $9\frac{3}{8}$ inches. If the width, w, is $\frac{7}{8}$ inch, what is the length, ℓ, of the rectangle?

Answers to Warm Ups   D. $d = \frac{6}{5}$ E. $3\frac{13}{16}$ inches

Exercises 5.7

OBJECTIVE: *Solve equations involving fractions.*

Solve.

A.

1. $2x - \dfrac{1}{5} = \dfrac{1}{5}$

2. $3x - \dfrac{2}{7} = \dfrac{1}{7}$

3. $\dfrac{1}{2}x + \dfrac{2}{3} = \dfrac{5}{3}$

4. $\dfrac{1}{3}x + \dfrac{3}{5} = \dfrac{8}{5}$

5. $\dfrac{2}{3}x + 1 = \dfrac{5}{3}$

6. $\dfrac{3}{5}x + 1 = \dfrac{11}{5}$

B.

7. $\dfrac{1}{5}a - \dfrac{1}{3} = \dfrac{1}{5}$

8. $\dfrac{1}{4}x - \dfrac{2}{3} = \dfrac{1}{2}$

9. $\dfrac{1}{6} + \dfrac{1}{2}y = \dfrac{2}{3}$

10. $\dfrac{5}{8} + \dfrac{1}{4}z = \dfrac{1}{2}$

11. $\dfrac{3}{7} + \dfrac{1}{4}c = \dfrac{1}{21}$

12. $\dfrac{4}{9} + \dfrac{1}{6}d = \dfrac{1}{3}$

13. $\dfrac{1}{4}y - \dfrac{3}{2} = \dfrac{1}{8}$

14. $\dfrac{1}{6}x - \dfrac{1}{4} = \dfrac{5}{12}$

15. $\dfrac{5}{7} - \dfrac{1}{2}x = \dfrac{3}{14}$

479

16. $\dfrac{5}{6} - \dfrac{1}{5} y = \dfrac{7}{30}$

17. $\dfrac{3}{10} y - \dfrac{4}{5} = \dfrac{7}{20}$

18. $\dfrac{5}{11} x - \dfrac{3}{22} = \dfrac{1}{2}$

C.

19. $\dfrac{a}{6} - \dfrac{a}{5} = \dfrac{2}{3} + \dfrac{1}{2}$

20. $\dfrac{x}{4} - \dfrac{x}{5} = \dfrac{1}{2} + \dfrac{3}{10}$

21. $\dfrac{a}{9} - \dfrac{1}{2} = \dfrac{5}{18} + \dfrac{2}{3}$

22. $\dfrac{y}{10} - \dfrac{7}{15} = \dfrac{5}{6} + \dfrac{1}{5}$

23. $\dfrac{3x}{10} + \dfrac{5x}{12} = \dfrac{5}{6} - \dfrac{1}{30}$

24. $\dfrac{5y}{7} + \dfrac{3y}{14} = \dfrac{5}{42} + \dfrac{1}{21}$

25. $\dfrac{13x}{20} - \dfrac{x}{3} = \dfrac{2}{5} - \dfrac{22}{30}$

26. $\dfrac{9y}{16} - \dfrac{3y}{8} = \dfrac{5}{4} - \dfrac{3}{2}$

27. $\dfrac{6}{7} c - 12 = \dfrac{3}{5} + 5$

28. $-\dfrac{4}{7} b + 3 = \dfrac{3}{4} - 7$

29. $-\dfrac{11}{12} b - \dfrac{2}{3} = -\dfrac{5}{6} + 3$

30. $-\dfrac{4}{9} a - 2 = \dfrac{5}{6} - \dfrac{5}{9}$

31. $\dfrac{2a}{18} + \dfrac{7a}{72} = \dfrac{1}{18} + \dfrac{7}{9}$

32. $\dfrac{6a}{11} + \dfrac{5a}{44} = \dfrac{9}{22} + \dfrac{5}{66}$

33. $\dfrac{9}{8} - \dfrac{7z}{20} = \dfrac{3}{50} + \dfrac{13}{24}$

34. $\dfrac{9}{14} - \dfrac{3y}{7} = \dfrac{5}{42} + \dfrac{2}{21}$

35. $\dfrac{3}{8} + \dfrac{5a}{6} - \dfrac{7a}{24} + \dfrac{5}{9} = 0$

36. $\dfrac{3}{5} + \dfrac{9b}{25} - \dfrac{7b}{75} + \dfrac{7}{10} = 0$

37. $\dfrac{1}{5}\left(\dfrac{8}{3} + \dfrac{4y}{15}\right) = \dfrac{11}{25}$

38. $\dfrac{1}{3}\left(\dfrac{7}{4} + \dfrac{5x}{12}\right) = \dfrac{13}{24}$

39. $\dfrac{2}{5}\left(\dfrac{2p}{3} - \dfrac{5}{8}\right) = -\dfrac{1}{20}$

40. $\dfrac{3}{5}\left(\dfrac{2a}{3} - \dfrac{1}{6}\right) = \dfrac{7}{50}$

41. $3x - \dfrac{7}{8} = 5x + \dfrac{2}{3}$

42. $\dfrac{16}{15} - \dfrac{4}{9}x = x + \dfrac{5}{6}$

43. $\dfrac{23}{24}x - \dfrac{2}{3}x = \dfrac{11}{12}x + 10$

44. $\dfrac{19}{30}y - \dfrac{4}{7} = \dfrac{33}{35}y$

45. $\dfrac{7}{18}c - \dfrac{5}{27} = -\dfrac{8}{9} - \dfrac{1}{9}c$

46. $\dfrac{9}{14}d + \dfrac{3}{7} = \dfrac{8}{21}d + \dfrac{5}{21}$

47. $\dfrac{1}{12}\left(\dfrac{17}{3}x + \dfrac{11}{4}\right) = \dfrac{1}{24}(13x + 7)$ **48.** $\dfrac{1}{3}\left(\dfrac{5}{3} + 2y\right) = \dfrac{1}{81}(32y + 36)$

49. The perimeter of a rectangle is $6\dfrac{2}{3}$ inches. If the length is $2\dfrac{1}{2}$ inches, what is the width of the rectangle?

50. The perimeter of a rectangle is $4\dfrac{3}{8}$ inches. If the length is $1\dfrac{3}{4}$ inches, what is the width of the rectangle?

51. The width of a rectangle is one-half the length. What is the length of the rectangle if the perimeter is 36 inches?

52. The width of a rectangle is one-third the length. What is the length of the rectangle if the perimeter is 72 inches?

53. A number is divided by 3. If the quotient is increased by $\frac{2}{5}$, the result is $1\frac{11}{15}$. Find the number.

54. A number is divided by 6. If the quotient is decreased by $\frac{5}{9}$, the result is $\frac{2}{3}$. Find the number.

55. Find the base of the triangle.

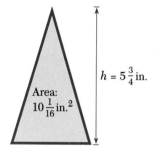

$h = 5\frac{3}{4}$ in.

Area: $10\frac{1}{16}$ in.^{2}

56. Find the height of a triangle whose base, b, is $7\frac{3}{8}$ feet and whose area, A, is $9\frac{7}{32}$ ft^2.

57. Find the height of the cylinder. Let $\pi \approx 3\frac{1}{7}$.

$r = 5$ cm

$V = 770$ cm^3

58. Find the height of a cone whose volume is 1100 ft^3 and whose radius is $2\frac{1}{2}$ feet. Let $\pi \approx \frac{22}{7}$.

Exercises 59–60 refer to the Chapter 5 Application. See pages 373 and 453.

59. According to a survey, 11 out of 20 women like a man's face best of all his physical attributes. In the same survey, 59 out of 100 women like a man's face or his hands best among his physical attributes. What is the probability that a woman chosen at random likes a man's hands best?

60. The same survey states that 1 in 4 men like a woman's face best of all her physical attributes whereas 3 in 10 men like a women's face or hair best. What is the probability that a man chosen at random prefers a woman's hair?

STATE YOUR UNDERSTANDING

61. Explain why multiplying an equation by the LCM of the denominators makes all the fractions disappear. Illustrate with an example.

62. Find the sum $\frac{5}{6} + \frac{7}{8}$ and solve the equation $\frac{x}{6} + \frac{7}{8} = 5$. Along with your solutions, write an explanation of each of the steps. At the end, write a brief description explaining why the procedures are different.

CHALLENGE

63. Finish writing the equation $\frac{3}{4}x - \frac{5}{12} = -\frac{1}{6} + \square$ so that $x = \frac{5}{8}$ is a solution.

64. Finish writing the equation $\frac{2}{3}y - \frac{5}{12} = 9 + \square$ so that $y = \frac{5}{8}$ is a solution.

GROUP ACTIVITY

65. Working separately, have each member of your group make up two equations containing fractions. Write them so that $x = \dfrac{5}{6}$ is the solution of one and $x = -\dfrac{3}{4}$ is the solution of the other. After they are written, get together to compare your problems, listing the differences and the similarities.

MAINTAIN YOUR SKILLS (Section 5.3)

66. Simplify: $\dfrac{25x^2}{45x}$

67. Simplify: $\dfrac{72xy^2}{81xy}$

68. Multiply: $\left(-\dfrac{1}{2}\right)\left(-\dfrac{4}{5}\right)\left(-\dfrac{10}{11}\right)$

69. Find the product of $-\dfrac{25y}{32}$ and $\dfrac{8y}{15}$.

70. Find the quotient of $-\dfrac{84ab}{25}$ and $\dfrac{35b}{27}$.

CHAPTER 5

OPTIONAL Group Project *(2–3 weeks)*

Your group is the estimation team for a company that sells and installs custom tile surfaces. It is your job to work from blueprints or other architectural drawings and produce for the client moderate and deluxe plans for materials and labor.

The materials list for standard tile installation includes the following:

- Thin set mortar—to fix the tiles to the floor or wall
- Mortar high strength additive—used for tiling walls
- Tiles
- Grout—to fill the joints between tiles
- Grout sealer

One 50-lb bag of thin set mortar covers 90 to 100 ft^2. If tiling walls, it is necessary to use a high strength additive with the mortar. Estimate one $2\frac{1}{2}$-gallon pail of additive per 50 lb of mortar. Nonsanded grout may be used for joints of $\frac{1}{8}$ in. or less. Sanded grout is recommended for joints of $\frac{1}{8}$ to $\frac{1}{4}$ in. The table below lists grout coverage for several combinations of tile and joint size (all measured in inches).

Tile Size	Joint Width	Coverage/lb
Nonsanded Grout		
$4\frac{1}{4} \times 4\frac{1}{4} \times \frac{1}{4}$	$\frac{1}{16}$ in.	16 ft^2
$4\frac{1}{4} \times 4\frac{1}{4} \times \frac{1}{4}$	$\frac{1}{8}$ in.	8 ft^2
$6 \times 6 \times \frac{1}{4}$	$\frac{1}{16}$ in.	18 ft^2
$6 \times 6 \times \frac{1}{4}$	$\frac{1}{8}$ in.	14 ft^2
Sanded Grout		
$4 \times 4 \times \frac{1}{4}$	$\frac{1}{4}$ in.	4 ft^2
$6 \times 6 \times \frac{1}{4}$	$\frac{1}{4}$ in.	3 ft^2

Your client would like to tile the floor, countertops, and the walls behind the tub in a master bathroom. The floor plan is below.

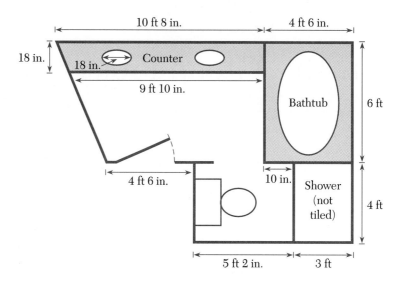

The shower is a fiberglass unit and does not need to be tiled. The tub is 5 ft long, 30 in. wide, and 20 in. high. The horizontal surface surrounding the tub is to be tiled. The walls surrounding the tub should each be tiled to a height of 6 ft off the floor. Include 4-in.-tiled backsplash around the counter. The client wants the tile on the counter, the tile surrounding the tub, and the tile on the walls to match. A coordinating tile should be selected for the floor.

Your group report will include a detailed list of the various surfaces to be tiled and their areas. Then go to a tile or home improvement store and get prices for all the materials needed. Prepare two cost lists for the client, one for a deluxe plan and one for a moderate plan. Figure the labor costs at $15 per ft². Finish your report with a recommendation to the client that includes a rationale for the decisions you made.

CHAPTER 5

True-False Concept Review

ANSWERS

Check your understanding of the language of algebra and arithmetic. Tell whether each of the following statements is True (always true) or False (not always true). For each statement you judge to be false, revise it to make a statement that is true.

1. _____

   1. If the numerator and denominator of a fraction are both positive, then the denominator is less than the numerator.

2. _____

   2. An improper fraction can be pictured with one or more unit regions.

3. _____

   3. A proper fraction such as $\dfrac{5}{6}$ cannot be changed to a mixed number.

4. _____

   4. The integer 7 is a multiple of 35.

5. _____

   5. Every whole number, greater than one (1), has at least two factors.

6. _____

   6. Every whole number is either a prime number or a composite number.

7. _____

   7. Every even number larger than 2 is a composite number.

8. _____

   8. When a fraction is simplified, its value is made smaller.

9. _____

   9. Every mixed number can be simplified.

10. _____

   10. It is possible for a number to have exactly five factors.

11. _____

   11. The numerator of the sum of two fractions is the sum of the two numerators.

12. _____

   12. Before simplifying, the numerator of the product of two fractions is the product of the two numerators.

13. _____

   13. It is possible for a number to have exactly five different prime factors.

14. _____

14. It is possible to multiply and divide mixed numbers by changing them into fractions first.

15. _____

15. To find the reciprocal of a mixed number, invert the numerator and denominator of the fractional part of the mixed number.

16. _____

16. Every division problem involving fractions can be changed to a multiplication problem.

17. _____

17. A rational number must be written in fraction form.

18. _____

18. The reciprocal of a fraction whose value is greater than one is a proper fraction.

19. _____

19. The opposite of a fraction has a negative numerator.

20. _____

20. The absolute value of a fraction is never negative.

21. _____

21. Building and simplifying fractions are both methods for giving fractions different names.

22. _____

22. Building fractions is useful only for adding and subtracting fractions.

23. _____

23. A fraction renamed by building has a larger value than the original fraction.

24. _____

24. It is possible for the LCM of two numbers to be the same as the product of the two numbers.

25. _____

25. The smallest divisor of the LCM of three numbers is the smallest of the three numbers.

26. _____

26. The LCM of four numbers is the smallest number that has each of the four numbers as a factor.

27. _____

27. The LCM of a list of numbers is always larger than any number in the list.

28. _____

28. The denominator of the sum of two fractions is the sum of the denominators of the two fractions.

29. _____

29. Unlike fractions cannot be added or subtracted.

30. _____

30. Every subtraction problem involving fractions can be changed to an addition problem.

31. _____

31. It is necessary for fractions to have a common denominator if they are to be subtracted.

32. _____

32. Mixed numbers can be added without first changing them to fractions.

33. _____

33. Mixed numbers can always be added without carrying.

34. _____

34. Subtracting mixed numbers sometimes involves the same kind of "borrowing" as subtracting whole numbers.

35. _____

35. It is possible to add and subtract mixed numbers by changing them into fractions first.

36. _____

36. The order of operations for rational numbers in fraction form is different from the order for whole numbers because the fractions must be eliminated first.

37. _____

37. The same properties of equations that are used for equations with whole numbers can be used for equations with fractions or mixed numbers.

38. _____

38. The multiplication property of equality can be used to eliminate the fractions in an addition problem.

39. _____

39. The multiplication property of equality can be used to eliminate the fractions in an equation.

40. _____

40. All rational numbers are written in fraction form.

41. _____

41. Two fractions are equivalent if their denominators are the same.

42. _____

42. Of two fractions, the fraction with the larger numerator has the larger value.

43. _____

43. The commutative and distributive properties are used to group the whole numbers and to group the fractions when adding mixed numbers.

44. _____

44. To add two fractions, add the numerators and add the denominators.

45. _____

45. The sum of two fractions is always larger than either of the fractions.

46. _____

46. To find the LCM of two or more numbers, find the product of the numbers.

47. _____

47. The LCM of the denominators of two or more fractions is the least common denominator of the fractions.

48. _____

48. The difference of two fractions can be larger than either of the two fractions.

CHAPTER 5

Review

Section 5.1 *Objective 1*

Write the fraction represented by the figure.

1.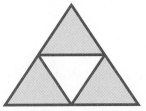

2.

3.

4.

One unit

5.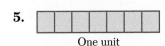

One unit

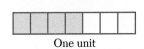

One unit

Section 5.1 *Objective 2*

6. Select the proper fractions from the following list: $\dfrac{7}{7}, -\dfrac{8}{9}, \dfrac{27}{27}, \dfrac{28}{29}, \dfrac{111}{112}, -\dfrac{149}{148}$

7. Select the improper fractions from the list in Exercise 6.

8. Select the fractions that represent the number one from the list in Exercise 6.

9. Select the improper fractions from the following list: $\dfrac{5}{6}, -\dfrac{6}{7}, \dfrac{7}{8}, \dfrac{8}{7}, -\dfrac{10}{9}, \dfrac{11}{10}, -\dfrac{12}{12}, \dfrac{13}{14}$

10. Select the proper fractions from the list in Exercise 9.

Section 5.1 *Objective 3*

Write the opposite of each of the following fractions:

11. $\dfrac{7}{9}$

12. $-\dfrac{19}{21}$

13. $\dfrac{-13}{7}$

14. $\dfrac{45}{9}$

15. $-\dfrac{8}{3}$

Section 5.1 *Objective 4*

Change each improper fraction to a mixed number:

16. $\dfrac{43}{6}$

17. $\dfrac{19}{4}$

18. $\dfrac{123}{5}$

19. $\dfrac{145}{7}$

20. $\dfrac{208}{17}$

Section 5.1 *Objective 5*

Write each of these numbers as an improper fraction:

21. $8\dfrac{2}{3}$

22. $13\dfrac{1}{12}$

23. $101\dfrac{4}{7}$

24. $24\dfrac{16}{17}$

25. $51\dfrac{4}{11}$

Section 5.2 *Objective 1*

List the first five multiples of each of the following numbers:

26. 9
27. 19
28. 62
29. 33
30. 71
31. Is 112 a multiple of 7?
32. Is 235 a multiple of 5?
33. Is 156 a multiple of 13?
34. Is 393 a multiple of 17?
35. Is 1197 a multiple of 19?

Section 5.2 *Objective 2*

Determine whether each number is divisible by 2, 3, 5, or 10.

36. 5230
37. 5231
38. 5232
39. 5233
40. 5324

Section 5.2 *Objective 3*

List all the factors (divisors) of each of the following numbers:

41. 124
42. 42
43. 630
44. 165
45. 384

Is each of the following numbers prime or composite?

46. 67
47. 213
48. 119
49. 181
50. 2111

Section 5.2 *Objective 4*

Write the prime factorization of each number:

51. 192
52. 84
53. 610
54. 880
55. 1224

Section 5.3 *Objective 1*

Reduce to lowest terms:

56. $-\dfrac{8}{24}$

57. $\dfrac{72}{96}$

58. $\dfrac{39}{65}a$

59. $-\dfrac{30ab}{105}$

60. $\dfrac{84xy}{35x}$

Section 5.3 *Objective 2*

61. Write four fractions equivalent to $\dfrac{6}{7}$ by multiplying by $\dfrac{2}{2}, \dfrac{3}{3}, \dfrac{4}{4}$, and $\dfrac{5}{5}$.

Find the missing numerator:

62. $\dfrac{4}{5} = \dfrac{?}{35}$

63. $\dfrac{5}{9} = \dfrac{?}{72}$

64. $\dfrac{11a}{12} = \dfrac{?}{96}$

65. $\dfrac{4y}{7x} = \dfrac{?}{21xz}$

Section 5.3 *Objective 3*

Multiply and reduce to lowest terms:

66. $\dfrac{12}{35} \cdot \dfrac{21}{30}$

67. $-\dfrac{36}{60} \cdot \dfrac{45}{63}$

68. $\left(\dfrac{-12}{27}x\right)\left(\dfrac{-9}{5}y\right)$

69. $\left(2\dfrac{1}{7}\right)\left(4\dfrac{3}{10}\right)$

70. If 82 bricks each measuring $6\dfrac{3}{4}$ inch in length are laid end to end, how long will the line of bricks be?

Section 5.3 *Objective 4*

Divide and reduce to lowest terms:

71. $\dfrac{16}{25} \div \dfrac{8}{15}$

72. $\dfrac{15}{32} \div \left(3\dfrac{1}{5}\right)$

73. $\dfrac{24}{49}d \div \dfrac{6}{7}$

74. $\dfrac{35bc}{9} \div \dfrac{7c}{12}$

75. If one turn of a screw sinks the screw $\dfrac{5}{16}$ inch, how many turns are needed to sink the screw $1\dfrac{1}{4}$ inches?

Section 5.4

Convert units as shown

76. 6 miles = ? feet

77. 124 pints = ? gallons

78. $\dfrac{55 \text{ miles}}{1 \text{ hour}} = \dfrac{? \text{ feet}}{1 \text{ second}}$

79. $\dfrac{40 \text{ km}}{1 \text{ hour}} = \dfrac{? \text{ meters}}{1 \text{ second}}$

80. $\dfrac{100 \text{ words}}{1 \text{ minute}} = \dfrac{? \text{ words}}{1 \text{ hour}}$

Section 5.5 *Objective 1*

Find the LCM of each group of numbers:

81. 5, 10, 40

82. 18, 24, 40

83. 15, 60, 90

84. 80, 100, 48, 24

85. 57, 95, 152, 190

Section 5.5 *Objective 2*

Compare these fractions, writing the result as an inequality.

86. $\dfrac{3}{7}$ and $\dfrac{4}{9}$

87. $\dfrac{8}{3}$ and $\dfrac{9}{4}$

88. $-\dfrac{5}{13}$ and $-\dfrac{11}{25}$

89. $\dfrac{15}{17}$ and $\dfrac{29}{33}$

90. $-\dfrac{42}{111}$ and $-\dfrac{65}{152}$

Section 5.5 *Objective 3*

Add:

91. $\dfrac{5}{17} + \dfrac{11}{17}$

92. $\dfrac{13}{24} + \dfrac{7}{24}$

93. $\dfrac{33}{40} + \left(-\dfrac{9}{40}\right)$

94. $\dfrac{7}{9}a + \dfrac{5}{9}a$

95. Find the sum of $\dfrac{12}{25}a$ and $-\dfrac{7}{25}a$.

96. $\dfrac{7}{10} + \dfrac{8}{15}$

97. $\dfrac{3}{4} + \dfrac{7}{18}$

98. $\dfrac{5}{27} + \left(-\dfrac{5}{36}\right)$

99. $\dfrac{17}{18}b + \dfrac{11}{45}b$

100. Find the sum of $\dfrac{3}{52}c$ and $-\dfrac{8}{65}c$.

Section 5.5 *Objective 4*

Subtract:

101. $\dfrac{23}{25} - \dfrac{6}{50}$

102. $\dfrac{13}{15} - \left(-\dfrac{2}{3}\right)$

103. $-\dfrac{5}{9} - \left(-\dfrac{1}{18}\right)$

104. $\dfrac{14}{33}x - \dfrac{5}{66}x$

105. Find the difference of $\dfrac{7}{20}y$ and $-\dfrac{1}{15}y$.

Section 5.5 *Objective 5*

Add:

106. $23\dfrac{4}{5}$

$+ \ \ 5\dfrac{2}{3}$

$\overline{}$

107.
$$9\frac{8}{9}$$
$$+\,12\frac{5}{6}$$

108.
$$54\frac{13}{20}$$
$$+\,27\frac{7}{12}$$

109. $6\frac{1}{8} + 19\frac{7}{12} + 42\frac{1}{15}$

110. Find the sum of $345\frac{31}{45}$, $75\frac{53}{81}$, and $121\frac{2}{5}$.

Subtract:

111.
$$6\frac{3}{8}$$
$$-\,3\frac{1}{8}$$

112.
$$9\frac{5}{6}$$
$$-\,6\frac{1}{15}$$

113.
$$8\frac{3}{7}$$
$$-\,5\frac{9}{14}$$

114. $13\frac{11}{18} - 6\frac{5}{27}$

115. Find the difference of $459\frac{17}{54}$ and $342\frac{41}{135}$.

Section 5.6 *Objective 1*

116. Evaluate $x - y + z$ if $x = \frac{3}{4}$, $y = \frac{1}{3}$, and $z = \frac{2}{5}$.

117. Evaluate $\frac{3}{4}a - \frac{2}{5}b + ab$ if $a = \frac{8}{9}$ and $b = \frac{5}{4}$.

118. Evaluate $\frac{4}{5}x^2 - \left(\frac{1}{4}y\right)^2$ if $x = \frac{1}{2}$ and $y = \frac{2}{3}$.

119. Evaluate the formula for A given $A = p + prt$ and $p = 6000$, $r = \frac{7}{100}$, and $t = \frac{3}{2}$.

120. Evaluate the formula for I given $I = \dfrac{E - e}{R}$ and $E = 5$, $e = \frac{4}{5}$, and $R = \frac{7}{10}$.

Section 5.6 *Objective 2*

Find the average of each set of numbers.

121. $\frac{1}{4}$, $\frac{3}{8}$, and $\frac{15}{16}$

122. $\frac{1}{5}$, $\frac{7}{10}$, and $\frac{17}{20}$

123. $\frac{1}{2}$, $\frac{1}{4}$, $\frac{9}{8}$, and $\frac{19}{16}$

124. $3\frac{3}{8}$, $5\frac{1}{4}$, and $6\frac{1}{2}$

125. $3\frac{1}{6}$, $5\frac{1}{2}$, $6\frac{3}{4}$, and $8\frac{5}{6}$

Section 5.6 *Objective 3*

Evaluate formulas containing π. Let $\pi \approx \frac{22}{7}$.

126. Find the circumference of a circle with diameter 16 inches.

127. Find the area of a circle with diameter 16 inches.

128. Find the volume of a cone with radius 8 feet and height 6 feet.

129. Find the volume of a sphere with diameter 25 meters.

130. A water tank is a cylinder that is 18 inches in diameter and 5 feet high. If there are 231 cubic inches in a gallon, how many gallons of water will the tank hold?

Section 5.7

Solve:

131. $\frac{3}{7}x - \frac{5}{14} = \frac{1}{2}$

132. $\frac{8}{9}y + \frac{2}{3} = \frac{5}{6}$

133. $\frac{a}{3} - \frac{11}{12} + \frac{5a}{6} = \frac{17}{12}$

134. $\frac{13d}{18} - \frac{5}{12} - \frac{7d}{9} + \frac{11}{24} = 0$

135. A number is divided by 21. If the quotient is decreased by $\frac{3}{7}$, the result is $-\frac{4}{21}$. Find the number.

CHAPTER 5

Test

ANSWERS

1. _____

    1. Find the average of $\frac{1}{5}$, $\frac{2}{3}$, $\frac{5}{6}$, and $\frac{7}{10}$.

2. _____

    2. Change to a mixed number: $\frac{81}{15}$

3. _____

    3. Change to an improper fraction: $5\frac{3}{8}$

4. _____

    4. Write the first five multiples of 13.

5. _____

    5. Is 72 a multiple of 8?

6. _____

    6. List all the factors of 72.

7. _____

    7. Is 113 a prime or a composite number?

8. _____

    8. Write the prime factorization of 260.

9. _____

    9. Simplify: $\frac{36b^2}{54b}$

10. _____

    10. Find the product of $\frac{3}{16}$, $-\frac{8}{9}$, and $\frac{15}{4}$.

11. _____

    11. Multiply: $\left(-\frac{18d}{45}\right)\left(-\frac{15d}{34}\right)$

12. _____

    12. Multiply: $\left(2\frac{3}{4}\right)\left(3\frac{3}{5}\right)$

495

13. _____

13. Divide: $\left(-\dfrac{30x}{25}\right) \div \left(-\dfrac{6x}{15}\right)$

14. _____

14. Divide: $\left(3\dfrac{5}{9}\right) \div \left(1\dfrac{1}{3}\right)$

15. _____

15. Find the missing numerator: $\dfrac{5}{8} = \dfrac{?}{64}$

16. _____

16. Find the least common multiple of 8, 27, and 36.

17. _____

17. True or false: $\dfrac{4}{5} < \dfrac{8}{9}$?

18. _____

18. Find the sum of $\dfrac{5}{12}a$ and $\dfrac{1}{12}a$.

19. _____

19. Find the sum of $\dfrac{4}{9}$, $-\dfrac{5}{12}$, and $\dfrac{3}{4}$.

20. _____

20. Add: $\dfrac{4}{7}y + \dfrac{2}{21}y + \dfrac{3}{14}y$

21. _____

21. Find the difference of $\dfrac{11}{15}$ and $\dfrac{1}{3}$.

22. _____

22. Subtract: $\dfrac{7}{18}ab - \left(-\dfrac{1}{4}ab\right)$

23. _____

23. Add: $5\dfrac{5}{6} + 3\dfrac{3}{10}$

24. _____

24. Subtract: $8\dfrac{3}{10} - 5\dfrac{8}{15}$

25. _____

25. Evaluate $xy - xz + y$ if $x = \dfrac{3}{4}$, $y = \dfrac{5}{8}$, and $z = \dfrac{1}{3}$.

26. _____

26. Solve: $\dfrac{2}{3}y - \dfrac{1}{6} = \dfrac{3}{4}$

27. _____

27. The formula for the area of a triangle is $A = \frac{1}{2}bh$, where b is the base and h is the height of the triangle.

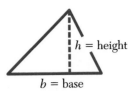

h = height

b = base

What is the area of a triangle that has a height of $2\frac{3}{4}$ inches and a base of $3\frac{3}{8}$ inches? Write the answer in square inches.

28. _____

28. A machinist takes $68\frac{1}{2}$ minutes to make 4 pins. How much time would it take him to make one pin?

29. _____

29. A railroad car contains $120\frac{1}{2}$ tons of baled hay. A truck that is being used to unload the hay can load $6\frac{3}{4}$ tons. How many truckloads of hay are in the railroad car?

30. _____

30. Jill buys a board that is 18 ft long. She needs a board that is $15\frac{3}{8}$ ft long. How much will she need to cut off the board that she bought?

Good Advice for Studying

Preparing for Tests

Testing usually causes the most anxiety for students. By studying more effectively, you can eliminate many of the causes of anxiety. But there are also other ways to prepare that will help relieve your fears.

If you are math anxious, you actually may study too much out of fear of failure and not allow enough time for resting and nurturing yourself. Every day, allow yourself some time to focus on your concerns, feelings, problems, or anything that might distract you when you try to study. Then, when these thoughts distract you, say to yourself, "I will not think about this now. I will later at _____ o'clock. Now I have to focus on math." If problems become unmanageable, make an appointment with a college counselor.

Nurturing is any activity that will help you recharge your energy. Choose an activity that makes you feel good such as going for a walk, daydreaming, reading a favorite book, doing yard work, taking a bubble bath, or playing basketball.

Other ways to keep your body functioning effectively under stress are diet and exercise. Exercise is one of the most beneficial means of relieving stress. Try to eat healthy foods and drink plenty of water. Avoid caffeine, nicotine, drugs, alcohol, and "junk food."

Plan to have all your assignments finished two days before the test, if possible. The day before the test should be completely dedicated to reviewing and practicing for the test.

Many students can do the problems, but cannot understand the instructions and vocabulary, so they do not know where to begin. Review any concepts that you have missed or any that you were unsure of or "guessed at."

When you feel comfortable with all of the concepts, you are ready to take the practice test at the end of the chapter. You should simulate the actual testing situation as much as possible. Have at hand all the tools that you will use on the real test: sharpened pencils, eraser, and calculator (if your instructor allows). Give yourself the same amount of time as you'll be given on the actual test. Plan a time for your practice test when you can be sure there will be no interruptions. Work each problem slowly and carefully. Remember, if you make a mistake by rushing through a problem and have to do it over, it will take more time than doing the problem carefully in the first place.

After taking the test, go back and study topics referenced with the answers you missed or that you feel you do not understand. You should now know if you are ready for the test. If you have been studying effectively and did well on the practice test, you should be ready for the real test. You are prepared!

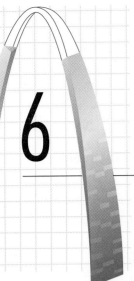

6 Rational Numbers: Decimals

Bruce Ayers/Tony Stone Images ©

APPLICATION

Sports hold a universal attraction. People all over the world enjoy a good game. For some sports it is relatively easy to determine which athlete is the best. In track and swimming for instance, each contestant races against the clock and the fastest time wins. In team sports, it is easy to tell which team wins, but sometimes difficult to determine how the individual athletes compare with one another. In order to make comparisons more objective, we often use sport statistics.

The easiest kind of statistic is simply to count how many times an athlete performs a particular feat in a single game. In basketball, for instance, it is usual to count the number of points scored, the number of rebounds made, and the number of assists for each player.

Consider the following statistics of members of the Houston Rockets in their four-game sweep of Orlando in the 1994–1995 National Basketball Association Championship.

Player	Points Scored	Rebounds	Assists
Olajuwon	131	46	22
Drexler	86	38	27
Horry	71	40	15
Elie	65	17	13
Cassell	57	7	12
Smith	30	7	16
Brown	12	11	0

(continued on next page)

1. Which player was the best in the championship series? Why?
2. Which two players have the best statistics that are closest to each other?
3. Which is more important in basketball, rebounds or assists?

6.1   Decimals: Reading, Writing, Rounding, and Inequalities

OBJECTIVES

1. Write word names from place value notation and place value notation from word names.
2. Change a decimal to a fraction.
3. List a set of decimals from smallest to largest.
4. Round a given decimal.

VOCABULARY

Decimal numbers, more commonly referred to as **decimals,** are another way of writing fractions and mixed numbers. The digits used to write whole numbers and a period called a **decimal point** are used to write place value names for these rational numbers.

The **number of decimal places** is the number of digits to the right of the decimal point. **Exact decimals** are decimals that show exact values. **Approximate decimals** are rounded values. **Equivalent decimals** are decimals that name the same number.

HOW AND WHY
Objective 1

Write word names from place value notation and place value notation from word names.

Decimals are written by using a standard place value in the same way we write whole numbers in place value. Numbers such as 12.65, 0.45, 0.795, -3.1267, 1306.94, and 19.36956 are examples of decimals. In general, the place value for decimals is

1. The same as whole numbers for digits to the left of the decimal point, and
2. A fraction whose denominator is 10, 100, 1000, and so on, for digits to the right of the decimal point.

The digits to the right of the decimal point have place values of

$$0.1 = \frac{1}{10}$$

$$0.01 = \frac{1}{100} = \frac{1}{10 \cdot 10} = \frac{1}{10^2}$$

$$0.001 = \frac{1}{1000} = \frac{1}{10 \cdot 10 \cdot 10} = \frac{1}{10^3}$$

$$0.0001 = \frac{1}{10,000} = \frac{1}{10 \cdot 10 \cdot 10 \cdot 10} = \frac{1}{10^4}$$

and so on, in that order from left to right.

Using the ones place as the central position (the place value of 10^0), the place values of a decimal are

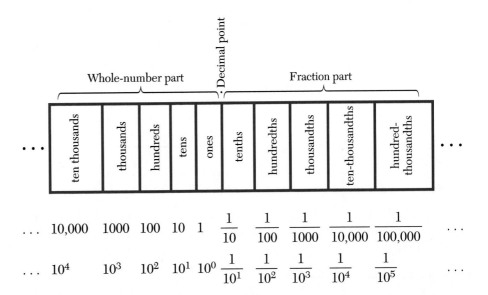

The chart reinforces our statement that the decimal point separates the whole number part from the fraction part. The place value notation for the digits in the fraction parts all end in "th." The "th" indicates that the place value of the digit is a fraction whose denominator is a power of 10.

▶ **To write the word name for a decimal**

1. Write the name for the whole number to the left of the decimal point.
2. Write the word "and" for the decimal point.
3. Write the whole number name for the number to the right of the decimal point.
4. Write the place value of the digit farthest to the right.

If the decimal has only zero or no number to the left of the decimal point, omit steps 1 and 2.

Here are some numbers and their corresponding word names:

Number	Word Name
10.21	Ten and twenty-one hundredths
0.723	Seven hundred twenty-three thousandths
0.00045	Forty-five hundred thousandths
6.006	Six and six thousandths

▶ **To write the place value notation from the word name**

1. Write the whole number. (The number before the word "and.")
2. Write a decimal point for the word "and."
3. Ignoring the place value notation, write the "whole number" after the word "and." Insert zeros, if necessary, between the decimal point and the digits following it to assure that the place on the far right has the given place value.

So, the place value notation for three hundred ten and sixty-five thousandths is

310	First write the whole number to the left of the word "and."
310.	Write a decimal point for the word "and."
310.065	The "whole number" after the word "and" is 65, a zero is inserted to place the "5" in the thousandths place.

Examples A–F   **Warm Ups A–F**

Directions: Write the word name.

Strategy: Write the word name for the whole number to the left of the decimal point. Then write "and." Last, write the word name for the number to the right of the decimal point followed by the place value of the digit farthest to the right.

A.   Write the word name for 0.58.

Fifty-eight	Write the word name for the number right of the decimal point.
Fifty-eight hundredths	Next, write the place of the digit 8.
	The word name for 0 in the ones place may be written or omitted. "Zero and fifty-eight hundredths" is correct but unnecessary.

A.   Write the word name for 0.31.

B.   Write the word name for 0.0034.

Thirty-four ten-thousandths

B.   Write the word name for 0.0089.

C.   Write the word name for 13.65.

Thirteen	Write the word name for the whole number left of the decimal point.
Thirteen and	Write "and" for the decimal point.
Thirteen and sixty-five	Write the word name for the number right of the decimal point.
Thirteen and sixty-five hundredths	Write the place value of the digit 5.

C.   Write the word name for 35.97.

D.   Janet called an employee to find the measurement of the outside diameter of a new wall clock the company is manufacturing. She asked the employee to check the plans. What is the word name the employee will read to her? The clock is shown below.

9.225 in.

D.   The measurement of the outside diameter of another clock is shown in the diagram below. What word name will the employee read?

11.375 in.

The employee will read "Nine and two hundred twenty-five thousandths inches."

Directions: Write the place value notation.

Strategy: Write the digit symbols for the corresponding words. Replace the word "and" with a decimal point.

E.   Fifteen hundred-thousandths

15	**First, write the number for fifteen.**
.00015	**The place value "hundred-thousandths" indicates five decimal places, so write three zeros before the numeral fifteen and then a decimal point.**
0.00015	**Since the number is between zero and one, we write a "0" in the ones place.**

F.   Write the place value notation for "four hundred five and four hundred five ten-thousandths."

405	**The whole number part is 405.**
405.	**Write the decimal point for "and."**
405.0405	**The "whole number" after the "and" is 405. A zero is inserted so the "5" is in the ten-thousandths place.**

E.   Twenty-nine thousandths

F.   Write the place value notation for "seven hundred three and three hundred seven ten-thousandths."

HOW AND WHY
Objective 2

Change a decimal to a fraction.

Most commonly used decimals can be written in the fraction form, $\frac{a}{b}$, where a and b are integers, and $b \neq 0$. For this reason, decimals are rational numbers.

The word name for the number to the right of the decimal point is the same as the word name of a fraction.

Place Value Notation	Word Name	Fraction
0.21	Twenty-one hundredths	$\frac{21}{100}$
0.3	Three tenths	$\frac{3}{10}$
0.125	One hundred twenty-five thousandths	$\frac{125}{1000} = \frac{1}{8}$
0.5	Five tenths	$\frac{5}{10} = \frac{1}{2}$

The number to the right of the decimal point is equivalent to a fraction whose denominator is a power of ten.

Answer to Warm Up   E. 0.029   F. 703.0307

▶ *To change a decimal to a fraction*

1. Read the word name.
2. Write the fraction equivalent to that word name.
3. Simplify.

| **Examples G–I** | **Warm Ups G–I** |

Directions: Change the decimal to a fraction.

Strategy: Write the fraction or mixed number equivalent to the word name. Simplify if possible.

G. Change to a common fraction: 0.83	G. Change to a common fraction: 0.51

    Eighty-three hundredths   **Word name.**

    $\dfrac{83}{100}$ **Write as a fraction.**

H. Write 0.475 as a fraction.	H. Write 0.625 as a fraction.

    Four hundred seventy-five thousandths   **Word name.**

    $\dfrac{475}{1000} = \dfrac{19}{40}$ **Write as a fraction and simplify.**

I. Write as a mixed number: 45.06	I. Write as a mixed number: 112.35

    Forty-five and six hundredths   **Word name.**

    $45\dfrac{6}{100} = 45\dfrac{3}{50}$ **Write as a mixed number and simplify.**

HOW AND WHY
Objective 3   **List a set of decimals from smallest to largest.**

Fractions can be listed in order, by comparing their numerators, when they have a common denominator. This idea can be extended to decimals when they have the same number of decimal places. For instance, $0.26 = \dfrac{26}{100}$ and $0.37 = \dfrac{37}{100}$ have a common denominator when written in fraction form. So 0.26 is less than 0.37, since 26 is less than 37.

$0.26 < 0.37$

    Graphically, we can see that $3.6 > 1.5$, because 3.6 is to the right of 1.5 on the number line.

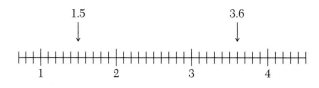

The decimals 0.3 and 0.15 have a common denominator when a zero is written after the 3. Thus,

$$0.3 = \frac{3}{10} \quad \text{and} \quad \frac{3}{10} = \frac{3}{10} \cdot \frac{10}{10} = \frac{30}{100}$$

so that

$$0.3 = \frac{30}{100} \quad \text{and} \quad 0.15 = \frac{15}{100}$$

Since $\frac{15}{100} < \frac{30}{100}$, we conclude that $0.15 < 0.3$.

There are many forms for decimal numbers that are equivalent. For example,

6.3 = 6.30 = 6.300 = 6.3000 = 6.30000
0.85 = 0.850 = 0.8500 = 0.85000 = 0.850000
45.982 = 45.9820 = 45.98200 = 45.982000 = 45.9820000

The zeros to the right of the decimal point following the last nonzero digit do not change the value of the decimal. Usually these extra zeros are not written, but they are useful when operating with decimals.

 To determine the order of a set of decimals

1. Make sure that all numbers have the same number of decimal places to the right of the decimal point by writing zeros to the right of the last digit when necessary.
2. Ignoring the decimal point, compare the numbers as if they were whole numbers.

Examples J–K   **Warm Ups J–K**

Directions: List the decimals from smallest to largest.

Strategy: Write zeros on the right so that all numbers have the same number of decimal places. Compare the numbers as if they are whole numbers and then remove the extra zeros.

J. 0.52, 0.537, 0.5139, 0.521   J. 0.65, 0.592, 0.648, 0.632

0.5200	**First, write all numbers with the same number of decimal places by inserting zeros on the right.**
0.5370	
0.5139	
0.5210	
0.5139, 0.5200, 0.5210, 0.5370	**Second, write the numbers in order as if they were whole numbers.**
0.5139, 0.52, 0.521, 0.537	**Third, remove the extra zeros.**

K. List 9.357, 9.361, 9.3534, and 9.358 from smallest to largest.

  9.357 = 9.3570   **Step 1.**

  9.361 = 9.3610

  9.3534 = 9.3534

  9.358 = 9.3580

  9.3534, 9.3570, 9.3580, 9.3610   **Step 2.**

  9.3534, 9.357, 9.358, 9.361   **Step 3.**

K. List 1.03, 1.0033, 1.0333, and 1.0303 from smallest to largest.

HOW AND WHY
Objective 4

Round a given decimal.

Decimals can be either *exact* or *approximate*. For example, decimals that count money are exact. The figure $23.95 shows an exact amount. Most decimals are approximations of measurements. For example 5.9 ft shows a person's height to the nearest tenth of a foot and 1.8 m shows the height to the nearest tenth of a meter, but neither is an exact measure. In measuring it is important that we know how to round decimals to a specific place.

Decimals are rounded using the same procedure as for whole numbers. Using a ruler, we round 2.563.

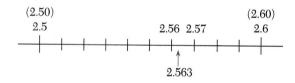

To the nearest tenth, we choose the larger number, 2.6, because 2.563 is closer to 2.6 than to 2.5. Rounded to the nearest hundredth, we choose the smaller number, 2.56, because 2.563 is closer to 2.56 than to 2.57.

To round 6.3265 to the nearest hundredth, without drawing a number line, draw an arrow under the hundredths place to identify the round off place.

6.3265

We must choose between 6.32 and 6.33. Since the digit to the right of the round-off position is 6, the number is more than halfway to 6.33. So, we choose the larger number.

6.3265 ≈ 6.33

To round a decimal number to a given place value

1. Draw an arrow under the given place value. (After enough practice, you will be able to round mentally and will not need the arrow.)

2. If the digit to the right of the arrow is 5, 6, 7, 8, or 9, add 1 to the digit above the arrow. That is, round to the larger number.

3. If the digit to the right of the arrow is 0, 1, 2, 3, or 4, keep the digit above the arrow. That is, round to the smaller number.

4. Write whatever zeros are necessary after the arrow so that the number above the arrow has the same place value as the original. See Example M.

This method is sometimes called the "four-five" rule. Although this rounding procedure is the most commonly used, it is not the only way to round. Many government agencies round by *truncation;* that is, by dropping the digit after the decimal point. Thus, $56.65 \approx $56. It is common for retail stores to round up for any amounts smaller than one cent. Thus, $1.333 \approx $1.34. There is also a rule for rounding numbers in science, which is sometimes referred to as the "even/odd" rule. You might learn and use a different round-off rule depending on what kind of work you are doing.

Examples L–N | **Warm Ups L–N**

Directions: Round as indicated.

Strategy: Draw an arrow under the round-off place. Examine the digit to the right of the arrow to determine whether to round up or down.

L. Round 0.3582 to the nearest hundredth.

$0.3582 \approx 0.36$ **The digit to the right of the round-off place is 8, so round up.**
↑

L. Round 0.3548 to the nearest thousandth.

M. Round 5582.9 to the nearest thousand.

$5582.9 \approx 6000$ **Three zeros must be written after the 6 to keep it in the thousands place.**
↑

M. Round 663.89 to the nearest ten.

N. Round 37.2828 and 3.9964 to the nearest unit, the nearest tenth, the nearest hundredth, and the nearest thousandth.

	Unit	Tenth	Hundredth	Thousandth
37.2828 ≈	37 ≈	37.3 ≈	37.28 ≈	37.283
3.9964 ≈	4 ≈	4.0 ≈	4.00 ≈	3.996

N. Round 12.8947 to the nearest unit, to the nearest tenth, to the nearest hundredth, and the nearest thousandth.

C A U T I O N **The zeros following the decimal in 4.0 and 4.00 are necessary to show that the original was rounded to the nearest tenth and hundredth, respectively.**

Exercises 6.1

OBJECTIVE 1: *Write word names from place value notation and place value notation from word names.*

A.

Write the word name.

1. 0.12

2. 0.34

3. 0.267

4. 0.712

5. 6.0004

6. 5.3002

Write the place value notation.

7. Eleven hundredths

8. Forty-five hundredths

9. One hundred eleven thousandths

10. Five hundred fourteen thousandths

11. Two and nineteen thousandths

12. One and six hundredths

B.

Write the word name.

13. 0.504

14. 5.04

15. 50.04

16. 5.004

17. 18.0205

18. 45.0051

Write the place value notation.

19. Twelve thousandths

20. Twelve thousand

21. Seven hundred and ninety-six thousandths

22. Six hundred and seven thousandths

23. Five hundred five and five thousandths

24. Five and five hundred five thousandths

OBJECTIVE 2: *Change a decimal to a fraction.*

A.

Change each decimal to a fraction and simplify if possible.

25. 0.33

26. 0.97

27. 0.75

28. 0.8

29. One hundred eleven thousandths

30. Five hundred thirteen thousandths

B.

31. 0.34

32. 0.98

33. 0.486

34. 0.504

35. Two hundred thousandths

36. Five hundred-thousandths

OBJECTIVE 3: *List a set of decimals from smallest to largest.*

A.

List the set of decimals from smallest to largest.

37. 0.6, 0.7, 0.1

38. 0.07, 0.03, 0.025

39. 0.05, 0.6, 0.07

40. 0.04, 0.1, 0.01

41. 4.16, 4.161, 4.159

42. 7.18, 7.183, 7.179

B.

43. 0.0729, 0.073001, 0.072, 0.073, 0.073015

44. 3.009, 0.301, 0.3008, 0.30101

45. 0.888, 0.88799, 0.8881, 0.88579

46. 8.36, 8.2975, 8.3599, 8.3401

47. 20.004, 20.04, 20.039, 20.093

48. 71.4506, 71.0456, 71.0546, 71.6405

Is the statement true or false?

49. 3.1231 < 3.1213

50. 4.1243 > 4.124

51. 13.1204 < 13.2014

52. 53.1023 > 53.1203

OBJECTIVE 4: *Round a given decimal.*

A.

Round to the nearest unit, tenth, and hundredth.

		Unit	Tenth	Hundredth
53.	15.888			
54.	51.666			
55.	477.774			
56.	344.333			
57.	0.7392			
58.	0.92937			

Round to the nearest cent.

59. $33.5374

60. $84.4167

61. $246.4936

62. $368.1625

513

B.

Round to the nearest ten, hundredth, and thousandth.

		Ten	Hundredth	Thousandth
63.	12.5532			
64.	21.3578			
65.	245.2454			
66.	118.0752			
67.	0.5536			
68.	0.9695			

Round to the nearest dollar.

69. $10.78 **70.** $15.49 **71.** $1129.38 **72.** $3178.48

C.

Use the graph to answer Exercises 73–80. The graph shows the precipitation for 4 weeks in a southern city.

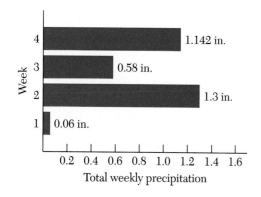

73. Write the word name for the number of inches of precipitation during the second week.

74. Write the word name for the number of inches of precipitation during the fourth week.

75. Write a simplified fraction to show the number of inches of precipitation in the first week.

76. Write a mixed number to show the number of inches of precipitation in the fourth week.

77. Round the amount of precipitation in week four to the nearest tenth of an inch.

78. Round the amount of precipitation in week one to the nearest tenth of an inch.

79. Which week has the least precipitation? The most precipitation?

80. List the weeks in order of the amount of precipitation from the most to the least.

81. Dan Ngo buys a deep fryer that has a marked price $64.79. What word name does he write on the check?

82. Fari Alhadet buys a truckload of organic fertilizer for her yard. The price of the load is $106.75. What word name does she write on the check?

83. The probability that a flipped coin will come up heads three times in a row is 0.125. Write this as a simplified fraction.

84. The probability that a flipped coin will come up heads twice and tails once out of three flips is 0.375. Write this as a simplified fraction.

Exercises 85–88 refer to the figure that follows.

85. What is the position of the arrow to the nearest hundredth?

86. What is the position of the arrow to the nearest thousandth?

87. What is the position of the arrow to the nearest tenth?

88. What is the position of the arrow to the nearest unit?

89. The Davis Meat Company bids 98.375¢ per pound to provide meat to the Beef and Bottle Restaurant. Circle K Meats puts in a bid of 98.35¢, and J & K Meats makes a bid of 98.3801¢. Which is the best bid for the restaurant?

90. Charles loses 2.165 pounds during the week. Karla loses 2.203 pounds and Mitchell loses 2.295 pounds during the same week. Who loses the most weight this week?

91. The computer at Grant's savings company shows that his account, including the interest he has earned, has a value of $1617.37099921. Round the value of the account to the nearest cent.

92. In doing her homework on a calculator, Catherine's calculator shows the answer to a division exercise is 34.78250012. If she is to round the answer to the nearest thousandth, what answer does she report?

 Exercises 93–95 relate to the chapter application.

93. In January of 1906 a Stanley car with a steam engine set a 1-mile speed record by going 127.659 miles per hour. Round this rate to the nearest tenth of a mile per hour.

94. In March of 1927 a Sunbeam set a 1-mile speed record by going 203.790 mph. What place value was this rate rounded to?

95. In October of 1970 a Blue Flame set a 1-mile speed record by going 622.407 mph. Explain why it is incorrect to round the rate to 622.5 mph.

STATE YOUR UNDERSTANDING

96. Explain the difference between an exact decimal value and an approximate decimal value. Give an example of each.

97. Explain how the number line can be a good visual aid for determining which of two numerals has the larger value.

98. Explain in words, the meaning of the values of 4 in the numeral 43.34. Include some comment on how and why the values of the digit 4 are alike and how and why they are different.

99. Consider the decimal represented by *abc.defg*. Explain how to round this number to the nearest hundredth.

CHALLENGE

100. Change 0.44, 0.404, and 0.04044 to fractions and simplify.

101. Determine whether each statement is true or false.

    a. $7.44 < 7\dfrac{7}{18}$ **b.** $8.6 > 8\dfrac{5}{9}$ **c.** $3\dfrac{2}{7} < 3.285$ **d.** $9\dfrac{3}{11} > 9.271$

102. Round 8.28282828 to the nearest thousandth. Is the rounded value less than or greater than the original value? Write an inequality to illustrate your answer.

GROUP ACTIVITY

103. Find a pattern, a vehicle manual, and/or a parts list whose measurements are given in decimal form. Change the measurements to fraction form.

104. Have each member of the group write one fraction and one decimal each with values between 3 and 4. Then as a group, list all of the fractions and decimal values from smallest to largest.

105. Discuss with the members of your group cases where you think rounding by "truncating" is the best way to round. Rounding by truncating means to drop all digit values to the right of the rounding position. For example, $2.77 \approx 2$, $\$19.33 \approx \$19$, $34,999 \approx 34,000$. Can each of the groups think of a situation where such rounding is actually used?

MAINTAIN YOUR SKILLS (Section 5.4)

Add and simplify.

106. $\dfrac{2}{5} + \dfrac{1}{3} + \dfrac{3}{8}$

107. $\dfrac{5}{6} + \left(-\dfrac{7}{8}\right)$

108. Find the sum of $\dfrac{3}{10}$ and $-\dfrac{1}{5}$.

109. Find the sum of $-\dfrac{1}{2}$, $-\dfrac{2}{3}$, and $-\dfrac{3}{4}$.

110. Find the sum of $\dfrac{2}{3}$, $\dfrac{5}{6}$, $\dfrac{7}{8}$, and $\dfrac{8}{9}$.

6.2 Adding and Subtracting Rational Numbers (Decimals)

OBJECTIVES
1. Add decimals
2. Subtract decimals.
3. Estimate the sum or difference of decimals.

HOW AND WHY
Objective 1

Add decimals.

The expanded form of a decimal shows the place value of each digit.

$$0.346 = \frac{3}{10} + \frac{4}{100} + \frac{6}{1000}$$

$$32.9 = 30 + 2 + \frac{9}{10}$$

An alternate expanded form uses words for the place values. So the number 0.346 can be written as:

3 tenths + 4 hundredths + 6 thousandths

What is the sum 6.3 + 2.5? We make use of the expanded form of the decimal to explain addition.

$$
\begin{array}{r}
6.3 = 6 \text{ ones} + 3 \text{ tenths} \\
+2.5 = 2 \text{ ones} + 5 \text{ tenths} \\
\hline
8 \text{ ones} + 8 \text{ tenths} = 8.8
\end{array}
$$

We use the same principle for adding decimals that we use for whole numbers. That is, we add like units. The vertical form gives us a natural grouping of the ones and tenths. By inserting zeros so all the numbers have the same number of decimal places, the addition 2.8 + 13.4 + 6.22 is written 2.80 + 13.40 + 6.22.

$$
\begin{array}{r}
2.80 \\
13.40 \\
+ \ 6.22 \\
\hline
22.42
\end{array}
$$

 To add decimals

1. Write in columns with the decimal points aligned. Place extra zeros on the right to help align the place values.
2. Add the decimals as if they were whole numbers.
3. Align the decimal point in the sum with those in the addends.

In algebraic sums, like terms with decimal coefficients are combined using the same rules as for integers and fractions. Add the numerical coefficients and multiply by the common variable factors.

$$9.34x + 3.67x = (9.34 + 3.67)x \quad \textbf{Distributive property.}$$
$$= 13.01x$$

Examples A–D

Warm Ups A–D

Directions: Add.

Strategy: Write each numeral with the same number of decimal places, align the decimal points, and add.

A.   Add: 1.3 + 21.41 + 32 + 0.05

<table>
<tr><td></td><td>1.30</td><td rowspan="5">Write each numeral with two decimal places. The extra zeros help line up the place values.</td></tr>
<tr><td></td><td>21.41</td></tr>
<tr><td></td><td>32.00</td></tr>
<tr><td>+</td><td>0.05</td></tr>
<tr><td></td><td>54.76</td></tr>
</table>

A.   Add: 2.4 + 37.52 + 19 + 0.08

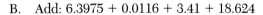

Calculator Example:

B.   Add: 6.3975 + 0.0116 + 3.41 + 18.624

Strategy: The extra zeros do not need to be inserted. The place values will be added correctly by the calculator.

The sum is 28.4431.

B.   Add:
8.4068 + 0.0229 + 4.56 + 34.843

C.   What is the total cost of an automobile tire if the retail price is $67.95, the federal excise tax is $2.72, the state sales tax is $5.44, and the local sales tax is $0.68?

Strategy: Add the retail price and the taxes.

$$\begin{array}{r} \$67.95 \\ 2.72 \\ 5.44 \\ +\underline{\quad 0.68} \\ \$76.79 \end{array}$$

The tire costs $76.79.

C.   What is the total cost of a pair of emerald earrings if the retail price is $103.95, the federal tax is $4.16, the state sales tax is $8.32, and the city sales tax is $0.83?

D.   Add: $1.05a + 0.723a + 72.6a$

$$1.05a + 0.723a + 72.6a = (1.05 + 0.723 + 72.6)a$$
$$= 74.373a$$

D.   Add: $35.7b + 0.32b + 0.925b$

HOW AND WHY
Objective 2   **Subtract decimals.**

What is the difference $6.59 - 2.34$? To find the difference of decimals, we use the same principle we used in addition. That is, we subtract like units. This is done by writing the numbers in column form aligning the decimal points. Now subtract as if they were whole numbers.

$$\begin{array}{r} 6.59 \\ -\underline{2.34} \\ 4.25 \end{array}$$ **The decimal point in the difference is aligned with those above.**

When necessary, we can regroup, or "borrow," as with whole numbers. What is the difference $6.271 - 3.845$?

$$\begin{array}{r} 6.271 \\ -3.845 \\ \hline \end{array}$$

We need to borrow one from the hundredth's column (1 hundredth = 10 thousandths) and we need to borrow one from the one's column (1 one = 10 tenths).

$$\begin{array}{r} \overset{5}{}\,\overset{12}{}\,\overset{6}{}\,\overset{11}{} \\ \cancel{6}.\cancel{2}\,\cancel{7}\,\cancel{1} \\ -3.8\,4\,5 \\ \hline 2.4\,2\,6 \end{array}$$

So, the difference is 2.426.

Sometimes it is necessary to write zeros on the right so the numbers have the same number of decimal places. See Example F.

▶ ***To subtract decimals***

1. Write the decimals in columns with the decimal points aligned. Place extra zeros on the right to align the place values.

2. Subtract the decimals as if they were whole numbers.

3. Align the decimal point in the difference with those in the problem.

Positive and negative decimals (rational numbers) are added and subtracted using the same rules as for integers. (See Sections 4.2 and 4.3.)

$-1.34 + 6.78 = 5.44$ **Since $|6.78| - |1.34| = 5.44$, and 6.78 has the larger absolute value.**

$-8.53 + 6.9 = -1.63$ **Since $|-8.53| - |6.9| = 1.63$ and -8.53 has the larger absolute value.**

$5.72 - 9.83 = 5.72 + (-9.83)$ **First rewrite as addition.**

$ = -4.11$ **Add.**

Like terms with decimal coefficients are combined using the same rules as for integers and fractions. Add or subtract the numerical coefficients and multiply by the common variable factors.

$9.5x - 3.45x - 0.51x = (9.5 - 3.45 - 0.51)x$ **Distributive property.**

$ = (6.05 - 0.51)x$ **Subtract: $9.5 - 3.45 = 6.05$**

$ = 5.54x$ **Subtract: $6.05 - 0.51 = 5.54$**

$75.025y^2 - 102.45y^2 = (75.025 - 102.45)y^2$ **Distributive property.**

$ = -27.425y^2$ **Subtract.**

Examples E–L   **Warm Ups E–L**

Directions: Subtract.

Strategy: Write each numeral with the same number of decimal places, align the decimal points, and subtract.

E. $5.831 - 0.287$ E. $7.946 - 0.378$

$$\begin{array}{r} 5.831 \\ -0.287 \\ \hline \end{array}$$
We need to borrow since we cannot subtract 7 thousandths from 1 thousandth.

$$\begin{array}{r} 5.8\overset{2\,11}{\cancel{3}\cancel{1}} \\ -0.28\,7 \\ \hline 4 \end{array}$$
Borrow 1 hundredth from the 3 in the hundredths place to add to the 1 in the thousandths place. (1 hundredth = 10 thousandths)

$$\begin{array}{r} \overset{7\,12}{5.\cancel{8}}\overset{2\,11}{\cancel{3}\cancel{1}} \\ -0.2\,8\,7 \\ \hline 5.5\,4\,4 \end{array}$$
Since we cannot subtract 8 hundredths from 2 hundredths, we regroup again. We borrow 1 tenth from the 8 in the tenths place to add to the 2 in the hundredths place. (1 tenth = 10 hundredths)

Check:

$$\begin{array}{r} 0.287 \\ +5.544 \\ \hline 5.831 \end{array}$$
Check by adding.

The difference is 5.544.

F.   Subtract 2.94 from 6.   F.   Subtract 5.736 from 7.

$$\begin{array}{r} 6.00 \\ -2.94 \\ \hline \end{array}$$
We write 6 as 6.00 so that both numerals will have the same number of decimal places.

$$\begin{array}{r} \overset{5\,\,10}{\cancel{6}.\cancel{0}\,0} \\ -2.9\,4 \\ \hline \end{array}$$
We need to borrow to subtract in the hundredths place. Since there is a 0 in the tenths place, we start by borrowing 1 from the ones place. (1 one = 10 tenths)

$$\begin{array}{r} \overset{9}{\overset{5\,\,10\,10}{\cancel{6}.\cancel{0}\,\cancel{0}}} \\ -2.9\,4 \\ \hline 3.0\,6 \end{array}$$
Now borrow 1 tenth to add to the hundredths place. (1 tenth = 10 hundredths) Subtract.

Check:

$$\begin{array}{r} 2.94 \\ +3.06 \\ \hline 6.00 \end{array}$$

The difference is 3.06.

G. Find the difference of 6.271 and 3.845. Round to the nearest tenth.

$$\begin{array}{r} \overset{5}{\cancel{6}}\overset{12}{.\cancel{2}}\overset{6}{\cancel{7}}\overset{11}{\cancel{1}} \\ -3.8\,4\,5 \\ \hline 2.4\,2\,6 \end{array}$$ **The check is left for the student.**

The difference is 2.4 to the nearest tenth

C A U T I O N   **Do not round before subtracting. Note the difference if we do:**

$$6.3 - 3.8 = 2.5$$

G. Find the difference of 9.382 and 5.736. Round to the nearest tenth.

Calculator Example:

H. Subtract: 145.9673 − 298.893

Strategy: The calculator lines up the decimal points.

The difference is −152.9257.

H. Subtract: 340.7445 − 445.895

I. Marta purchases a small radio for $33.89. She gives the clerk two $20 bills to pay for the radio. How much change does she get?

Strategy: Since two $20 bills are worth $40, subtract the cost of the radio from $40.

$$\begin{array}{r} \$40.00 \\ -33.89 \\ \hline \$\ 6.11 \end{array}$$

Marta gets $6.11 in change.

Clerks sometimes make change by counting backwards, that is, by adding to $33.89 the amount necessary to equal $40.

$33.89 + a penny = $33.90
$33.90 + a dime   = $34.00
$34.00 + 1 dollar = $35.00
$35.00 + 5 dollars = $40.00

So, the change is $0.01 + $0.10 + $1 + $5 = $6.11.

I. Mickey buys a video for $18.69. She gives the clerk a $20 bill to pay for the cassette. How much change does she get?

J. Subtract: 3.56 − 11.74 − 0.78

3.56 − 11.74 − 0.78

$= (3.56) + (-11.74) + (-0.78)$ **Rewrite as addition.**

$= 3.56 + (-12.52)$ **The sum of two negative numbers is negative.**

$= -8.96$ **Subtract the absolute value of the numbers and use the sign of the one with the larger absolute value.**

J. Subtract: 18.19 − 23.8 − 1.01

Answers to Warm Ups   G. 3.6   H. −105.1505   I. $1.31   J. −6.62

K. Combine terms: $-7.6y - 8.5y + 3.21y - 7.4y$

$-7.6y - 8.5y + 3.21y - 7.4y$

$= (-7.6 - 8.5 + 3.21 - 7.4)y$	**Distributive property.**
$= [-7.6 + (-8.5) + 3.21 + (-7.4)]y$	**Rewrite as addition.**
$= -20.29y$	**Add.**

K. Combine terms: $3.76a^2 - 13a^2 - 5.69a^2 + 0.453a^2$

L. Subtract $(3b - 0.43c + 0.56)$ from $(1.2b - 4.5c - 3.4)$.

$(1.2b - 4.5c - 3.4) - (3b - 0.43c + 0.56)$

$= (1.2b - 4.5c - 3.4) + [-(3b - 0.43c + 0.56)]$
Rewrite as addition.

$= (1.2b - 4.5c - 3.4) + (-3b + 0.43c - 0.56)$
Write the opposite.

$= [1.2b + (-4.5c) + (-3.4)] + [-3b + 0.43c + (-0.56)]$
Rewrite as addition.

$= [1.2b + (-3b)] + [(-4.5c) + 0.43c] + [(-3.4) + (-0.56)]$
Group like terms.

$= -1.8b + (-4.07c) + (-3.96)$ **Add.**

$= -1.8b - 4.07c - 3.96$ **Rewrite as subtraction.**

L. Subtract $(1.6m - 3.9n - 18.4)$ from $(-5.6m - 2n + 2.07)$.

HOW AND WHY
Objective 3   **Estimate the sum or difference of decimals.**

The sum or difference of decimals can be estimated by rounding each to the largest nonzero place value of the numbers and then adding or subtracting the rounded values. For instance,

0.756	0.8	**The largest place value is tenths, so round each**
0.092	0.1	**number to the nearest tenth.**
0.0072	0.0	
$+0.0205$	$+0.0$	
	0.9	

The estimate of the sum is 0.9. One use of the estimate is to see if the sum of the group of numbers is correct. If the calculated sum is not close to the estimated sum, 0.9, you should check the addition by re-adding. In this case the exact sum, 0.8757, is close to the estimate.

Example M   **Warm Up M**

Directions: Estimate the difference. Then subtract and compare.

Strategy: Round each number to the largest nonzero place value of the numbers. Then, subtract and compare.

M. $0.0569 - 0.845$ M. $0.00782 - 0.00298$

$0.0569 - 0.845 \approx 0.1 - 0.8$ **Round to the largest
nonzero place value.**

$\approx 0.1 + (-0.8)$ **Write as addition.**

≈ -0.7

Now subtract and compare.

$0.0569 - 0.845 = 0.0569 + (-0.845)$

$= -0.7881$

The difference is -0.7881 and is close to the estimation, -0.7.

Exercises 6.2

OBJECTIVE 1: *Add decimals.*

Add.

A.

1. $0.4 + 0.3$

2. $0.8 + 0.3$

3. $2.5 + 1.3$

4. $6.7 + 2.1$

5. $1.4 + 2.1 + 4.2$

6. $3.2 + 1.1 + 2.4$

7. $23.3 + 4.13$

8. $17.7 + 2.28$

9. To add 4.5, 6.78, 9.342, and 23, first rewrite each with _____ decimal places.

10. The sum of 6.7, 8.93, 5.4321, and 45.72 has _____ decimal places.

B.

11.
$$\begin{array}{r} 8.3 \\ +5.541 \\ \hline \end{array}$$

12.
$$\begin{array}{r} 7.6 \\ +6.44 \\ \hline \end{array}$$

13.
$$\begin{array}{r} 8.28 \\ 0.28 \\ 12.3 \\ + \ 2.54 \\ \hline \end{array}$$

14.
$$\begin{array}{r} 9.06 \\ 0.82 \\ 11.5 \\ + \ 4.35 \\ \hline \end{array}$$

15. $0.438 + 0.834 + 1.483$

16. $1.254 + 1.425 + 0.524$

17. $0.0017 + 1.007 + 7 + 1.071$

18. $1.0304 + 1.4003 + 1.34 + 0.403$

527

19. $37.008 + 38.007 + 3.87 + 3.708$

20. $82.005 + 8.25 + 2.085 + 28.55$

21.
$$
\begin{array}{r}
7.5 \\
14.378 \\
+\,33.6583 \\
\hline
\end{array}
$$

22.
$$
\begin{array}{r}
10.03 \\
223.231 \\
+\,5603.3056 \\
\hline
\end{array}
$$

23.
$$
\begin{array}{r}
43.524 \\
12.8 \\
+\,774.943 \\
\hline
\end{array}
$$

24.
$$
\begin{array}{r}
314.143 \\
712.217 \\
+\,333.444 \\
\hline
\end{array}
$$

25. Find the sum of 9.76, 9.6, 0.581, and 7.04.

26. Find the sum of 0.9855, 4.913, 6.72, and 3.648.

OBJECTIVE 2: *Subtract decimals.*

A.

Subtract.

27. $0.9 - 0.6$

28. $2.7 - 2.2$

29. $5.7 - 2.3$

30. $0.25 - 0.12$

31.
$$
\begin{array}{r}
8.31 \\
-\,3.21 \\
\hline
\end{array}
$$

32.
$$
\begin{array}{r}
17.48 \\
-\,6.23 \\
\hline
\end{array}
$$

33.
$$
\begin{array}{r}
19.05 \\
-\,12.64 \\
\hline
\end{array}
$$

34.
$$
\begin{array}{r}
6.28 \\
-\,1.19 \\
\hline
\end{array}
$$

35. Subtract 8.11 from 16.20.

36. Find the difference of 18.477 and 9.2.

B.

Subtract.

37.　0.612
　　　$-\,0.155$

38.　3.457
　　　$-\,2.509$

39.　2.712
　　　$-\,1.148$

40.　7.303
　　　$-\,0.178$

41. $5.678 - 3.069$

42. $4.823 - 1.167$

43. $134.98 - 67.936$

44. $405.4057 - 316.316$

45. Subtract 9.34 from 12.1.

46. Subtract 3.576 from 8.4.

47. Find the difference of 8.642 and 8.573.

48. Find the difference of 71.505 and 69.948.

OBJECTIVE 3: *Estimate the sum or difference of decimals.*

A.

Estimate the sum or difference.

49. $0.09346 + 0.0371 + 0.0444$

50. $0.00567 + 0.003211 + 0.00123$

529

51. $0.678 - 0.351$

52. $0.074 - 0.056$

53. $3.895 + 2.045 + 0.34 + 0.045$

54. $0.45 + 0.0021 + 0.0043 + 0.1001$

55. $0.075 - 0.0023$

56. $0.00675 - 0.000984$

57.
```
   0.542
   0.125
   0.32974
+ 0.7421
```

58.
```
   0.0346
   0.067
   0.02389
+ 0.06002
```

B.

Estimate the sum or difference and add or subtract.

59.
```
   0.0764
- 0.03621
```

60.
```
   0.0056982
- 0.003781
```

61. $0.0342 + 0.00687 + 0.057294 + 0.00843$

62. $2.005 + 0.8741 + 0.006723 + 5.0555$

63. $11.9867 - 27.3074$

64. $0.7382 - 0.88914$

65. $0.067 + (-0.456) + (-0.0964) + 0.5321 + (-0.112)$

66. $4.005 + (-0.875) + (-3.96) + 7.832 + (-4.009)$

67. $0.0456 - 0.7834 - 0.456 + 0.3097 - 0.5067$

68. $3.079 - 7.935 - 0.983 + 3.115 - 7.22$

C.

Perform the indicated operations.

69. $135.904 - (-34.651) - 78.45$

70. $-87.16 - (-1.976) - (-52.871)$

71. $3.17t + 3t + 0.5t + 0.8t$

72. $20.03a + 112a + 0.21a + 0.25a$

73. $1.21x - 9.34x$

74. $0.845y - 0.087y$

75. $-23.4a - (-9.5a)$

76. $-54.1t - (-23.4t)$

77. $6.788x - (-3.408) - (-5.009x) - 8.1$

78. $5.782m - (-7.601) - 6.78m + (-5.3332)$

79. $-34.98b - (-10.05c) - (-8.04b) + 3.2c$

80. $-61.884x - 24.97y + 12.308x + 9.005y$

81. Find the sum of $-34.98x - 10.5y$ and $18.04x + 47y$.

82. Find the sum of $-61.44a - 23.08b$ and $-14.44a - 6.08b$.

83. Find the difference of $-5a^2 - 3.46a$ and $0.34a + 1.87a^2$.

84. Find the difference of $3.07ab - 2.6cd$ and $8.05ab - 0.45cd$.

85. On a vacation trip, Manuel stops for gas four times. The first time he bought 9.2 gallons. At the second station he bought 11.9 gallons, and at the third he bought 15.4 gallons. At the last stop he bought 12.6 gallons. How much gas did he buy on the trip?

86. Heather wrote five checks in the amounts of $45.78, $23.90, $129.55, $7.75, and $85. She has $295.67 in her checking account. Does she have enough money to cover the five checks?

87. Find the sum of 45.984, 134.6, 98.992, 89.56, and 102.774. Round the sum to the nearest tenth.

88. Find the sum of 235.98, 785.932, 6.94432, 11.116, and 8.0034. Round the sum to the nearest thousandth.

▼ Exercises 89–93 relate to the chapter application.

89. Muthoni swims 50 meters in 12.16 seconds, whereas Sera swims 50 meters in 11.382 seconds. How much faster is Sera?

90. A skier posts a race time of 1.257 minutes. A second skier posts a time of 1.32 minutes. The third skier completes the race in 1.2378 minutes. Find the difference between the fastest and the slowest times.

91. A college men's 4 × 100-meter relay track team has runners with individual times of 9.35 sec, 9.91 sec, 10.04 sec, and 9.65 sec. What is the time for the relay?

92. A high school girl's swim team has a 200-yd freestyle relay in which its members have times of 21.79 sec, 22.64 sec, 22.38 sec, and 23.13 sec. What is the time for the relay?

93. A high school women's track coach knows that the rival school's team in the 4 × 100-meter relay has a time of 52.78 sec. If the coach knows that her top three sprinters have times of 12.83, 13.22, and 13.56 sec, how fast does the fourth sprinter need to be in order to beat the rival school's relay team?

Use the chart for Exercises 94–96. The chart shows the annual annuity sales in billions of dollars.

**ANNUITY SALES IN
BILLIONS OF DOLLARS**

1985	$4.5	1990	$12
1986	$8.1	1991	$17.3
1987	$9.3	1992	$28.5
1988	$7.2	1993	$46.6
1989	$9.8	1994	$50.4

94. Find the total sales for the ten years.

95. How many more dollars were invested in 1992 than in 1989?

96. How many more dollars were invested in the '90s than in the '80s?

97. Doris makes a gross salary (before deductions) of $2796 per month. She has the following monthly deductions: Federal Income Tax, $254.87; State Income Tax, $152.32; Social Security, $155.40; Medicare, $35.61; retirement contribution, $82.45; union dues, $35; and health insurance, $134.45. Find her actual take-home (net) pay.

98. Jack goes shopping with $75 in cash. He pays $14.99 for a T-shirt, $9.50 for a CD, and $25.75 for a sweater. On the way home he buys $18.55 worth of gas. How much money does he have left?

99. In 1996 the average interest on a 30-year home mortgage dropped from 8.23% to 7.88% in 1 week. What was the drop in interest rate?

100. What is the total cost of a cart of groceries that contains bread for $1.88, bananas for $2.12, cheese for $5.87, cereal for $3.57, coffee for $7.82, and meat for $7.89?

101. How high from the ground level is the top of the smokestack in the drawing below? Round to the nearest foot.

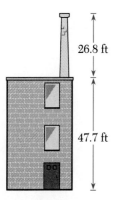

26.8 ft

47.7 ft

102. Find the length of the piston skirt (A) shown in the following drawing if the other dimensions are as follows: B = 0.3125 in., C = 0.250 in., D = 0.3125 in., E = 0.250 in., F = 0.3125 in., G = 0.375 in., H = 0.3125.

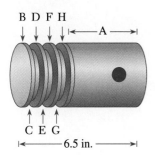

B D F H

A

C E G

6.5 in.

103. What is the center-to-center distance, A, between the holes in the diagram below?

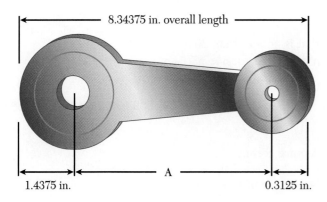

8.34375 in. overall length

1.4375 in.

A

0.3125 in.

104. What is the total length of the connecting bar shown below?

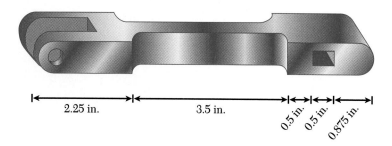

2.25 in.   3.5 in.   0.5 in.   0.5 in.   0.875 in.

STATE YOUR UNDERSTANDING

105. Explain the procedure for adding 2.005, 8.2, 0.00004, and 3.

106. Explain the procedure for subtracting 8.9215 from 11.2.

107. Explain the similarities between subtracting decimals and subtracting fractions.

108. Copy the table and fill it in.

Operation	Procedure	Example
Addition		
Subtraction		

CHALLENGE

109. How many 8.75s must be added to have a sum that is greater than 200?

110. Find the missing number in the sequence: 0.4, 0.8, 1.3, _____, 2.6, 3.4, 4.3, 5.3.

111. Find the missing number in the sequence: 0.2, 0.19, 0.188, _____, 0.18766, 0.187655.

112. Which number in the following group is 11.1 less than 989.989: 999.999, 989.999, 988.889, 979.889, or 978.889?

113. Write the difference between $6\dfrac{9}{16}$ and 4.99 in decimal form.

114. Round the sum of 9.8989, 8.9898, 7.987, and 6.866 to the nearest tenth.

GROUP ACTIVITY

115. As a group, review the multiplication of fractions. Have each member make up a pair of fractions whose denominators are in the list: 10, 100, 1000, and 10,000. Find the product of each pair and change it to decimal form. In group discussion, make up a rule for multiplying decimals.

MAINTAIN YOUR SKILLS (Sections 4.5, 5.3)

116. Find the product of -5, 27, and -300.

117. Find the product of $\dfrac{7}{8}$ and $-\dfrac{2}{3}$.

118. Find the product of $-\dfrac{3}{11}$ and $-\dfrac{22}{9}$.

119. Multiply: $(4x)(-13x)(x^2)$

120. Multiply: $(-3y^2)(-10y)(-25y^2)$

6.3 Multiplying Rational Numbers (Decimals)

OBJECTIVES

1. Multiply decimals.

2. Estimate the product of decimals.

HOW AND WHY
Objective 1   **Multiply decimals.**

The "multiplication table" for decimals is the same as for whole numbers. In fact, decimals are multiplied the same way as whole numbers with one exception. The exception is the location of the decimal point in the product. Change the decimals to fractions to find the product.

Decimals	Fractions	Product of Fractions	Product as a Decimal	Number of Decimal Places in Product
0.3×0.8	$\dfrac{3}{10} \times \dfrac{8}{10}$	$\dfrac{24}{100}$	0.24	Two
11.2×0.07	$\dfrac{112}{10} \times \dfrac{7}{100}$	$\dfrac{784}{1000}$	0.784	Three
0.02×0.13	$\dfrac{2}{100} \times \dfrac{13}{100}$	$\dfrac{26}{10000}$	0.0026	Four

From the chart we see that the product in decimal form has the same number of decimal places as the total number of places in the decimal factors.

The shortcut is to multiply the numbers and insert the decimal point. If necessary, insert zeros so that there are enough decimal places. The product 0.2×0.3 has two decimal places, since tenths multiplied by tenths yields hundredths.

$$0.2 \times 0.3 = 0.06 \qquad \text{since} \qquad \frac{2}{10} \times \frac{3}{10} = \frac{6}{100}$$

 To multiply decimals

1. Multiply the numbers as if they were whole numbers.

2. Locate the decimal point in the product by counting the number of decimal places (to the right of the decimal point) in both factors. The sum of these two counts is the number of decimal places the product must have.

3. If necessary, zeros are inserted at the *left of the numeral* so there are enough decimal places (see Example D).

When multiplying decimals it is not necessary to align the decimal points in the decimals being multiplied.

Positive and negative decimals (rational numbers) are multiplied using the same rules as for integers and fractions. The product of two positive or of two negative numbers is positive. The product of a positive number and a negative number is negative.

$$-2.03(5.6) = -11.368 \qquad \text{and} \qquad (-4.1)(-0.22) = 0.902$$

Terms with coefficients that are rational numbers are multiplied using the same rules as for integers and fractions.

$$7.2t(1.25t) = 9t^2$$

| **Examples A–G** | **Warm Ups A–G** |

Directions: Multiply.

Strategy: First multiply the numbers, ignoring the decimal points. Place the decimal point in the product by counting the number of decimal places in the two factors. Insert zeros if necessary to produce the number of required places.

A. $(0.7)(11)$	A. $(9)(0.6)$

$(0.7)(11) = 7.7$ **Multiply 7 and 11.**

The total number of decimal places in both factors is one (1), so there is one decimal place in the product.

B. Find the product of 0.8 and 0.21.	B. Find the product of 0.7 and 0.33.

$(0.8)(0.21) = 0.168$ **There are three decimal places in the product as the total number of places in the factors is three.**

C. Find the product of (-5.67) and (-3.8).	C. Find the product of (-16.9) and (0.34).

$$
\begin{array}{r}
5.67 \\
\times\ \ 3.8 \\
\hline
4536 \\
1701\ \ \\
\hline
21.546
\end{array}
$$
Multiply the absolute values of the numbers. There are three decimal places in the product.

$(-5.67)(-3.8) = 21.546$ **The product of two negatives is positive.**

D. Multiply 1.2 times 0.0004.	D. Multiply 0.05 times 0.015.

$$
\begin{array}{r}
1.2 \\
\times\ 0.0004 \\
\hline
0.00048
\end{array}
$$
Since 1.2 has one decimal place and 0.0004 has four decimal places, the product must have five decimal places. We must insert three zeros at the left so there are enough places in the answer.

Calculator Example:

E. Find the product: $(12.87)(-64.862)$	E. Find the product: $-23.84(79.035)$

Strategy: The calculator will automatically place the decimal point in the correct position.

The product is -834.77394.

Answers to Warm Ups   A. 5.4   B. 0.231   C. -5.746 D. 0.00075   E. -1884.1944

F. Multiply: $0.38w(0.43w^2)$

$0.38w(0.43w^2) = (0.38 \times 0.43)(w \times w^2)$
$= 0.1634w^3$

F. Multiply: $7.8xy(3.765y^2)$

G. If exactly eight strips of metal, each 3.875 inches wide, are to be cut from a piece of sheet metal, what is the smallest (in width) piece of sheet metal that can be used? Ignore the waste in cutting the sheet metal.

Strategy: To find the width of the piece of sheet metal we multiply the width of one of the strips by the number of strips needed.

$$\begin{array}{r} 3.875 \\ \times 8 \\ \hline 31.000 \end{array}$$ **The extra zeros can be dropped.**

The piece must be 31 inches wide.

G. If 12 strips, each 6.45 centimeters wide, are to be cut from a piece of sheet metal, what is the narrowest piece of sheet metal that can be used?

HOW AND WHY
Objective 2

Estimate the product of decimals.

The product of two decimals can be estimated by rounding each number to the largest nonzero place in each number and then multiplying the rounded numbers. For instance,

$$\begin{array}{r} 0.0673 \\ \times 0.79 \end{array} \qquad \begin{array}{r} 0.07 \\ \times 0.8 \\ \hline 0.056 \end{array} \qquad \begin{array}{l} \text{\textbf{Round to the nearest hundredth.}} \\ \text{\textbf{Round to the nearest tenth.}} \\ \text{\textbf{Multiply.}} \end{array}$$

The estimate of the product is 0.056. One use of the estimate is to see if the product is correct. If the calculated product is not close to 0.056, you should check the multiplication. In this case the exact product is 0.053167, which is close to the estimate.

 To estimate the product of two decimals.

1. Round each number to its largest nonzero place.
2. Multiply the rounded numbers.

| **Example H** | **Warm Up H** |

Directions: Estimate the product. Then find the product and compare.

Strategy: Round each number to its largest nonzero place. Multiply the rounded numbers. Multiply the original numbers.

H. 0.316 and 9.107

$$
\begin{array}{r}
0.3 \\
\times\ \ 9 \\
\hline
2.7
\end{array}
$$
    Round to the nearest tenth.
    Round to the nearest one.

$$
\begin{array}{r}
0.316 \\
\times 9.107 \\
\hline
2\ 212 \\
31\ 60\ \ \\
2\ 844\ \ \ \ \\
\hline
2.877812
\end{array}
$$

The product is 2.877812 and is close to the estimated product of 2.7.

H. 3.871 and 0.0051

Exercises 6.3

OBJECTIVE 1: *Multiply decimals.*

A.

Multiply.

1. 0.4
  × 8

2. 0.6
  × 7

3. 1.5
  × 6

4. 2.1
  × 8

5. 3 × 0.09

6. 0.07 × 8

7. 0.5 × 0.4

8. 0.4 × 0.2

9. 0.03 × 0.5

10. 0.7 × 0.008

11. 0.16 × (−0.4)

12. −1.2 × (−0.05)

13. The number of decimal places in the product of 9.456 and 4.23 is _____.

14. In the product of 0.034 × ? = 0.0408, the number of decimal places in the missing factor is _____.

B.

Multiply.

15. 7.45
  ×0.002

16. 1.13
  ×0.005

17. 1.45
  × 4.6

18. 4.23
  × 3.2

19. 7.84
  ×0.53

20. 45.8
  ×0.12

21. 0.346
  × 7.8

22. 0.073
  × 9.6

23. Find the product of 8.52 and -3.54.

24. Find the product of -6.79 and 1.34.

25. Multiply: $6.5(-0.6)(0.03)$

26. Multiply: $3.6(-1.6)(-0.012)$

Multiply.

27. $(0.06x)(1.2x)$

28. $(0.08w)(3.4w^2)$

29. $(7.5t^2)(-0.3t^2)$

30. $(-6.4v^2)(-0.7v)$

OBJECTIVE 2: *Estimate the product of decimals.*

A.

Estimate the product.

31. $32(0.845)$

32. $680(0.00231)$

33. $0.045(0.0672)$

34. $0.00389(0.0912)$

35. $(16.95)(0.0781)$

36. $(0.00439)(31.9)$

37. $\begin{array}{r} 0.0875 \\ \times\ \underline{0.021} \end{array}$

38. $\begin{array}{r} 0.2753 \\ \times \underline{0.9631} \end{array}$

B.

Estimate the product and then multiply.

39.   23.5
      ×0.47

40.   18.6
      ×0.32

41.   0.356
      ×0.067

42.   0.832
      ×0.041

43.   0.0975
      ×   3.92

44.   0.00732
      ×   7.05

45.   0.825
      ×0.0054

46.   0.575
      ×0.00378

C.

Use the table for Exercises 47–51. The table shows the amount of gas purchased by Grant and the price he paid per gallon for five fill-ups.

Number of Gallons	Price per Gallon
20.7	$1.375
20.4	$1.405
19.3	$1.447
18.9	$1.393
18.4	$1.523

47. What is the total number of gallons of gas that Grant purchased?

48. To the nearest cent, how much did he pay for the second fill-up?

49. To the nearest cent, how much did he pay for the fifth fill-up?

50. To the nearest cent, what is the total amount he paid for the five fill-ups?

51. At which price per gallon did he pay the least for his fill-up?

Estimate the product and then multiply.

52. (223.6)(8.45)

53. (98.67)(3.52)

54. (343.17)(8.73)

55. (12.6)(760.02)

Multiply.

56. (9.58)(5.63)(23.22), Round to the nearest hundredth.

57. (9.86)(146.3)(14.83), Round to the nearest thousandth.

58. (2.15)(1.8)(0.54)(−13.5), Round to the nearest tenth.

59. (−5.7)(0.57)(−5.07)(50.7), Round to the nearest hundredth.

60. $7.2a(3a + 1.5)$

61. $5.3b(4b + 3.1)$

62. $-0.6x(1.8x - 4.3)$

63. $-4.5y(2.1y - 0.9)$

64. $0.3t^2(0.6t^2 - 3.4t)$

65. $0.4b^2(3b - 0.4c)$

66. Joe earns \$9.85 per hour. How much does he earn if he works 30.25 hours in one week? Round to the nearest cent.

67. If upholstery fabric costs \$37.83 per yard, how much will Joanne pay for 14.7 yards? Round to the nearest cent.

Use the chart for Exercises 68–71. The chart shows the cost of renting a car from a local agency.

Type of Car	Cost per Day	Price per Mile Driven
Compact	\$25.95	\$0.25
Mid-size	\$39.72	\$0.32
Luxury	\$47.85	\$0.42

68. What does it cost to rent a compact car for four days if it is driven 324 miles?

69. What does it cost to rent a mid-size car for 3 days if it is driven 312 miles?

70. What does it cost to rent a luxury car for 6 days if it is driven 453 miles?

71. Which costs less, renting a mid-size car for 3 days or a compact car for 5 days if both are driven 345 miles? How much less does it cost?

72. Tiffany can choose any of the following ways to finance her new car. Which method is the least expensive in the long run?

$650 down and $305.54 per month for 5 years
$350 down and $343.57 per month for 54 months
$400 down and $386.42 per month for 4 years

73. A new refrigerator-freezer is advertised at three different stores as follows:

Store 1: $75 down and $85.95 for 18 months
Store 2: $125 down and $63.25 per month for 24 months
Store 3: $300 down and $109.55 per month for 12 months

Which store is selling the refrigerator-freezer for the least total cost?

74. An order of 31 bars of steel is delivered to a machine shop. Each bar is 19.625 ft long. Find the total linear feet of steel in the order.

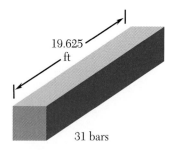

19.625 ft

31 bars

75. From a table in a machinist's handbook, it is determined that hexagon steel bars 1.125 in. across weigh 3.8 lbs per running foot. Using this constant, find the weight of a 1.125 in. hexagon steel bar that is 18.875 ft long.

Exercises 76–78 relate to the chapter application.

In Olympic diving, seven judges rate each dive using a whole or half number between 0 and 10. The high and low scores are thrown out and the remaining scores are added together. The sum is then multiplied by 0.6 and by the difficulty factor of the dive to obtain the total points awarded.

76. A diver does a reverse $1\frac{1}{2}$ somersault with $2\frac{1}{2}$ twists, a dive with a difficulty factor of 2.9. She receives scores of 6.0, 6.5, 6.5, 7.0, 6.0, 7.5, and 7.0. What are the total points awarded for the dive?

77. Another diver also does a reverse $1\frac{1}{2}$ somersault with $2\frac{1}{2}$ twists. This diver receives scores of 7.5, 6.5, 7.5, 8.0, 8.0, 7.5, and 8.0. What are the total points awarded for the dive?

78. A cut through reverse $1\frac{1}{2}$ somersault has a difficulty factor of 2.6. What is the highest number of points possible with this dive?

79. In 1970 the per capita consumption of red meat was 132 pounds. In 1980 the consumption was 126.4 pounds. In 1990, the amount consumed was 112.3 pounds per person. Compute the total weight of red meat consumed by a family of four using the rates for each of these years. Discuss the reasons for the change in consumption.

80. Older models of toilets use 5.5 gallons of water per flush. Models made in the 1970's use 3.5 gallons per flush. The new low-flow models use 1.55 gallons per flush. Assume each person flushes the toilet an average of five times per day. Determine the amount of water used in a town with a population of 34,782 in one day for each type of toilet. How much water is saved using the low-flow model as opposed to the pre-1970's model?

81. The annual property-tax bills arrive in early November in some states. The McNamara house is assessed for $97,700. The annual tax rate is $0.199 per hundred dollars of assessment. Find what they owe in taxes. Round to the nearest cent.

82. Find the property tax on the Gregory Estate which is assessed at $1,895,750. The tax rate in the area is $1.235 per hundred dollars of assessment. Round to the nearest dollar.

STATE YOUR UNDERSTANDING

83. Explain how to determine the number of decimal places needed in the product of two decimals.

84. Suppose you use a calculator to multiply $(0.005)(3.2)(68)$ and get 10.88. Explain, using placement of the decimal point in a product, how you can tell that at least one of the numbers was entered incorrectly. How can such errors occur? How can you estimate the answers before using the calculator so you can avoid such errors?

CHALLENGE

85. What is the smallest whole number you can multiply 0.66 by to get a product that is greater than 55?

86. What is the largest whole number you can multiply 0.78 by to get a product that is less than 58.9?

87. Find the missing number in the following sequence: 1.8, 0.36, 0.108, 0.0432, _____.

88. Find the missing number in the following sequence: 3.1, -0.31, _____, -0.0000031, 0.0000000031.

GROUP ACTIVITY

89. Visit a grocery store or use newspaper ads to "purchase" the items listed. Have each member of your group use a different store or chain. Which members of your group "spent" the most? least? Which group "spent" the most? least?

Three 12-packs of Diet Pepsi	Five 6-roll packs of toilet paper
Eight gallons of 2% milk	Seven pounds of butter
Five pounds of hamburger	72 hamburger rolls
Four cans of the store-brand creamed corn	12 large boxes of Cheerios

MAINTAIN YOUR SKILLS (Section 1.4)

Multiply or divide as indicated.

90. 803×10^3

91. $67,000 \div 10^2$

92. $380,000 \div 10^4$

93. 23×10^7

94. $4,210,000,000 \div 10^5$

6.4   Multiplying and Dividing by Powers of Ten and Scientific Notation

OBJECTIVES

1. Multiply or divide a number by a power of ten.

2. Write a number in scientific notation or change a number in scientific notation to its place value notation.

VOCABULARY

A **power of ten** is a number that can be written as 10^a where a is an integer. **Scientific notation** is a special way to write numbers as a product using a number between one and ten and a power of ten.

HOW AND WHY
Objective 1

Multiply or divide a number by a power of ten.

The shortcut used in Section 1.4 for multiplying and dividing by ten or a power of ten works in a similar way with decimals. Consider the following products:

$$
\begin{array}{ccc}
0.5 & 0.23 & 5.67 \\
\underline{10} & \underline{10} & \underline{10} \\
0 & 0 & 0 \\
\underline{5\,0} & \underline{2\,30} & \underline{56\,70} \\
5.0 = 5 & 2.30 = 2.3 & 56.70 = 56.7
\end{array}
$$

Notice in each case that multiplying a decimal by 10 has the effect of moving the decimal point one place to the right.

Since $100 = 10 \cdot 10$, multiplying by 100 is the same as multiplying by 10 two times in succession. So, multiplying by 100 has the effect of moving the decimal point two places to the right. For instance,

$$(0.53)(100) = 0.53(10 \cdot 10) = (0.53 \cdot 10) \cdot 10 = 5.3 \cdot 10 = 53$$

Since $1000 = 10 \cdot 10 \cdot 10$, the decimal point will move three places to the right when multiplying by 1000. Since $10,000 = 10 \cdot 10 \cdot 10 \cdot 10$, the decimal point will move four places to the right when multiplying by 10,000, and so on in the same pattern:

$$(0.08321)(10,000) = 832.1$$

Zeros may have to be placed on the right in order to move the correct number of decimal places:

$$(2.3)(1000) = 2.300 = 2300$$

In this problem, two zeros are placed on the right.

Since multiplying a decimal by 10 has the effect of moving the decimal point one place to the right, dividing a number by 10 must move the decimal point one place to the left. Again, we are using the fact that multiplication and division are inverse operations. Division by 100 will move the decimal point two places to the left, and so on. Thus,

$$347.1 \div 100 = 347.1 = 3.471$$

$$0.763 \div 1000 = 0.000763$$

Three zeros are placed on the left so that the decimal point may be moved three places to the left.

To multiply a number by a power of ten

Move the decimal point to the right. The number of places to move is shown by the number of zeros in the power of ten.

 To divide a number by a power of ten

Move the decimal point to the left. The number of places to move is shown by the number of zeros in the power of ten.

Examples A–G	**Warm Ups A–G**

Directions: Multiply or divide as indicated.

Strategy: To multiply by a power of ten, move the decimal point to the right. To divide by a power of ten, move the decimal point to the left. The exponent of ten specifies the number of places to move the decimal point.

A. 3.828(10)	A. 22.58(10)
3.828(10) = 38.28 **Multiplying by 10 moves the decimal point one place to the right.**	
B. Multiply: 0.428(100)	B. Multiply: 0.017(100)
0.428(100) = 42.8 **Multiplying by 100 moves the decimal point two places to the right.**	
C. Find the product: $264.3(10^3)$	C. Find the product: $31.45(10^3)$
$264.3(10^3) = 264{,}300$ **Multiplying by 10^3 moves the decimal point three places to the right. Two zeros must be placed on the right to make the move.**	
D. Divide: 27.3 ÷ 10	D. Divide: 334 ÷ 10
27.3 ÷ 10 = 2.73 **Dividing by 10 moves the decimal point one place to the left.**	
E. Divide: 3.12 ÷ 100	E. Divide: 5.267 ÷ 100
3.12 ÷ 100 = 0.0312 **Dividing by 100 moves the decimal point two places to the left.**	
F. Find the quotient: $47.8 ÷ 10^4$	F. Find the quotient: $68.935 ÷ 10^4$
$47.8 ÷ 10^4 = 0.00478$ **Move the decimal point four places to the left.**	
G. A stack of sheet metal contains 100 sheets. The stack is 6.25 inches high. How thick is each sheet of metal?	G. One hundred sheets of clear plastic are 0.05 inch thick. How thick is each sheet? (This is the thickness of some household plastic wrap.)
6.25 ÷ 100 = 0.0625 **To find the thickness of each sheet, divide the height by the number of sheets.**	
Each sheet is 0.0625 inch thick.	

Answers to Warm Ups   A. 225.8   B. 1.7   C. 31,450   D. 33.4   E. 0.05267   F. 0.0068935   G. 0.0005 inch

HOW AND WHY
Objective 2

Write a number in scientific notation or change a number in scientific notation to its place value notation.

Scientific notation is widely used in science, technology, and industry to write large and small numbers. Every "scientific calculator" has a key for entering numbers in scientific notation. This notation makes it possible for a calculator or computer to deal with much larger or smaller numbers than those that take up eight, nine, or ten spaces on the display.

Scientific Notation

A number in scientific notation is written as the product of two numbers. The first number is between one and ten (including one but not ten) and the second number is a power of ten. You must use an " $\times$ " to show the multiplication.

For example,

Word Form	Place Value (Numeral Form)	Scientific Notation	Calculator or Computer Display
One million	1,000,000	1×10^6	1. 06 or 1 E 6
Five billion	5,000,000,000	5×10^9	5. 09 or 5 E 9
One trillion, three billion	1,003,000,000,000	1.003×10^{12}	1.003 12 or 1.003 E 12

Small numbers are shown by writing the power of ten using a negative exponent. (You will learn more about this when you take a course in algebra.) For now, remember that multiplying by a negative power of ten is the same as *dividing* by a power of ten, which means you will be moving the decimal point to the left.

Word Form	Place Value (Numeral Form)	Scientific Notation	Calculator or Computer Display
Seven thousandths	0.007	7×10^{-3}	7. −03 or 7 E −3
Six ten-millionths	0.0000006	6×10^{-7}	6. −07 or 6 E −7
Fourteen hundred-billionths	0.00000000014	1.4×10^{-10}	1.4 −10 or 1.4 E −10

▶ *To write a number in scientific notation*

1. Move the decimal point right or left so that only one nonzero digit remains to the left of the decimal point. The result will be a number between one and ten. If the choice is one or ten itself, use one.

2. Multiply the decimal found in step 1 by a power of ten. The exponent of ten to use is one that will make the new product equal to the original number.

   a. If you had to move the decimal to the left, multiply by the same number of tens as the number of places moved.

   b. If you had to move the decimal to the right, divide (by writing a negative exponent) by the same number of tens as the number of places moved.

To change from scientific notation to place-value notation

1. If the exponent of ten is positive, multiply by as many tens (move the decimal to the right as many places) as the exponent shows.

2. If the exponent of ten is negative, divide by as many tens (move the decimal to the left as many places) as the exponent shows.

For numbers larger than 1:

Place-Value Notation:	12,000	3,400,000	12,300,000,000,000	
Number Between 1 and 10:	1.2	3.4	1.23	Move the decimal (which is after the units place) to the left until the number is between 1 and 10 (one digit to the left of the decimal).
Scientific Notation:	1.2×10^4	3.4×10^6	1.23×10^{13}	Multiply each by a power of ten that shows how many places left the decimal moved, or how many places you would have to move to the right to recover the original number.

For numbers smaller than 1:

Place-Value Notation:	0.000033	0.00000007	0.0000000000345	
Number Between 1 and 10:	3.3	7.	3.45	Move the decimal to the right until the number is between 1 and 10.
Scientific Notation:	3.3×10^{-5}	7×10^{-8}	3.45×10^{-11}	Divide each by the power of ten that shows how many places right the decimal moved. Show this division by a negative power of 10.

It is important to note that scientific notation is not rounding. The scientific notation has exactly the same value as the original name.

Examples H–J   **Warm Ups H–J**

Directions: Write in scientific notation.

Strategy: Move the decimal point so that there is one nonzero digit to the left. Multiply this number by the appropriate power of ten so the value is the same as the original number.

H. 782,000,000   H. 13,000,000

  7.82 is between 1 and 10   **Move the decimal eight places to the left.**

  $7.82 \times 100,000,000$ is 782,000,000   **Moving the decimal left is equivalent to dividing by 10 for each place.**

Answers to Warm Ups   H. 1.3×10^7

$782{,}000{,}000 = 7.82 \times 10^8$	**To keep the values the same, we multiply by 10 eight times.**

I.   Write 0.0000000092 in scientific notation.

9.2 is between 1 and 10	**Move the decimal nine places to the right.**
$9.2 \div 1{,}000{,}000{,}000$ is 0.0000000092	**Moving the decimal right is equivalent to multiplying by 10 for each place.**
$0.0000000092 = 9.2 \times 10^{-9}$	**To keep the values the same, we divide by 10 nine times.**

I.   Write 0.00000774 in scientific notation.

J.   Approximately 12,000,000 people in the United States have type II diabetes. Write this number in scientific notation.

1.2 is between 1 and 10.

$1.2 \times 10{,}000{,}000 = 12{,}000{,}000$

$12{,}000{,}000 = 1.2 \times 10^7$

In scientific notation, the number of people with type II diabetes is 1.2×10^7.

J.   The age of a 22-year-old student is approximately 694,000,000 seconds. Write this number in scientific notation.

Answers to Warm Ups   I. 7.74×10^{-6} J. 6.94×10^8

Exercises 6.4

OBJECTIVE 1: *Multiply or divide a number by a power of ten.*

A.

Multiply or divide.

1. $4.25 \div 10$

2. $56.98 \div 10$

3. $(3.67)(100)$

4. $28.9(100)$

5. $(0.62833)(1000)$

6. $(34.6211)(1000)$

7. $\dfrac{569.2}{1000}$

8. $\dfrac{9568.3}{1000}$

9. $\dfrac{5645}{100}$

10. $\dfrac{3459}{1000}$

11. 0.87×10^4

12. 4.3×10^5

13. To multiply 4.56 by 10^5 move the decimal point five places to the _____.

14. To divide 4.56 by 10^3 move the decimal point three places to the _____.

B.

Multiply or divide.

15. $(6.274)(1000)$

16. $8.75(100)$

17. $1.85 \div 10$

18. $912.5 \div 1000$

19. $36.9(1000)$

20. $0.6783(10)$

21. $\dfrac{6895.3}{10,000}$

22. $\dfrac{213.775}{100,000}$

23. $14.78\,(100,000)$

24. $5.732\,(1,000,000)$

25. $1367.94 \div 100$

26. $78.94 \div 1000$

27. $45.8 \div 100,000$

28. $2.789 \div 1,000,000$

OBJECTIVE 2: *Write a number in scientific notation or change a number in scientific notation to its place value notation.*

A.

Write in scientific notation.

29. 230,000

30. 4700

31. 0.00035

32. 0.0000521

33. 467.95

34. 1245.6

Write the place value notation.

35. 6×10^4

36. 7×10^2

37. 8×10^{-3}

38. 2×10^{-5}

39. 4.78×10^3

40. 9.02×10^5

B.

Write in scientific notation.

41. 780,000

42. 4,520,000

43. 0.0000345

44. 0.0007432

45. 0.0000000000821

46. 0.00000002977

47. 3567.003

48. 56.8004

Write the place value notation.

49. 1.345×10^{-6}

50. 8.031×10^{-5}

51. 7.11×10^9

52. 8.032×10^8

53. 4.44×10^{-7}

54. 3.9×10^{-10}

55. 5.6723×10^2

56. 7.892111×10^5

C.

57. Ken's Shoe Store buys 100 pairs of shoes that cost $22.29 per pair. What is the total cost of the shoes?

58. If Mae's Shoe Store buys 100 pairs of shoes for a total cost of $4897.50, what is the cost of each pair of shoes?

59. Ms. James buys 100 acres of land at a cost of $985 per acre. What is the total cost of her land?

60. If 1000 bricks weigh 5900 pounds, how much does each brick weigh?

61. The total land area of the Earth is approximately 52,000,000 square miles. What is the total area written in scientific notation?

62. A local computer store offers a small computer with 1152K (1,152,000) bytes of memory. Write the number of bytes in scientific notation.

 Exercises 63–65 relate to the chapter application.

In baseball, a hitter's batting average is calculated by dividing the number of hits by the number of times at bat. Mathematically, this number is always between zero and one.

63. In 1988 Wade Boggs led the American League with a batting average of 0.366. However, players and fans would say that Boggs has an average of "three hundred sixty-six." Mathematically, what are they doing to the actual number?

64. Explain why the highest possible batting average is 1.0.

65. The major league player with the highest season batting average in this century is Roger Hornsby of St. Louis. In 1924 he batted 424. Change this to the mathematically calculated number of his batting average.

66. The length of a red light ray is 0.000000072 cm. Write this length in scientific notation.

67. The time it takes light to travel one kilometer is approximately 0.0000033 second. Write this time in scientific notation.

68. The speed of light is approximately 1.116×10^7 miles per minute. Write this speed in place value notation.

69. The table below gives the distances from the sun to the planets in the solar system. Write the place value notation for each distance.

Planet	Mean Distance in Miles
Mercury	3.6×10^7
Venus	6.724×10^7
Earth	9.296×10^7
Mars	1.4164×10^8
Jupiter	4.8364×10^8
Saturn	8.87×10^8
Uranus	1.783×10^9
Neptune	2.795×10^9
Pluto	3.666×10^9

70. The shortest wavelength of visible light is approximately 4×10^{-5} centimeter. Write this length in place value notation.

71. A sheet of paper is approximately 1.3×10^{-3} inch thick. Write the thickness in place value notation.

72. A family in the Northeast used 3.276×10^8 BTUs of energy during 1989. A family in the Midwest used 3.312×10^8 BTUs in the same year. A family in the South used 3.933×10^8 BTUs and a family in the West used 1.935×10^8 BTUs. Write the total energy usage for the four families in place value notation.

73. In 1990 the per capita consumption of fish was 15.5 pounds. In the same year the per capita consumption of poultry was 63.6 pounds and of red meat was 112.3 pounds. Write the total amount in each category consumed by 100,000 people in scientific notation.

74. The population of Cabot Cove was approximately 10,000 in 1995. During the year, the community consumed a total of 276,000 gallons of milk. What was the per capita consumption of milk in Cabot Cove in 1995?

75. In 1980, $24,744,000,000 was spent on air pollution abatement. Ten years later, $26,326,000,000 was spent. In scientific notation, how much more money was spent in 1990 than in 1980? What is the average amount of increase per year during the period?

STATE YOUR UNDERSTANDING

76. Find a pair of numbers whose product is larger than ten trillion. Explain how scientific notation makes it possible to multiply these factors on a calculator. Why is it not possible without scientific notation?

CHALLENGE

77. A Parsec is a unit of measure used to determine distance between stars. One Parsec is approximately 206,265 times the average distance of the Earth from the sun. If the average distance from the Earth to the sun is approximately 93,000,000 miles. Find the approximate length of one Parsec. Write the length in scientific notation. Round the number in scientific notation to the nearest hundredth.

78. Light will travel approximately 5,866,000,000,000 miles in one year. Approximately how far will light travel in 8 years? Write the distance in scientific notation. Round the number in scientific notation to the nearest thousandth.

Simplify.

79. $\dfrac{(3.25 \times 10^{-3})(2.4 \times 10^{3})}{(4.8 \times 10^{-4})(2.5 \times 10^{-3})}$

80. $\dfrac{(3.25 \times 10^{-7})(2.4 \times 10^{6})}{(4.8 \times 10^{4})(2.5 \times 10^{-3})}$

GROUP ACTIVITY

81. Find the 1990 population for the ten largest cities and the five smallest cities in your state. Round these numbers to the nearest thousand. Find the total number of pounds of fruit, at the rate of 92.3 pounds per person, and the total number of pounds of vegetables, at the rate of 11.2 pounds per person, consumed in each of these 15 cities.

MAINTAIN YOUR SKILLS (Section 1.3)

82. $38\overline{)18{,}050}$

83. $76\overline{)8208}$

84. $103\overline{)21{,}527}$

85. Find the quotient of 15,023 and 25.

86. Find the quotient of 231,876 and 203.

6.5   Dividing Rational Numbers (Decimals)

OBJECTIVES

1. Divide decimals.

2. Change fractions to decimals.

3. Estimate the quotient of decimals.

**HOW AND WHY
Objective 1**

Divide decimals.

Division of decimals is the same as division of whole numbers, with one exception. The exception is the location of the decimal point in the quotient.

$$\begin{array}{r} 0.0019 \\ 20\overline{)0.0380} \\ \underline{20} \\ 180 \\ \underline{180} \\ 0 \end{array}$$

When dividing by a whole number, you can correctly place the decimal point for the quotient by writing it above the decimal point in the dividend.

Check:

$$\begin{array}{r} 0.0019 \\ \times 20 \\ \hline 0.0380 \end{array}$$

It may be necessary to insert zeros to do the division. See Example B.

When a decimal is divided by 7, the division process may not have a remainder of zero at any step:

$$\begin{array}{r} 0.33 \\ 7\overline{)2.34} \\ \underline{2\,1} \\ 24 \\ \underline{21} \\ 3 \end{array}$$

At this step we can write zeros to the right of the digit 4, since $2.34 = 2.340 = 2.3400 = 2.34000 = 2.340000$.

$$\begin{array}{r} 0.33428 \\ 7\overline{)2.34000} \\ \underline{2\,1} \\ 24 \\ \underline{21} \\ 30 \\ \underline{28} \\ 20 \\ \underline{14} \\ 60 \\ \underline{56} \\ 4 \end{array}$$

It appears that we might go on inserting zeros and continue endlessly. This is indeed what happens. Such decimals are called "nonterminating, repeating decimals." For example, this one is sometimes written

$$0.33428571428571\ldots \quad \text{or} \quad 0.33\overline{428571}$$

The bar written above the sequence of digits, 428571, indicates that these digits are repeated endlessly.

In practical applications we stop the division process one place value beyond the accuracy required by the situation and then round. Therefore,

$2.34 \div 7 \approx 0.33$ to the nearest hundredth

$2.34 \div 7 \approx 0.3343$ to the nearest ten-thousandth

If the divisor contains a decimal point, we change the problem:

$0.7\overline{)2.338} = 0.7\,\overline{)2.338} = 7\overline{)23.38}$ because $\dfrac{2.338}{0.7} \cdot \dfrac{10}{10} = \dfrac{23.38}{7}$

$0.014\overline{)7.8} = 0.014\,\overline{)7.800} = 14\overline{)7800}$ because $\dfrac{7.8}{0.014} \cdot \dfrac{1000}{1000} = \dfrac{7800}{14}$

$0.23\overline{)7.2} = 0.23\,\overline{)7.20} = 23\overline{)720}$ because $\dfrac{7.2}{0.23} \cdot \dfrac{100}{100} = \dfrac{720}{23}$

> ### To divide two numbers
>
> 1. If the divisor is not a whole number, move both decimal points to the right the same number of decimal places until the divisor is a whole number. In other words, multiply both the divisor and the dividend by the same power of ten so the divisor is a whole number.
> 2. Place the decimal point in the quotient above the decimal point in the dividend.
> 3. Divide as if both numbers were whole numbers.
> 4. Round to the given place value. (If no round-off place is given, divide until the remainder is zero or round as appropriate in the problem. For instance, in problems with money, round to the nearest cent.)

Examples A–H   **Warm Ups A–H**

Directions: Divide. Round as indicated.

Strategy: If the divisor is not a whole number, move the decimal point in both the divisor and the dividend to the right the number of places necessary to make the divisor a whole number. The decimal point in the quotient is found by writing it directly above the decimal (as moved) in the dividend.

A. $13\overline{)13.026}$ A. $15\overline{)45.105}$

$$
\begin{array}{r}
1.002 \\
13\overline{)13.026} \\
\underline{13} \\
00 \\
\underline{0\,0} \\
02 \\
\underline{00} \\
26 \\
\underline{26} \\
0
\end{array}
$$

The numerals in the answer are lined up in columns that have the same place value as those in the dividend. Check by multiplying 13×1.002.

$$
\begin{array}{r}
1.002 \\
\times 13 \\
\hline
3\,006 \\
10\,02 \\
\hline
13.026
\end{array}
$$

The quotient is 1.002.

CAUTION **Write the decimal point for the quotient directly above the decimal point in the dividend.**

B.   Find the quotient of 1.88 and 8.

$$\begin{array}{r} 0.23 \\ 8\overline{)1.88} \\ \underline{1\,6} \\ 28 \\ \underline{24} \\ 4 \end{array}$$

Here the remainder is not zero, so the division is not complete. We write a zero on the right (1.880) without changing the value of the dividend and continue dividing.

$$\begin{array}{r} 0.235 \\ 8\overline{)1.880} \\ \underline{1\,6} \\ 28 \\ \underline{24} \\ 40 \\ \underline{40} \\ 0 \end{array}$$

Both the quotient (0.235) and the rewritten dividend (1.880) have three decimal places. Check by multiplying 8×0.235.

$$\begin{array}{r} 0.235 \\ \times\quad 8 \\ \hline 1.880 \end{array}$$

The quotient is 0.235.

B.   Find the quotient of 2.16 and 16.

C.   Divide 486.5 by 23 and round quotient to the nearest hundredth.

$$\begin{array}{r} 21.152 \\ 23\overline{)486.500} \\ \underline{46} \\ 26 \\ \underline{23} \\ 3\,5 \\ \underline{2\,3} \\ 1\,20 \\ \underline{1\,15} \\ 50 \\ \underline{46} \\ 4 \end{array}$$

It is necessary to place two zeros on the right in order to round to the hundredths place, since the division must be carried out one place past the place to which you wish to round.

The quotient is approximately 21.15.

C.   Divide 241.3 by 21 and round quotient to the nearest hundredth.

D.   Divide: $1.32 \div 0.7$ and round to the nearest hundredth.

$$0.7\overline{)1.32}$$

$$\begin{array}{r} 1.8 \\ 7\overline{)13.2} \\ \underline{7} \\ 6\,2 \\ \underline{5\,6} \\ 6 \end{array}$$

First, move both decimals one place to the right so the divisor is the whole number, 7. The same result is obtained by multiplying both divisor and dividend by 10.

$$\frac{1.32}{0.7} \times \frac{10}{10} = \frac{13.2}{7}$$

D.   Divide: $2.48 \div 0.7$ and round to the nearest hundredth.

(continued on next page)

$$\begin{array}{r} 1.885 \\ 7\overline{)13.200} \\ \underline{7} \\ 6\,2 \\ \underline{5\,6} \\ 60 \\ \underline{56} \\ 40 \\ \underline{35} \\ 5 \end{array}$$

The number of zeros you place on the right depends on either the directions for rounding or your own choice of the number of places. Here we find the approximate quotient rounded to the nearest hundredth.

The quotient is approximately 1.89.

E. Divide 0.47891 by 0.072 and round to the nearest thousandth.

$$0.072\overline{)0.47891}$$

$$\begin{array}{r} 6.6515 \\ 72\overline{)478.9100} \\ \underline{432} \\ 46\,9 \\ \underline{43\,2} \\ 3\,71 \\ \underline{3\,60} \\ 110 \\ \underline{72} \\ 380 \\ \underline{360} \\ 20 \end{array}$$

Move both decimals three places to the right.

The quotient is approximately 6.652.

E. Divide 0.75593 by 0.043 and round to the nearest thousandth.

Calculator Example:

F. Find the quotient of 78.1936 and −8.705 and round to the nearest thousandth.

$$78.1936 \div -8.705 \approx -8.9826077$$

The quotient is −8.983, to the nearest thousandth.

F. Find the quotient of −103.843 and 4.088 and round to the nearest thousandth.

G. Divide: $\dfrac{5.6t}{-1.4}$

$$\dfrac{5.6t}{-1.4} = -4t \qquad \textbf{Divide the coefficients and multiply times the variable factor.}$$

G. Divide: $\dfrac{-4.9n}{-0.7}$

H. What is the cost per ounce of a 12-ounce can of root beer that costs 45¢? This is called the "unit price" and is used for comparing prices. Many stores are required to show this price for the food they sell.

H. What is the unit price of potato chips if a 7-ounce bag costs $1.26?

Answers to Warm Ups   E. 17.580   F. −25.402   G. 7n   H. 18¢ per ounce

$$
\begin{array}{r}
3.75 \\
12\overline{)45.00} \\
36 \\
\hline
9\,0 \\
8\,4 \\
\hline
60 \\
60 \\
\hline
\end{array}
$$

To find the unit price (cost per ounce), we divide the cost by the number of ounces.

The root beer costs 3.75¢ per ounce.

HOW AND WHY
Objective 2   **Change fractions to decimals.**

Every fraction can be thought of as a division problem, $\dfrac{2}{5} = 2 \div 5 = 0.4$. Therefore, a method for changing fractions to decimals is division. As you discovered earlier, many division problems with decimals do not divide evenly. Whenever a fraction is changed to a decimal, we have one of two possibilities:

1. The quotient is exact, as in $\dfrac{2}{5} = 0.4$. This decimal is a "terminating decimal."

2. The quotient is nonterminating, repeating decimal, as in

$$\dfrac{2}{3} = 0.666666\ldots = 0.\overline{6}$$

The bar over the 6 indicates that the decimal repeats the number 6 forever. In the exercises for this section, round the division to the desired decimal place.

  To change a fraction to a decimal

Divide the numerator by the denominator.

Examples I–M	**Warm Ups I–M**

Directions: Change the fraction or mixed number to a decimal.

Strategy: Divide the numerator by the denominator. Round as necessary. If the number is a mixed number, add the decimal to the whole number.

I. $\dfrac{9}{20}$

$$
\begin{array}{r}
0.45 \\
20\overline{)9.00} \\
8\,0 \\
\hline
1\,00 \\
1\,00 \\
\hline
0 \\
\end{array}
$$

Divide the numerator 9 by the denominator, 20.

Therefore, $\dfrac{9}{20} = 0.45$.

I. $\dfrac{11}{20}$

J. $7\dfrac{7}{50}$

$$\begin{array}{r} 0.14 \\ 50\overline{)7.00} \\ \underline{5\,0} \\ 2\,00 \\ \underline{2\,00} \\ 0 \end{array}$$

Divide the numerator, 7, by the denominator, 50.

$7\dfrac{7}{50} = 7.14$

Add the decimal to the whole number: $7 + 0.14$

or

$\dfrac{7}{50} = \dfrac{7}{50} \cdot \dfrac{2}{2} = \dfrac{14}{100} = 0.14$

Often a fraction with a denominator that has only 2's and 5's for prime factors can be changed to a decimal more easily by building than by dividing.

So, $7\dfrac{7}{50} = 7 + 0.14 = 7.14$.

J. $4\dfrac{17}{25}$

K. $\dfrac{11}{16}$

$$\begin{array}{r} 0.6875 \\ 16\overline{)11.0000} \\ \underline{9\,6} \\ 1\,40 \\ \underline{1\,28} \\ 120 \\ \underline{112} \\ 80 \\ \underline{80} \\ 0 \end{array}$$

Divide the numerator by the denominator. This fraction can be changed by building to a denominator of 10,000 but the factor is not easily recognized, except by calculator.

$\dfrac{11}{16} = \dfrac{11}{16} \cdot \dfrac{625}{625} = \dfrac{6875}{10,000} = 0.6875$

Therefore, $\dfrac{11}{16} = 0.6875$.

C A U T I O N **Most fractions cannot be represented by terminating decimals. They are represented by repeating decimals. It is common practice to round these to an approximate decimal.**

K. $\dfrac{23}{40}$

L.   Change $\dfrac{7}{12}$ to a decimal and round to the nearest hundredth.

$$\begin{array}{r} 0.583 \\ 12\overline{)7.000} \\ 6\,0 \\ \hline 1\,00 \\ 96 \\ \hline 40 \\ 36 \\ \hline 4 \end{array}$$

Divide 7 by 12. Carry out the division to three decimal places and round to the nearest hundredth.

So, $\dfrac{7}{12} \approx 0.58$.

L.   Change $\dfrac{2}{7}$ to a decimal and round to the nearest hundredth.

Calculator Example:

M.   Change $\dfrac{78}{136}$ to a decimal and round to the nearest thousandth.

$\dfrac{78}{136} \approx 0.5735294118$

So, $\dfrac{78}{136} \approx 0.574$ to the nearest thousandth.

M.   Change $\dfrac{87}{143}$ to a decimal and round to the nearest thousandth.

HOW AND WHY
Objective 3

Estimate the quotient of decimals.

The quotient of two decimals can be estimated by rounding each number to its largest nonzero place value and then dividing the rounded numbers. For instance,

$0.27\overline{)0.006345}$ $0.3\overline{)0.006}$ **Round each to its largest nonzero place.**

$$\begin{array}{r} 0.02 \\ 3\overline{)0.06} \\ 6 \\ \hline 0 \end{array}$$

Multiply by 10 so the divisor is a whole number. Divide.

The estimate of the quotient is 0.02. One use of the estimate is to see if the quotient is correct. If the calculated quotient is not close to 0.02, you should check the division. In this case the exact quotient is 0.0235, which is close to the estimate.

The estimated quotient will not always come out even. When this happens, round the first partial quotient and use it for the first nonzero entry in the quotient. For instance,

$7.41\overline{)42.95577}$ $7\overline{)40}$ **Round each number to its largest nonzero place.**

$$\begin{array}{r} 6 \\ 7\overline{)40} \end{array}$$

Divide 40 by 7 for the first partial quotient. Since $5(7) = 35$ and $6(7) = 42$, we choose the closer value, 6.

The estimated quotient is 6. The exact quotient is 5.797.

To estimate the quotient of decimals

1. Round each decimal to its largest nonzero place.

2. Divide the rounded numbers.

3. If the first partial quotient has a remainder, choose the digit that gives the closer value when multiplied by the divisor.

Example N   **Warm Up N**

Directions: Estimate the quotient and find the quotient to the nearest ten-thousandth.

Strategy: Round each number to its largest nonzero place and then divide the rounded numbers. If the first partial quotient has a remainder, choose the digit that gives the closer value when multiplied by the divisor.

N.   Divide 0.05682 by 0.908.

$$0.908\overline{)0.05682}$$ $$0.9\overline{)0.06}$$ **Round.**

$$\begin{array}{r} 0.07 \\ 9\overline{)0.60} \end{array}$$
$9(0.06) = 0.54$
$9(0.07) = 0.63$
So choose 0.07 as the estimate.

$$\begin{array}{r} 0.06257 \\ 908\overline{)56.82} \\ \underline{5448} \\ 2340 \\ \underline{1816} \\ 5240 \\ \underline{4540} \\ 7000 \\ \underline{6356} \\ 644 \end{array}$$

The estimated quotient is 0.07 and the quotient rounded to the nearest ten-thousandth is 0.0626.

N.   Divide 0.0039 by 0.723.

Exercises 6.5

OBJECTIVE 1: *Divide decimals.*

A.

Divide.

1. $7\overline{)3.5}$

2. $6\overline{)4.8}$

3. $2\overline{)19.6}$

4. $3\overline{)19.2}$

5. $0.1\overline{)18.31}$

6. $0.1\overline{)2.72}$

7. $242.4 \div 0.12$

8. $337.7 \div 0.11$

9. To divide 2.65 by 0.05 we first multiply both the dividend and the divisor by 100 so we are dividing by a _____.

10. To divide 0.4763 by 0.287 we first multiply both the dividend and the divisor by _____.

B.

Divide.

11. $80\overline{)1008}$

12. $80\overline{)104.8}$

13. $-16.64 \div 32$

14. $3.936 \div (-32)$

Divide and round to the nearest tenth.

15. $7\overline{)8.96}$

16. $8\overline{)0.912}$

17. $1.3\overline{)11.778}$

18. $6.7\overline{)34.562}$

Divide and round to the nearest thousandth.

19. $2.2\overline{)34.22}$

20. $24.9\overline{)60.363}$

OBJECTIVE 2: *Change fractions to decimals.*

A.

Change the fraction or mixed number to a decimal.

21. $\dfrac{3}{4}$

22. $\dfrac{7}{10}$

23. $\dfrac{3}{8}$

24. $\dfrac{5}{8}$

25. $\dfrac{11}{16}$

26. $\dfrac{1}{32}$

27. $3\dfrac{11}{20}$

28. $6\dfrac{13}{20}$

29. $11\dfrac{3}{125}$

30. $12\dfrac{7}{50}$

B.

Change to a decimal rounded to the indicated place value.

		Tenth	**Hundredth**
31.	$\dfrac{3}{7}$		
32.	$\dfrac{2}{9}$		
33.	$\dfrac{5}{11}$		

		Tenth	Hundredth
34.	$\dfrac{5}{6}$		
35.	$\dfrac{2}{13}$		
36.	$\dfrac{3}{14}$		
37.	$\dfrac{8}{15}$		
38.	$\dfrac{11}{19}$		
39.	$5\dfrac{17}{18}$		
40.	$11\dfrac{9}{17}$		

OBJECTIVE 3: *Estimate the quotient.*

A.

41. $4\overline{)0.0782}$ **42.** $7\overline{)0.0359}$ **43.** $0.2\overline{)2.67}$

44. $0.3\overline{)5.823}$ **45.** $0.468 \div 0.523$ **46.** $0.2489 \div 0.1943$

47. $34 \div 0.0756$ **48.** $76 \div 0.08659$

B.

Estimate the quotient. Divide and round to the nearest hundredth.

49. $64\overline{)6211.84}$ **50.** $32\overline{)201.824}$ **51.** $2.97\overline{)0.2267}$

52. $3.46\overline{)0.5699}$ **53.** $0.12 \div 0.007$ **54.** $0.15 \div 0.0083$

C.

Divide.

55. $\dfrac{8.4x}{-0.4}$ **56.** $\dfrac{-8.4x}{-0.4}$ **57.** $\dfrac{-0.45m}{1.2}$ **58.** $\dfrac{4.5m}{-0.12}$

59. Find the quotient of -9.19 and 0.11, and round to the nearest hundredth.

60. Find the quotient of -0.481 and -94, and round to the nearest hundredth.

Use the table for Exercises 61–67. The table shows some prices from a grocery store.

Item	Quantity	Price	Item	Quantity	Price
Oranges	3 lb	$4.42	Potatoes	15 lb	$1.16
Strawberries	6 pints	$4.77	Rib Steak	3.24 lb	$9.62
Syrup	36 ounces	$2.77	Ham	4.6 lb	$16.05

61. Find the unit price (price per pound) of oranges. Round to the nearest tenth of a cent.

62. Find the unit price (price per ounce) of syrup. Round to the nearest tenth of a cent.

63. Find the unit price of rib steak. Round to the nearest tenth of a cent.

64. Find the unit price of potatoes. Round to the nearest tenth of a cent.

65. Using the unit price find the cost of 7 lb of ham. Round to the nearest cent.

66. Using the unit price find the cost of 11 pints of strawberries. Round to the nearest cent.

67. Using unit pricing, find the total cost of 6 lb of rib steak, 2 pints of strawberries, 4 lbs of potatoes, and 5 lb of oranges.

Divide.

68. $\dfrac{8.5x - 3.5}{0.5}$

69. $\dfrac{3.8a - 5.7}{1.9}$

70. $\dfrac{-0.9x^2 + 0.36x}{1.8x}$

71. $\dfrac{-2.16a^2 - 0.0096a}{0.24a}$

Change each of the following fractions to decimals to the nearest indicated place value.

		Hundredth	**Thousandth**
72.	$\dfrac{15}{43}$		
73.	$\dfrac{12}{31}$		
74.	$\dfrac{28}{57}$		
75.	$\dfrac{85}{91}$		

Change to a repeating decimal.

76. $\dfrac{5}{13}$

77. $\dfrac{11}{12}$

78. $\dfrac{6}{7}$

79. $\dfrac{23}{26}$

80. Ninety-seven alumni of Tech U. donated $7635 to the university. To the nearest cent, what was the average donation?

81. Vern bought a pair of green socks. The socks are on sale at three pairs for $12.45. How much does he pay for the socks?

82. June drove 354.6 miles on 12.3 gallons of gas. What is her mileage (miles per gallon)? Round to the nearest mile.

83. A 65-gallon drum of cleaning solvent in an auto repair shop is being used at the rate of 1.94 gallons per day. At this rate, how many days will the drum last? Round to the nearest day.

84. The Williams Construction Company uses cable that weighs 3.5 pounds per foot. A partly filled spool of the cable is weighed. The cable itself weighs 813 pounds after taking off for the weight of the spool. To the nearest foot, how many feet of cable are on the spool?

85. A plumber connects four building's sewers to the public sewer line like the one shown in the diagram. The total bill for the job is $6355.48. What is the average cost for each connection?

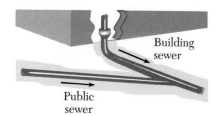
Building sewer

Public sewer

86. A carpenter works 195.5 hours in one month. How many hours did she work each day if the month contained 23 work days? (Assume that she worked the same number of hours each day.)

87. A 15-inch I beam weighs 32.7 pounds. What is the length of a beam weighing 630.6 pounds? Find the length to the nearest tenth of a foot.

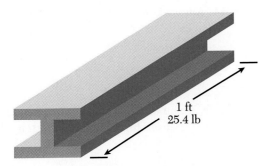

1 ft
25.4 lb

88. A contractor estimates the labor cost of pouring 75 cubic yards of concrete to be $4050. How much does he allow for each cubic yard of concrete?

89. What is the cost of the illustrated clay flue lining if 56 feet cost $312.60? Round to the nearest cent.

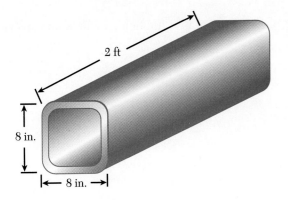

2 ft

8 in.

|← 8 in. →|

90. Allowing 0.125 inch of waste for each cut, how many bushings, which are 1.25 inches in length, can be cut from an 8-inch length of bronze? What is the length of the piece that is left?

Exercises 91–94 relate to the chapter application.

In baseball, a pitcher's earned run average (ERA) is calculated by dividing the number of earned runs by the quotient of the number of innings pitched and 9. The lower a pitcher's ERA the better.

91. Suppose a pitcher allowed 26 earned runs in 80 innings of play. Calculate his ERA and round to the nearest hundredth.

92. A pitcher allows 15 earned runs in 100 innings. Calculate his ERA, rounding to the nearest hundredth.

93. A runner's stolen base average is the quotient of the number of bases stolen and the number of attempts. As with the batting averages, this number is usually rounded to the nearest thousandth. Calculate the stolen base average of a runner who stole 18 bases in 29 attempts.

94. A good stolen base average is 0.700 or higher. Express this as a fraction and say in words what the fraction represents.

STATE YOUR UNDERSTANDING

95. Describe a procedure for determining the placement of the decimal in a quotient. Include an explanation for the justification of the procedure.

96. Explain how to find the quotient of $-4.1448 \div 0.0012$.

97. Copy the table below and fill it in.

Operation	Procedure	Example
Division		

CHALLENGE

98. Which is larger, 0.0036 or $\dfrac{2}{625}$?

99. Which is larger, 2.5×10^{-4} or $\dfrac{3}{2000}$?

100. What will be the value of $3000 invested at 6% (0.06) interest compounded quarterly at the end of one year? (Compounded quarterly means that the interest earned for the quarter is added to the principal and then earns interest for the next quarter.) How much more is earned by compounding quarterly instead of annually?

GROUP ACTIVITY

101. Determine the distance each member of your group travels to school each day. Find the average distance to the nearest hundredth of a mile for your group. Compare these results with the class. Find the average distance for the entire class. Recalculate the class average after throwing out the longest and the shortest distances. Are the averages different? Why?

102. In diving, a back $3\dfrac{1}{2}$ somersault, tuck position has a difficulty factor of 3.3. What do the judges need to score this dive in order for it to be worth the same number of points as a perfect cut through reverse $1\dfrac{1}{2}$ somersault that has a difficulty factor of 2.6? See Exercises 76–78, Section 6.3.

MAINTAIN YOUR SKILLS (Sections 5.5, 6.2)

Perform the indicated operations.

103. $2\dfrac{1}{4} - \dfrac{3}{4} + 1\dfrac{5}{8}$

104. $3 + 2\dfrac{3}{4} - 1\dfrac{5}{16}$

105. $17 - 5\dfrac{7}{9}$

106. $23 - 19\dfrac{15}{16}$

107. Mr. Lewis buys 350 books for $60 at an auction. He sells two fifths of them for $25, 26 books at $1.50 each, 45 books at $1 each, and gives away the rest. How many books does he give away? What is his total profit if his handling cost is $15?

6.6   Another Look at Conversion of Units

Convert units of measure.

HOW AND WHY
Objective

Convert units of measure.

In this section we convert units of measure that require decimal representation. Most of these conversions are approximate, so we round to an indicated place value. For instance, convert 4.1 ft to yards.

$$4.1 \text{ ft} = \frac{4.1 \text{ ft}}{1} \cdot \frac{1 \text{ yd}}{3 \text{ ft}} = \frac{4.1 \times 1}{1 \times 3} \text{ yd} = \frac{4.1}{3} \text{ yd} \approx 1.37 \text{ yd (to the nearest hundredth)}$$

Decimal representation provides us with an alternate method for conversion within the metric system. This method utilizes the fact that the metric system uses the base-ten place value system. The following conversion chart, based on the prefixes and the base unit, will help. We include in the chart three seldom used prefixes, h for hecto, 100; da for deka, 10; and d for deci, 0.1. For example, 1 hectogram = 100 grams.

b
a
s
e
k   h   da   d   c   m
u
n
i
t

To convert from one metric measure to another, move the decimal point the same number of places and in the same direction as you do to go from the original prefix to the new one on the chart. This is the same process we used to divide or multiply by powers of ten. For instance,

			base			
k	h	da	unit	d	c	m
0	0	0	0	0	4.	5

4.5 cg = ? kg

The "k" prefix is five places to the left of the "c" prefix, so move the decimal point five places to the left.

4.5 cg = 0.000045 kg

Also,

56 kℓ = ? mℓ

56 kℓ = 56,000,000 mℓ

The "m" prefix is six places to the right of the "k" prefix, so move the decimal point six places to the right.

Three other widely used prefixes are

mega: one million — 1,000,000
giga: one billion — 1,000,000,000
nano: one-billionth — 0.000000001

Using these, we have 80 megabytes, which means 80(1,000,000) bytes or 80,000,000 bytes, and 15 nanoseconds, which means 15(0.000000001) seconds or 0.000000015 seconds. These large and small numbers are used in the sciences and world of computers.

The world is almost entirely metric and the United States is moving in that direction. We are seeing comparisons of the two systems by double listing. This is apparent on marked packages of food products (ounces and grams) and on some highway signs (miles per hour and kilometers per hour).

We perform conversions between the systems (English and metric) using the basic units. Since the systems were developed independently, there are no exact comparisons. All conversion units are approximate.

English–Metric Conversions	Metric–English Conversions
1 inch ≈ 2.54 centimeters	1 centimeter ≈ 0.3937 inch
1 foot ≈ 0.3048 meter	1 meter ≈ 3.281 feet
1 yard ≈ 0.9144 meter	1 meter ≈ 1.094 yards
1 mile ≈ 1.609 kilometers	1 kilometer ≈ 0.6214 mile
1 quart ≈ 0.946 liter	1 liter ≈ 1.057 quarts
1 gallon ≈ 3.785 liters	1 liter ≈ 0.2642 gallon
1 ounce ≈ 28.35 grams	1 gram ≈ 0.0353 ounce
1 pound ≈ 453.59 grams	1 gram ≈ 0.0022 pound

With these tables of conversion we can convert between the two systems. For example, convert 45 inches to centimeters.

$$45 \text{ inches} \approx \frac{45 \text{ in.}}{1} \cdot \frac{2.54 \text{ cm}}{1 \text{ in.}}$$

Multiply by the conversion factor 2.54 cm/in. from the table.

$$\approx \frac{45(2.54)}{(1)(1)} \text{ cm}$$

$$\approx 114.3 \text{ cm}$$

So, 45 inches is approximately 114.3 centimeters.

C A U T I O N

1. Because the conversions in the chart are all rounded to the nearest hundredth, thousandth, or ten-thousandth, we cannot expect closer accuracy when using them.

2. Since there are two conversion factors that can be used for each conversion, it is possible for answers to vary slightly, depending upon which factor is chosen.

All conversions between the systems in the examples and exercises are rounded to the nearest tenth. Since these conversions are approximate, the results may not always be accurate. For greater accuracy, conversion units with more decimal places are required.

For conversions involving area, this chart is useful.

Area Conversions	
$1 \text{ ft}^2 = 144 \text{ in.}^2$	$1 \text{ cm}^2 = 100 \text{ mm}^2$
$1 \text{ yd}^2 = 9 \text{ ft}^2$	$1 \text{ m}^2 = 10{,}000 \text{ cm}^2$
$1 \text{ mi}^2 = 3{,}097{,}600 \text{ yd}^2$	$1 \text{ km}^2 = 1{,}000{,}000 \text{ m}^2$
$1 \text{ in.}^2 \approx 6.4516 \text{ cm}^2$	$1 \text{ cm}^2 \approx 0.155 \text{ in.}^2$
$1 \text{ ft}^2 \approx 0.0929 \text{ m}^2$	$1 \text{ m}^2 \approx 10.765 \text{ ft}^2$
$1 \text{ mi}^2 \approx 2.590 \text{ km}^2$	$1 \text{ km}^2 \approx 0.3861 \text{ mi}^2$

Examples A–J   Warm Ups A–J

Directions: Convert the units of measure. If the conversion is not exact, round to the nearest tenth.

Strategy: Multiply the given unit(s) of measure by fractions formed from equivalent measures to get the desired unit(s). Simplify.

A.   Convert 6.85 pints to gallons.

$$6.85 \text{ pt} = \frac{6.85 \text{ pt}}{1} \cdot \frac{1 \text{ qt}}{2 \text{ pt}} \cdot \frac{1 \text{ gal}}{4 \text{ qt}}$$

Multiply by the conversion factors.

$$= \frac{6.85(1)(1)}{1(2)(4)} \text{ gal}$$

$$= 0.85625 \text{ gal}$$

So, 6.85 pints is 0.85625 gallons.

A.   Convert 0.32 mi to feet.

B.   Convert 34.8 m to kilometers.

Strategy: Use the conversion chart to move the decimal point. This is an exact conversion, so we will not round.

$$34.8 \text{ m} = 0.0348 \text{ km}$$

The "k" unit is three places to the left of the base unit, "m." Move the decimal point three places left.

So, 34.8 m = 0.0348 km.

B.   Convert 0.43 g to centigrams.

C.   Convert 3 pints to liters.

Strategy: First change pints to quarts, as we have a conversion from quarts to liters.

$$3 \text{ pt} \approx \frac{3 \text{ pt}}{1} \cdot \frac{1 \text{ qt}}{2 \text{ pt}} \cdot \frac{0.946 \, \ell}{1 \text{ qt}}$$

Multiply by the conversion factors.

$$\approx \frac{3(1)(0.946)}{1(2)(1)} \ell$$

$$\approx 1.419 \, \ell$$

So, 3 pints are approximately 1.4 liters

C.   Convert 12 ounces to grams.

Answers to Warm Ups   A. 1689.6 ft   B. 43 cg   C. 340.2 grams

D. Convert 55 miles per hour to kilometers per hour.

$$\frac{55 \text{ mi}}{\text{hour}} \approx \frac{55 \text{ mi}}{\text{hour}} \cdot \frac{1 \text{ km}}{0.6214 \text{ mi}}$$ **Multiply by the conversion factor.**

$$\approx \frac{55(1)}{1(0.6214)} \frac{\text{km}}{\text{hr}}$$

$$\approx 88.5 \frac{\text{km}}{\text{hr}}$$

So, 55 miles per hour is approximately 88.5 kilometers per hour.

D. Convert 64 feet per minute to meters per minute.

E. Convert 2 kilograms to pounds.

$$2 \text{ kg} \approx \frac{2 \text{ kg}}{1} \cdot \frac{1000 \text{ g}}{1 \text{ kg}} \cdot \frac{0.0022 \text{ lb}}{1 \text{ g}}$$ **Multiply by the conversion factors.**

$$\approx \frac{2(1000)(0.0022)}{1(1)(1)} \text{ lb}$$

$$\approx 4.4 \text{ lb}$$

So, 2 kilograms is approximately 4.4 pounds.

E. Convert 2 pounds to kilograms.

F. Convert 8.3 liters to quarts.

$$8.3 \ \ell \approx \frac{8.3 \ \ell}{1} \cdot \frac{1.057 \text{ qt}}{1 \ \ell}$$ **Multiply by the conversion factor.**

$$\approx \frac{8.3(1.057)}{1(1)} \text{ qt}$$

$$\approx 8.7731 \text{ qt}$$

So, 8.3 liters is approximately 8.8 quarts.

F. Convert 8.95 kiloliters to gallons.

G. Convert 25 centigrams per centimeter to ounces per inch. Round to the nearest hundredth.

$$\frac{25 \text{ cg}}{1 \text{ cm}} \approx \frac{25 \text{ cg}}{1 \text{ cm}} \cdot \frac{1 \text{ g}}{100 \text{ cg}} \cdot \frac{0.0353 \text{ oz}}{1 \text{ g}} \cdot \frac{1 \text{ cm}}{0.3937 \text{ in.}}$$

$$\approx \frac{25(1)(0.0353)(1)}{1(100)(1)(0.3937)} \frac{\text{oz}}{\text{in.}}$$

$$\approx 0.02242 \frac{\text{oz}}{\text{in.}}$$

So, 25 centigrams per centimeter is approximately 0.02 ounces per inch.

G. Convert 1.3 pounds per foot to grams per meter. Round to the nearest hundredth.

H. Mary is driving an old car on a road in Canada where the speed limit is posted as 80 km/hr. The speedometer is registering 45 mi/hr. Is she driving within the speed limit?

H. Mary is traveling on another road in Canada where the speed limit is posted as 115 km/hr. Can she drive 70 mi/hr and be within the speed limit?

Answers to Warm Ups   D. 19.5 meters/minute   E. 0.9 kilogram   F. 2364.6 gallons   G. 1934.60 grams/meter   H. yes

Strategy:　Change the posted speed limit to mi/hr and compare with her speed.

$$\frac{80 \text{ km}}{1 \text{ hr}} \approx \frac{80 \text{ km}}{1 \text{ hr}} \cdot \frac{1 \text{ mi}}{1.609 \text{ km}}$$

$$\approx \frac{(80)(1)}{1(1.6090)} \frac{\text{mi}}{\text{hr}}$$

$$\approx 49.72 \frac{\text{mi}}{\text{hr}}$$

The speed limit is approximately 49.7 mi/hr, so Mary is within the limit.

I.　Convert 504.8 in.^{2} to yd^2.

$$504.8 \text{ in.}^2 = \frac{504.8 \text{ in.}^2}{1} \cdot \frac{1 \text{ ft}^2}{144 \text{ in.}^2} \cdot \frac{1 \text{ yd}^2}{9 \text{ ft}^2}$$

Multiply by the conversion factors.

$$= \frac{504.8(1)(1)}{1(144)(9)} \text{ yd}^2$$

$$\approx 0.4 \text{ yd}^2$$

So, 504.8 in.^{2} is approximately 0.4 yd^2.

I.　Convert 0.45 yd^2 to in.^{2}.

J.　Convert 3.6 mi^2 to km^2.

$$3.6 \text{ mi}^2 \approx \frac{3.6 \text{ mi}^2}{1} \cdot \frac{2.590 \text{ km}^2}{\text{mi}^2}$$

$$\approx \frac{3.6(2.590)}{1(1)} \text{ km}^2$$

$$\approx 9.3 \text{ km}^2$$

So, 3.6 mi^2 is approximately 9.3 km^2.

J.　Convert 7.8 m^2 to ft^2.

Answers to Warm Ups　I. 583.2 in.^{2}　J. 84.0 ft^2

Exercises 6.6

OBJECTIVE: *Convert units of measure.*

A.

Convert the units as shown.

1. 8 ounces = ? pounds

2. 30 quarts = ? gallons

3. 15 mm = ? cm

4. 600 grams = ? kilograms

5. 550 mℓ = ?ℓ

6. 456 mm = ? km

7. 6.9 ft = ? yd

8. 36 minutes = ? hour

9. 3.02 ft = ? in.

10. 0.7 pint = ? cup

B.

11. 1.83 m = ? mm

12. 4.756 cg = ? mg

13. 4.56 mi = ? ft

14. 3.62 hours = ? minutes

Convert and round to the nearest hundredth. (Answers may vary depending upon the conversion factors used.)

15. 16 in. = ? cm

16. 15 lb = ? kg

17. 4.5 ℓ = ? qt

18. 20 yd = ? m

19. 37.9 cm = ? in.

20. 0.85 ℓ = ? qt

21. 9.5 sq in. = ? sq ft

22. 16.85 sq ft = ? sq yd

23. 53400 cm^2 = ? m^2

24. 67930 m^2 = ? km^2

C.

Convert and round to the nearest tenth.

25. 14.8 ft = ? m

26. 19.8 yd = ? m

27. 3.2 kg = ? lb

28. 67.4 km = ? mi

29. 8235 m = ? mi

30. 9.67 mi = ? m

31. 7.5 cups = ? cℓ

32. 0.65 kℓ = ? pints

33. 14.5 ft^2 = ? m^2

34. 78.5 cm^2 = ? in.^{2}

Convert and round to the nearest hundredth.

35. $\dfrac{1.5 \text{ lb}}{\text{ft}} = \dfrac{? \text{ g}}{\text{m}}$

36. $\dfrac{55 \text{ miles}}{\text{hour}} = \dfrac{? \text{ meters}}{\text{second}}$

37. $\dfrac{5.2 \text{ lb}}{\text{ft}^2} = \dfrac{? \text{ g}}{\text{cm}^2}$

38. $\dfrac{45 \text{ ft}}{\text{sec}} = \dfrac{? \text{ m}}{\text{min}}$

39. $\dfrac{525 \text{ g}}{\ell} = \dfrac{? \text{ lb}}{\text{qt}}$

40. $\dfrac{\$1.52}{\text{lb}} = \dfrac{? \text{ \$}}{\text{kg}}$

41. A box of Wheat Bran cereal weighs 18.9 ounces. To the nearest tenth, how many grams does it weigh?

42. A farmer's harvest of strawberries averages 10 tons per acre. To the nearest tenth, express the harvest in kilograms per acre.

43. The Heatherton Corporation reimburses its employees 30¢ per mile when they use their private car on company business. The company is opening a plant in Canada. What reimbursement per kilometer should the company pay, in U.S. dollars, to the nearest cent per kilometer?

44. The Williams Company advertises that a water pump will deliver water at 35 gallons per minute. Nyen needs to put an ad in a Canadian paper. How many liters per minute should he put in the ad, to the nearest liter per minute?

45. A newborn baby elephant at the Oxnard Zoo weighs 295 pounds. Express the weight in kilograms.

46. The Japanese fishing fleet has a weekly catch of 110,000 kilograms of whiting. Express the catch in tons.

47. The Georgia Pacific Corporation grows trees for poles. A typical pole measures 95 feet in length. Express the length in meters.

48. The average length of a salmon returning to the Oakridge Hatchery is 37 inches. Express the length in centimeters.

49. The Coffee Company packages 3 kilograms of coffee that sells in Canada for $18.95. The company also packages 6 pounds of coffee that sells for $17.95 in the United States. Assuming both dollar amounts are in American currency, which is the better buy?

50. Laura drives 67 miles to work each day. Her cousin Mabel, who lives in a foreign country, drives 115 kilometers to work each day. Who has the shorter drive?

51. The average rainfall in Freeport, Florida, is 26.83 inches. Express the average to the nearest tenth of a centimeter.

52. The Toyo Tire Company recommends that its premium tire be inflated to 1.5 kilograms per square centimeter. In preparation to sell the tire in America, Mishi must convert the pressure to pounds per square inch. What measure, to the nearest pound per square inch, must she use?

Exercises 53–56 relate to the chapter application.

In swimming in the United States, there is a short course season that takes place in 25-yd pools, and a long course season that takes place in 50-m pools. Consequently, swimmers gather both yard and meter times in their events. Occasionally, for purposes of qualification for major meets, it is necessary to convert a yard time into its equivalent meter time.

53. Doug can swim the 100-yd breaststroke in 1 min 10 sec. Express this rate as sec per yard, then convert to sec per meter.

54. Using your results, what should Doug's time be for the 100-m breaststroke?

55. In doing yard to meter swimming conversions, there is an extra consideration, the number of turns in the race. In a 100-yd race, there are four lengths of the pool and three turns. By contrast in a 100-m race, there are two lengths of the pool and only one turn. When converting times, some allowance must be made for the extra turns in the yard pool. Since turning is faster than straight swimming, it is usual to add 1 sec to the yard time for each extra turn, and then convert to meters. Following this rule, what is Doug's expected time for the 100-m breaststroke?

56. In 1995 Miguel Indurain won the Tour de France, completing the 3635-km course in a total time of 92 hr, 44 min, 59 sec. Calculate his average speed in km/hr and in mi/hr.

Exercises 57–60. The following chart lists common mariners' measures.

1 fathom = 6 ft
1 nautical mile ≈ 6076 ft
1 knot = 1 nautical mile per hour

57. How many km are in a nautical mile? Round to the nearest thousandth.

58. Due to flooding, a river level is rising .3 fathoms/day. Convert this to ft/hr.

59. How many nautical miles are in a statute (land) mile? Round to the nearest thousandth.

60. Convert 85 knots to statute miles per hour. Round to the nearest tenth.

STATE YOUR UNDERSTANDING

61. Write your opinion about which measurement system, English or metric, is easier to use and why?

62. Explain why you might get different answers when converting 25 meters to feet using conversion units from the chart in this section.

63. Explain how to convert 40 mph to km/min.

CHALLENGE

64. Convert $\dfrac{110 \text{ lb}}{\text{ft}^2}$ to $\dfrac{\text{kg}}{\text{m}^2}$.

65. Mary has 5 gallons 3 quarts 1 pint of blackberry juice. To the nearest tenth, how many liters does she have?

66. The Candy Basket receives an order of candy from Europe that costs $21 per kilogram. The store plans to sell the candy for $14.50 per pound. To the nearest dollar, what is the profit on the sale of 100 pounds of candy?

GROUP ACTIVITY

63. Write a brief history of the metric system, including the controversy over the change in Canada, Great Britain, and the United States. Present your findings as an oral report in class.

MAINTAIN YOUR SKILLS (Section 5.6)

64. Evaluate $4x - 3y$ if $x = \dfrac{7}{12}$ and $y = \dfrac{5}{8}$.

65. Evaluate $2xyz - 3xz$ if $x = \dfrac{7}{12}$, $y = \dfrac{5}{8}$, and $z = \dfrac{3}{5}$.

66. Evaluate $\dfrac{4x + 2z}{z}$ if $x = \dfrac{7}{12}$ and $z = \dfrac{3}{5}$.

67. Is $x = -\dfrac{2}{3}$ a solution of $2x^2 + 17x + 8 = 0$?

68. Is $y = -\dfrac{3}{10}$ a solution of $20y^2 - 22y - 9 = 0$?

6.7 Evaluating Algebraic Expressions and Formulas with Rational Numbers (Decimals)

OBJECTIVE

Evaluate algebraic expressions and formulas with decimals.

HOW AND WHY
Objective

Evaluate algebraic expressions and formulas with decimals.

The order of operations and the method for evaluating expressions and formulas are the same as for whole numbers, integers, and fractions.

Order of Operations

In a number expression with two or more operations, perform the operations in the following order:

1. Grouping symbols—perform operations included within the grouping symbols first.
2. Exponents—perform operations indicated by exponents.
3. Multiply and divide—perform multiplication and division from left to right as you come to them.
4. Add and subtract—perform addition and subtraction from left to right as you come to them.

Evaluate the expression $3x + 2y$ if $x = 5.23$ and $y = 12.04$.

$$
\begin{aligned}
3x + 2y &= 3(5.23) + 2(12.04) && \textbf{Substitute 5.23 for } x \textbf{ and 12.04 for } y. \\
&= 15.69 + 24.08 && \textbf{Multiply first.} \\
&= 39.77 && \textbf{Add.}
\end{aligned}
$$

Many formulas involve the number pi, represented by the symbol π. (See Section 5.6.) Pi is found by dividing the circumference of a circle by its diameter. Pi has no exact decimal representation. We have already used $\dfrac{22}{7}$ to approximate pi.

Using decimals we can get a better approximation, $\pi \approx 3.14159$. For the purposes of this text we will round the value of pi to the nearest hundredth, so we will let $\pi \approx 3.14$. Scientific calculators have a key, $\boxed{\pi}$, that gives π to eight or ten decimal places. When using your calculator, use the π key for greater accuracy.

Examples A–F

Directions: Evaluate.

Strategy: Substitute the given values in the expression or formula and simplify.

A.   Evaluate $a - 17.93$ if $a = -7.64$.

$$a - 17.93 = -7.64 - 17.93 \qquad \textbf{Replace } a \textbf{ with } -7.64.$$
$$= -7.64 + (-17.93) \qquad \textbf{Rewrite as addition.}$$
$$= -25.57$$

A.   Evaluate $b - 14.7$ if $b = -8.21$.

B.   Evaluate $6.82 - \dfrac{x}{y}$ if $x = 12.45$ and $y = -1.5$.

$$6.82 - \frac{x}{y} = 6.82 - \frac{12.45}{-1.5} \qquad \textbf{Substitute.}$$
$$= 6.82 - (-8.3) \qquad \textbf{Divide first.}$$
$$= 6.82 + 8.3 \qquad \textbf{Rewrite as addition.}$$
$$= 15.12$$

B.   Evaluate $10.34 - \dfrac{a}{b}$ if $a = 32.4$ and $b = -1.6$.

C.   Find the value of $-4x^2 + 17.4y$ when $x = 3.2$ and $y = -6.3$.

$$-4x^2 + 17.4y = -4(3.2)^2 + 17.4(-6.3) \qquad \textbf{Substitute.}$$
$$= -4(10.24) + 17.4(-6.3) \qquad \textbf{Exponents first.}$$
$$= -40.96 + (-109.62) \qquad \textbf{Multiply.}$$
$$= -150.58$$

C.   Find the value of $-3y^2 - 0.5w$ when $y = -4.2$ and $w = -6.7$.

D.   Is 4.62 a solution of $5x - 45.3 - 10x = -68.4$?

Does $5(4.62) - 45.3 - 10(4.62) = -68.4$?   **Substitute.**

Does $\qquad 23.1 - 45.3 - 46.2 = -68.4$?

Does $\qquad\qquad\qquad -68.4 = -68.4$?   **Yes.**

Yes, 4.62 is a solution of $5x - 45.3 - 10x = -68.6$.

D.   Is 6.24 a solution of $5t - 3.45 = -33.65 + 10t$?

Calculator Example:

E.   The formula for the volume of one-half a cylindrical tank is $V = \dfrac{1}{2}\pi r^2 h$. These tanks are often used for watering cattle or for feed troughs. Find the volume if $r = 2.25$ ft, and $h = 12.35$ ft. Round to the nearest tenth.

$$V \approx 0.5(\pi)(2.25)^2(12.35) \qquad \textbf{Substitute; } \frac{1}{2} = 0.5 \textbf{ and}$$
$$\approx 98.209 \qquad\qquad\qquad \textbf{use the } \boxed{\pi} \textbf{ key.}$$

So, the volume is approximately 98.2 ft³.

E.   A second trough in a feed lot has a radius, r, of 3.75 ft and a length of 17.82 ft. Using the formula in Example E, find the volume of this trough. Round to the nearest tenth.

Answers to Warm Ups   A. −22.91   B. 30.59   C. −49.57   D. no   E. 393.6 ft³

F. A shopping mall is to have a central patio in the form of a circle. It will be constructed with decorative bricks with a concrete base. If the circle is to have a diameter of 75 feet, how many square feet of bricks will the patio have? Round to the nearest square foot.

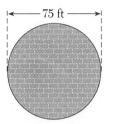

⊢——— 75 ft ———⊣

Formula:

$A = \pi r^2$ **Area of a circle.**

$\approx 3.14\left(\dfrac{75}{2}\right)^2$ **Substitute, r is one half the diameter. Let $\pi \approx 3.14$.**

$\approx 3.14(37.5)^2$

$\approx 3.14(1406.25)$

≈ 4415.625

The mall patio has about 4416 square feet of brick. Using the $\boxed{\pi}$ key, the patio has 4418 square feet of brick.

F. A mall is to have a central area in the form of a circle. It will be constructed with decorative bricks with a concrete base. If the circle is to have a diameter of 92 feet, how many square feet of bricks will the central area have? Round to the nearest square foot.

Answer to Warm Up F. 6644 ft² of bricks or 6648 ft² of bricks.

Exercises 6.7

OBJECTIVE: *Evaluate algebraic expressions and formulas.*

A.

Evaluate.

1. ac, $a = 0.5$ and $c = -3$

2. $a + b$, $a = 0.45$ and $b = 1.6$

3. abx, $a = 0.6$, $b = 3$, and $x = 0.01$

4. cyx, $c = -9$, $y = 0.5$, and $x = 0.001$

5. $b - y$, $b = -3.56$ and $y = 7.11$

6. $a - z$, $a = 4.89$ and $z = 2.25$

7. x^3, $x = 0.2$

8. m^2, $m = 0.03$

9. $\dfrac{a}{b}$, $a = -0.44$ and $b = -1.1$

10. $\dfrac{c}{d}$, $c = 6.4$ and $d = -0.002$

11. Is 40.4 a solution of $x + 3.3 = 43.73$?

12. Is 23.73 a solution of $y - 5.42 = 18.31$?

B.

Evaluate if $a = -0.3$, $b = 100$, $c = -0.26$, $x = 2.02$, $y = 0.4$, and $z = -0.45$.

13. $a^2 x^2$

14. $a^2 + x^2$

15. $ab + yz$

16. $ay + bz$

17. $bc - by$

18. $ab - ay$

19. $cbx - ay$

20. $acz - bx$

21. $b^2(a + y)$

22. $c^2(b + x)$

23. $(a + x)(b + y)$

24. $(a + y)(b + x)$

25. Is 0.283 a solution of $4x + 3 = 4.132$?

26. Is 0.73 a solution of $8x - 5 = 0.84$?

C.

27. Find the area of a triangle, $A = \dfrac{bh}{2}$, if $b = 6.5$ and $h = 3.9$.

28. Find the area of a rectangle, $A = \ell w$, if $\ell = 32.67$ and $w = 31.78$. Estimate the area and then calculate and round to the nearest tenth.

29. Find the volume of a cube, $V = s^3$, if $s = 3.56$. Round to the nearest tenth.

30. Find the area of a circle, $A = \pi r^2$, if $\pi \approx 3.14$ and $r = 12.6$. Round to the nearest thousandth.

Evaluate if $a = 0.15$, $b = 55$, $c = -0.342$, $x = -4.53$, $y = 0.6$, and $z = -0.74$. Round to the nearest thousandth.

31. $\dfrac{a + c}{b}$

32. $\dfrac{y - z}{x}$

33. $\dfrac{(a + b)^2}{b}$

34. $\dfrac{x^2 - z^2}{c}$

35. $5x^2 - ac$

36. $10y^2 - bz$

37. Is 2.3 a solution of $3w - 7 = -0.1$?

38. Is 1.04 a solution of $3y + 22 = 25.12$?

39. Is 0.34 a solution of $7.64 - 2m = 0.68$?

40. Is 0.47 a solution of $6.71 - 4p = 52.3$?

41. The distance a vehicle travels is given by the formula $d = rt$, where d = distance, r = rate, and t = time. What distance does a vehicle travel in $4\frac{1}{2}$ hours if it travels at a rate of 63.45 miles per hour?

42. How far does a plane travel at the rate of 377 miles per hour in $9\frac{1}{4}$ hours? $(d = rt)$

43. The stage at an outdoor amphitheater is in the shape of a semicircle of radius 125 feet. Find the area (A) of the stage, $A = \dfrac{\pi r^2}{2}$ and $\pi \approx 3.14$. Round to the nearest ft².

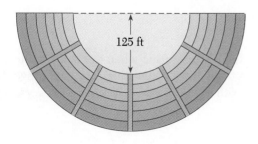

125 ft

44. The plaza at the new mall in downtown Seattle is in the shape of a square with a semicircle at one end. If the side of the square is 60 feet, how many square feet are in its area? $\left(A = s^2 + \dfrac{\pi r^2}{2}, \pi \approx 3.14, s = 60, \text{ and } r = 30\right)$ Round to the nearest ft^2.

45. A real estate agent receives a commission of 5% (0.05) on the first $40,000 of a sale and 3% (0.03) for anything above $40,000. How much commission C is earned from the sale of a $135,850 home? (Formula: $C = 2000 + 0.03(h - 40,000)$, where h is the sale price of the home.)

46. The real estate agent in Exercise 45 sells another home for $265,900. How much commission is earned from the sale?

47. A painter needs to find the area of the surface of a cylindrical drum. If the drum has radius $r = 7.85$ feet and a height of 10.5 feet, what is the surface area of the drum to the nearest tenth? If a gallon of industrial paint covers 125 sq ft, how many gallons of paint must he buy? (Formula: $A = 2\pi rh + 2\pi r^2$, let $\pi \approx 3.14$)

48. The original height (h) of the Great Pyramid of Egypt was 480 feet and the area of the base (B) was 583,696 square feet. Find the original volume of the pyramid in cubic yards and cubic meters. $\left(\text{Formula: } V = \dfrac{hB}{3}\right)$

 Exercises 49 and 50 relate to the chapter application.

In football, placekickers can score field goals that are worth 3 points each or point after touchdowns that are worth 1 point. Therefore, a formula for the total number of points by a placekicker is $T = 3F + P$, where T is the total points scored, F is the number of field goals scored, and P is the number of point after touchdowns.

49. In 1944, John Carey of San Diego led the American Football Conference in scoring with 33 point after touchdowns and 34 field goals. How many points did he score for the season?

50. A running back's rushing average is calculated by adding up all the yardage made when the ball is in his possession, and dividing by the total number of possessions. Write a formula for calculating rushing average. Use your formula to calculate the rushing average of a running back with carries of 4 yd, 12 yd, -3 yd, 7 yd, 6 yd, 22 yd, 1 yd, and 3 yd.

Exercises 51–52. The following graph gives the number (in millions) of farms in the United States.

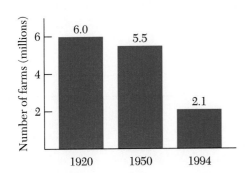

51. What was the average number of farms lost per year between 1920 and 1950? Round to the nearest thousand farms.

52. What was the average number of farms lost per year between 1950 and 1994? Round to the nearest thousand farms.

 Exercises 53–55 relate to the chapter application. On a two day cycling trip, the fastest cyclist in a group averaged 19 miles per hour. The slowest cyclist averaged 16 miles per hour.

53. Complete the following table.

Time (hours)	1	2	3	4	5
Total Distance of Fastest Cyclist					
Total Distance of Slowest Cyclist					

54. How many miles behind is the slowest cyclist after 6 hours? Express the answer as a signed number.

55. If the group left at 6:30 am and stopped to camp for the night after 150 miles, what time did the fastest cyclist get to camp. The slowest cyclist? Round to the nearest minute.

56. Explain how to evaluate the expression $\dfrac{(a+c)^2}{b}$, when $a = 0.03$, $b = 100$, and $c = 5.6$.

CHALLENGE

57. The horsepower developed in one end of the cylinder of a steam engine is found by the formula

$$\text{horsepower} = \frac{PLAN}{33{,}000}$$

where P = mean effective pressure in pounds per square inch

L = length of stroke in feet

A = area of end of piston in square inches

N = number of strokes per minute

Find the horsepower when $P = 75$ lbs per square inch, $L = 2.5$ feet, $A = 30.05$ square inches, and $N = 75$. Round your answer to the nearest tenth.

58. Using the formula in Problem 57, find the horsepower when $P = 95$ pounds per square inch, length is 54 inches, $A = 98.8$ square inches, and $N = 90$. Round your answer to the nearest tenth.

59. If the formula for finding the area of a circle is $A \approx 3.14r^2$, write a formula for finding the shaded area of the ring shown in the following illustration. Use the formula to find the area shaded of a ring with an outside radius (R) of 4.5 inches and an inside radius (r) of 2.5 inches. Round to the nearest hundredth.

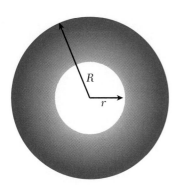

GROUP ACTIVITY

60. With your group, review the solution of equations studied in previous chapters. Apply these procedures to equations involving decimals. Will these procedures remain unchanged? Report your findings to the class.

MAINTAIN YOUR SKILLS (Sections 4.9, 5.7)

Solve:

61. $5x + 33 = 13$

62. $12w + 26 = 206$

63. $3 - 5y = 63$

64. $\dfrac{1}{2}x + \dfrac{7}{8} = \dfrac{9}{4}$

65. $\dfrac{3}{8}b + \dfrac{6}{5} = \dfrac{3}{2}$

6.8   Solving Equations Involving Rational Numbers (Decimals)

OBJECTIVE   Solve equations involving decimals.

HOW AND WHY
Objective   **Solve equations involving decimals.**

Equations with decimals are solved using the same properties of equations as were used with whole numbers and fractions. The addition, subtraction, multiplication, and division properties of equality are used to isolate the variable.

When an equation contains fractions we can "clear the fractions" by using the multiplication property of equality. Similarly, when an equation contains decimals we can "clear the decimals" using the same property.

$0.4x - 1.2 = -3.6$	
$10(0.4x - 1.2) = 10(-3.6)$	**Multiply both sides of the equation by 10 to clear the decimals. The factor 10 is chosen because the greatest number of decimal places in any number is one (in this case they all have one).**
$10(0.4x) - 10(1.2) = 10(-3.6)$	**On the left, apply the distributive property.**
$4x - 12 = -36$	**The equation now looks exactly like the equations of earlier chapters and can be solved as before.**
$4x = -24$	**Add 12 to both sides.**
$x = -6$	**Divide both sides by 4.**

Check:

$0.4x - 1.2 = -3.6$	**Original equation.**
$0.4(-6) - 1.2 = -3.6$	**Substitute.**
$-2.4 - 1.2 = -3.6$	**Multiply.**
$-3.6 = -3.6$	**Combine on the left side. The equation checks.**

The solution is $x = -6$.

It is not necessary to clear the decimals in an equation. See Example C. Each student may decide whether or not to "clear the decimals."

> ▶ *To solve an equation involving decimals*
>
> **1.** Multiply both sides of the equation by a power of 10 so that all numbers in the equation are integers.
> **2.** Solve the resulting equation using the properties of equality to isolate the variable.

Examples A–E   **Warm Ups A–E**

Directions: Solve the equations.

Strategy: Clear the decimals by multiplying both sides by a power of 10 and then solve.

A.   Solve: $3b + 8.3 = 6.29$

$100(3b + 8.3) = 100(6.29)$ Since 6.29 has the most decimal places, multiply both sides by 100.

$300b + 830 = 629$ Multiply.

$300b = 629 - 830$ Subtract 830 from each side.

$300b = -201$

$b = -0.67$ Divide each side by 300.

Check:

$3b + 8.3 = 6.29$ Original equation.

$3(-0.67) + 8.3 = 6.29$ Substitute.

$-2.01 + 8.3 = 6.29$

$6.29 = 6.29$

The solution is $b = -0.67$.

A.   Solve: $4c + 7.5 = 4.78$

B.   Solve: $0.8t - 4.51 - 0.06t = 4$

$100(0.8t - 4.51 - 0.06t) = 100(4)$ Multiply both sides by 100.

$80t - 451 - 6t = 400$ Multiply.

$80t + (-451) + (-6t) = 400$ Rewrite as addition.

$74t + (-451) = 400$ Combine like terms.

$74t = 400 + 451$ Add 451 to each side.

$74t = 851$

$t = \dfrac{851}{74}$ Divide each side by 74.

$t = 11.5$

Check:

$0.8t - 4.51 - 0.06t = 4$ Original equation.

$0.8(11.5) - 4.51 - 0.06(11.5) = 4$ Substitute.

$9.2 - 4.51 - 0.69 = 4$

$4 = 4$

The solution is $t = 11.5$.

B. $0.25m - 3.824 - 0.05m = -3$

Answers to Warm Ups   A. $c = -0.68$ B. $m = 4.12$

C.   Solve: $x - 4.5 = -0.875$

Strategy: Solve without "clearing" the decimals. Students can decide which method works best for them.

$$x - 4.5 = -0.875 \qquad \textbf{Add 4.5 to both sides.}$$
$$x = -0.875 + 4.5$$
$$x = 3.625$$

Check:

$$x - 4.5 = -0.875 \qquad \textbf{Always check in the}$$
$$\textbf{original equation.}$$
$$3.625 - 4.5 = -0.875 \qquad \textbf{Substitute.}$$
$$-0.875 = -0.875$$

The solution is $x = 3.625$.

D.   Solve: $0.33y - 9.7 = -0.76y - 123.8$, and round answer to the nearest hundredth.

$$100(0.33y - 9.7) = 100(-0.76y - 123.8) \qquad \textbf{Multiply both sides}$$
$$\textbf{by 100 to clear the}$$
$$\textbf{decimals.}$$
$$33y - 970 = -76y - 12380$$
$$33y + 76y - 970 = -12380 \qquad \textbf{Add 76y to each}$$
$$\textbf{side.}$$
$$109y - 970 = -12380$$
$$109y = -12380 + 970 \qquad \textbf{Add 970 to each}$$
$$\textbf{side}$$
$$109y = -11410$$
$$y \approx -104.6788991$$

Since the solution is approximate, $y \approx -104.68$, the check will be approximate.

$$0.33y - 9.7 = -0.76y - 123.8 \qquad \textbf{Original}$$
$$\textbf{equation.}$$
$$0.33(-104.68) - 9.7 \approx -0.76(-104.68) - 123.8$$
$$-34.5444 - 9.7 \approx 79.5568 - 123.8$$
$$-44.2444 \approx -44.2432$$

The two sides of the equation are not exactly equal. However, to the nearest hundredth, -44.24, they are equal.

The solution is $y \approx -104.68$ to the nearest hundredth.

E.   Latoya purchases a new car for \$1183.75 down, including license and registration fees. She signs a contract to pay off the balance,

---

C.   Solve: $y - 2.7 = -0.654$

D.   Solve: $-0.9r + 7.3 = 4.3r - 1.245$, and round answer to the nearest hundredth.

E.   Benny, a friend of Latoya, also purchases a new car. He pays \$2407.65 down and signs a contract to pay off the balance, including interest,

(continued on next page)

Answers to Warm Ups   C. $y = 2.046$ D. $r \approx 1.64$

including interest, in 48 monthly payments. If the total she pays is $18,207.91, what is her monthly payment? Use the formula, $C = D + nP$, where C is the total cost, D the down payment, n the number of payments, and P the monthly payment.

in 36 months. If the total he pays is $21,965.73, what is his monthly payment? Use the formula in Example E.

Formula:

$C = D + nP$ **Identify the value of the variables.**

$C = 18207.91$

$D = 1183.75$

$n = 48$

$P = ?$ **Substitute.**

Subtract 1183.75 from each side.

$18207.91 = 1183.75 + 48P$

$17024.16 = 48P$ **Divide each side by 48. The check is left for the student.**

$354.67 = P$

The monthly payment is $354.67.

Exercises 6.8

OBJECTIVE: *Solve equations involving rational numbers (decimals).*

A.

Solve.

1. $x + 4.52 = 8.93$

2. $y + 16.66 = 25.97$

3. $a - 0.005 = 5.342$

4. $b - 0.342 = 8.116$

5. $3x = -9.036$

6. $-4y = 12.488$

7. $c - (-4.75) = -5.43$

8. $d - 6.32 = -9.45$

9. $-6b = -36.066$

10. $-7y = 7.0014$

11. $0.02c = 3.4$

12. $0.003t = -1.8$

B.

13. $4b - 0.66 = 32.34$

14. $7b - 0.65 = 44.43$

15. $0.7c - 22.6 = 16.6$

16. $0.8c - 12.9 = 23.9$

17. $-0.11x = 0.3872$

18. $-0.24y = -2.52$

19. $y - (-24) = 15.67$

20. $z - (-32) = 19.73$

21. $w + 5.67 = -1.567$

22. $n + 4.772 = -3.007$

23. $0.2y - 34.78 = 15.43$

24. $0.03t - 45.09 = -16.5$

25. $4.8b - 0.83 = -9.95$

26. $8.4b - 0.38 = -16.34$

27. $8.7 = 7.8c + 13.77$

28. $9.4 = 4.9c + 29.735$

C.

29. The formula for the balance of a loan (D) is $D = B - nP$, where P represents the monthly payment, n represents the number of payments made, and B represents the amount of money borrowed including interest. Find the balance on a loan of \$785 if seven payments of \$45.95 have been made.

30. Using the formula in Exercise 29, find the balance of a $45,890 loan, including interest, if fifteen payments of $1268.56 have been made.

31. Using the formula in Exercise 29, find the number of payments needed to pay off a loan of $7100, including interest, if the monthly payment is $258.67. Find the amount of the final payment if it is not $258.67.

32. If a loan of $6500, not including interest, is paid off in 16 years with a monthly payment of $60.13, what is the amount of interest paid?

33. If a house loan of $125,000, not including interest, is paid off in 30 years with a monthly payment of $931.68, what is the amount of interest paid?

Solve. Round to the nearest tenth.

34. $56 + 9.23y = 7.832$

35. $78.43 + 6.81t = -4.56$

36. $3y + 8.67 - 1.03y = 2.5553$

37. $5.7a - 9.54 - 3.23a = 4.872$

38. $104.53 = -6.7r + 17.832 + 3.56r$

39. $55.78 = 9.83t - 34.78 - 13.59t$

40. $-0.0056 + 0.075a = -0.09432 + 0.346a$

41. $0.0356 - 0.0236z = 0.0976z + 0.00421$

42. If the perimeter of a rectangle is 24.7 cm and its length is 4.5 cm, what is its width?

43. If the perimeter of a rectangle is 48.96 m and its width is 5.62 m, what is its length?

44. The velocity (v_1) of a falling object at any time t with initial velocity v_o is given by the formula $v_1 = v_0 + 32.2t$. What is the initial velocity if the velocity after 5.5 seconds is 188 feet per second?

45. The velocity (v_1) of a falling object at any time t with initial velocity v_o is given by the formula $v_1 = v_0 + 32.2t$. What is the initial velocity if the velocity after 8.2 seconds is 305.5 feet per second?

46. The formula for finding the area of a triangle is $A = 0.5bh$. Find the height (h) of a triangle that has a base (b) of 15.35 inches and an area (A) of 63.95 square inches. Round to the nearest hundredth.

47. Using the formula in Exercise 46, find the base of a triangle that has an area of 243.6 cm^2 and a height of 12.6 cm. Round to the nearest hundredth.

6.9   **Square Roots and the Pythagorean Formula**

OBJECTIVES

1. Find the square root or the approximate square root of a number.

2. Find the missing side of a right triangle.

3. Determine if a triangle is a right triangle given the lengths of its sides.

VOCABULARY

A **perfect square** is a whole number or fraction that is the square of another whole number or fraction.

A **square root** of a positive number is one of the two equal factors of the number. The symbol for the positive square root is called a **radical sign:** $\sqrt{}$. In this text, we work problems in which only the positive square root of a number is used. Thus, we use the phrase "the square root" to mean the "positive square root."

A number such as 2 is not a perfect square. The square root of 2 is not a rational number. In these cases, we use an **approximate square root.** Some calculators display 1.414213562 for the square root of 2. Like π this is a never-ending decimal so we use an approximation for its value. Hence is it not uncommon to say that $\sqrt{2} \approx 1.414$ correct to the nearest thousandth. The **Pythagorean formula** describes a relationship between the sides of right triangles. A **right triangle** is a triangle that contains one right (90°) angle. The sides a and b, which form the right angle, are called the **legs.** The longest side, c, which is opposite the right angle, is called the **hypotenuse.** Right triangle ABC with right angle at C is shown in Figure 6.1. (For all triangles in this section, we assume the right angle is at C.)

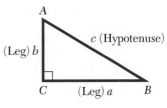

Figure 6.1

HOW AND WHY
Objective 1

Find the square root or the approximate square root of a number.

The whole number 81 is the square of 9, so 9 is the square root of 81. We write $9^2 = 81$ and $\sqrt{81} = 9$.

If a whole number is a perfect square, the square root can be found by trial and error, or by referring to a square root table, or by using a calculator.

To find $\sqrt{196}$ by trial and error, we discover that

10 is too small, since $10(10) = 10^2 = 100$
12 is too small, since $12(12) = 12^2 = 144$
15 is too large, since $15(15) = 15^2 = 225$
14 is the square root, since $14(14) = 14^2 = 196$

So $\sqrt{196} = 14$.

Most whole numbers are not perfect squares. To find $\sqrt{95}$, we discover that

9 is too small, since $9(9) = 9^2 = 81$
10 is too large, since $10(10) = 10^2 = 100$
So $\sqrt{95}$ is between 9 and 10.

The number 95 is not a perfect square and thus we cannot write an exact decimal value for its square root. The number $\sqrt{95}$ can be approximated by using a calculator or a table. (See the inside the back cover.)

Approximations			
Square Root	**Tenth**	**Hundredth**	**Thousandth**
$\sqrt{95}$	9.7	9.75	9.747
$\sqrt{3}$	1.7	1.73	1.732
$\sqrt{111}$	10.5	10.54	10.536

It is assumed for the following exercises that you have access to a calculator or will use the square root table to find the approximations. In the table n is the square root of n^2.

n	n^2	$\sqrt{n}$
185	34,225	13.601

From the table we can see that

$$\sqrt{34225} = 185 \text{ and } \sqrt{185} \approx 13.601$$

Examples A–D	**Warm Ups A–D**

Directions: Find the square roots.

Strategy: Use a calculator, a table, or trial and error

A. Find $\sqrt{64}$ and $-\sqrt{25}$.

$\sqrt{64} = 8$ Since $8(8) = 8^2 = 64$
$-\sqrt{25} = -5$ Since $5(5) = 5^2 = 25$

A. Find $\sqrt{36}$ and $-\sqrt{225}$.

B. Find $\sqrt{\dfrac{9}{25}}$ and $\sqrt{\dfrac{16}{121}}$.

$\sqrt{\dfrac{9}{25}} = \dfrac{3}{5}$ Since $\dfrac{3}{5}\left(\dfrac{3}{5}\right) = \dfrac{9}{25}$

$\sqrt{\dfrac{16}{121}} = \dfrac{4}{11}$ Since $\dfrac{4}{11}\left(\dfrac{4}{11}\right) = \dfrac{16}{121}$

B. Find $\sqrt{\dfrac{4}{49}}$ and $\sqrt{\dfrac{81}{64}}$.

Answers to Warm Ups   A. $6, -15$ B. $\dfrac{2}{7}, \dfrac{9}{8}$

▦ Calculator example:

C.   Find the square root of 2925 using a calculator. Round the result to the nearest tenth.

$$\sqrt{2925} \approx ?$$ **Depending on your calculator, first enter the number or the radical sign.**

$$\sqrt{2925} \approx 54.08326913$$

So $\sqrt{2925} \approx 54.1$ **Round.**

C.   Find $\sqrt{3529}$. Round the result to the nearest tenth.

D.   The formula to find the approximate number of seconds that it takes a free-falling body to fall a given distance is

$$t = \sqrt{\dfrac{d}{16}}$$

where t is the time in seconds and d is the distance in feet. Find the approximate number of seconds it will take for a ball to drop 10,000 ft.   **Formula.**

$$t = \sqrt{\dfrac{d}{16}}$$ **Substitute 10,000 for d.**

$$t = \sqrt{\dfrac{10000}{16}}$$ **Simplify.**

   The square root of 625 is 25.

$$t = \sqrt{625}$$

$$t = 25$$

Therefore, it will take about 25 seconds for the ball to fall 10,000 ft.

D.   The formula to find the approximate number of seconds that it takes a free-falling body to fall a given distance is

$$t = \sqrt{\dfrac{d}{16}}$$

where t is the time in seconds and d is the distance in feet. Find the approximate number of seconds it will take for a ball to drop 28,224 ft.

HOW AND WHY
Objective 2   **Find the missing side of a right triangle.**

The Pythagorean formula states that in any right triangle, $c^2 = a^2 + b^2$.

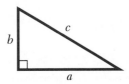

Figure 6.2

If any two sides of a right triangle are known, the third side can be found by substituting the known values into the formula.

Given that one leg of a right triangle is 6 inches and the other leg is 8 inches, find the length of the hypotenuse.

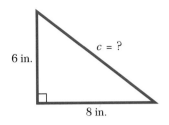

$$c^2 = a^2 + b^2$$
$$c^2 = 6^2 + 8^2$$
$$c^2 = 36 + 64$$
$$c^2 = 100$$
$$c = \sqrt{100}$$
$$c = 10$$

6 in.

$c = ?$

8 in.

The hypotenuse is 10 inches long.

Examples E–F	**Warm Ups E–F**

Directions: Find the missing side of each right triangle.

Strategy: Substitute the known values into the formula $c^2 = a^2 + b^2$ and solve.

E. Find the length of the leg of a right triangle whose other leg is 12 cm and whose hypotenuse is 15 cm.

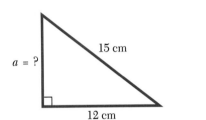

15 cm

$a = ?$

12 cm

$$c^2 = a^2 + b^2$$
$$15^2 = a^2 + 12^2 \qquad \textbf{Substitute 15 for } c \textbf{ and 12 for } b \textbf{ and}$$
$$\textbf{simplify.}$$
$$225 = a^2 + 144$$
$$81 = a^2 \qquad\qquad \textbf{Subtract 144 from both sides.}$$
$$a^2 = 81$$
$$a = \sqrt{81} \qquad\quad \textbf{Solve for } a.$$
$$a = 9$$

Side a has length 9 cm.

E. Find the length of the leg of a right triangle whose other leg is 12 cm and whose hypotenuse is 13 cm.

F. Find the length of the hypotenuse of the right triangle whose legs are 5 ft and 6 ft respectively. Round the result to the nearest hundredth.

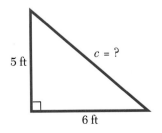

5 ft

$c = ?$

6 ft

F. Find the length of the hypotenuse of the right triangle whose legs are 5 ft and 8 ft respectively. Round the result to the nearest hundredth.

Answers to Warm Ups   E. 5 cm   F. 9.43 ft

$$c^2 = a^2 + b^2$$

$$c^2 = 5^2 + 6^2$$ **Substitute 5 for a and 6 for b and simplify.**

$$c^2 = 25 + 36$$

$$c^2 = 61$$

$$c = \sqrt{61}$$

$$c \approx 7.81$$

The hypotenuse is approximately 7.81 ft.   **The approximate value was found using a calculator**

HOW AND WHY

Objective 3   **Determine if a triangle is a right triangle given the lengths of its sides.**

Given the lengths of three sides of a triangle we can determine if it is a right triangle by using the Pythagorean formula. If it is a right triangle, the square of the longest side must equal the sum of the squares of the other two sides. Is the triangle with sides of 15 in., 12 in., and 9 in. a right triangle?

$$c^2 = a^2 + b^2$$

Does $(15)^2 = (12)^2 + (9)^2$?

Does $225 = 144 + 81$?   **Yes**

The triangle is a right triangle.

Example G   **Warm Up G**

Directions: Determine whether a triangle is a right triangle.

Strategy: Substitute the known values into the formula $c^2 = a^2 + b^2$. If the result is true the triangle is a right triangle, otherwise it is not.

G. Is the triangle whose sides are 12 ft, 18 ft, and 24 ft a right triangle?

$$c^2 = a^2 + b^2$$ **Pythagorean formula. The longest side, 24, is c. Substitute 24 for c, 12 for a, and 18 for b.**

Does $24^2 = 12^2 + 18^2$?

Does $576 = 144 + 324$?

Does $576 = 468$?   **No.**

Since the lengths of the sides do not satisfy the Pythagorean relationship, the triangle is not a right triangle.

G. Is the triangle whose sides are 6 m, 8 m, and 10 m a right triangle?

Answer to Warm Up   G. Yes

Exercises 6.9

OBJECTIVE 1: *Find either the square root or the approximate square root of a number.*

A.

Find the square root.

1. $\sqrt{81}$

2. $\sqrt{64}$

3. $\sqrt{121}$

4. $\sqrt{144}$

5. $\sqrt{\dfrac{4}{25}}$

6. $\sqrt{\dfrac{9}{49}}$

7. $\sqrt{\dfrac{121}{144}}$

8. $-\sqrt{\dfrac{144}{49}}$

B.

9. $\sqrt{484}$

10. $\sqrt{576}$

Find the approximate square root. Round to the indicated place value.

11. $\sqrt{116}$ (tenth)

12. $\sqrt{150}$ (tenth)

13. $-\sqrt{78}$ (hundredth)

14. $-\sqrt{56}$ (hundredth)

15. $\sqrt{21}$ (tenth)

16. $\sqrt{210}$ (hundredth)

17. $\sqrt{0.675}$ (hundredth)

18. $\sqrt{0.0453}$ (hundredth)

OBJECTIVE 2: *Find the missing side of a right triangle.*

A.

Find the missing side of each of the right triangles.

19. $a = ?, b = 8, c = 17$

20. $a = 9, b = 12, c = ?$

21. $a = 12, b = 5, c = ?$

22. $a = 8, b = ?, c = 10$

23. $a = ?, b = 16, c = 20$

24. $a = 10, b = 24, c = ?$

25. $a = 40, b = 30, c = ?$

26. $a = 27, b = ?, c = 45$

B.

Find the missing side of each of the right triangles. Round answers to the nearest hundredth.

27. $a = 2, b = 3, c = ?$

28. $a = ?, b = 11, c = 16$

29. $a = 5, b = ?, c = 7.7$

30. $a = 2.5, b = 3.6, c = ?$

31. $a = 110, b = ?, c = 175$

32. $a = ?, b = 210, c = 312$

33. $a = 5.7, b = 13.2, c = ?$

34. $a = 10.4, b = 19.6, c = ?$

35. $a = 37, b = 55, c = ?$

36. $a = ?, b = 37, c = 72$

OBJECTIVE 3: *Determine if a triangle is a right triangle given the length of its sides.*

A.

Determine if the triangle whose sides are given is a right triangle.

37. 16, 30, and 34

38. 6, 8, and 9

39. 9, 12, and 16

40. 10, 24, and 26

B.

41. 4, 2.4, and 3.2

42. 2.6, 1, and 2.4

43. 8.4, 9.2, and 3.5

44. 6, 7.1, and 3.2

C.

45. The formula for finding the approximate number of seconds that it takes a free-falling body to fall a given distance is

$$t \approx \sqrt{\frac{d}{16}}$$

where t is the time in seconds and d is the distance in feet. Find the approximate number of seconds, to the nearest tenth, that it takes a free falling body to fall 50 feet.

46. Using the formula in Exercise 45, find the approximate number of seconds, to the nearest tenth, that it takes a free-falling body to fall 200 feet.

47. Given a right triangle with the two shorter sides of length 25 feet and 39 feet, find the length of the third side. Find to the nearest tenth of a foot.

48. Given a right triangle with sides 67 cm and 72 cm, find the length of the shortest side. Find to the nearest tenth of a centimeter.

Complete the following chart.

		Nearest Tenth	Nearest Hundredth	Nearest Thousandth	Nearest Ten-thousandth
49.	$\sqrt{365.96}$				
50.	$\sqrt{0.9682}$				
51.	$\sqrt{20.037}$				
52.	$\sqrt{0.03985}$				

53. Evaluate the formula $t = 2\pi\sqrt{\dfrac{\ell}{g}}$ for t if $\ell = 49$, $g = 36$, and $\pi = \dfrac{22}{7}$.

54. Evaluate the formula $t = 2\pi\sqrt{\dfrac{\ell}{g}}$ for t if $\ell = 49$, $g = 121$, and $\pi = \dfrac{22}{7}$.

55. To meet the city code, the attic of a new house needs a vent with a minimum area of 706 square inches. What is the radius of a circular vent that will meet the code requirement? The radius (r) of a circle in terms of the area (A) is given by the formula $r \approx .564\sqrt{A}$. Compute the length of the radius to one decimal place.

56. What is the perimeter of a square field whose area is 13,240 square feet? (Find the perimeter to the nearest tenth of a foot.)

57. What is the length of a rafter that has a rise of 6 feet and a run of 12 feet? Round the answer to the nearest hundredth of a foot. (The rise and run are the legs of a right triangle where the rafter is the hypotenuse.)

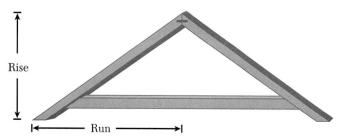

Rise

Run

58. What is the length of a rafter that has a rise of 4 feet and a run of 13 feet? Round the answer to the nearest hundredth of a foot.

59. What is the length of a cable needed to replace a brace that goes from the top of a 50-foot power pole to a ground-level anchor that is 35 feet from the base of the pole? Round the answer to the nearest tenth of a foot.

60. What is the length of a cable needed to replace a brace that goes from the top of a 175-foot power pole to a ground-level anchor that is 42 feet from the base of the pole? Round the answer to the nearest tenth of a foot.

61. A plane is flying south at a speed of 200 miles per hour. The wind is blowing from the west at a rate of 50 miles per hour. To the nearest tenth of a mile, how many miles does the plane actually fly in one hour, in a southeasterly direction? See the accompanying figure.

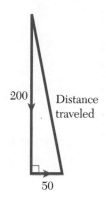

62. A plane is flying south at a speed of 310 miles per hour. The wind is blowing from the west at a rate of 32 miles per hour. To the nearest tenth of a mile, how many miles does the plane actually fly in one hour, in a southeasterly direction?

63. A baseball "diamond" is actually a square that is 90 feet on each side (between the bases). To the nearest tenth of a foot, what is the distance the catcher must throw when attempting to tag out a runner who is attempting to steal second base?

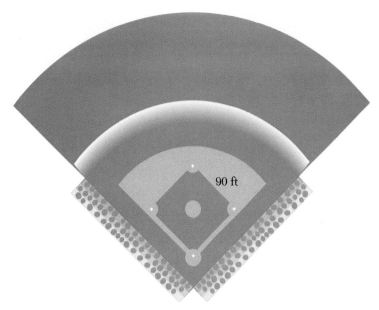

90 ft

64. A 10-foot ladder is leaning against a building. The bottom of the ladder is 3.6 feet from the building. To the nearest tenth of a foot, how high up the building will the ladder reach?

65. Carlotta is 8.3 miles east of the ranger station and Jorge is 5.9 miles south of the station. What is the shortest land distance between them, to the nearest tenth of a mile?

STATE YOUR UNDERSTANDING

66. Explain the difference between the phrases "the square of 36" and "the square root of 36." Describe the symbols for each. Is there an easy way to keep them straight?

67. Describe how to approximate the $\sqrt{7}$ to three decimal places without using a table or calculator.

68. Describe two ways to identify which of the sides of a right triangle is the hypotenuse.

CHALLENGE

Perform the indicated operations.

69. $\sqrt{\dfrac{1}{9}} - \sqrt{0.36} + \sqrt{\dfrac{9}{49}} + (-\sqrt{32.49}) + \sqrt{3.24}$

70. $\sqrt{\dfrac{81}{100}} + \sqrt{\dfrac{196}{225}} - \sqrt{39.69} + \sqrt{54.76} - \sqrt{5.76}$

71. $a\sqrt{b}$ means $a(\sqrt{b})$. Simplify: $3\sqrt{16} + 4\sqrt{25} - 8\sqrt{4}$

72. A square field contains 3 acres. Find the cost, to the nearest dollar, of fencing the field if the fence costs $0.95 per linear foot installed. Use 1 acre = 4840 sq yd and round the perimeter to the nearest foot.

GROUP ACTIVITY

73. The number 8 is a perfect cube because $2(2)(2) = 8$ and we write $\sqrt[3]{8} = 2$. Find the value of $\sqrt[3]{64}$, $\sqrt[3]{125}$, $\sqrt[3]{1000}$, $\sqrt[4]{16}$, $\sqrt[4]{81}$, and $\sqrt[5]{32}$. Define $\sqrt[n]{x}$.

MAINTAIN YOUR SKILLS (Sections 1.4, 1.6)

Evaluate the following.

74. 8^3

75. $4^2 + 5^2$

76. $9^2 - 4^3$

77. $2^2 + 3^2 + 4^2$

78. $5^2 + 6^2 + 7^2$

CHAPTER 6

Group Project *(3–4 weeks)*

a. Each person in the group selects three stocks and tracks their progress for two full weeks. Record the beginning price, the daily change, and closing price.

b. For each stock, calculate the net change for the 2 weeks and the average daily change.

c. The entire group has $10,000 to invest for 1 week, with the objective being to make as much money as possible. The group must buy at least two different stocks but not more than four different ones. All stocks purchased must be from the ones tracked by the members of the group. Once the decision is made, the group's stocks are tracked for 1 week.

d. The group issues a final report. The report includes the data gathered, analysis of the data, a clear rationale for the investment decision, and the final results.

CHAPTER 6

True-False Concept Review

ANSWERS

Check your understanding of the language of algebra and arithmetic. Tell whether each of the following statements is True (always true) or False (not always true). For each statement that you judge to be False, revise it to make a statement that is True.

1. _____

2. _____

3. _____

4. _____

5. _____

6. _____

7. _____

8. _____

9. _____

10. _____

11. _____

12. _____

1. It is common to write the decimal 6.7 in words: six point seven.

2. Some fractions cannot be written as terminating decimals.

3. A decimal such as 0.47 is an alternate way of writing a fraction.

4. A decimal point is the separation of the tens place from the tenths place.

5. The last word in the word name for a decimal is the name of the place value of the last digit.

6. To the nearest ten, 74.49 rounds to 80.

7. To the nearest tenth, 74.49 rounds to 74.5.

8. The rounded value of a decimal is always smaller or larger than the original decimal.

9. The sum of 8 and 0.5 is 8.5.

10. The commutative and associative properties of addition and multiplication also apply to decimals.

11. The same properties for adding positive and negative integers also apply to adding positive and negative decimals.

12. It is not possible to subtract 9.4 from 6.

13. _____

13. The product of 0.6 and 0.5 is 0.3.

14. _____

14. The quotient of 0.6 and 0.5 is 1.1.

15. _____

15. Scientific notation is used only by scientists who deal with large numbers (astronomers) and small numbers (physicists).

16. _____

16. It is always necessary to move the decimal point when dividing decimals.

17. _____

17. It is always necessary to round off when dividing decimals.

18. _____

18. The rules for solving equations involving whole numbers can also be used to solve equations involving decimals.

19. _____

19. The procedure for dividing rational numbers in decimal form is the same as the procedure for dividing rational numbers in fraction form.

20. _____

20. Decimals can be added, subtracted, multiplied, and divided using the same procedures as for whole numbers except for the placement of the decimal point in the result.

21. _____

21. The quotient of two decimals must be smaller than either of the two decimals.

22. _____

22. There is one basic rule for rounding.

23. _____

23. The sum of 8.15 and 0.3 is 8.18.

24. _____

24. The product of two decimals has more decimal places than either decimal.

25. _____

25. The smaller of two decimals has the greater number of decimal places.

26. _____

26. Every repeating decimal can be written as a fraction.

27. _____

27. There are two decimal places in the product of 7.61 and 4.63.

28. _____

28. The square root of a number is less than the original number.

29. _____

29. The Pythagorean formula can be used to find the third side of any triangle if two sides are known.

30. _____

30. The Pythagorean formula can be used to tell whether or not a triangle with three known sides is a right triangle.

CHAPTER 6

Review

Section 6.1 *Objective 1*

1. Write the place value notation for seven hundred twenty-one thousandths.
2. Write the place value notation for fifty-seven ten-thousandths.
3. Write the place value notation for eighteen and six hundred two ten-thousandths.
4. Write the word name for 0.047.
5. Write the word name for 344.00082.

Section 6.1 *Objective 2*

Change the decimal to a fraction.

6. 0.15
7. 0.34
8. 0.125
9. 0.875
10. 0.95

Section 6.1 *Objective 3*

List the decimals from smallest to largest.

11. 0.345, 0.3409, 0.2998, 0.3426
12. 1.2336, 1.2328, 1.23, 1.223, 1.233
13. 0.0976, 0.0909, 0.09099, 0.090733, 0.0907

True or false?

14. $0.0034 < 0.00044$
15. $83.25 > 83.035$

Section 6.1 *Objective 4*

16. Round to the nearest tenth: 45.758
17. Round to the nearest hundredth: 0.0445
18. Round to the nearest hundredth: 255.016
19. Round to the nearest thousandth: 2632.9378
20. Round to the nearest thousand: 2632.9384

Section 6.2 *Objective 1*

21. Add: $6.994 + 12.536 + 4.64 + 0.8993$
22. Add: $0.7761 + 2.83825 + 37.8817 + 2.66$
23. Add:
    $5.16 + 0.006824 + 11.20367 + 3.55546 + 9.9$

24. Add: $5.25x + 7.4x + 0.007x + 22.067x$
25. Add: $-8.3t + 6.04t + (-0.55t) + 1.925t$

Section 6.2 *Objective 2*

26. Subtract: $36.0354 - 18$
27. Subtract: $39 - 17.0354$
28. Subtract: $2507.7443 - 785.48$
29. Subtract: $33v - 0.961v$
30. Subtract:
    $-2.91w - (-0.3w) - 7.05w - (-3.55w)$

Section 6.2 *Objective 3*

Estimate the sum or difference.

31. $0.05679 + 0.00965 + 0.07321 + 0.021$
32. $0.00934 - 0.005609$
33.
```
  0.6078
  0.2409
  0.007
  0.0467
+ 0.1
```
34.
```
  0.07845
- 0.03499
```
35. $0.0678 - 0.782 - 0.5601 + 0.34$

Section 6.3 *Objective 1*

36. Multiply: $0.006(5.45)$
37. Multiply: $4.2(0.484)$
38. Multiply: $0.64(22.8)$
39. Multiply: $1.022(0.025)$
40. Multiply: $8.5(5.8)(2.58)$

Section 6.3 *Objective 2*

Estimate the product.

41. $(0.0412)(0.76)$
42. $(-0.0954)(0.62)$
43. $(0.045)(0.321)(0.000374)$
44. $(-3.09)(0.00783)$
45. $(0.9854)(0.8723)(0.0438)$

Section 6.4 *Objective 1*

Multiply or divide.

46. $35.8(100)$
47. $0.0092(1000)$
48. $12.637 \div 100$
49. $3.467 \div 10^{-6}$
50. 83.92×10^4

Section 6.4 *Objective 2*

51. Write 40,000 in scientific notation.
52. Write 347.8 in scientific notation.
53. Write 0.00000007 in scientific notation.
54. Change 9.2×10^{-5} to place value notation.
55. Change 6.78×10^6 to place value notation.

Section 6.5 *Objective 1*

56. Divide: $15\overline{)35.265}$
57. Divide: $32\overline{)400.96}$
58. Divide: $19\overline{)2.879}$, and round to the nearest thousandth.
59. Divide: $2.41\overline{)5.63}$, and round to the nearest thousandth.
60. Divide: $0.019\overline{)0.36567}$, and round to the nearest thousandth.

Section 6.5 *Objective 2*

Change each fraction to a decimal:

61. $\dfrac{7}{16}$

62. $\dfrac{12}{25}$

63. $\dfrac{23}{64}$

Change each fraction to a decimal, rounded as indicated:

64. $\dfrac{12}{19}$; thousandths

65. $\dfrac{19}{29}$; hundredths

Section 6.5 *Objective 3*

Estimate the quotient.

66. $0.45\overline{)0.008682}$
67. $3.4\overline{)0.87902}$
68. $0.005623 \div 0.216$
69. $4.89 \div 0.0673$
70. $0.00823 \div 30.6$

Section 6.6

71. Convert 6.7 cm to km.
72. Convert 5.78 ft to in.

632

73. Convert 4.7 oz to g.
74. Convert 3.4 m to in.
75. Convert 34 mph to meters per minute.

Section 6.7

76. Evaluate $aw + y^2$ if $a = 0.06$, $w = 43.5$, and $y = 1.2$.
77. Evaluate $a(w + y)^2$ if $a = 0.06$, $w = 43.5$, and $y = 1.2$.
78. Evaluate $\dfrac{p - q}{r^2}$ if $p = 77.7$, $q = 19.44$, and $r = 2.5$.
79. Evaluate the formula $P = 2\pi r$ if $\pi \approx 3.14$ and $r = 6.6$.
80. Is 2.45 a solution of $16x - 4.78 - 3.2x = 26.58$?

Section 6.8

81. Solve: $2.25x + 3.9 = 4.125$
82. Solve: $0.03y - 13.5 = 2.22$
83. Solve: $1.3z + 1.466 = 1.57$
84. Solve: $2.8t - 1.75 - 2.3t = 0.015$
85. Solve: $4.1s + 51.6 + 0.7s = 150$

Section 6.9 *Objective 1*

Find the square roots.

86. $\sqrt{625}$
87. $\sqrt{900}$
88. $\sqrt{484}$
89. Approximate $\sqrt{236}$ to the nearest hundredth.
90. What is the perimeter of a square whose area is 3025 sq ft?

Section 6.9 *Objective 2*

Find the missing side of each of the following right triangles. Round decimal answers to the nearest hundredth.

91. $a = ?, b = 16, c = 34$
92. $a = 18, b = 24, c = ?$
93. $a = 32, b = ?, c = 43$
94. $a = ?, b = 90, c = 120$
95. What is the length of a cable needed to replace a brace that goes from the top of a 500-foot tower to a ground-level anchor that is 375 feet from the base of the tower?

Section 6.9 *Objective 3*

In each of the following, determine if the triangle whose sides are given is a right triangle.

96. 4.5, 6, and 7.5
97. 11, 26.4, and 28.6

CHAPTER 6

Test

ANSWERS

1. _____

1. Divide and round to the nearest thousandth: $0.79\overline{)0.6523}$

2. _____

2. Change $\dfrac{12}{17}$ to a decimal rounded to the nearest hundredth.

3. _____

3. List the decimals from smallest to largest: 0.0673, 0.06893, 0.067, 0.06729

4. _____

4. Write the word name for 106.00408.

5. _____

5. Multiply: $\begin{array}{r} 7.89 \\ \times\, 3.45 \\ \hline \end{array}$

6. _____

6. Evaluate the formula $S = \pi r \ell$ if $\pi \approx 3.14$, $r = 2.4$, and $\ell = 0.8$.

7. _____

7. Round to the nearest hundredth: 4.998

8. _____

8. Change 0.024 to a fraction and simplify.

9. _____

9. Subtract: $12 - 6.7834$

10. _____

10. Solve: $0.05x - 0.008 = 0.082$

11. _____

11. Write 0.0000071 in scientific notation.

12. _____

12. Is 1 a solution of $3.84 + 0.45y - 4.5 = 4.24 - 1.45y$?

13. _____

13. Solve: $11.3 - 2.4a = 9.16 + 3.6a - 11$

14. _____

14. Estimate the difference and then subtract:
$$\begin{array}{r} 8.773 \\ -3.4891 \\ \hline \end{array}$$

15. _____

15. Find $\sqrt{1622}$ to the nearest hundredth.

16. _____

16. Write 43820 in scientific notation.

17. _____

17. True or false: $0.00567 < 0.0056099$?

18. _____

18. Estimate the quotient and divide, round to the nearest ten-thousandth: $3.332 \div 5.3$.

19. _____

19. Estimate the product and then multiply: $7.26(-0.0085)$

20. _____

20. Estimate the sum and then add:
$$\begin{array}{r} 6.9 \\ 0.08 \\ 21.64 \\ 13.89 \\ 9.84 \\ +\ 0.234 \\ \hline \end{array}$$

21. _____

21. Divide: $0.026 \div 100{,}000$

22. _____

22. Combine terms: $-6.3a - 8.4 - 5.6a + 1.23 - 0.87a + 0.03$

23. _____

23. For each of the four days that Juanita drove her delivery route last week, she drove 198.7 miles, 203.5 miles, 386.2 miles, and 187.6 miles. What is the average number of miles she drove each day?

24. _____

24. In one second, a computer can do 6.3×10^4 calculations. Write this number of calculations in place value notation.

25. _____

25. During a canned vegetable sale, Ted bought 21 cans of assorted vegetables. If the sale price was 3 cans for $2.12, how much did Ted pay for the canned vegetables?

26. _____

26. Find the missing side of a right triangle given that $a = 320$ m and $c = 400$ m.

CHAPTERS 1-6

Cumulative Review

1. Find the quotient of 287,064 and 36.

2. Round 20,049 to the nearest hundred.

3. Find the product of 72,198 and 302.

4. Find the difference between 87,234 and 61,345.

5. Is the statement "963 > 936" true or false?

6. Simplify: $b + 8b + 12b$

7. Multiply: $3x(4x + 3y - 8z)$

8. Solve: $9x - 18 = 72$

9. Solve: $8a - 3 = 29$

10. Solve: $15y + 22 = 142$

11. Add: 3 yd 2 ft
            1 yd 2 ft
          +2 yd 1 ft

12. Change 27 feet to inches.

13. June harvested five fields of wheat. The production on each field is given in the chart.

Number of Acres	Bushels per Acre
100	65
150	70
85	55
125	83
62	90

What is the average yield per acre on the five fields? (To the nearest bushel.)

14. Find the perimeter of the following figure.

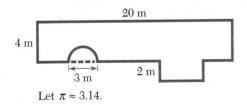

Let $\pi \approx 3.14$.

15. Find the area of the following figure.

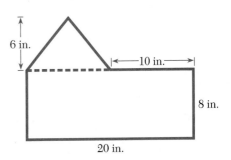

16. Find the volume of a cylinder that has a diameter of 18 cm and a height of 35 cm. Let $\pi \approx 3.14$.

For Exercises 17–21, use the following bar graph which shows the average points per quarter by an NBA basketball team.

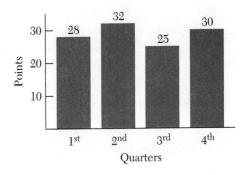

17. Which quarter has the highest point production?

18. What is the average number of points scored in the second half?

19. How many fewer points are scored, on the average, in the first quarter than the fourth quarter?

20. Which quarter has the poorest point production?

21. What is the average number of points per game?

22. Find the sum of $-467, -321, 899, -45, 901,$ and -237.

23. Subtract: $-67 - (-34) - 13 - (-72) - 53$
24. Add: $(13x^2 - 14x + 56) + (3x^2 + 11x - 41)$
25. What is 15 times the difference of -34 and -54?
26. Solve: $35 = 15y + 44 - 18y$
27. In the cooler of a convenience store there are 24 cans of cola and 30 cans of other soft drinks. What fraction of the cans are cola?
28. How many quarter-pound hamburgers can be made from $35\frac{3}{4}$ pounds of hamburger?
29. List the first five multiples of 235.
30. Is 1999 a multiple of 47?
31. Is 1999 prime or composite?
32. What is the smallest prime number that is larger than 777?
33. Simplify: $\dfrac{36xy}{54x}$
34. Find the product of $\dfrac{2x}{7}$, $\dfrac{14y}{3}$, and $\dfrac{6x}{7}$.
35. Find the quotient of $-\dfrac{5}{21}$ and $\dfrac{3}{20}$.
36. Add: $-\dfrac{11t}{96} + \dfrac{35t}{72}$
37. Add. $8\frac{3}{5} + 2\frac{3}{10} + 9\frac{3}{20}$
38. Subtract: $\dfrac{25}{48} - \dfrac{15}{64}$
39. Subtract: $45\frac{7}{15} - 21\frac{5}{6}$
40. Is $\dfrac{12}{25}$ a solution of $\dfrac{7}{10} - x = \dfrac{11}{50}$?
41. Blake poured $8\frac{3}{10}$ yards of cement for a fountain. Elena's fountain took $5\frac{3}{8}$ yards of cement. How much more cement was needed for the larger fountain?
42. Evaluate $3x^2z$ if $x = \dfrac{2}{3}$ and $z = -\dfrac{5}{8}$.
43. Solve: $3x - \dfrac{3}{4} = 18$
44. Solve: $\dfrac{1}{3} - x = \dfrac{3}{5}x + \dfrac{7}{9}$
45. The perimeter of a rectangle is 17 feet. If the length is $6\frac{4}{9}$ feet, what is the width?
46. The average of $2\frac{1}{2}$, $8\frac{9}{10}$, and a third number is 5. What is the third number?
47. Write the place value notation for sixteen and twelve thousandths.

48. Write the word name for 8.004.
49. Change 0.48 to a fraction and simplify.
50. True or false: $0.098 < 0.0099$?
51. Round 8.9049 to the nearest hundredth.
52. Find the sum of 8.2, 3.097, 0.809, 0.0083, and 21.32.
53. Subtract: $8.9053 - 6.9144$
54. Multiply: $8.09(-2.04)$
55. Find the product of 1.0025 and 100,000.
56. Multiply: 0.007934×1000
57. Divide: $823,467 \div 1000$
58. Divide: $0.3456 \div 100$
59. The distance from the Earth to the sun is approximately 93,000,000 miles. Write this distance in scientific notation.
60. The distance from the Earth to the sun is approximately 4.9×10^{11} feet. Write this distance in place value notation.
61. Divide: $0.55\overline{)2.8765}$
62. Divide: $12\overline{)0.01188}$
63. Change $\dfrac{17}{64}$ to a decimal.
64. Evaluate the formula $A = \pi r^2$ if $\pi \approx 3.14$ and $r = 7.5$.
65. Evaluate the formula $V = \dfrac{4}{3}\pi r^3$ if $\pi \approx 3.14$ and $r = 0.25$.
66. Solve: $0.4x + 38.2 = 10.75$
67. During the 1996 presidential election, in one precinct in California, the following votes were recorded: Clinton, 674; Dole, 593; Perot, 127; Nader, 57; others, 52. How many votes were cast for the presidency at that precinct?
68. During a meeting of the Hartford Community Foundation it was decided to contact 720 prominent citizens for donations. The trustees agreed to write to the following fraction of the citizens: Trustee A, one-tenth; Trustee B, one-sixth; Trustee C, one-fifth; Trustee D, one-fifteenth; Trustee E, one-fourth. How many names are left to be divided between the remaining four trustees?
69. Catherine is planning her vacation. The motel costs are $72 per night. She allots $23 per day for food. She plans to drive 1350 miles. Her car averages 28 miles per gallon and gas costs $1.495 per gallon. If her vacation lasts 6 days and 5 nights, how much should she budget for food, lodging, and car expenses? If she has budgeted $800 for her vacation, how much will she have to spend on entertainment?
70. How much will it cost to carpet a 20 ft by 35 ft room, wall to wall, if the carpet and pad costs $37.75 per square yard installed?

Good Advice for Studying

Low-Stress Tests

It's natural to be anxious before an exam. In fact, a little anxiety is actually good: it keeps you alert and on your toes. Obviously, too much stress over tests is not good. Here are some proven tips for taking low-stress tests.

1. Before going to the exam, find a place on campus where you can physically and mentally relax. Don't come into the classroom in a rush.

2. Arrive at the classroom in time to arrange all the tools you will need for the test: sharpened pencils, eraser, plenty of scratch paper, and a water bottle. Try to avoid talking with classmates about the test. Instead, concentrate on deep breathing and relaxation.

3. Before starting the test, on a separate piece of paper, write all the things you may forget while you are busy at work: formulas, rules, definitions, and reminders to yourself. Doing so relieves the load on your short-term memory.

4. Read all of the test problems and mark the easiest ones. Don't skip reading the directions. Note point values so that you don't spend too much time on problems that count only a little, at the expense of problems that count a lot.

5. Do the easiest problems first; do the rest of the problems in order of difficulty.

6. Estimate a reasonable answer before you make calculations. When you finish the problem, check to see that your answer agrees with your estimate.

7. If you get stuck on a problem, mark it and come back to it later.

8. When you have finished trying all the problems, go back to the problems you didn't finish and do what you can. Show all steps because you may get partial credit even if you cannot complete a problem.

9. When you are finished, go back over the test to see that all the problems are as complete as possible and that you have indicated your final answer. Use all of the time allowed, unless you are sure there is nothing more that you can do.

10. Turn in your test and be confident that you did the best job you could. Congratulate yourself on a low-stress test!

If you find yourself feeling anxious during the test, it may help you to have a 3×5 "calming"card. It may include the following: (a) a personal coping statement such as "I have studied hard and prepared well for this test, I will do fine."; (b) a brief description of your peaceful scene; and (c) a reminder to stop, breathe, and relax your tense muscles.

By now, you should be closer to taking control over math instead of allowing math to control you. You are avoiding learned helplessness (believing that other people or influences control your life). Saying "Why try?" and lack of motivation are indicators of this type of attitude. Perfectionism, procrastination, fear of failure, and blaming others are also ineffective attitudes that block your power of control. Take responsibility, and believe that you have the power within to control your life situation.

7 Ratio, Proportion, and Percent

Alan Levenson/Tony Stone Images ©

APPLICATION

The price we pay for everyday items such as food and clothing is theoretically simple. The manufacturer of the item sets the price based on how much it costs to produce and adds a small profit. The manufacturer then sells the item to a retail store which in turn marks it up and sells it to you the consumer. But as you know, it is rarely as simple as that. The price you actually pay for an item also depends on the time of year, the availability of raw materials, the amount of competition between manufacturers of comparable items, the economic circumstances of the retailer, the geographic location of the retailer, and many other factors.

Group Activity

Select a common item whose price is affected by the factors listed below. Discuss how the factor varies and how the price of the item is affected. For each factor, make a plausible bar graph which shows the change in price as the factor varies. (You may estimate specific price levels.)

a. Time of year
b. Economic circumstances of the retailer
c. Competition of comparable products

7.1 Ratio and Rate

OBJECTIVES

1. Write a fraction that shows a ratio comparison of two like measurements.

2. Write a fraction that shows a rate comparison of two unlike measurements.

3. Write a unit rate.

VOCABULARY

A **ratio** is a comparison of two quantities by division.

Like measurements have the same unit of measure.

A **rate** is a comparison of two unlike quantities by division.

A **unit rate** is a rate with a denominator of one unit.

HOW AND WHY
Objective 1

Write a fraction that shows a ratio comparison of two like measurements.

Two numbers can be compared by subtraction or by division. If we compare 12 and 3 we could say that since $12 - 3 = 9$,

12 is 9 more than 3

And since $12 \div 3 = 4$, we say

12 is 4 times larger than 3

The indicated division, $12 \div 3$, is called a *ratio*. These are common ways to write the ratio to compare 12 and 3:

$$12{:}3 \qquad 12 \div 3 \qquad 12 \text{ to } 3 \qquad \frac{12}{3}$$

Since we are comparing 12 to 3, 12 is written first or placed in the numerator of the fraction.

Here we write ratios as fractions. Since a ratio is a fraction, it can be simplified. The ratio $\frac{4}{6}$ is simplified to $\frac{2}{3}$. If the ratio contains two like measurements, it can be simplified in the same way as a fraction.

$$\frac{\$12}{\$25} = \frac{12}{25} \qquad \textbf{The units, \$, are dropped since they are the same.}$$

$$\frac{15 \text{ miles}}{25 \text{ miles}} = \frac{3}{5} \qquad \textbf{The common units are dropped and the fraction simplified.}$$

Examples A–C

Warm Ups A–C

Directions: Write a ratio in simplified form.

Strategy: Write the ratio as a simplified fraction.

A. Write the ratio of 84 to 105.

$$\frac{84}{105} = \frac{4}{5} \qquad \textbf{Write 84 in the numerator and simplify.}$$

The ratio of 84 to 105 is $\frac{4}{5}$.

A. Write the ratio of 16 to 20.

B. Write the ratio of the length of a room to its width if the room is 24 feet by 18 feet.

$$\frac{24 \text{ feet}}{18 \text{ feet}} = \frac{24}{18} = \frac{4}{3}$$ **Simplify the fraction.**

The ratio of the length to the width is $\frac{4}{3}$.

B. Write the ratio of the length of a room to its width if the room is 36 feet by 28 feet.

C. Write the ratio of 2 dimes to 5 quarters. (Compare in cents.)

$$\frac{2 \text{ dimes}}{5 \text{ quarters}} = \frac{20 \text{ cents}}{125 \text{ cents}} = \frac{20}{125} = \frac{4}{25}$$

The ratio of 2 dimes to 5 quarters is $\frac{4}{25}$.

C. Write the ratio of 45 mm to 2 m. (Compare in mm.)

HOW AND WHY
Objective 2   **Write a fraction that shows a rate comparison of two unlike measurements.**

Fractions are also used to compare unlike measurements. The rate of $\frac{31 \text{ children}}{10 \text{ families}}$ compares the unlike measurements "31 children" and "10 families." A common application of a rate is computing gas mileage. For example, if a car runs 208 miles on 8 gallons of gas, we compare miles to gallons by writing $\frac{208 \text{ miles}}{8 \text{ gallons}}$. This rate can be simplified as long as the units are stated, not dropped.

$$\frac{208 \text{ miles}}{8 \text{ gallons}} = \frac{104 \text{ miles}}{4 \text{ gallons}} = \frac{26 \text{ miles}}{1 \text{ gallon}} = 26 \text{ miles per gallon} = 26 \text{ mpg}$$

C A U T I O N   **When units are different, they are *not* dropped.**

Examples D–F

Warm Ups D–F

Directions: Write a rate in simplified form.

Strategy: Write the simplified fraction and retain the unlike units.

D. Write the rate of 10 chairs to 11 people.

$$\frac{10 \text{ chairs}}{11 \text{ people}}$$ **The units must be kept since they are different.**

D. Write the rate of 15 people to 8 tables.

E. Write the rate of 8 cars to 6 homes.

$$\frac{8 \text{ cars}}{6 \text{ homes}} = \frac{4 \text{ cars}}{3 \text{ homes}}$$

E. Write the rate of 12 TV's to 8 homes.

Answers to Warm Ups   B. $\frac{9}{7}$ C. $\frac{9}{400}$ D. $\frac{15 \text{ people}}{8 \text{ tables}}$ E. $\frac{3 \text{ TV's}}{2 \text{ homes}}$

F.   An urban environmental committee urged the local citizens to plant deciduous trees around their homes as a means of conserving energy. (The leaves provide shade in the summer and the fallen leaves allow the sun to shine through the limbs in the winter.) The committee provided the trees to the citizens at cost. After the program had been completed, they determined that 825 oak trees and 675 birch trees had been sold.

1.   What is the rate of the number of oak trees to the number of birch trees sold?

2.   What is the rate of the number of oak trees to the total number of trees sold?

1.   **Strategy:**   Write the first unit, 825 oaks, in the numerator, and the second unit, 675 birch trees, in the denominator.

$$\frac{825 \text{ oak trees}}{675 \text{ birch trees}} = \frac{11 \text{ oak trees}}{9 \text{ birch trees}} \qquad \textbf{Simplify.}$$

The rate of oak trees sold to birch trees is 11 to 9, that is, 11 oak trees were sold for every 9 birch trees sold.

2.   **Strategy:**   Write the first unit, 825 oaks, in the numerator, and the second unit, total number of trees, in the denominator.

$$\frac{825 \text{ oak trees}}{1500 \text{ trees total}} = \frac{11 \text{ oak trees}}{20 \text{ trees total}} \qquad \textbf{Simplify.}$$

The rate of oak trees sold to the total is 11 to 20, that is, 11 out of every 20 trees sold were oak trees.

F.   The following spring the committee repeated the tree program. This time they sold 770 oak trees and 440 birch trees.

1.   What is the rate of oak trees to birch trees sold?

2.   What is the rate of birch trees to the total number of trees?

HOW AND WHY
Objective 3   **Write a unit rate.**

When a rate is simplified so that the denominator is one unit, then we have a *unit rate*. For example,

$$\frac{208 \text{ miles}}{8 \text{ gallons}} = \frac{26 \text{ miles}}{1 \text{ gallon}} \qquad \textbf{Read "26 miles per gallon."}$$

Simplifying rates can lead to statements such as "There are 3.1 children to a family," since

$$\frac{31 \text{ children}}{10 \text{ families}} = \frac{3.1 \text{ children}}{1 \text{ family}}$$

The last rate is a comparison, not a fact, since no family has 3.1 children.

   To write a unit rate given a rate

1.   Do the indicated division.
2.   Retain the units.

Answers to Warm Up   F.   1. $\dfrac{7 \text{ oak trees}}{4 \text{ birch trees}}$ 2. $\dfrac{4 \text{ birch trees}}{11 \text{ trees total}}$

Examples G–I

Directions: Write as a unit rate.

Strategy: Simplify the rate so that the denominator is one unit.

G. Write the unit rate for $\dfrac{\$2.34}{3 \text{ cans of peas}}$.

Strategy: Do the division by 3. Retain the units.

$$\frac{\$2.34}{3 \text{ cans of peas}} = \frac{\$0.78}{1 \text{ can of peas}}$$

The rate is 78¢ per can.

G. Write the unit rate for $\dfrac{492 \text{ pounds}}{12 \text{ square inches}}$.

H. Write the unit rate for $\dfrac{270 \text{ miles}}{12.5 \text{ gallons}}$.

$$\frac{270 \text{ miles}}{12.5 \text{ gallons}} = \frac{21.6 \text{ miles}}{1 \text{ gallon}}$$

Divide numerator and denominator by 12.5.

The unit rate is 21.6 miles per gallon.

H. Write the unit rate for $\dfrac{301 \text{ miles}}{14 \text{ gallons}}$.

Calculator Example:

I. The population density of a region is a unit rate. The rate is the number of people per one square mile of area. Find the population density of Stone County if the population is 13,550 and the area of the county is 1700 square miles. Round your answer to the nearest tenth.

Strategy: Write the rate and divide numerator and denominator by 1700 using your calculator.

$$\text{Density} = \frac{13,550 \text{ people}}{1700 \text{ square miles}}$$

$$= \frac{7.970588234 \text{ people}}{1 \text{ mi}^2} \quad \textbf{Divide.}$$

$$\approx \frac{8.0 \text{ people}}{1 \text{ mi}^2} \quad \textbf{Round to the nearest tenth.}$$

The density is 8.0 people per square mile, to the nearest tenth.

I. What was the estimated population density of Los Angeles in 1995 if the population was estimated at 10,414,000 and the area is 1110 mi^2? Round your answer to the nearest whole number.

Answers to Warm Ups   G. 41 pounds per in.^{2} H. 21.5 miles per gallon   I. 9382 people per mi^2

Exercises 7.1

OBJECTIVE 1: *Write a fraction that shows a ratio comparison of two like measurements.*

A.

Write as a ratio in simplified form.

1. 7 to 35

2. 9 to 54

3. 12 meters to 10 meters

4. 12 feet to 9 feet

5. 40 cents to 45 cents

6. 30 dimes to 54 dimes

B.

7. 1 dime to 4 nickels
(compare in cents)

8. 3 quarters to 5 dimes
(compare in cents)

9. 16 inches to 2 feet
(compare in inches)

10. 3 feet to 3 yards
(compare in feet)

11. 200 cm to 3 km
(compare in centimeters)

12. 200 yards to 5 miles
(compare in yards)

OBJECTIVE 2: *Write a fraction that shows a rate comparison of two unlike measurements.*

A.

Write a rate and simplify.

13. 8 people to 11 chairs

14. 6 families to 18 children

15. 110 miles in 2 hours

16. 264 kilometers in 3 hours

17. 63 miles to 3 gallons

18. 100 kilometers to 4 gallons

19. 88 pounds to 33 feet

20. 36 buttons to 24 bows

B.

21. 18 trees to 63 feet

22. 91 TV's to 52 houses

23. 38 books to 95 students

24. 750 people for 3000 tickets

25. 765 people to 27 rooms

26. 8780 households to 6 cable companies

27. 345 pies to 46 sales

28. $17.85 per 34 pounds of apples

OBJECTIVE 3: *Write a unit rate.*

A.

29. 50 miles to 2 hours

30. 60 miles to 4 minutes

31. 36 feet to 9 seconds

32. 75 meters to 3 minutes

33. 132 yards to 2 billboards

34. 60 yards to 15 posts

35. 90¢ per 10 pounds of potatoes

36. 117¢ per 3 pounds of broccoli

B.

Write a unit rate. Round to the nearest tenth.

37. 825 miles per 22 gallons

38. 13,266 kilometers per 220 gallons

39. 1000 feet to 12 seconds

40. 1000 yards to 15 minutes

41. 12,095 pounds to 45 square miles

42. 5486 kilograms to 290 square centimeters

43. 225 gallons per 14 minutes

44. 850 liters per 14 minutes

C.

45. The parking lot in the lower level of the Senter Building has 18 spaces for compact cars and 24 spaces for larger cars.

   a. What is the ratio of compact spaces to larger spaces?

   b. What is the ratio of compact spaces to the total number of spaces?

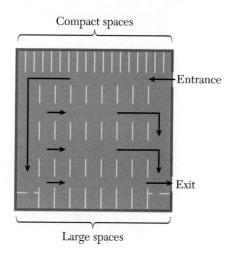

Compact spaces

Entrance

Exit

Large spaces

46. The Reliable Auto Repair Service building has eight stalls for repairing automobiles and four stalls for repairing small trucks.

   a. What is the ratio of the number of stalls for small trucks to the number of stalls for automobiles?

   b. What is the ratio of the number of stalls for small trucks to the total?

47. One section of the country has 3500 TV sets per 1000 houses. Another section has 500 TV sets per 150 houses. Are the rates of the TV sets to the number of houses the same in both parts of the country?

48. In Oakville there are 5000 automobiles per 3750 households. In Firland there are 6400 automobiles per 4800 households. Are the rates of the number of automobiles to the number of households the same?

49. A store bought a sofa for a cost of $175 and sold it for $300. What is the ratio of the cost to the selling price?

50. In Exercise 49, what is the ratio of the markup to the cost?

51. A coat is regularly priced at $99.99, but during a sale its price is $66.66. What is the ratio of the sale price to the regular price?

52. In Exercise 51, what is the ratio of the discount to the regular price?

53. What is the population density of the city of Dryton if there are 22,450 people and the area is 230 square miles? Simplify to a unit comparison, rounded to the nearest tenth.

54. What is the population density of Struvaria if 950,000 people live there and the area is 18,000 square miles? Simplify to a unit comparison, rounded to the nearest tenth.

55. What was the population density of your city in 1997?

56. What was the population density of your state in 1997?

57. In the United States, four people use an average of 250 gallons of water per day. One hundred gallons are used to flush the toilet, 80 gallons in baths/showers, 35 gallons doing laundry, 15 gallons washing dishes, 12 gallons for cooking and drinking, and 8 gallons in the bathroom sink.

   a. Write the ratio of laundry use to toilet use.

   b. Write the ratio of bathing/showering use to dish-washing use.

58. In Exercise 57,

   a. Write the ratio of cooking/drinking use to dish-washing use.

   b. Write the ratio of laundry use per person.

59. Drinking water is considered to be polluted when a pollution index of 0.05 mg of lead per liter is reached. At that rate, how many mg of lead are enough to pollute 25 liters of drinking water?

60. Data indicate that 3 of every 20 rivers in the United States showed an increase in water pollution from 1974 to 1983. Determine how many rivers are in your state. At the same rate, determine how many of those rivers had an increased pollution level during the same period.

Exercises 61–66 relate to the Chapter 7 Application.

It is often difficult to compare the price of various food items, many times because of the packaging. Is a 14-oz can of pears for $0.89 a better buy than a 16-oz can of pears for $1.00? To help consumers compare, *unit pricing* is often posted. Mathematically, we write the information as a rate and simplify to a one-unit comparison.

61. Write a ratio for a 14-oz can of pears that sells for $0.89 and simplify it to price per 1 oz of pears. Do the same with the 16-oz can of pears for $1.00. Which is the better buy?

62. Which is the best buy: a 15-oz box of Cheerios for $2.49, a 20-oz box for $3.29, or a 2-lb 3-oz box for $5.39?

63. Which is the better buy: 5 lb of granulated sugar on sale for $4.95 or 25 lb of sugar for $24.90?

Some food items have the same unit price regardless of the quantity purchased. Other food items have a decreasing unit price as the size of the container increases. In order to determine which category a food falls in, find the unit price for each item.

64. Is the unit price of hamburger the same if 2.35 lb costs $4.44 and 3.52 lb costs $6.65?

65. Is the unit price of frozen orange juice the same if a 12-oz can costs $1.09 and a 16-oz can costs $1.30?

66. List five items that usually have the same unit price regardless of quantity purchased, and five that do not. What circumstances could cause an item to change categories?

STATE YOUR UNDERSTANDING

67. Write a short paragraph explaining why ratios are useful ways to compare measurements.

68. Explain the difference between a ratio, a rate, and a unit rate.

CHALLENGE

69. Give an example of a ratio that is not a rate. Give an example of a rate that is not a ratio.

70. The ratio of noses to persons is $\dfrac{1}{1}$ or one-to-one. Find three examples of two-to-one ratios and three examples of three-to-one ratios.

71. Each gram of fat contains 9 calories. Chicken sandwiches at various fast-food places contain the following total calories and grams of fat.

		Calories	**Grams of Fat**
a.	RB's Light Roast Chicken Sandwich	276	7
b.	KB's Broiler Chicken Sandwich	267	8
c.	Hard B's Chicken Filet	370	13
d.	LJS's Baked Chicken Sandwich	130	4
e.	The Major's Chicken Sandwich	482	27
f.	Mickey's Chicken	415	19
g.	Tampico's Soft Chicken Taco	213	10
h.	Winston's Grilled Chicken Sandwich	290	7

Find the ratio of fat calories to total calories for each sandwich.

GROUP ACTIVITY

72. Have each member of your group select a country other than the United States. Each member is to use the library or other resources to find the population and area of the country selected. Calculate the population density for each country and compare your findings. Which country has the greatest population density? The least?

73. The golden ratio of 1.618 to 1 has been determined by artists to be very pleasing aesthetically. The ratio has been discovered to occur in nature in many places, including the human body. In particular, the ratio applies to successive segments of the fingers.

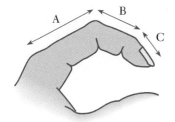

Measure as accurately as possible at least three fingers of everyone in the group. Calculate the ratio of successive segments, and fill in the table.

Name	**Finger**	**A**	**B**	**C**	$\dfrac{A}{B}$	$\dfrac{B}{C}$

Whose fingers come closest to the golden ratio? Can you find other body measures that have this ratio?

Solve.

74. $\dfrac{x}{2} = \dfrac{3}{4}$

75. $\dfrac{x}{2} - \dfrac{2}{3} = \dfrac{3}{4}$

76. $\dfrac{y}{10} = -\dfrac{3}{2}$

77. $\dfrac{y}{10} + \dfrac{6}{25} = -\dfrac{3}{2}$

78. $\dfrac{1}{2} = \dfrac{2w}{3}$

7.2 Solving Proportions

OBJECTIVES

1. Determine whether a proportion is true or false.

2. Solve a proportion.

VOCABULARY

A **proportion** is a statement that two ratios are equal.

In a proportion, **cross multiplication** means multiplying the numerator of each ratio times the denominator of the other.

Cross products are the products obtained from cross multiplication.

Solving a proportion means finding a missing number that will make a proportion true.

HOW AND WHY
Objective 1

Determine whether a proportion is true or false.

The statement $\dfrac{14}{8} = \dfrac{35}{20}$ is a proportion. A proportion states that two rates or ratios are equal. To check whether the proportion is true or false we use "cross multiplication."

The proportion $\dfrac{14}{8} = \dfrac{35}{20}$ is true if the cross products are equal.

$$\frac{14}{8} = \frac{35}{20}$$

$20(14) = 280$ and $8(35) = 280$, so $20(14) = 8(35)$.

This test is based on the multiplication property of equality. Multiply both sides of the equation (proportion) by the product of the denominators to eliminate the fractions.

$$160\left(\frac{14}{8}\right) = 160\left(\frac{35}{20}\right)$$
$$20(14) = 8(35)$$
$$280 = 280$$

When we cleared fractions in Chapter 5, we multiplied both sides by the LCM of the denominators. To cross multiply, we just use the product of the denominators that is also a common denominator but not necessarily the LCM.

▶ *To check whether a proportion is true or false*

1. Check that the ratios or rates have the same units.

2. Cross multiply.

3. If the cross products are equal, the proportion is true.

Examples A–C | **Warm Ups A–C**

Directions: Determine whether a proportion is true or false.

Strategy: Check the cross products. If they are equal the proportion is true.

A. Is $\dfrac{6}{5} = \dfrac{72}{60}$ true or false?

$\dfrac{6}{5} = \dfrac{72}{60}$ **Find the cross products.**

$6(60) = 5(72)$

$360 = 360$ **True.**

The proportion is true.

A. Is $\dfrac{7}{8} = \dfrac{42}{48}$ true or false?

B. Is $\dfrac{4.1}{7.1} = \dfrac{4}{7}$ true or false?

$\dfrac{4.1}{7.1} = \dfrac{4}{7}$ **Find the cross products.**

$4.1(7) = 7.1(4)$

$28.7 = 28.4$ **False.**

The proportion is false.

B. Is $\dfrac{5.6}{6.3} = \dfrac{5}{6}$ true or false?

C. Is $\dfrac{1 \text{ dollar}}{3 \text{ quarters}} = \dfrac{8 \text{ dimes}}{12 \text{ nickels}}$ true or false?

Strategy: The units in the rates are not the same. We change all units to cents and simplify.

$\dfrac{1 \text{ dollar}}{3 \text{ quarters}} = \dfrac{8 \text{ dimes}}{12 \text{ nickels}}$

$\dfrac{100 \text{ cents}}{75 \text{ cents}} = \dfrac{80 \text{ cents}}{60 \text{ cents}}$

$\dfrac{100}{75} = \dfrac{80}{60}$ **Like units may be dropped.**

$100(60) = 75(80)$

$6000 = 6000$ **True.**

The proportion is true.

C. Is $\dfrac{1 \text{ dollar}}{2 \text{ quarters}} = \dfrac{16 \text{ nickels}}{4 \text{ dimes}}$ true or false?

HOW AND WHY
Objective 2   **Solve a proportion.**

Proportions are used to solve many problems in science, technology, and business. There are four numbers or measures in a proportion. If three of the numbers are given, we can find the missing number.

Answers to Warm Ups   A. true   B. false   C. true

For example,

$$\frac{x}{5} = \frac{15}{25}$$

$25x = 5(15)$ **Cross multiply.**

$25x = 75$

$x = 3$ **Divide each side by 25.**

We can also multiply both sides by the LCM, 25, to get the same result.

$$25\left(\frac{x}{5}\right) = 25\left(\frac{15}{25}\right)$$

$5x = 15$

$x = 3$

The missing number is 3.

 To solve a proportion:

1. Cross multiply.

2. Solve the resulting equation.

Examples D–H

Directions: Solve the proportion.

Strategy: Cross multiply, then divide.

Examples D–H	**Warm Ups D–H**
D. Solve: $\dfrac{4}{9} = \dfrac{8}{x}$	D. Solve: $\dfrac{5}{9} = \dfrac{10}{y}$
$4x = 9(8)$ **Cross multiply.**	
$4x = 72$ **Simplify.**	
$x = 18$ **Divide both sides by 4.**	
The missing number is 18.	
E. Solve: $\dfrac{0.6}{t} = \dfrac{1.2}{0.84}$	E. Solve: $\dfrac{0.5}{c} = \dfrac{1.5}{0.75}$
$0.6(0.84) = 1.2t$ **Cross multiply.**	
$0.504 = 1.2t$ **Simplify.**	
$0.42 = t$ **Divide both sides by 0.42.**	
The missing number is 0.42.	
F. Solve: $\dfrac{\frac{3}{4}}{1\frac{2}{3}} = \dfrac{\frac{1}{2}}{x}$	F. Solve: $\dfrac{\frac{3}{4}}{\frac{5}{8}} = \dfrac{\frac{1}{2}}{w}$

(continued on page 656)

Answers to Warm Ups   D. $y = 18$ E. $c = 0.25$ F. $w = \dfrac{5}{12}$

$$\frac{3}{4}x = \left(1\frac{2}{3}\right)\left(\frac{1}{2}\right) \qquad \textbf{Cross multiply.}$$

$$\frac{3}{4}x = \left(\frac{5}{3}\right)\left(\frac{1}{2}\right)$$

$$\frac{3x}{4} = \frac{5}{6} \qquad \textbf{Simplify. Remember that } \frac{3}{4}x = \frac{3x}{4}.$$

$$18x = 20 \qquad \textbf{Cross multiply again.}$$

$$x = \frac{20}{18} = \frac{10}{9} \qquad \textbf{Divide both sides by 18.}$$

The missing number is $\frac{10}{9}$.

Calculator Example:

G.   Solve $\dfrac{3}{z} = \dfrac{9.6}{7.32}$ and round to the nearest hundredth.

$$3(7.32) = 9.6z \qquad \textbf{Cross multiply.}$$

$$\frac{3(7.32)}{9.6} = z \qquad \textbf{Before simplifying, divide both sides by 9.6.}$$

$$2.2875 = z \qquad \textbf{First multiply, then divide using a calculator.}$$

$$2.29 \approx z \qquad \textbf{Round.}$$

The missing number is 2.29 to the nearest hundredth.

G.   Solve $\dfrac{8}{y} = \dfrac{1.82}{21.24}$ and round to the nearest hundredth.

H.   If $\dfrac{1}{5}$ of people bought products with no plastic wrapping $\dfrac{1}{5}$ of the time, 144 tons of plastic would be eliminated from our landfills each year. At the same rate, how many tons would be eliminated if $\dfrac{1}{5}$ of people purchased products with no plastic wrapping $\dfrac{3}{4}$ of the time?

Strategy: The ratios of the fractions and the ratio of the tons is the same.

$$\frac{\frac{1}{5}}{\frac{3}{4}} = \frac{144 \text{ tons}}{x \text{ tons}}$$

$$\frac{1}{5}x = \frac{3}{4}(144)$$

$$\frac{1}{5}x = 108$$

$$x = 108 \div \frac{1}{5}$$

$$x = 540$$

So, 540 tons of plastic would be eliminated from our landfills.

H.   If $\dfrac{1}{5}$ of people bought products with no plastic wrapping $\dfrac{11}{20}$ of the time, how much plastic would be eliminated from our landfills each year? Assume the same rate as in Example H.

Answers to Warm Ups   G. $y \approx 93.36$ H. **396 tons eliminated**

Exercises 7.2

OBJECTIVE 1: *Determine whether a proportion is true or false.*

A.

True or false.

1. $\dfrac{6}{3} = \dfrac{16}{8}$

2. $\dfrac{2}{3} = \dfrac{10}{15}$

3. $\dfrac{6}{8} = \dfrac{9}{12}$

4. $\dfrac{6}{9} = \dfrac{8}{12}$

5. $\dfrac{3}{2} = \dfrac{9}{4}$

6. $\dfrac{3}{4} = \dfrac{9}{16}$

B.

7. $\dfrac{18}{15} = \dfrac{12}{10}$

8. $\dfrac{16}{24} = \dfrac{10}{15}$

9. $\dfrac{35}{22} = \dfrac{30}{20}$

10. $\dfrac{24}{32} = \dfrac{36}{38}$

11. $\dfrac{27}{45} = \dfrac{30}{60}$

12. $\dfrac{36}{45} = \dfrac{20}{25}$

13. $\dfrac{2.8125}{3} = \dfrac{15}{16}$

14. $\dfrac{9.375}{3} = \dfrac{25}{8}$

OBJECTIVE 2: *Solve a proportion.*

A.

Solve.

15. $\dfrac{1}{2} = \dfrac{a}{18}$

16. $\dfrac{1}{3} = \dfrac{b}{18}$

17. $\dfrac{2}{6} = \dfrac{c}{18}$

18. $\dfrac{2}{9} = \dfrac{x}{18}$

19. $\dfrac{2}{a} = \dfrac{5}{10}$

20. $\dfrac{8}{b} = \dfrac{2}{5}$

21. $\dfrac{14}{28} = \dfrac{5}{c}$

22. $\dfrac{8}{12} = \dfrac{6}{d}$

23. $\dfrac{2}{3} = \dfrac{12}{x}$

24. $\dfrac{3}{6} = \dfrac{8}{y}$

25. $\dfrac{w}{7} = \dfrac{2}{28}$

26. $\dfrac{z}{2} = \dfrac{2}{12}$

B.

27. $\dfrac{x}{7} = \dfrac{3}{2}$

28. $\dfrac{y}{5} = \dfrac{3}{4}$

29. $\dfrac{2}{z} = \dfrac{5}{11}$

30. $\dfrac{12}{x} = \dfrac{16}{3}$

31. $\dfrac{16}{24} = \dfrac{y}{16}$

32. $\dfrac{9}{11} = \dfrac{z}{15}$

33. $\dfrac{15}{16} = \dfrac{12}{a}$

34. $\dfrac{28}{7} = \dfrac{50}{b}$

35. $\dfrac{0.1}{c} = \dfrac{0.2}{1.2}$

36. $\dfrac{0.5}{a} = \dfrac{0.2}{0.6}$

37. $\dfrac{\frac{3}{5}}{b} = \dfrac{8}{5}$

38. $\dfrac{\frac{2}{3}}{c} = \dfrac{\frac{8}{9}}{1\frac{7}{9}}$

39. $\dfrac{0.9}{4.5} = \dfrac{0.05}{x}$

40. $\dfrac{1.2}{2.7} = \dfrac{3.4}{y}$

41. $\dfrac{7}{42} = \dfrac{w}{3}$

42. $\dfrac{80}{30} = \dfrac{b}{2}$

43. $\dfrac{y}{3} = \dfrac{9}{\frac{1}{8}}$

44. $\dfrac{s}{40} = \dfrac{\frac{3}{4}}{5}$

45. $\dfrac{t}{24} = \dfrac{3\frac{1}{2}}{10\frac{1}{2}}$

46. $\dfrac{w}{4\frac{1}{4}} = \dfrac{3\frac{1}{3}}{2\frac{1}{2}}$

Solve. Round to the nearest tenth.

47. $\dfrac{3}{11} = \dfrac{w}{5}$

48. $\dfrac{3}{11} = \dfrac{x}{15}$

49. $\dfrac{9}{35} = \dfrac{14}{y}$

50. $\dfrac{9}{35} = \dfrac{24}{z}$

Solve. Round to the nearest hundredth.

51. $\dfrac{1.5}{5.5} = \dfrac{a}{0.8}$

52. $\dfrac{1.5}{5.5} = \dfrac{b}{2.8}$

53. $\dfrac{\frac{3}{7}}{c} = \dfrac{9}{20}$

54. $\dfrac{\frac{3}{7}}{d} = \dfrac{9}{32}$

C.

Fill in the boxes so the statement is true. Explain your answer.

55. If $\dfrac{\square}{70} = \dfrac{x}{7}$, then $x = 1$.

56. If $\dfrac{\square}{35} = \dfrac{y}{7}$, then $y = 4$.

57. Find the errro(s) in the statement: If $\dfrac{2}{5} = \dfrac{x}{19}$, then $2x = 5(19)$. Correct the statement. Explain how you would avoid this error.

58. Find the error(s) in the statement: If $\dfrac{3}{x} = \dfrac{7}{9}$, then $3x = 7(9)$. Correct the statement. Explain how you would avoid this error.

Exercises 59–61 relate to the Chapter 7 Application.

59. If 2.4 pounds of bananas cost $1.66, how much would you expect 4.5 pounds of bananas to cost?

60. If 6 ounces of dried cranberries cost $2.25, how much would you expect $1\dfrac{1}{2}$ pounds of dried cranberries to cost?

61. A box of Tide that is sufficient for 18 loads costs $3.65. What is the most that a store brand of detergent can cost if the box is sufficient for 25 loads and is cheaper to use than Tide?

STATE YOUR UNDERSTANDING

62. Explain how to solve $\dfrac{3.5}{\frac{1}{4}} = \dfrac{7}{y}$.

63. Write a short paragraph explaining the similarities and differences of the methods of solution for proportions and equations.

64. Look up the word "proportion" in the dictionary and write two definitions that differ from the mathematical definition of the word. Write three sentences, using the word "proportion," that illustrate each of the meanings.

CHALLENGE

Solve.

65. $\dfrac{9+3}{9+6} = \dfrac{8}{a}$

66. $\dfrac{5(9)-2(5)}{8(6)-3(2)} = \dfrac{8(5)}{b}$

Solve. Round to the nearest thousandth.

67. $\dfrac{7}{w} = \dfrac{18.92}{23.81}$

68. $\dfrac{7}{t} = \dfrac{18.81}{23.92}$

GROUP ACTIVITY

69. Five ounces of decaffeinated coffee contain approximately 3 mg of caffeine whereas 5 ounces of regular coffee contain an average of 120 mg of caffeine. Five ounces of tea brewed for 1 minute contain an average of 21 mg of caffeine. Twelve ounces of regular cola contain an average of 54 mg of caffeine. Six ounces of hot cocoa contain an average of 11 mg of caffeine. Twelve ounces of iced tea contain an average of 72 mg of caffeine. Determine the total amount of caffeine each member of your group consumed yesterday. Make a chart to illustrate this information. Combine this information with the other groups in your class to make a class amount. Make a class chart to illustrate this information. Determine the average amount of caffeine consumed by each member of the group and then by each member of the class. Compare these averages by making ratios. Discuss the similarities and the differences.

MAINTAIN YOUR SKILLS (Sections 5.7, 6.2, 6.4)

70. Find the difference of 620.3 and 499.9781.

71. Solve: $\dfrac{x}{9} - \dfrac{4}{18} = \dfrac{1}{9}$

72. Solve: $\dfrac{y}{15} + \dfrac{11}{5} = \dfrac{8}{3}$

73. Multiply: 4.835(10,000)

74. Divide: 4.835 ÷ 1000

7.3   **Applications of Proportions**

OBJECTIVE   Solve word problems using proportions.

HOW AND WHY
  Objective   **Solve word problems using proportions.**

If the ratio of two quantities is constant, the ratio is used to find the missing part of a second ratio. For instance, if two pounds of bananas cost $0.48, what will 12 pounds of bananas cost?

	Case I	Case II
Pounds of bananas	2	12
Cost in dollars	0.48	

In the table, the cost in Case II is missing. Call the missing value y.

	Case I	Case II
Pounds of bananas	2	12
Cost in dollars	0.48	y

Write the proportion using the ratios as shown in the chart.

$$\frac{2 \text{ lb of bananas}}{\$0.48} = \frac{12 \text{ lb of bananas}}{\$y}$$

Cross multiplying gives us

$(2 \text{ lb of bananas})(\$y) = (12 \text{ lb of bananas})(\$0.48)$

The units are the same on each side of the equation, so we can drop them and have

$2y = 12(0.48)$
$2y = 5.76$
$\ y = 2.88$ **Divide both sides by 2.**

So, 12 pounds of bananas will cost $2.88.

Using a table forces the units of a proportion to match. Therefore we usually do not write the units in the proportion itself. We always use the units in the answer.

Examples A–D	Warm Ups A–D

Directions: Solve the following problems using proportions.

Strategy: Make a table with two columns and two rows. Label the columns Case I and Case II, and the rows with the units in the problem. Fill in the table with the quantities given, and assign a variable to the unknown quantity. Write the proportion contained in the table, and solve it.

A.   If 3 cans of tuna fish sell for $3.73, what is the cost of 48 cans of tuna fish?

	Case I	**Case II**
Cans	3	48
Cost	$3.73	C

Make a table.

$$\frac{3}{3.73} = \frac{48}{C}$$ Write the proportion.

$3C = (3.73)(48)$ Cross multiply.

$3C = 179.04$

$C = 59.68$

The cost of 48 cans of tuna fish is $59.68.

A.   A sporting goods store advertises golf balls at 6 for $7.25. At this rate, what will 3 dozen balls cost?

B.   Mary Alice pays $1650 property tax on a house valued at $55,000. At the same rate, what would be the property tax on a house valued at $82,000?

	Case I	**Case II**
Tax	$1650	T
Value	$55,000	$82,000

Make a table.

$$\frac{1650}{55,000} = \frac{T}{82,000}$$ Write the proportion.

$1650(82,000) = 55,000T$ Cross multiply.

$135,300,000 = 55,000T$

$2460 = T$

The tax on the $82,000 house is $2460.

B.   A house has a property tax of $1260 and is valued at $45,000. At the same rate, what will be the property tax on a house valued at $56,000?

C.   On a roadmap of Texas, $\frac{1}{4}$ inch represents 50 miles. How many miles are represented by $1\frac{1}{2}$ inches?

C.   On a road map of Jackson County, $\frac{1}{4}$ inch represents 25 miles. How

	Case I	Case II
Inches	$\dfrac{1}{4}$	$1\dfrac{1}{2}$
Miles	50	N

Make a table.

many miles are represented by $2\dfrac{1}{2}$ inches?

$$\frac{\frac{1}{4}}{50} = \frac{1\frac{1}{2}}{N}$$ **Write the proportion.**

$\dfrac{1}{4}N = 1\dfrac{1}{2}(50)$ **Cross multiply.**

$\dfrac{1}{4}N = \dfrac{3}{2}(50)$ **Change to improper fraction.**

$\dfrac{1}{4}N = \dfrac{150}{2} = 75$

$\dfrac{N}{4} = \dfrac{75}{1}$

$N = 300$ **Cross multiply again.**

$1\dfrac{1}{2}$ inches on the map represents 300 miles.

D. Paula decided to start a savings account. Her weekly take-home pay without overtime is $278. She decides to save $13.90 of this each week. One week she works overtime and her take-home pay is $320. If she wants to save the same ratio from this check, how much should she save?

D. Paula receives a raise and now is taking home $330 without overtime. If she wants to continue saving in the same ratio as in Example D, how much should she save each week?

	Case I	Case II
Pay	$278	$320
Savings	$13.90	x

Make a table.

$$\frac{278}{13.90} = \frac{320}{x}$$ **Write the proportion.**

$278x = 13.90(320)$ **Cross multiply.**

$278x = 4448$

$x = 16$

Paula should save $16 from her paycheck.

Exercises 7.3

OBJECTIVE: *Solve word problems using proportions.*

A.

A photograph that measures 6 inches wide and 4 inches high is to be enlarged so that the width will be 15 inches. What will be the height of the enlargement?

15 in.

x in.

6 in.

4 in.

	Case I	Case II
Width (in.)	(a)	(c)
Height (in.)	(b)	(d)

1. What goes in box (a)?

2. What goes in box (b)?

3. What goes in box (c)?

4. What goes in box (d)?

5. What is the proportion for the problem?

6. What is the height of the enlargement?

If a fir tree is 30 feet tall and casts a shadow of 18 feet, how tall is a tree that casts a shadow of 48 feet?

	First Tree	Second Tree
Height (ft)	(1)	(3)
Shadow (ft)	(2)	(4)

7. What goes in box (1)?

8. What goes in box (2)?

9. What goes in box (3)?

10. What goes in box (4)?

11. What is the proportion for the problem?

12. How tall is the second tree?

Jean and Jim are building a fence around their yard. From past experience they know that they are able to build 48 feet in 8 hours. If they work at the same rate, how many hours will it take them to complete the job if the perimeter of the yard is 288 feet?

	Case I	Case II
Time (hr)	(5)	(7)
Length of fence (ft)	(6)	(8)

13. What goes in box (5)?

14. What goes in box (6)?

15. What goes in box (7)?

16. What goes in box (8)?

17. What is the proportion for the problem?

18. How many hours will it take to build the fence?

B.

The Midvale Junior High School expects a fall enrollment of 910 students. The district assigns teachers at the rate of 3 teachers for every 65 students. The district currently has 38 teachers assigned to the school. How many additional teachers does the district need to assign to the school?

	Case I	Case II
Teachers	3	(e)
Students	65	(f)

19. What goes in box (e)?

20. What goes in box (f)?

21. What is the proportion for the problem?

22. How many teachers will be needed at the school next year?

23. How many additional teachers will need to be assigned?

The average restaurant in Universeville produces 30 pounds of garbage in $1\frac{1}{2}$ days. How many pounds of garbage do they produce in 2 weeks (14 days)? (Use x for the missing number of pounds.)

	Case I	Case II
Days		
Garbage (lb)		

24. What goes in each of the four boxes?

25. What proportion should be used to solve this problem?

26. How many pounds of garbage do they have at the end of 2 weeks?

C.

27. Merle is knitting a sweater. The knitting gauge is eight rows to the inch. How many rows must she knit to complete $12\frac{1}{2}$ inches of the sweater?

	Case I	Case II
Rows		
Inches		

28. For every 2 hours a week that Helen is in class, she plans to spend 5 hours a week doing her homework. If she is in class 15 hours each week, how many hours will she plan to be studying each week?

29. If 16 lb of fertilizer will cover 1500 ft^2 of lawn, how much fertilizer is needed to cover 2500 ft^2?

30. If 30 lb of fertilizer covers 1500 ft^2 of lawn, how many ft^2 will 50 lb of fertilizer cover?

Exercises 31–33. The Logan Community College basketball team won 12 of its first 15 games. At this rate how many games will they win if they play a 30-game schedule?

	Case I	Case II
Games Won		
Games Played		

31. What goes in each of the four boxes?

32. What is the proportion for the problem?

33. How many games should they win with a 30-game schedule?

34. John must do 25 hours of work to pay for the tuition for three college credits at the local university. If John is going to take 15 credits in the fall, how many hours will he need to work to pay for his tuition?

35. If John (see Exercise 34) works 40 hours per week, how many weeks will he need to work to pay for his tuition? (Any part of a week counts as a full week.)

36. Larry sells men's clothing at the University Men's Shop. If he sells $100 worth of clothing, he makes $15. How much does he make if he sells $340 worth of clothes?

37. Hazel sells automobiles at the Quality Used Car Company. If she sells an automobile for $1200 she is paid $60. If she sells an automobile for $2900, how much is she paid?

38. If gasoline sells for $1.229 per gallon, how many gallons can be purchased for $24.58?

39. If 44 ounces of soap powder costs $4.84, what does 20 ounces cost?

40. In Jean's Vegetable Market, onions are priced at 2 pounds for $0.63. If Mike buys 6 pounds, what is his cost?

41. Twenty-five pounds of tomatoes cost $23.70 at the local market. At this rate, what is the cost of 10 pounds?

42. A new car travels 369 miles in 8.2 hours. At the same rate, how long does it take to go 900 miles?

43. A brine solution is made by dissolving 1.5 pounds of salt in one gallon of water. At this rate, how many gallons of water are needed when 9 pounds of salt are used?

44. Celia earns a salary of $900 per month from which she saves $45 each month. Her salary is increased to $980 per month. How much must she save each month to save at the same rate?

45. Ginger and George have a room in their house that needs to be carpeted. It is determined that a total of 33 yards of carpet are needed for the job. Hickson's Carpet Emporium will install the 33 yards of carpet for $526.35. If Ginger and George decide to have a second room of their house carpeted and the room will need 22 yards of carpet, at the same rate, how much will it cost to have the second room carpeted?

46. A 16-ounce can of pears costs $0.98 and a 29-ounce can costs $1.69. Is the price per ounce the same in both cases? If not, then what should be the price of the 29-ounce can to equalize the price per ounce?

47. A doctor requires that Ida, the nurse, give 8 milligrams of a certain drug to a patient. The drug is in a solution that contains 20 milligrams in 1 cubic centimeter. How many cubic centimeters should Ida use for the injection?

48. If a 24-foot beam of structural steel contracts 0.0036 inch for each drop of 5 degrees in temperature, then, at the same rate, how much does a 50-foot beam of structural steel contract for a drop of 5 degrees in temperature?

49. If a package of gumdrops weighing 1.5 ounces costs 45¢, at the same rate what is the cost of 1 pound (16 ounces) of the gumdrops?

50. The ratio of boys to girls taking math is 5 to 4. How many boys are in a math class of 81 students? (*Hint:* Fill in the rest of the table.)

	Case I	Case II
Number of boys		
Number of students	9	81

51. Betty prepares a mixture of nuts that has cashews and peanuts in a ratio of 3 to 7 How many pounds of each will she need to make 40 pounds of the mixture?

52. The Local Health-Food Store is making a cereal mix that has nuts to cereal in a ratio of 2 to 7. If they want to make 126 ounces of the mix, how many ounces of nuts will they need?

53. Debra is making green paint by using 3 quarts of blue paint for every 4 quarts of yellow paint. How much blue paint will she need to make 98 quarts of green paint?

54. A concrete mix takes three bags of cement for every two bags of sand and every three bags of gravel. How many bags of cement are necessary if 80 bags of the concrete mix are needed?

55. Mario makes meatballs for his famous spaghetti sauce by using 10 pounds of ground round to 3 pounds of additives. How many pounds of ground round should he buy for 91 pounds of meatballs?

56. When $1 is worth 210 drachma (Greek currency) and a used refrigerator costs $247, what is the cost in drachmas?

57. When $1 is worth £0.65 (British pound) and a computer costs $2300, what is the cost in pounds?

673

58. When $1 is worth 1455 lire (Italian currency), a pair of shoes costs 69,840 lire. What is the cost in dollars?

59. Auto batteries are sometimes priced proportionally to the number of years they are expected to last. If a $35.85 battery is expected to last 36 months, what is the comparable price of a 60-month battery?

60. In 1960 only 6.7 of every 100 pounds of waste was recovered. In 1970, this rose to 7.1 pounds. By 1980, the amount was 9.7 pounds. In 1990 the amount was up to 13.1 pounds. Determine the amount of waste recovered from 56,000,000 pounds of waste in each of these years.

61. The amount of ozone contained in 1 cubic meter of air may not exceed 235 mg or the air is considered to be polluted. What is the greatest amount of ozone that can be contained in 12 cubic meters of air and not be considered polluted?

Exercises 62–63 relate to the Chapter 7 Application.

62. A 5.5-ounce can of Alpo cat food is priced at three cans for $1.00. A 13-ounce can is $0.59. The store manager wants to put the smaller cans on sale so that they are the same unit price as the larger cans. What price should the smaller cans be marked?

63. A large box of brownie mix that makes four batches of brownies costs $4.79 at a warehouse store. A box of brownie mix that makes one batch costs $1.29 in a grocery store. By how much should the grocery store reduce each box so that their prices are competitive with the warehouse store?

STATE YOUR UNDERSTANDING

64. What is a proportion? Write three examples of situations that are proportional.

65. Look on the label of any food package to find the number of calories in one serving. Use this information to create a problem that can be solved by a proportion. Write the solution of your problem in the same way as the examples in this section are written.

66. From a consumer's viewpoint, explain why it is not always an advantage for costs of goods and services to be proportional.

CHALLENGE

67. In 1982, approximately 25 California condors were alive. This low population was caused by losses from hunting, habitat loss, and poisoning. The U.S. Fish and Wildlife Service instituted a program that resulted in 73 condors alive in 1992. If this increase continues proportionally, predict how many condors will be alive in 2017.

68. The tachometer of a sports car shows the engine speed is 2800 revolutions per minute. The transmission ratio (engine speed to drive shaft speed) for the car is 2.5 to 1. Find the drive shaft speed.

69. Two families rented a mountain cabin for 19 days at a cost of $1905. The Santini family stayed for 8 days and the Nguyen family stayed for 11 days. How much did it cost each family? Round the rents to the nearest dollar.

GROUP ACTIVITY

70. A $13\frac{1}{2}$-ounce bag of Cheetos costs $2.69 and a 24-ounce bag costs $4.09. You have a coupon for $0.50 that the store will double. Divide the group into two teams. One team will formulate an argument that using the coupon on the smaller bag results in a better value. The other team will formulate an argument that using the coupon on the larger bag is better. Present your arguments to the whole group and select the one that is most convincing. Share your results with the rest of the class.

71. List all of the types of recycling done by you and your group members. Determine how many people participate in each type of recycling. Determine ratios for each kind of recycling. Find the population of your city or county. Using your class ratios, determine how many people in your area are recycling each type of material. Make a chart to illustrate your findings. Contact your local recycling center to see how your ratios compare to their estimates. Explain the similarities and differences.

72. Round 37.4145 to the nearest thousandth.

73. Round 37.4145 to the nearest hundredth.

74. Compare the decimals 0.00872 and 0.011. Write the result as an inequality.

75. Compare the decimals -0.06 and -0.15. Write the result as an inequality.

76. What is the total cost of 11.9 gallons of gasoline that costs \$1.379 per gallon? Round to the nearest cent.

7.4 Percent, Decimals, and Fractions

OBJECTIVES

1. Write a fraction as a percent.

2. Write a decimal as a percent.

3. Write a percent as a decimal or a fraction.

VOCABULARY

A **percent comparison** or just the **percent** is a ratio with a base unit of 100.

The **base unit** is the denominator in a ratio.

**HOW AND WHY
Objective 1**

Write a fraction as a percent.

The word "percent" means "by the hundred." It is from the Roman word *percentum.* In Rome, taxes were collected by the hundred. For example, if you had 100 cattle, the tax collector might take 14 of them in payment of your taxes. Hence, 14 per 100, or 14 percent, was the tax rate.

Using $\dfrac{1}{100}$ or % to represent per hundred, we can write these percents.

24 parts per hundred: $\dfrac{24}{100} = 24 \cdot \dfrac{1}{100} = 24\%$

100 parts per hundred: $\dfrac{100}{100} = 100 \cdot \dfrac{1}{100} = 100\%$

115 parts per hundred: $\dfrac{115}{100} = 115 \cdot \dfrac{1}{100} = 115\%$

The ratio of any two numbers, written as a fraction, can be changed to a percent. Compare 7 to 20. The ratio comparison of 7 to 20 is $\dfrac{7}{20}$. To find the percent comparison, we find the equivalent ratio with denominator, 100.

$\dfrac{7}{20} = \dfrac{35}{100} = 35 \cdot \dfrac{1}{100} = 35\%$

Not all fractions have denominators that are factors of 100. Consequently, to change a fraction, such as $\dfrac{5}{12}$, to a percent, we solve a proportion.

$\dfrac{5}{12} = \dfrac{x}{100}$ **Percent means "per hundred."**

$500 = 12x$ **Cross multiply.**

$x = \dfrac{500}{12} = 41\dfrac{2}{3}$ or $41.\overline{6}$

So, $\dfrac{5}{12} = 41\dfrac{2}{3}\%$. If we round to the nearest tenth of a percent we have $\dfrac{5}{12} \approx 41.7\%$.

To change a fraction to a percent

1. Write a proportion of the form: fraction $= \dfrac{x}{100}$.
2. Solve the proportion.
3. Write the value of x followed by the % symbol.

Examples A–D   **Warm Ups A–D**

Directions: Change the fraction to a percent.

Strategy: Build the fraction to a denominator of 100 or write and solve a proportion.

A. Change $\dfrac{4}{5}$ to a percent.

$\dfrac{4}{5} = \dfrac{4(20)}{5(20)}$ **Build so denominator is 100.**

$= \dfrac{80}{100}$

$= 80 \cdot \dfrac{1}{100}$

$= 80\%$

So, $\dfrac{4}{5} = 80\%$.

A. Change $\dfrac{3}{5}$ to a percent.

B. Change $\dfrac{7}{16}$ to a percent.

$\dfrac{7}{16} = \dfrac{x}{100}$ **Write the related proportion.**

$7(100) = 16x$ **Cross multiply.**

$700 = 16x$

$43\dfrac{3}{4} = x$ or $43.75 = x$ **Solve for x.**

So, $\dfrac{7}{16} = 43\dfrac{3}{4}\%$ or 43.75%.

B. Change $\dfrac{1}{8}$ to a percent.

C. Change $1\dfrac{5}{32}$ to a percent.

$\dfrac{37}{32} = \dfrac{x}{100}$ **Change the mixed number to a fraction and write the related proportion.**

$37(100) = 32x$

$\dfrac{3700}{32} = x$

C. Change $1\dfrac{9}{32}$ to a percent.

Answers to Warm Ups   A. 60%   B. $12\dfrac{1}{2}\%$ or 12.5%   C. $128\dfrac{1}{8}\%$ or 128.125%

$$\frac{925}{8} = x \quad \text{or} \quad 115\frac{5}{8} = x \quad \text{or} \quad 115.625 = x$$

So, $1\frac{5}{32} = 115\frac{5}{8}\%$ or 115.625%.

D. Jen buys a pair of jeans that were \$3 off the regular price of \$32.99. What percent of saving does this represent?

Strategy: We are comparing 3 to 32.99 as a percent.

$$\frac{3}{32.99} = \frac{x}{100}$$

$$3(100) = 32.99x$$

$$\frac{300}{32.99} = x$$

$$9 \approx x \quad \textbf{Rounded to the nearest percent.}$$

Jen had a saving of about 9%.

D. Tom buys a designer sweatshirt for \$5 off the regular price of \$37.95. What percent of saving does this represent?

HOW AND WHY
Objective 2

Write a decimal as a percent.

To write a decimal such as 0.3462 as a percent, the factor $\frac{1}{100}$ is introduced. If we simply multiply by $\frac{1}{100}$, the value is changed. To keep the value unchanged, multiply by $100\left(\frac{1}{100}\right)$, which has the value 1.

$$0.3462(100)\left(\frac{1}{100}\right) = 34.62\left(\frac{1}{100}\right) \qquad \textbf{Multiplying by 100 moves the decimal point two places to the right.}$$

$$= 34.62\% \qquad \textbf{Replace } \frac{1}{100} \textbf{ by \%.}$$

Study the examples in the chart.

Decimal Form	Multiply by 1 $100\left(\frac{1}{100}\right)$	Multiply	Percent Form
0.45	$0.45(100)\left(\frac{1}{100}\right)$	$45\left(\frac{1}{100}\right)$	45%
0.2	$0.2(100)\left(\frac{1}{100}\right)$	$20\left(\frac{1}{100}\right)$	20%
5	$5(100)\left(\frac{1}{100}\right)$	$500\left(\frac{1}{100}\right)$	500%

Answers to Warm Ups   D. about 13%

In each case the decimal point is moved two places to the right and the percent symbol, %, is written on the right.

 To change a decimal to a percent

1. Move the decimal point two places to the right. Write zeros on the right if necessary.
2. Write the percent symbol, %, on the right.

Examples E–H	**Warm Ups E–H**
Directions: Change the decimal to a percent.	
Strategy: Move the decimal point two places to the right and write the percent symbol.	
E. Change 0.217 to a percent. 0.217 = 21.7% **Move the decimal two places to the right and write the % symbol.**	E. Change 0.905 to a percent.
F. Change 0.008 to a percent. 0.008 = 0.8%	F. Change 0.001 to a percent.
G. Change 8 to a percent. 8 = 8.00 = 800%	G. Change 9 to a percent.
H. If the tax rate on a building lot is given as 0.03, what is the rate expressed as a percent? 0.03 = 03% = 3% **Move the decimal two places to the right and write the % symbol.**	H. If the tax rate on a commercial building is given as 0.045, what is the rate expressed as a percent?

HOW AND WHY
Objective 3   Write a percent as a decimal or a fraction.

To change a percent to a decimal, replace the percent sign by $\dfrac{1}{100}$ (one hundredth) and multiply. Observe that multiplying by $\dfrac{1}{100}$ is the same as dividing by 100, so we move the decimal point two places to the left.

$$19\% = 19\left(\dfrac{1}{100}\right) = 0.19$$

The decimal form of 19% is 0.19.

Answers to Warm Ups   E. 90.5%   F. 0.1%   G. 900%   H. 4.5%

To change a percent to a decimal

1. Move the decimal point two places to the left. Write zeros on the left if necessary.
2. Drop the percent symbol, %.

In the business world, it is not uncommon to see a percent written using a mixed number. For example, a bank might advertise an interest rate of $4\frac{3}{4}\%$. The mixed number should be written as a decimal and rounded as desired before changing the percent to a decimal.

$$4\frac{3}{4}\% = 4.75\% = 0.0475$$

 To change a percent to a fraction

Replace the percent symbol, %, by $\frac{1}{100}$, multiply, and simplify.

$$12\% = 12\left(\frac{1}{100}\right) = \frac{12}{100} = \frac{3}{25}$$

$$7\frac{1}{4}\% = 7\frac{1}{4}\left(\frac{1}{100}\right) = \frac{29}{4}\left(\frac{1}{100}\right) = \frac{29}{400}$$

$$6.65\% = 6\frac{65}{100}\% = 6\frac{13}{20}\% = 6\frac{13}{20}\left(\frac{1}{100}\right) = \frac{133}{20}\left(\frac{1}{100}\right) = \frac{133}{2000}$$

Examples I–M   **Warm Ups I–M**

Directions: Change the percent to a decimal or fraction.

Strategy: To change to a decimal, move the decimal point two places to the left and drop the percent sign. To change to a fraction, replace the percent sign by $\frac{1}{100}$ and multiply.

I. Change 14.5% to a decimal. 14.5% = 0.145 **Move the decimal point two places to the left and drop the percent symbol.**	I. Change 89.4% to a decimal.
J. Change 238% to a decimal. 238% = 2.38 **Move the decimal point two places to the left and drop the percent symbol.**	J. Change 1001% to a decimal.
K. Change 0.005% to a decimal. 0.005% = 0.00005 **Move the decimal point two places to the left and drop the percent symbol.**	K. Change 0.34% to a decimal.

Answers to Warm Ups   I. 0.894   J. 10.01   K. 0.0034

L. Change 1.22% to a fraction.

$$1.22\% = 1.22\left(\frac{1}{100}\right)$$ **Replace the % by $\frac{1}{100}$.**

$$= \frac{61}{50}\left(\frac{1}{100}\right)$$ $1.22 = 1\frac{22}{100} = 1\frac{11}{50} = \frac{61}{50}.$

$$= \frac{61}{5000}$$

L. Change 89.4% to a fraction.

M. Change $3\frac{5}{8}\%$ to a fraction.

$$3\frac{5}{8}\% = 3\frac{5}{8}\left(\frac{1}{100}\right)$$ **Replace the % by $\frac{1}{100}$.**

$$= \frac{29}{8}\left(\frac{1}{100}\right)$$

$$= \frac{29}{800}$$

M. Change $66\frac{2}{3}\%$ to a fraction.

Answers to Warm Ups   L. $\frac{447}{500}$ M. $\frac{2}{3}$

Exercises 7.4

OBJECTIVE 1: *Write a fraction as a percent.*

A.

Change to a percent.

1. $\dfrac{19}{100}$

2. $\dfrac{82}{100}$

3. $\dfrac{1}{2}$

4. $\dfrac{1}{4}$

5. $\dfrac{7}{10}$

6. $\dfrac{9}{10}$

7. $\dfrac{9}{25}$

8. $\dfrac{13}{25}$

B.

9. $\dfrac{2}{5}$

10. $\dfrac{4}{5}$

11. $\dfrac{18}{25}$

12. $\dfrac{23}{25}$

13. $\dfrac{3}{8}$

14. $1\dfrac{7}{8}$

OBJECTIVE 2: *Write a decimal as a percent.*

A.

Change to a percent.

15. 0.37

16. 0.81

17. 2.34

18. 3.61

19. 1.4

20. 9.2

21. 2

22. 3

B.

23. 0.052

24. 0.023

25. 0.1025

26. 0.0775

27. 0.001

28. 0.005

29. 0.0001

30. 0.0009

Write each fraction as a decimal rounded to the nearest thousandth. Change the decimal to a percent.

31. $\dfrac{2}{3}$

32. $\dfrac{5}{6}$

33. $\dfrac{5}{12}$

34. $\dfrac{11}{12}$

35. $\dfrac{4}{7}$

36. $\dfrac{2}{7}$

OBJECTIVE 3: *Write a percent as a decimal or a fraction.*

A.

Change to a decimal.

37. 17%

38. 79%

39. 2.35%

40. 1.08%

41. 315%

42. 525%

43. 0.12%

44. 0.9%

45. 0.25%

46. 0.60%

47. 1%

48. 100%

Change to a simplified fraction.

49. 37%

50. 49%

51. 10%

52. 35%

B.

Change to a decimal.

53. 4.756%

54. 2.94%

55. $3\dfrac{4}{5}\%$

56. $4\dfrac{1}{5}\%$

57. $46\dfrac{3}{8}\%$

58. $85\dfrac{5}{8}\%$

Change to a simplified fraction.

59. $33\dfrac{1}{3}\%$

60. $66\dfrac{2}{3}\%$

61. $15\dfrac{2}{3}\%$

62. $17\dfrac{1}{3}\%$

63. 0.02%

64. 0.50%

C.

65. In an algebra class, 28 students out of 40 took arithmetic before taking algebra. What percent took arithmetic?

66. Serafina got 21 problems correct on a 25-problem test. What percent were correct?

67. If the state income tax rate on Kyle's income is 0.022, what is the rate expressed as a percent?

68. A class has completed 62% of a group project. Express this as a decimal.

69. A solution in the chemistry lab contains 18 parts of acid to 100 parts of the solution. Write the comparison as a percent.

70. A bank pays $5.25 interest per year for every $100 in savings. What is the annual interest rate?

71. If 0.375 of the contestants in a race withdraw, what is this expressed as a percent?

72. When bidding for a job, an estimator adds 10% to cover unexpected expenses. What decimal part is this?

73. Jonna is paid a 7.5% rate of commission. What decimal is used to compute the amount of her commission?

74. Huong is paid a 9.3% rate of commission. What decimal is used to compute the amount of her commission?

75. During the blizzard of 1977, the price of a snow blower increased by a factor of 1.87. Express this as a percent.

76. Mary Ellen measured the July rainfall. At the end of the year she found that it was 0.235 of the year's total. Express this as a percent.

77. Unemployment is down 0.3%. Express this as a decimal.

78. In industrialized countries, 60% of river pollution is due to agricultural runoff. Change this to a decimal.

79. The cost of living rose 0.6% during November. Express this as a decimal.

80. In 1993 Americans had 4% less blood cholesterol than in 1981. Change this to a decimal.

81. The price of gasoline at Colexo Service Station is currently three times the price it was 10 years ago. What percent of the price 10 years ago is the price today?

82. In the United States, one out of every two women is on a diet at any given time. What percent of women are on a diet?

83. A grilled chicken sandwich contains 7 grams of fat. Each gram of fat contains 9 calories. If the entire sandwich contains 290 calories, what percent of the calories come from the fat content? Round to the nearest percent.

84. In 1979, one area of California had 188 smoggy days. What was the percent of smoggy days rounded to the nearest tenth of a percent? In 1991 there were only 129 smoggy days. What was the percent that year rounded to the nearest tenth of a percent? Compare these percents and discuss the possible reasons for the change.

85. Jorge spends 42% of his salary on his car each month. What fractional part is spent on Jorge's car?

86. The enrollment at City Community College this year is 118% of last year's enrollment. What fractional increase in enrollment took place this year?

STATE YOUR UNDERSTANDING

87. What is a percent? How is it related to fractions and decimals?

88. Explain why not all fractions can be changed to a whole number percent. What is special about those fractions that can be changed to a whole-number percent?

89. Every percent can be written as a decimal. Write two sentences that illustrate the use of percent. Rewrite each sentence replacing the percent with the equivalent decimal. Which form best communicates the intent of your sentences?

CHALLENGE

90. Change $\dfrac{67}{500}$ to both a decimal and a percent.

91. Change $24\dfrac{7}{8}\%$ to both a decimal and a fraction.

92. Change 0.00025 to both a percent and a fraction.

GROUP ACTIVITY

93. Use your library resources to find some history of the use of the percent symbol (%). What is the earliest date you can find for its use?

MAINTAIN YOUR SKILLS (Sections 6.2, 6.8)

94. Combine terms: $-3.07a - 2.77a - (-0.5a) - 3.9a$

95. Combine terms: $7.653t^2 + 89.603t + (-1.863t^2) + (-27.536t)$

96. Solve: $0.03t - 0.19t + 4.5 = 1.5$

97. Solve: $1.25y + 3.07 = 7.57$

98. Solve: $7.006 - 0.3x + 0.2x = 6.951$

7.5   Solving Percent Problems

OBJECTIVE   Solve percent problems.

VOCABULARY   In the statement "*R* of *B* is *A*":

R is the **rate** of percent.

B is the **base** unit and follows the words "percent of."

A is the **amount** that is compared to *B*.

HOW AND WHY
Objective

Solve percent problems.

To solve a percent problem we need to identify *A*, *R*, and *B*. For example, in the sentence "30% of 60 is 18,"

R is 30% since it is the percent and includes the percent symbol.

B is 60 since it follows the word "of."

A (sometimes called the *percentage*) is 18, the number compared to *B*.

The formula for percent is

$$RB = A$$

where we understand the percent symbol to be part of *R*. The formula is based on the fact that the word "of" in mathematics usually indicates multiplication and the word "is" indicates equality.

 To solve a percent problem using the formula $RB = A$

1. Identify the values of the variables in the formula that are given. You need to know two of the values to find the third value.
2. Substitute the given values in the formula. Use the decimal form of the percent when substituting for *R*.
3. Solve the resulting equation.

For example, 25% of 60 is what number?

$R = 25\% = 0.25$	**R is the percent. Change to decimal form.**
$B = 60$	**B follows the word "of."**
$A = ?$	**A is the unknown number.**
$RB = A$	**Basic formula.**
$0.25(60) = A$	**Substitute 0.25 for R and 60 for B.**
$15 = A$	**Simplify.**

Thus, 25% of 60 is 15.

The following examples illustrate other situations.

| **Examples A–D** | **Warm Ups A–D** |

Directions: Solve a percent problem using the formula.

Strategy: Write the percent formula, $RB = A$. Substitute the known values and solve.

A.   45% of what number is 9?

$R = 45\% = 0.45$ **R is the percent. Change to decimal form.**

$B = \,?$ **The base (follows "of") is unknown.**

$A = 9$ **We're comparing 9 to B.**

$0.45B = 9$ **Substitute in the formula $RB = A$.**

$B = 20$ **Divide both sides by 0.45.**

So, 45% of 20 is 9.

A.   75% of what number is 9?

B.   5 is what percent of 55? Round to the nearest whole percent.

$R = \,?$ **The percent is unknown.**

$B = 55$ **The base follows "of."**

$A = 5$

$R(55) = 5$ **Substitute in the formula $RB = A$.**

$R = \dfrac{5}{55} = \dfrac{1}{11}$ **Divide both sides by 55.**

$R \approx 0.09$ **Round to the nearest hundredth.**

So, 5 is approximately 9% of 55.

B.   11 is what percent of 55?

C.   78% of 36 is what?

$0.78(36) = A$ **Substitute 0.78 for R and 36 for B in the formula $RB = A$.**

$28.08 = A$

Thus, 78% of 36 is 28.08.

C.   48% of 98 is what?

D.   50 is what percent of 180? Round to the nearest tenth of a percent.

Strategy: Round the decimal value of R to the nearest thousandth because $\dfrac{1}{10}$ of 1% is $\dfrac{1}{10}\left(\dfrac{1}{100}\right) = \dfrac{1}{1000}$.

$R(180) = 50$ **Substitute 180 for B and 50 for A.**

$R = \dfrac{50}{180}$ **Divide both sides by 180.**

$R \approx 0.278$ **Round to the nearest thousandth.**

$R \approx 27.8\%$ **Change the decimal to percent.**

Therefore, 50 is 27.8% of 180 to the nearest tenth of a percent.

D.   78 is what percent of 162? Round to the nearest tenth of a percent.

Answers to Warm Ups   A. 12   B. 20%   C. 47.04   D. 48.1%

Exercises 7.5

OBJECTIVE: *Solve percent problems.*

A.

Solve.

1. 9 is 50% of _____?_____ .

2. 9 is 90% of _____?_____ .

3. 3 is _____?_____ % of 1.

4. 8 is _____?_____ % of 4.

5. _____?_____ % of 60 is 30.

6. _____?_____ % of 65 is 13.

7. 70% of _____?_____ is 28.

8. 80% of _____?_____ is 28.

9. 80% of 45 is _____?_____ .

10. _____?_____ is 80% of 25.

11. 64 is _____?_____ % of 80.

12. _____?_____ % of 56 is 14.

13. 19% of _____?_____ is 19.

14. 16 is _____?_____ % of 16.

15. 0.5% of 200 is _____?_____ .

16. 0.25% of 800 is _____?_____ .

17. 70% of 40 is _____?_____ .

18. 40% of 70 is _____?_____ .

© 1997 Saunders College Publishing.

B.

19. 140% of ___?___ is 56.

20. 175% of ___?___ is 66.5.

21. 9.5% of 40 is ___?___ .

22. 8.5% of 40 is ___?___ .

23. 0.4 is ___?___ % of 40.

24. 0.6 is ___?___ % of 20.

25. 78 is 65% of ___?___ .

26. 45 is 25% of ___?___ .

27. 10 is ___?___ % of 1000.

28. 1 is ___?___ % of 1000.

29. 48% of 45 is ___?___ .

30. 74% of 80 is ___?___ .

31. 37% of ___?___ is 222.

32. 76% of ___?___ is 380.

33. 145% of ___?___ is 66.7.

34. 165% of ___?___ is 95.7.

35. 66 is ___?___ % of 125.

36. 92 is ___?___ % of 160.

C.

37. 14.4% of 75 is ___?___ .

38. 26.3% of 80 is ___?___ .

39. 3.6% of 0.8 is ___?___ .

40. ___?___ is 6.4% of 0.5.

41. ___?___ is $\frac{1}{2}$% of 0.4.

42. ___?___ is $\frac{1}{4}$% of 0.2.

43. $11\frac{1}{9}$% of 1548 is ___?___ .

44. $22\frac{2}{9}$% of 1458 is ___?___ .

45. $16\frac{2}{3}$% of 4002 is ___?___ .

46. $83\frac{1}{3}$% of 1002 is ___?___ .

Solve. Write the answer as shown.

47. 1.25% of 1250 is ___?___ . (nearest tenth)

48. 3.85% of 750 is ___?___ . (nearest tenth)

49. 16% of ___?___ is 9.3. (nearest tenth)

50. 23% of ___?___ is 10.5. (nearest tenth)

51. ___?___ % of 75 is 8. (nearest tenth of one percent)

52. ___?___ % of 92 is 56. (nearest tenth of one percent)

53. 38 is ___?___ % of 28. (nearest whole-number percent)

54. 47.4 is ___?___ % of 11.6. (nearest whole-number percent)

55. $5\frac{1}{3}$% of $6\frac{1}{2}$ is ___?___ . (as a fraction)

56. $6\frac{1}{5}$% of $8\frac{3}{4}$ is ___?___ . (as a fraction)

STATE YOUR UNDERSTANDING

57. Explain how to find fifteen percent of a number. Explain how to find what percent of a number is thirty-eight.

58. Explain the errors in the following statements: "Slewhind Manufacturing charges $4.00 for a tool that cost them $1.00 to make. They are making 4% profit on each tool."

CHALLENGE

59. $\frac{1}{2}$% of $33\frac{1}{3}$ is what fraction?

60. $\frac{2}{5}$% of $66\frac{2}{3}$ is what fraction?

61. $\frac{9}{10}$% of what fraction is $2\frac{1}{2}$?

62. $\frac{4}{3}$% of what fraction is $\frac{7}{8}$?

GROUP ACTIVITY

63. Divide up the task of computing these percents: 45% of 37; 37% of 45; 18% of 80; 80% of 18; 130% of 22; 22% of 130; 0.6% of 5.5; 5.5% of 0.6. Compare your answers and together write up a statement about the answers.

64. Divide up the task of computing these percents: 30% of the number that is 80% of 250; 80% of the number that is 30% of 250; 60% of the number that is 20% of 340; 20% of the number that is 60% of 230; 150% of the number that is 200% of 40; 200% of the number that is 150% of 40. Compare your answers and together write up a statement about the answers.

MAINTAIN YOUR SKILLS (Sections 6.3, 6.5, 6.8)

65. Find the product of 27.82 and -3.096.

66. Find the product of -2.6, 0.92, and -17.5.

67. Find the quotient: $0.016\overline{)0.5792}$

68. Round the quotient of 17 and 143 to the nearest thousandth.

69. Solve: $16t - 18.4t = 5.76$

7.6 Applications of Percents

OBJECTIVES

1. Solve percent word problems.

2. Solve business and personal finance percent problems.

3. Read data from a circle graph or construct a circle graph from data.

VOCABULARY

The **percent of increase** is the value of R when a quantity, B, is increased by an amount, A.

The **percent of decrease** is the value of R when a quantity, B, is decreased by an amount, A.

A **circle graph** or **pie chart** illustrates a whole unit or a total divided into parts or percents. Each of the parts or percents is represented by a **sector** (pie-shaped piece) of the graph. The entire circle represents 100%.

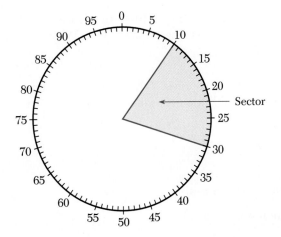

HOW AND WHY
Objective 1

Solve percent word problems.

When a word problem is translated to the simpler word form, "What percent of what is what?" or "R of B is A", the unknown value can be found using the percent formula. For instance, "What percent of the loan is the interest?" See Example A.

Examples A–F   **Warm Ups A–F**

Directions: Solve the percent word problem.

Strategy: Write the problem in the word form, "What percent of what is what?" or "R of B is A." Fill in the known values and solve for the unknown.

A.   Terry bought a motorcycle with a 15% one-year loan. The interest payment is $75. How much is his loan?

   15% of what is $75?   **The $75 interest is 15% of the loan.**

   $0.15B = 75$ **Substitute 0.15 for R and 75 for A**
                          in the formula RB = A.
   $B = \dfrac{75}{0.15}$

   $B = 500$

   Terry's loan is for $500.

A.   James buys a motorcycle with a 14% one-year loan. If the interest payment is $112, how much is the loan?

B.   The population of Lane County is 160% of what it was 10 years ago. The population 10 years ago was 117,000. What is the population now?

Strategy: The current population is 160% of the population 10 years ago. So the *base* is the old population and the *amount* is the new population.

   160% of 117,000 is what?

   $1.60(117,000) = A$

   $187,200 = A$ **Simplify.**

   The population is now 187,200.

B.   The cost of a car is now 140% of what it was four years ago. What is the cost today of a car that sold for $6850 four years ago?

C.   The Goliath Bakery has 500 loaves of day-old bread it wants to sell. If the price was originally $1.19 a loaf and is reduced to 71¢ a loaf, what percent discount, based on the original price, should the baker advertise?

Strategy: The discount is the difference between the original price and the sale price. The problem can be stated: "What percent of the original price is the discount?" The *base* is the original price and the *amount* is the discount.

   What percent of 1.19 is 0.48?   **The discount is**
   $R(1.19) = 0.48$ **$1.19 − $0.71 = $0.48.**

   $R = \dfrac{0.48}{1.19} \approx 0.403361$ **An approximate decimal**
                                      form of the percent.

   $R \approx 40.3\%$ **Change to percent and round**
                                   to the nearest tenth.

   The bakery will probably advertise "40% off."

C.   The Goliath Bakery has 200 packages of day-old buns it wants to sell. If the price was originally $1.23 per package and it is reduced to 87¢ per package, what percent discount, based on the original price, should the bakery advertise?

D. The student newspaper polls a group of students. Four of them say they walk to school, 14 others say they ride the bus, 20 drive in car pools, and 6 drive their own cars. What percent of the group ride the bus?

Strategy: There are 44 students in the group so the base is 44 and 14 ride the bus so the amount is 14.

What percent of 44 is 14?

$R(44) = 14$

$R = \dfrac{14}{44} \approx 0.318$ **Round to the nearest thousandth.**

$R \approx 31.8\%$ **Change to percent.**

About 32% of the group ride the bus.

D. A list of the grades in a math class reveals that 7 students received A's, 15 received B's, 23 received C's, and 5 received D's. What percent of the students received a grade of B?

E. Find the percent of increase from 360 to 432.

Strategy: The difference, 72, is the *amount* of increase from 360 to 432. The percent of increase is the increase, 72, compared to the base, which is 360, the initial quantity.

What percent of 360 is 72?

$R(360) = 72$ **Substitute.**

$R = \dfrac{72}{360} = 0.2$

$R = 20\%$

There is a 20% increase from 360 to 432.

E. What is the percent of increase from 360 to 450?

F. Find the percent of decrease from 556 to 361.4.

Strategy: The difference, 194.6, is the *amount* of decrease from 556 to 361.4. The percent of decrease is the decrease, 194.6, compared to the base, which is 556, the initial quantity.

What percent of 556 is 194.6?

$R(556) = 194.6$ **Substitute.**

$R = \dfrac{194.6}{556} = 0.35$

$R = 35\%$

There is a 35% decrease from 556 to 361.4.

F. What is the percent of decrease from 450 to 360?

HOW AND WHY
Objective 2   **Solve business and personal finance percent problems.**

Businesses and individuals use percent in a variety of ways. Among these are percent of markup, percent of discount, depreciation, interest rates, taxes, salary increases, and commissions. The new words and phrases are explained in the next set of examples.

Examples G–K

Warm Ups G–K

Directions: Solve the business-related word problem.

Strategy: Write the problem in the word form, "What percent of what is what?" or "*R* of *B* is *A*." Fill in the known values and solve for the unknown.

G. The cost of an electric iron is $18.50. The markup is 30% of the cost. What is the selling price of the iron?

Strategy: *Markup* is the number added to the cost of an article so the store can pay its expenses and make a profit. Let *M* represent the markup.

30% of the cost is what?

30%($18.50) = *M*

0.30($18.50) = *M*

$5.55 = *M*

The markup is $5.55

S.P. = cost + markup   **Add the cost and markup to find the selling price.**

S.P. = $18.50 + $5.55

S.P. = $24.05

The selling price is $24.05.

G. The cost of a coffee maker is $28.50. The markup is 40% of the cost. What is the selling price of the coffee maker?

H. The regular price of a personal stereo is $29.98. What is the sale price of the stereo if it is discounted 30%? Round to the nearest cent.

Strategy: The *amount of discount, D,* is the amount subtracted from the regular price. The *percent of discount* is 30%. Find the discount and then the sale price.

30% of the regular price is the discount.

30%($29.98) = *D*

0.30($29.98) = *D*

$8.994 = *D*

The amount of discount is $8.994.

Sale price = $29.98 − $8.994

= $20.986 ≈ $20.99

The sale price is $20.99.

H. The price of a CD player is $179.95. What is the sale price of the player if the percent of discount is 18%? Round to the nearest cent.

I. A toaster-oven is priced to sell for $29.95. The markup is 40% of the selling price. What is the cost of the oven?

Strategy: *Cost* is the amount the store pays for an item, not what the customer pays. Let *M* be the markup.

I. A camera is priced to sell for $189.95. If the markup is 40% of the selling price, what is the cost of the camera?

Answers to Warm Ups   G. $39.90   H. $147.56   I. $113.97

40% of the selling price is the markup.

40%($29.95) = M

0.40($29.95) = M

$11.98 = M

The markup is $11.98

Cost = $S.P.$ − markup

  = 29.95 − 11.98

  = 17.97

The stores' cost for the toaster-oven is $17.97.

J.   A phone company buys a new truck that costs $15,000. During the first year, it will depreciate $12\frac{1}{2}$% of its original value. What will its value be at end of the year?

Strategy: *Depreciation* is the name given to the amount of decrease in value caused by age or use. Let D represent the depreciation.

$12\frac{1}{2}$% of the cost is the depreciation.

$12\frac{1}{2}$%($15,000) = D

0.125($15,000) = D

   $1875 = D

The depreciation is $1875.

Value = $15,000 − $1875 = $13,125

The value of the truck at year's end is $13,125.

J.   The phone company buys a new van for $28,530. If the depreciation rate for the van is $11\frac{1}{9}$%, what will its value be at the end of the year?

K.   Jean's rate of pay is $8.48 per hour. She gets time and one-half for each hour over 40 hours worked in 1 week. What are her earnings if she works for 46.5 hours 1 week?

Strategy: Time and one-half means that she will earn 1.5 times or 150% of her regular hourly wage.

150% of the regular pay is the overtime.

150%($8.48) = overtime

1.50($8.48) = overtime

   $12.72 = overtime

Earnings = 40($8.48) + 6.5($12.72)   **Add the overtime pay to the pay for the first 40 hours.**

Earnings = $339.20 + $82.68

Earnings = $421.88

Jean earns $421.88 for the week.

K.   What are Jean's earnings for a week in which she works 49.75 hours after her pay is raised to $9.50 per hour?

Answers to Warm Ups   J. $25,360   K. $518.94

Read data from a circle graph or construct a circle graph from data.

A circle graph or pie chart is used to show how a whole unit is divided into parts. The area of the circle represents the entire unit, and each subdivision is represented by a sector. Percents are often used as the unit of measure of the subdivision. Consider the following pie chart.

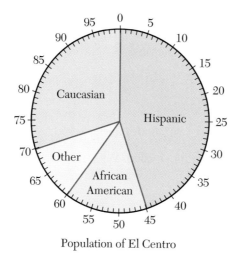

Population of El Centro

From the circle graph we can conclude:

1. The largest ethnic group in El Centro is Hispanic.

2. The Caucasian population is twice the African-American population.

3. The African-American and Hispanic populations are 60% of the total.

If the population of El Centro is approximately 125,000, we can also compute the approximate number in each group. For instance, the number of Hispanics is found by:

$$RB = A$$
$$45\%(125{,}000) = A$$
$$0.45(125{,}000) = A$$
$$56{,}250 = A$$

There are approximately 56,250 Hispanics in El Centro.

To construct a circle graph, determine what fractional part or percent each subdivision is compared to the total. Then draw a circle and divide it accordingly. We can draw a pie chart of the data in the following table.

Age-groups	0–21	22–50	over 50
Population	14,560	29,120	14,560

Begin by adding another row and column to the data table.

Age-groups	0–21	22–50	over 50	Total
Population	14,560	29,120	14,560	58,240
Fractional part or percent	$\dfrac{1}{4}$ 25%	$\dfrac{1}{2}$ 50%	$\dfrac{1}{4}$ 25%	1 100%

The third row is computed by writing each age-group as a fraction of the total population and simplifying. For example, the 0–21 age-group is

$$\frac{14{,}560}{58{,}240} = \frac{1456}{5824} = \frac{364}{1456} = \frac{1}{4} \quad \text{or} \quad 25\%$$

Now draw the circle graph and label it.

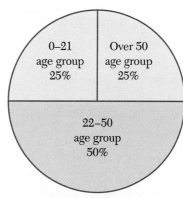

Group distribution by age

Example L   **Warm Up L**

Directions: Answer the questions associated with the graph.

Strategy: Examine the graph to determine the size of the related sector.

L. The sources of City Community College's revenue are displayed in the circle graph.

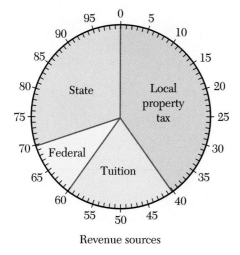

Revenue sources

L. The sales of items at Grocery Mart are displayed in the circle graph.

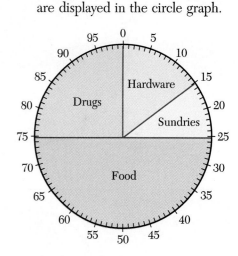

1. What percent of the revenue is from the federal government?

2. What percent of the revenue is from tuition and property taxes?

3. What percent of the revenue is from the federal and state governments?

1. 10%   Read directly from the graph.

2. 60%   Add the percents for tuition and taxes.

3. 40%   Add the percents for state and federal sources.

1. What is the area of highest sales?

2. What percent of total sales is from sundries and drugs?

3. What percent of total sales is from food and hardware?

Answers to Warm Up   L. (1) food; (2) 35%; (3) 65%

| **Example M** | **Warm Up M** |

Directions: Draw a circle graph to display the given data.

Strategy: Determine the percent, to the nearest whole percent, that is represented by each given quantity. Divide the circle based on the percents. Label the graph.

M. The Florida Panthers hockey team had a record of 33 wins, 25 losses, and 17 ties at one point in the 1997 NHL season. Display their record in a circle graph.

Strategy: Use a chart to organize the data.
Round to nearest whole percent.

	Wins	Losses	Ties	Total
Number	33	25	17	75
Fraction	$\frac{33}{75}$	$\frac{25}{75}$	$\frac{17}{75}$	$\frac{75}{75}$
Percent	44%	33%	23%	100%

Draw the circle graph and label it.

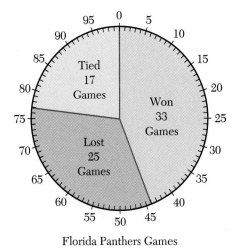

Florida Panthers Games

M. The Ottawa Senators hockey team had a record of 25 wins, 34 losses, and 15 ties at one point in the 1997 NHL season. Display their record in a circle graph.

Answer to Warm Up   M.

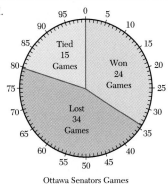

Ottawa Senators Games

Exercises 7.6

OBJECTIVE 1: *Solve percent word problems.*

---

1. If there is a 4% sales tax on a television set costing $129.95, how much is the tax?

2. Dan bought a used motorcycle for $955. He made a down payment of 18%. How much cash did he pay as a down payment?

3. Last year Joan had 14% of her salary withheld for taxes. If the total amount withheld was $2193.10, what was Joan's yearly salary?

4. The manager of a fruit stand lost $16\frac{2}{3}$% of his bananas to spoilage and sold the rest. He discarded four boxes of bananas in 2 weeks. How many boxes did he have in stock?

5. Maria's base rate of pay is $7.82 per hour. She receives time and a half for all hours over 40 that she works in 1 week. What were her total earnings if last week she worked a total of 47 hours?

6. Truong's base rate of pay is $6.72 per hour. He receives time and a half for all hours over 40 that he works in 1 week. What were his earnings if last week he worked a total of 42 hours?

7. John got 25 problems correct on a 30-problem test. What was his percent score to the nearest whole-number percent?

8. To pass a test to qualify for a job interview, Whitney must score at least 70%. If there are 40 questions on the test, how many must she get correct to score 70%?

9. Eddie and his family went to a restaurant for dinner. The dinner check was $33.45. He left the waiter a tip of $5. What percent of the check was the tip, to the nearest whole-number percent?

10. Adams High School's basketball team finished the season with a record of 15 wins and 9 losses. What percent of the games played were won?

11. The town of Verboort has a population of 15,560, which is 45% male. Of the men, 32% are 40 years or older. How many men are there in Verboort who are younger than 40?

12. If you use a 35¢ coupon to buy a $2.89 box of ice cream sandwiches, what percent off is this, to the nearest whole-number percent?

13. A store advertises $20 off every coat in stock. What percent savings is this on a coat that is regularly $72.99, to the nearest tenth of 1%?

14. Good driving habits can increase mileage and save on gas. If good driving causes a car's mileage to go from 31.5 mpg to 35 mpg, what is the percent of increased mileage? Round to the nearest whole-number percent.

15. According to the Bureau of Labor, there were 90,000 physical therapists in the United States in 1992. Physical therapy is one of the fastest growing occupations, with the Census Bureau predicting an 88% increase over 1992 levels by the year 2005. How many physical therapists are predicted by the year 2005? Round to the nearest thousand.

16. Human services workers are growing at an even faster rate than physical therapists. The Census Bureau is predicting a 135% increase over 1992 levels by the year 2005. How many human services workers are predicted in the year 2005 if there were 189,000 in the United States in 1992? Round to the nearest thousand.

17. For customers who use a bank's credit card, there is a $1\frac{3}{4}$% finance charge on monthly accounts that have a balance of $400 or less. Merle's finance charge for August was $2.80. What was the amount of her account for August?

18. A fast-food hamburger has 342 calories in fat content. This is approximately 54.3% of the total number of calories in the hamburger. How many calories are in one of the hamburgers? Round to the nearest calorie.

19. In 1980, 64% of all workers drove to work alone. In 1990, 73% drove alone. Find the population of your community and compute the number of single-car drivers for both of these years.

20. Find the percent of the students in your class who drove alone to class today. Compare this percent with the percents in Exercise 19. How does the class compare to the national average for both years?

OBJECTIVE 2: *Solve business and personal finance percent problems.*

---

21. An article that costs the store owner $16.80 is marked up $5.04. What is the percent of markup based on the cost?

22. A tool that costs a hardware merchant $8.40 is marked up 30% of the cost. What is the selling price?

23. An article is priced to sell at $39.99. If the markup is 28% of the selling price, how much is the markup to the nearest cent?

24. The Bright TV Store regularly sells a television set for $329.95. An advertisement in the paper shows that it is on sale at a discount of 25%. What is the sale price to the nearest cent?

25. A competitor of the store in Exercise 24 has the same TV set on sale. The competitor normally sells the set for $335.95 and has it advertised at a 27% discount. To the nearest cent, what is the sale price of the TV? Which is the better buy and by how much?

26. A salesman earns a 9% commission on all of his sales. How much did he earn last week if his total sales were $5482?

27. If the salesman in Exercise 26 received a 12% commission on all sales, and his total sales for one week were $4725, what were his earnings for the week?

28. The Top Company offered a 6% rebate on the purchase of their best model of canopy. If the regular price is $398.98, what is the amount of the rebate to the nearest cent?

29. The taxes on a piece of property are $2238.30, and the property has an assessed value of $82,900. What is the percent tax rate in the district where the property is located?

Exercises 30–37 relate to the Chapter Application.

30. Corduroy overalls that are regularly $34.99 are on sale for 25% off. What is the sale price of the overalls?

31. A pair of Reebok cross trainers, which regularly sells for $59.99, goes on sale for $47.99. What percent off is this? Round to the nearest whole number percent.

32. A bag of Tootsie Rolls is marked "20% more free—14.5 oz for the price of 12 oz." Assuming there has not been a change in price, is the claim accurate? Explain.

33. A department store puts a blazer, which was originally priced at $139.95, on sale for 20% off. At the end of the season, the store has an "additional 40% off everything that is already reduced" sale. What is the price of the blazer? What percent saving does this represent over the original price?

34. A shoe store advertises "Buy one pair, get 50% off a second pair of lesser or equal value." The mother of twin boys buys a pair of basketball shoes priced at $36.99 and a pair of hikers priced at $27.99. How much did she pay for the two pairs of shoes? What percent savings is this to the nearest tenth of a percent?

35. The Klub House advertises on the radio that all merchandise is on sale at 25% off. When you go in to buy a set of golf clubs that originally sold for $279.95, you find that the store is giving an additional 10% discount off the original price. What is the price you will pay for the set of clubs?

36. In Exercise 35, if the salesperson says that the 10% discount can only be applied to the sale price, what is the price of the clubs?

37. A store advertises "30% off all clearance items." A boy's knit shirt is on a clearance rack that is marked 20% off. How much is saved on a knit shirt that was originally priced $14.99?

38. Three people resolve to become business associates. The first associate pays 36% for a share of a franchise, the second associate pays 29%, and the third associate pays for the remaining 35% of the franchise. The profits are to be divided among the three according to each associate's share of the ownership. The profits for last month were $5274. How much does each associate receive?

39. In 1990, use of office paper resulted in over 11 tons of solid waste. Of this amount, 15% was recycled. How many tons were recycled?

40. A labor union negotiated a contract with the Swift Copier Company. The average wage under the new contract is $4\frac{3}{4}$% higher. The average wage under the old contract was $9.24 per hour. What is the average wage under the new contract?

OBJECTIVE 3: *Read data from a circle graph or construct a circle graph from data.*

For Exercises 41–42. The circle graph below shows ethnic distribution of children enrolled in Head Start.

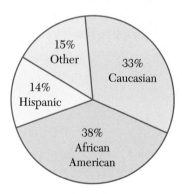

41. Which group has the smallest number of children in Head Start?

42. What percent of children in Head Start are non-Caucasian?

For Exercises 43–46, the following circle graph shows how dollars are spent in a particular industry.

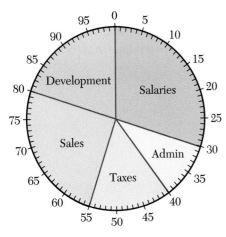

43. What item has the greatest expenditure of funds?

44. Which is more costly, development or taxes?

45. If the total expenditures of the industry were $2,500,000, how much was spent on development?

46. Using the total expenditures in Problem 45, how much was spent on salaries?

47. In a family of three children, there are eight possibilities of boys and girls. One possibility is that they are all girls. Another possibility is that they are all boys. There are three ways for the family to have two girls and a boy. There are also three ways for the family to have two boys and a girl. Make a circle graph to illustrate these possibilities.

48. According to one state official, 48% of all households have no firearms, 12% of all households store firearms unloaded and ammunition locked up, 10% of all households keep firearms loaded and unlocked, and 30% store firearms another way. Make a circle graph to illustrate these possibilities.

49. The major causes of death worldwide are listed below. Make a circle graph to illustrate this information.

Infectious and parasitic diseases	32%
Heart, circulatory diseases, & stroke	19%
Unknown causes	16%
Cancer	12%
Accidents and violence	8%
Infant death	6%
Chronic lung diseases	6%
Other causes	less than 1%

STATE YOUR UNDERSTANDING

50. Explain why the percent of increase from 500 to 750 (50%) is not the same as the percent of decrease from 750 to 500 $\left(33\frac{1}{3}\%\right)$.

51. Explain why a 10% increase in $100 followed by a 10% decrease does *not* yield $100.

52. Write a few sentences clarifying the following statements: (a) "a price increase of 100% is double the original price," and (b) "an enlargement of a floor area by 200% of the original is triple the original area."

CHALLENGE

53. The markup on furniture at NuMart is 30% based on the selling price. The cost of a lounge chair is $58. What is the selling price?

54. A retailer buys a shipment of athletic shoes for $34.67 a pair and sells them for $49.99 a pair. On the next shipment the cost is increased 8%, and they, in turn, increase the selling price to $54.99. Is the percent of markup more or less after the price increase? By how much, rounded to the nearest tenth of a percent?

55. During a 6-year period, the cost of maintaining a diesel engine averages $2650 and the cost of maintaining a gasoline engine averages $4600. What is the percent of savings of the diesel compared with the gasoline engine? Round to the nearest whole percent.

56. Carol's baby weighed $7\frac{1}{2}$ pounds when he was born. On his first birthday he weighed $23\frac{3}{4}$ pounds. What was the percent of increase during the year? Round to the nearest whole percent.

GROUP ACTIVITY

57. During one month in 1991, the interest rate on a 30-year home mortgage averaged 9.29%. A 15-year mortgage averaged 8.98%. Two years later, the interest rate on a 30-year mortgage averaged 6.5%. Find the difference in simple interest costs for 1 year on houses that cost $85,000, $97,000, $115,000, and $166,000. Find the assessed value of your own dwelling and then compute the costs at these different rates. Determine the average interest for the members of your group and for the class. Compare the values with each other, with the other groups, and with the values for 1991 and 1993.

MAINTAIN YOUR SKILLS (Section 6.8)

Solve.

58. $2.1z - 5.6 + 3.4z = -3.4$

59. $9.54 + 4t - 3.6t = 3.58$

60. $2.1 - 2.5w - 3w = -5.6$

61. $0.34y + 0.55 + 0.76y - 0.3y = -0.32$

62. $0.11x - 1.01 + 0.01x = 0.202$

CHAPTER 7

OPTIONAL Group Project *(1–2 weeks)*

Much research has been conducted in the area of human nutrition, and yet there are still many unanswered questions. Most nutritionists agree, however, that the American diet has too much fat and refined carbohydrates (primarily sugar).

You are about to become a nutritional consultant to John and Jane Doe. John is a 32-year-old male, 5'10", 165 lb, and consumes 2800 calories per day. Jane is a 31-year-old female, 5'5", 130 lb, and consumes 2000 calories per day. Both exercise regularly and try to be careful about their diet. However, neither of them likes to cook, and so they eat out a lot.

Experts disagree about the exact ratio of carbohydrates, protein, and fat in an optimum diet. Some recommend a 40% carbohydrate, 30% protein, 30% fat ratio. Others would shift the amounts to 50%, 25%, and 25%, respectively. Almost no one recommends more than 30% fat.

All food sold in grocery stores is required by the government to include nutritional labeling. Among other things, the label specifies the total calories per serving and the number of grams of carbohydrates, protein, and fat per serving. Most fast-food chains also provide this information. A gram of carbohydrate and a gram of protein each contain about 4 calories. A gram of fat contains 9 calories.

Your job is to identify exactly what John and Jane will eat for a day so that they meet their nutritional requirements entirely with food purchased from fast-food restaurants. Make a second day's plan for each of them with food purchased from a grocery, bearing in mind that the only cooking they do is heating up in a microwave oven.

Include in your final report:

a. A rationale for the ratio of carbohydrates, protein, and fat selected

b. Two days of menus for John and Jane

c. A table for each person and each day that shows the total calories of carbohydrates, protein, and fat and the final percentages

CHAPTER 7

True-False Concept Review

ANSWERS

Check your understanding of the language of algebra and arithmetic. Tell whether each of the following statements is True (always true) or False (not always true). For each statement you judge to be false, revise it to make a statement that is true.

1. _____

  1. A fraction can be regarded as a ratio.

2. _____

  2. Cross multiplying has the same effect as using the multiplication law of equality.

3. _____

  3. $\dfrac{7 \text{ miles}}{1 \text{ gallon}} = \dfrac{14 \text{ miles}}{2 \text{ hours}}$

4. _____

  4. To solve a proportion, we must know the values of three of the four numbers.

5. _____

  5. If $\dfrac{7}{3} = \dfrac{t}{4}$, then $t = \dfrac{3}{28}$.

6. _____

  6. A percent comparison is always based on the number 100.

7. _____

  7. Fractions with a denominator of 6, such as $\dfrac{7}{6}$, cannot be written as percents.

8. _____

  8. $1.2 = 120\%$

9. _____

  9. $7\dfrac{1}{2} = 7.5\%$

10. _____

  10. $0.009\% = 0.9$

11. _____

  11. If 0.3% of B is 84, then $B = 280$.

12. _____

  12. If $R\%$ of 12 is 24, then $R = 50$.

13. _____

  13. If $2\dfrac{4}{5}\%$ of 300 is A, then $A = 8.4$.

14. _____

14. Two consecutive decreases of 15% is the same as a decrease of 30%.

15. _____

15. If Selma is given a 10% raise on Monday but has her salary cut 10% on Wednesday, her salary is the same as it was Monday before the raise.

16. _____

16. It is possible to increase a city's population by 110%.

17. _____

17. In a proportion, two ratios are equal.

18. _____

18. Twelve inches and one foot are like measures.

19. _____

19. Ratios always compare like units.

20. _____

20. If the price of stock increases 100% each of 3 years, the value of $1 of stock is worth $8 at the end of 3 years.

21. _____

21. $\frac{1}{2}\% = 0.5$.

22. _____

22. A 50% growth in population is the same as 150% of the original population.

23. _____

23. Every rational number can be expressed as a percent.

24. _____

24. To change a decimal to a percent, move the decimal two places to the right and write the percent symbol on the right.

25. _____

25. To change a fraction to a percent, we must build the fraction to have a denominator of 100.

CHAPTER 7

Review

Section 7.1   *Objective 1*

Write a ratio and simplify.

1. 63 feet to 18 feet
2. 102 meters to 68 meters
3. 2 dimes to 1 dollar (compare in nickels)
4. 1 mile to 4000 feet (compare in feet)
5. 18 ounces to 2 pounds (compare in ounces)

Section 7.1   *Objective 2*

Write a rate and simplify.

6. 1295¢ per 14 pounds of meat
7. 1898¢ per 6 boxes
8. 12 printers to 90 computers
9. 234 VCR's to 36 CD players
10. A supermarket parking lot has 36 spaces for compact cars and 62 spaces for larger cars. What is the ratio of compact spaces to the total number of spaces.

Section 7.1   *Objective 3*

Write a unit rate. Round as necessary.

11. $12.95 per 14 pounds of meat
12. $18.98 per 6 boxes
13. 44,896 people per 460 square miles
14. 35,854 miles per 1820 gallons
15. 31,535 nautical miles per 1802 gallons

Section 7.2   *Objective 1*

Are the following proportions true or false?

16. $\dfrac{15}{18} = \dfrac{75}{90}$

17. $\dfrac{21}{36} = \dfrac{105}{170}$

18. $\dfrac{5.7}{8.3} = \dfrac{14.82}{21.58}$

19. $\dfrac{9.6}{5.7} = \dfrac{6.72}{4.05}$

20. $\dfrac{3\frac{1}{2}}{7\frac{1}{3}} = \dfrac{1\frac{3}{4}}{3\frac{2}{3}}$

Section 7.2   *Objective 2*

Solve these proportions.

21. $\dfrac{x}{8} = \dfrac{13}{4}$

22. $\dfrac{6}{x} = \dfrac{3}{7}$

23. $\dfrac{6.8}{1.5} = \dfrac{x}{4.5}$

24. $\dfrac{\frac{3}{7}}{5} = \dfrac{6}{x}$

25. $\dfrac{0.024}{14} = \dfrac{x}{7}$

Section 7.3

26. Taxes on a home with an assessed value of $84,000 are $2450. How much will the taxes be on a home valued at $64,000 in the same district? (Round to the nearest dollar.)

27. A farmer estimates her profit at $110 on a 5-acre plot of land. At this rate what will be her profit on 460 acres?

28. The Carpetorium advertises it will carpet a 20 ft by 25 ft room for $445. At the same rate, what will it cost to carpet a room that measures 30 ft by 30 ft?

29. A map of Australia is scaled so that $1\frac{1}{2}$ inches represent 100 miles. How many miles are represented by 4 inches on the map? (Round to the nearest mile.)

30. The Praline Shop prepares a mixture of nuts that has almonds and hazelnuts in a ratio of 3 to 4. How many pounds of each will be needed to make 84 pounds of the mixture?

Section 7.4 *Objective 1*

Write each fraction as a percent:

31. $\dfrac{1}{8}$

32. $\dfrac{19}{25}$

33. $\dfrac{15}{16}$

34. $\dfrac{18}{29}$ (Round to the nearest tenth of a percent.)

35. $\dfrac{101}{85}$ (Round to the nearest tenth of a percent.)

Section 7.4 *Objective 2*

Write each decimal as a percent:

36. 3.067
37. 0.0893
38. 46.9
39. 0.0009
40. 67

Section 7.4 *Objective 3*

Write each of the following as a decimal:

41. 0.375%
42. 4.68%
43. 134.9%
44. 0.008%
45. 0.08642%

Section 7.5

Solve:

46. 78% of 92 is ___?___ .
47. 72 is ___?___ % of 160.
48. 212% of ___?___ is 99.68. (Round to the nearest hundredth.)
49. ___?___ is 73.4% of 123.7. (Round to the nearest tenth.)
50. 37.8 is ___?___ % of 63.8. (Round to the nearest tenth of a percent.)

Section 7.6 *Objective 1*

51. Forty-three of 210 people screened for cholesterol had readings above 225. What percent of those screened had readings above 225? Round to the nearest tenth of a percent.
52. On July 1, Henrietta will get a 7% raise. If Henrietta's current annual salary is $48,500, what will be her annual salary after the raise?

53. After a 12% increase in population, the town of Gaston had a population of 35,280. What was the former population?
54. A new piece of equipment was purchased by Exaco Electronics for a $22,500. At the end of the first year it had depreciated to a value of $18,900. What was the rate of depreciation in the first year?
55. Hazel spends 23% of her monthly income for rent. If her monthly rent is $350, what is her monthly income? Round to the nearest dollar.

Section 7.6 *Objective 2*

56. An article is priced to sell at $79.95. If the markup is 40% of the selling price, how much is the markup?
57. A salesperson earns an 11% commission on all sales. What was the commission earned on sales of $45,600?
58. The retail price (selling price) of a sofa and chair set is $975. If the markup is 50% of the cost, what is the cost?
59. For customers who use a bank's credit card, there is a 1.25% finance charge on monthly accounts that have a balance of $700 or less. Leslie's finance charge for April was $6.10. What was the balance of her account for that month?
60. Belinda bought a house of $129,000. She made a down payment of 16%. How much did she pay as the down payment?

Section 7.6 *Objective 3*

The following graph displays the way the Andrews spend their monthly income.

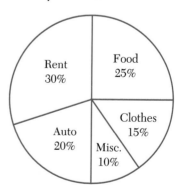

61. What percent of their income is spent on cars?
62. What percent of their income is spent on rent?
63. Do the Andrews spend more money per month on automobile expenses or clothes?
64. What percent of the Andrews' income is spent on clothes and miscellaneous items?
65. What percent of the Andrews' income is spent on cars and rent?

CHAPTER 7

Test

ANSWERS

1. _____

1. Write a ratio to compare 15 to 75 and simplify.

2. _____

2. Is the proportion $\dfrac{16}{34} = \dfrac{24}{51}$ true or false?

3. _____

3. Write as a percent: 1.6

4. _____

4. Write as a percent: $\dfrac{1}{5}$

5. _____

5. Write as a decimal: 78%

6. _____

6. Write a ratio to compare 3 quarts to 1 gallon (in quarts) and simplify.

7. _____

7. Write as a fraction: 32.5%

8. _____

8. Solve the proportion: $\dfrac{2.1}{9} = \dfrac{0.56}{w}$

9. _____

9. 62% of what number is 58.9?

10. _____

10. Write as a percent: 0.003

11. _____

11. Write a ratio to compare 10 pounds to 35 pounds.

12. _____

12. Write as a decimal: 0.13%

13. _____

13. What number is 11.5% of 212?

14. _____

14. Solve the proportion: $\dfrac{9}{24} = \dfrac{x}{28}$

15. _____

15. Write as a percent: $\dfrac{7}{23}$. Round to the nearest tenth of a percent.

16. _____

16. What percent of 23 is 9? Round to the nearest tenth of a percent.

17. _____

17. If 20 lb of mixed nuts contains 6 lb of cashews, how many pounds of cashews may be expected in 100 lb of mixed nuts?

18. _____

18. There were 10,685 tickets sold for a local rock concert. If 475 of the tickets were not used, what percent of the ticket holders were no shows? Round to the nearest whole percent.

19. _____

19. A TV set regularly sells for $225. During a sale the dealer discounts the price by $67.50. What is the percent of the discount?

20. _____

20. What is the selling price of a suit if the markup is 75% of the cost and the cost is $255?

Good Advice for Studying

Evaluating Your Performance

Pick up your graded test as soon as possible, while the test is still fresh in your mind. No matter what your score, time spent reviewing errors you may have made can help improve future test scores. Don't skip the important step of "evaluating your performance." Begin by categorizing your errors:

1. *Careless errors.* These occur when you know how to do the problem correctly, but don't. Careless errors happen when you read the directions incorrectly, make computational errors, or forget to do the problem.

2. *Concept errors.* These occur because you don't grasp the concept fully or accurately. Even if you did the problem again, you would probably make the same error.

3. *Study errors.* These occur when you do not spend enough time studying the material most pertinent to the test.

4. *Application errors.* These usually occur when you are not sure what concept to use.

If you made a careless error, ask yourself if you followed all the suggestions for better test-taking. If not, vow to do so on the next test. List what you will do differently next time so that you can minimize these careless errors in the future.

Concept errors need more time to correct. You must review what you didn't understand or you will repeat your mistakes. It is important to grasp the concepts because you will use them later. You may need to seek help from your instructor or a tutor for these kinds of errors.

Avoid study errors by asking the teacher before the test what concepts are most important. Also, pay careful attention to the section objectives, because these clarify what you are expected to know.

You can avoid application errors by doing as many of the word problems as you can. Reading the strategies for word problems can help you think about how to start a problem. It also helps to mix up the problems between sections and to do even-numbered problems, for which you do not have answers. It is especially important to try to estimate a reasonable answer before you start calculations.

As a final step in evaluating your performance, take time to think about what you said to yourself during the test. Was this self-talk positive or negative? If the talk was helpful, remember it. Use it again. If not, here are a few suggestions. Be certain that you are keeping a detailed record of your reactions to anxiety. Separate your thoughts from your feelings. Use cue words as signals to begin building coping statements. Your coping statements must challenge your negative belief patterns and they must be believable, not merely pep talk such as "I can do well, I won't worry about it." Make the statements brief and state them in the present tense.

Learn from your mistakes so that you can use what you have learned on the next test. If you are satisfied with the results of the test overall, congratulate yourself. You have earned it.

8

Equations in Two Variables

K. Giese/SuperStock ©

APPLICATION

Emma Raine Bless was born August 19, 1986, weighing 7 lb 13 oz. The proud grandparents, Joe and Hazel, decided to buy a US Savings bond every year and save them for Emma's college fund. On September 1 of each year, they purchase a bond with a face value of $1000. They buy the bond for $500 and it will mature to face value in 10–12 years, depending upon the current interest rates. When the bonds mature, Joe and Hazel will cash them in and deposit the proceeds in a special account for Emma. Fill in the following Table.

Year	Total Amount Invested	Face Value of Fund
1986		
1987		
1988		
1989		
1990		
1991		
1992		
1993		
1994		
1995		
1996		

8.1 The Rectangular Coordinate System

OBJECTIVES

1. Check whether an ordered pair is a solution of a linear equation in two variables.
2. Plot a point given its coordinates.
3. Identify the coordinates of a point given its graph.

VOCABULARY

A **linear equation in two variables** is any equation that can be written in the form

$$ax + by = c$$

where a, b, and c are numbers and a and b are not zero.

The **rectangular coordinate system** is formed by two number lines, one vertical and one horizontal.

The **x-axis** is the horizontal number line.

The **y-axis** is the vertical number line.

The **origin** is the intersection at zero on each line. See Figure 8.1.

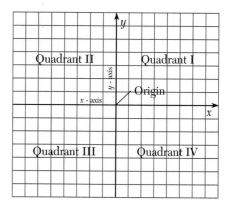

Figure 8.1

Points in the rectangular coordinate system are identified by ordered pairs of numbers, (x, y). The pairs are called **coordinates**. The pairs are also called **ordered pairs** because the x-value is always written first. The x-value is called the **abscissa** and the y-value is called the **ordinate** of the point.

The **solution** of an equation in two variables is written as an ordered pair of numbers, (x, y).

HOW AND WHY
Objective 1

Check whether an ordered pair is a solution of a linear equation in two variables.

A solution of an equation in two variables is a pair of values, one value for each variable, that makes the equation true. For example, for the two-variable equation $4x + 3y = 17$, replacing x with 2 and y with 3 yields a true statement:

$$4x + 3y = 17$$
$$4(2) + 3(3) = 17 \qquad \textbf{Replace } x \textbf{ with 2 and } y \textbf{ with 3.}$$
$$17 = 17 \qquad \textbf{Simplify.}$$

We write the solution using the **ordered pair notation** (2, 3), thereby showing that the x-value is 2 and the y-value is 3. The value of x is always given first; the value of y is always given second. The values in an ordered pair (x, y) may be called by any of these names:

(x-value, y-value)

(abscissa, ordinate)

(x-coordinate, y-coordinate)

(independent variable, dependent variable)

The ordered pair (2, 3) is not the only solution of the equation $4x + 3y = 17$. In the next section we see how to find an unlimited number of solutions of this equation. For now, we concentrate on verifying whether a particular ordered pair is a solution.

 To check whether an ordered pair is a solution of an equation

1. Replace each variable with its value in the ordered pair.

2. Simplify. If the resulting equation is a true statement, the ordered pair is a solution.

Examples A–D   **Warm Ups A–D**

Directions: Check whether the ordered pair is a solution of the equation.

Strategy: Replace each variable with its value in the ordered pair.

A.   Is $(0, -3)$ a solution of $3x - 4y = 12$?

$$3(0) - 4(-3) = 12 \qquad \textbf{Replace } x \textbf{ with 0}$$
$$\textbf{and } y \textbf{ with } -3.$$
$$0 - (-12) = 12$$
$$12 = 12 \qquad \textbf{True.}$$

The pair $(0, -3)$ is a solution.

A.   Is $(2, 0)$ a solution of $5x - y = 10$?

B.   Is $(8, 3)$ a solution of $3x - 4y = 12$?

$$3(8) - 4(3) = 12 \qquad \textbf{Replace } x \textbf{ with 8}$$
$$\textbf{and } y \textbf{ with 3.}$$
$$24 - (12) = 12$$
$$12 = 12 \qquad \textbf{True.}$$

The pair $(8, 3)$ is a solution.

B.   Is $(4, 10)$ a solution of $5x - y = 10$?

Answers to Warm Ups   A. $(2, 0)$ is a solution.   B. $(4, 10)$ is a solution

C. Is $(2, -\dfrac{3}{2})$ a solution of $3x - 4y = 12$?

$3(2) - 4(-\dfrac{3}{2}) = 12$ **Replace x with 2 and y with $-\dfrac{3}{2}$.**

$6 - (-6) = 12$

$12 = 12$ **True.**

The pair $(2, -\dfrac{3}{2})$ is a solution.

C. Is $(\dfrac{3}{5}, -7)$ a solution of $5x - y = 11$?

D. Is $(2, -5)$ a solution of $3y + 7x = 2$?

$3(-5) + 7(2) = 2$ **Replace x with 2 and y with -5.**

$-15 + 14 = 2$

$-1 = 2$ **False.**

The pair $(2, -5)$ is not a solution.

D. Is $(-4, 10)$ a solution of $4y + 9x = 4$?

HOW AND WHY

Objective 2 **Plot a point given its coordinates.**

We can *graph* an ordered pair, (x, y) by constructing a **rectangular coordinate system,** as shown in Figure 8.2.

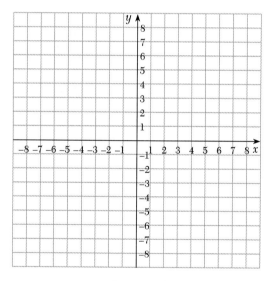

Figure 8.2

The horizontal axis (x-axis) is the number line that we have worked with before; the vertical axis (y-axis) is a number line with the positive numbers above the x-axis and the negative numbers below the x-axis. We associate each point in this coordinate system with an ordered pair (x, y). The x-value measures the distance of the point from the y-axis, and the y-value measures the distance of the point from the x-axis.

Answers to Warm Ups C. $(\dfrac{3}{5}, -7)$ is not a solution D. $(-4, 10)$ is a solution

To graph the ordered pair (6, 3), for example, we count 6 units to the right of the origin on the *x*-axis and 3 units up. We end up at the corner point of a rectangle, the corner opposite the origin. This rectangle is shown in Figure 8.3. (This is the reason we call it a *rectangular* coordinate system.) We say that the **coordinates** of the point are the ordered pair (6, 3). Graphing an ordered pair is called **plotting** a point.

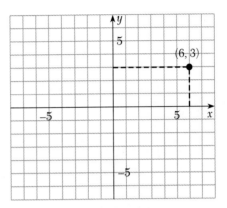

Figure 8.3

> ### To plot a point given its coordinates (x, y)
>
> **1.** From the origin, count along the *x*-axis right or left (positive or negative) according to the *x*-coordinate.
>
> **2.** From this point on the *x*-axis, count up or down (positive or negative) according to the *y*-coordinate.
>
> **3.** Draw a dot at this point and label it with the coordinates (*x*, *y*).

In many real-life situations, the *x*- and *y*-axes do not have the same scale. See Section 8.2.

Example E

Directions: Plot the points with given coordinates on the same rectangular coordinate system.

Strategy: For each point, count right or left according to the *x*-coordinate. From this point count up or down according to the *y*-coordinate. Label the point.

E.   (3, −2), (−5, 3), (−4, −1), (1.5, 5), (3, −4.5), (0, 0)

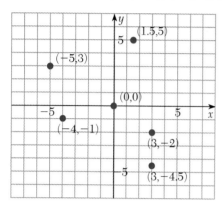

The points are shown as dots on the graph at the left.

E.   (3, 2), (5, 1), (−2, 5), (−1, 3), (−4.5, −2), (−7, 3.5), (2, −4)

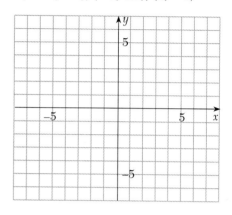

HOW AND WHY
Objective 3

Identify the coordinates of a point given its graph.

To find the coordinates of a point when its graph is given, find the number of units the point is from the *x*- and *y*-axes, as shown in Figure 8.4.

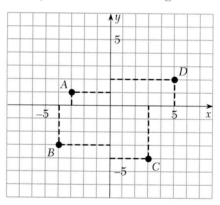

Figure 8.4

Answer to Warm Up   E.

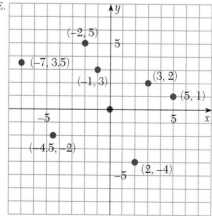

The point A is 3 units to the left and 1 unit above the origin: the coordinates of A are $(-3, 1)$. Point B is 4 units to the left and 3 units below the origin: the coordinates of B are $(-4, -3)$. Similarly, the coordinates of C are $(3, -4)$, and the coordinates of D are $(5, 2)$.

 To identify the coordinates of a point given its graph

1. Find the distance of the point from the y-axis. This is the x-value. It is positive if the point is to the right of the y-axis and negative if it is to the left.

2. Find the distance of the point from the x-axis. This is the y-value. It is positive if the point is above the x-axis and negative if it is below.

3. Write the ordered pair (x, y).

Examples F–G   **Warm Ups F–G**

Directions: Identify the coordinates of the points on the graph.

Strategy: Find the distance, positive or negative, that the point is from each axis and write the ordered pair.

F.

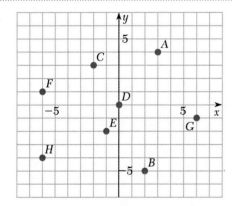

Read the values from the graph. The coordinates are A, $(3, 4)$; B, $(2, -5)$; C, $(-2, 3)$; D, $(0, 0)$; E, $(-1, -2)$; F, $(-6, 1)$; G, $(7, -1)$; and H, $(-6, -4)$.

F.

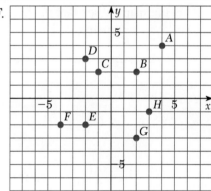

Answers to Warm Ups   F. The coordinates are A, $(4, 4)$; B, $(2, 2)$; C, $(-1, 2)$; D, $(-2, 3)$; E, $(-2, -2)$; F, $(-4, -2)$; G, $(2, -3)$; and H, $(3, -1)$.

G. Maps commonly use letters rather than numbers for one axis. Maps also use the coordinates to identify a rectangular area rather than a single point. What are the coordinates of Mount St. Helens, Washington, shown on the map?

G. What are the coordinates of Vancouver, Washington, shown on the map?

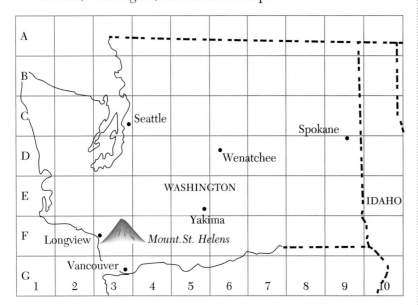

On the map we see that the horizontal coordinate is 3, while the vertical coordinate is *F*. The coordinates of Mount St. Helens are (3, *F*).

Exercises 8.1

OBJECTIVE 1: *Check whether an ordered pair is a solution of a linear equation.*

A.

Check whether the ordered pair is a solution of the equation.

1. $(0, -3); 2x - y = 3$

2. $(4, 0); 2x - y = 8$

3. $(4, -5); x + y = 9$

4. $(-3, 5); x - y = 8$

5. $(3, -4); x + y = -1$

6. $(2, 8); x - y = -6$

7. $(1, -1); 2x - y = 3$

8. $(1, -1); 5x + y = 4$

9. $(0, -5); 3x - 2y = 10$

10. $(-4, 0); 3x - 7y = 12$

B.

11. $(-10, 7); 2x + 4y = 8$

12. $(15, -3); 4x + 5y = 45$

13. $(-4, -12); 3x - 7y = -72$

14. $(-6, -8); 5x - 5y = 10$

15. $(-4, 9); 2x - 7y + 71 = 0$

16. $(-6, -6); 6x - 5y + 6 = 0$

17. $\left(-\dfrac{1}{2}, -\dfrac{6}{5}\right)$; $4x - 5y = -4$

18. $\left(-\dfrac{7}{3}, -\dfrac{8}{5}\right)$; $6x - 10y = 2$

19. $(2.25, 0.6)$; $60x - 40y = 111$

20. $(-1.5, 0.75)$; $3x - 2y = -6$

OBJECTIVE 2: *Plot a point given its coordinates.*

A.

Construct a rectangular coordinate system and plot the following points.

21. $(3, 4)$ **22.** $(-3, -2)$ **23.** $(-1, -4)$

24. $(5, -2)$ **25.** $(-4, 5)$ **26.** $(0, -3)$

27. $(3, 0)$ **28.** $(-1, 6)$ **29.** $(8, -6)$

30. $(7, 7)$

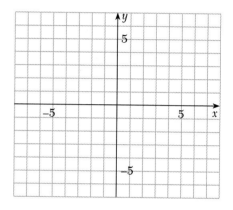

B.

Construct a rectangular coordinate system and plot the following points.

31. $(-1, -9)$ **32.** $(9, 7)$ **33.** $(1.5, 2)$

34. $(2.5, 1)$ **35.** $(-1.5, -2)$ **36.** $(-2, 2.5)$

37. $\left(4\dfrac{1}{2}, -1\right)$ **38.** $\left(3\dfrac{1}{2}, -3\right)$ **39.** $(-3, 3.5)$

40. $(0, -0.5)$

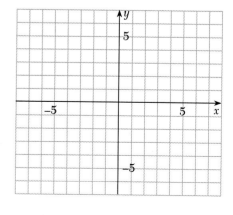

OBJECTIVE 3: *Identify the coordinates of a point given its graph.*

A.

Identify the points on the following graph.

41. *A*   **42.** *B*

43. *C*   **44.** *D*

45. *E*   **46.** *F*

47. *G*   **48.** *H*

49. *I*   **50.** *J*

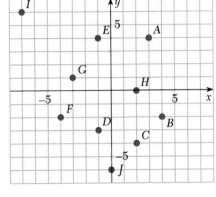

B.

Identify the points on the following graph.

51. *A*   **52.** *B*

53. *C*   **54.** *D*

55. *E*   **56.** *F*

57. *G*   **58.** *H*

59. *I*   **60.** *J*

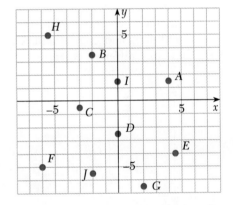

C.

Identify the coordinates of the following cities on the map of Washington in Example G on page 732.

61. Yakima   **62.** Seattle   **63.** Wenatchee

64. Spokane

65. What city on the map of Washington has the same coordinates as Mount St. Helens?

Identify the coordinates of the cities on the map of Texas.

66. Austin   **67.** Houston

68. El Paso   **69.** Dallas

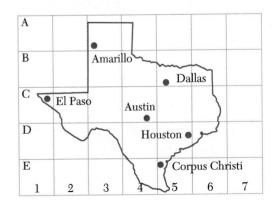

735

70. Consider the following picture.

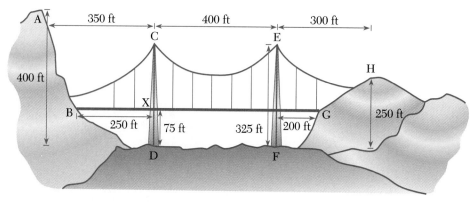

Superimpose a coordinate axis system with the origin at X. Determine the coordinates of the points A – H.

 Exercise 71 refers to the Chapter 8 Application. See page 725.

71. Set up a coordinate system with the years on the horizontal axis and dollars on the vertical axis. Select a scale so that all the entries in the savings bond table are visible. Use the first two columns of the table as eleven ordered pairs and graph them. Use the first and third columns in the table as eleven ordered pairs and graph them in a different color.

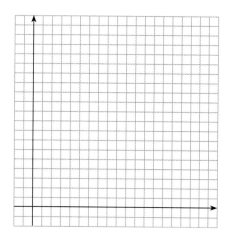

STATE YOUR UNDERSTANDING

72. Describe the procedure used to determine the coordinates of a point given its graph.

73. Explain how you would plot the point (−4, 7).

CHALLENGE

74. Is $(4, 0)$ a solution of $x = 4$?

75. Is $\left(-\dfrac{27}{5}, -\dfrac{18}{7}\right)$ a solution of $10x - 182y + 167 = 0$?

76. If a point is in Quadrant I, both coordinates are positive, indicated by $(+, +)$. Identify the coordinates in the other three quadrants by their signs.

GROUP ACTIVITY

77. Given the equation $3x - 8y = 24$, find five solutions to the equation. Explain how you arrive at the solutions.

MAINTAIN YOUR SKILLS (Sections 4.9, 5.7, 6.8)

Solve.

78. $3x - 14 - x = -76$

79. $15y - 4y + 17 = -60$

80. $\dfrac{5x}{12} - \dfrac{3}{4} + \dfrac{2x}{3} = -4$

81. $0.35a - 6.7 = 0.25a + 8.93$

82. $3.47b - 8.32 - 4.32b = 9.19$

8.2 Solving Linear Equations in Two Variables

OBJECTIVE Find solutions of a linear equation in two variables.

HOW AND WHY
Objective **Find solutions of a linear equation in two variables.**

A linear equation in two variables has many solutions. Each of these equations can be written in the form

$$y = mx + b$$

by solving the equation for y. We will use this form to find solutions of the equation by substituting values for x. For instance, to find a solution of

$$y = 3x - 5$$

we can pick any value for x and solve for y. The easiest value to use is $x = 0$.

$$y = 3(0) - 5$$
$$= 0 - 5$$
$$= -5$$

So, one solution is $(0, -5)$.

We can see from the general form, $y = mx + b$, that one solution is always $(0, b)$. Substituting $x = 0$ in the general form we have

$$y = mx + b$$
$$= m(0) + b$$
$$= 0 + b$$
$$= b$$

So, the solution is $(0, b)$.

Other solutions to $y = 3x - 5$ can be found by substituting integer values for x.

Let $x = 3$.   Let $x = -3$.   Let $x = 5$.

$y = 3(3) - 5$ $y = 3(-3) - 5$ $y = 3(5) - 5$
$\quad = 9 - 5$ $\quad = -9 - 5$ $\quad = 15 - 5$
$\quad = 4$ $\quad = -14$ $\quad = 10$

$(3, 4)$ $(-3, -14)$ $(5, 10)$

There are an unlimited number of solutions of a linear equation in two variables. We can list some of these solutions in a table as well as by pairs. Here are two equations and some of their solutions.

Equation	Solution in a Table		Solutions as Ordered Pairs
$y = -2x + 6$	**x**	**y**	
	0	6	$(0,\ 6)$
	4	-2	$(4,\ -2)$
	-3	12	$(-3,\ 12)$
	-5	16	$(-5,\ 16)$
	7	-8	$(7,\ -8)$
$y = \dfrac{2}{3}x - 1$	**x**	**y**	
	0	-1	$(0,\ -1)$
	3	1	$(3,\ 1)$
	-3	-3	$(-3,\ -3)$
	5	$\dfrac{7}{3}$	$(5,\ \dfrac{7}{3})$
	-1	$-\dfrac{5}{3}$	$(-1,\ -\dfrac{5}{3})$

In the equation $y = \dfrac{2}{3}x - 1$, you can avoid fraction values for y by choosing values for x that are multiples of 3. For example, let $x = 9$,

$$y = \frac{2}{3}(9) - 1$$
$$= 2(3) - 1$$
$$= 6 - 1$$
$$= 5$$

So a solution is $(9,\ 5)$.

It is impossible to list all the solutions, either in a table or as ordered pairs. Later in this chapter we draw a graph of the equation that is a "picture" that illustrates the solutions.

  To find a solution of an equation in two variables, written in the form $y = mx + b$, given a value for x

1. Substitute the value for x in the equation.
2. Solve the equation for y.
3. Write the solution as an ordered pair, $(x,\ y)$.

| **Examples A–C** | **Warm Ups A–C** |

Directions: Find three solutions of the linear equation in two variables.

Strategy: Substitute the given values in the equation and solve. Write the solutions as ordered pairs.

A. $y = 4x - 4, x = 1, x = -2, x = 0$

$y = 4x - 4$

$y = 4(1) - 4$ **Substitute 1 for x.**

$y = 4 - 4$

$y = 0$

So $(1, 0)$ is a solution.

$y = 4(-2) - 4$ **Substitute -2 for x.**

$y = -8 - 4$

$y = -12$

So $(-2, -12)$ is a solution.

$y = 4(0) - 4$ **Substitute 0 for x.**

$y = 0 - 4$

$y = -4$

So, $(0, -4)$ is a solution.

The three solutions are $(1, 0)$, $(-2, -12)$, and $(0, -4)$.

A. $y = -2x + 3, x = -1, x = 2, x = 3$

B. $y = -\dfrac{1}{3}x + 5, x = 0, x = 3, x = -5$

$y = -\dfrac{1}{3}x + 5$

$y = -\dfrac{1}{3}(0) + 5$ **Substitute 0 for x.**

$y = 5$

So $(0, 5)$ is a solution.

$y = -\dfrac{1}{3}x + 5$

$y = -\dfrac{1}{3}(3) + 5$ **Substitute 3 for x.**

$y = -1 + 5$

$y = 4$

So, $(3, 4)$ is a solution.

B. $y = \dfrac{4}{5}x + 10, x = 0, x = 5,$
$x = -8$

(continues next page)

$$y = -\frac{1}{3}x + 5$$

$$y = -\frac{1}{3}(-5) + 5 \qquad \text{Substitute } -5 \text{ for } x.$$

$$y = \frac{5}{3} + 5$$

$$y = \frac{5}{3} + \frac{15}{3}$$

$$y = \frac{20}{3}$$

So $(-5, \frac{20}{3})$ is a solution.

The three solutions are $(0, 5)$, $(3, 4)$, and $(-5, \frac{20}{3})$.

C. An apparel department is having a sale on jeans and tops. The jeans will sell for $15 a pair and the tops for $12 each. One sale comes to $66. If the customer buys three tops, how many pairs of jeans does she buy?

Write a simpler word form using the cost and the total sale.

Simpler word form:

Cost of jeans + cost of tops = 66

Select variables:

Organize the data in a table. Let J represent the number of jeans bought and T represent the number of tops bought.

	Cost of Each	Number Sold	Total Cost of Each
Jeans	15	J	$15J$
Tops	12	T	$12T$

Translate to algebra:

$15J + 12T = 66$

Since we are given a value for T, $T = 3$, we could solve the equation for J to write the equation in the form $y = mx + b$. In this case it is easier to just replace T with 3 in the equation and then solve for J.

Solve:

$$15J + 12(3) = 66$$
$$15J + 36 = 66 \qquad \text{Substitute 3 for } T.$$
$$15J = 30$$
$$J = 2$$

Therefore, the customer buys two pairs of jeans.

C. Another sale of jeans and tops comes to $117. If the customer buys three pairs of jeans, how many tops does she buy?

Answers to Warm Ups C. 6 tops

Exercises 8.2

OBJECTIVE: *Find solutions of linear equations in two variables.*

A.

Given $y = -x + 18$, find the solution whose x-value is given.

1. $x = 6$ **2.** $x = 3$ **3.** $x = 13$

4. $x = 18$ **5.** $x = -3$ **6.** $x = -5$

Given $y = -\dfrac{1}{4}x - 7$, find the solution whose x-value is given.

7. $x = 8$ **8.** $x = -12$ **9.** $x = 0$

10. $x = -4$ **11.** $x = 16$ **12.** $x = -16$

B.

Given $y = \dfrac{2}{7}x + 6$, find the solution whose x-value is given.

13. $x = 0$ **14.** $x = 7$ **15.** $x = -3$

16. $x = -28$ **17.** $x = -15$ **18.** $x = -20$

Given $y = 0.8x + 10$, find the solution whose x-value is given.

19. $x = 0$ **20.** $x = -6$ **21.** $x = -1.5$

22. $x = 5.5$ **23.** $x = -12.5$ **24.** $x = -30$

C.

25. A building purchased for \$45,000 is depreciated over a 12.5-year period to no value. This is represented by the equation

$$y = 45{,}000 - 3600x$$

where y is the value of the building and x is the number of years from 0 to 12. Find the value of the building at the end of 4 and 8 years.

26. The length (ℓ) and the width (w) of a series of rectangles whose perimeters are 36 inches is given by $2w + 2\ell = 36$. Find the length of the rectangles whose widths are 2, 5, and 9 inches.

27. A first number is two less than twice a second ($x = 2y - 2$). If the first number is 47, what is the second number?

28. A first number is three less than eight times a second number. If the second number is 21, what is the first?

Given $y = \dfrac{3}{2}x + \dfrac{15}{2}$, find the solution whose x-value is given.

29. $x = 3$ **30.** $x = 12$ **31.** $x = -\dfrac{16}{3}$

32. $x = -0.5$ **33.** $x = -4.2$ **34.** $x = -6.2$

35. A load of cartons, some of which weigh 6 lb each and some 7 lb each, are loaded on a truck. The total weight of the boxes is 129 lb. If there are nine 7-lb boxes, how many 6-lb boxes are in the load?

36. In Exercise 35, if there are three 7-lb boxes, how many 6-lb boxes are in the load?

Given $y = -\dfrac{7}{3}x + 7$, find the solution whose x- value is given.

37. $x = 3$ **38.** $x = 6$ **39.** $x = -5$

40. $x = 1$ **41.** $x = \dfrac{2}{7}$ **42.** $x = -\dfrac{1}{3}$

43. Frosty Flakes cereal provides 28 g of carbohydrate per serving; Cheerios provides 23 g of carbohydrates per serving. Gaston wants 171 g of carbohydrates (from cereal) per day. If he has already eaten two servings of Frosty Flakes, how many servings of Cheerios should he eat?

44. The end of a flower box is a trapezoid with an area of 66 in.^{2}. The formula for the area is $A = 0.5(b_1 + b_2)h$. If the lengths of the two bases are 10 in. and 12 in., find the height.

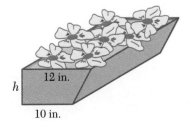

© 1997 Saunders College Publishing.

45. The perimeter of an isosceles triangle (a triangle with two equal sides) is given by the formula $P = 2a + b$. Find a if $P = 38$ cm and $b = 10$ cm.

46. The cost of renting a circular saw is \$7 for the first day and \$4 for each additional day the saw is kept. This can be expressed by the equation $y = 4x + 3$, where y is the total rental fee and x is the number of days the saw is kept. Find the cost of renting the saw for 3, 6, and 10 days.

47. The equation $T = 21.75s$ gives the total amount of money, T, that a newspaper carrier has collected for s subscriptions. How much money does the carrier have for 8 subscriptions? for 25 subscriptions? What is the cost of one subscription?

48. The equation $D = 55t$ gives the distance, D, that a train has traveled after t hours. How far has the train gone after 3.5 hours? How long does it take the train to travel 385 miles?

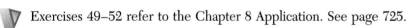 Exercises 49–52 refer to the Chapter 8 Application. See page 725.

49. Consider the equation $I = 500t$. Let $t = 1, 2, 5,$ and 10 and solve for corresponding values of I. Write each solution as an ordered pair of the form (t, I).

50. Relate your solutions to Exercise 49 to the savings bond table.

51. Consider the equation $V = 1000t$. Let $t = 1, 2, 3, 6,$ and 10 and solve for corresponding values of V. Write each solution as an ordered pair of the form (t, V).

52. Relate your solution to Exercise 51 to the savings bond table.

STATE YOUR UNDERSTANDING

53. Find the error in the following problem. What is the correct answer?

Find the solution for $y = \dfrac{5}{2}x - 5$ when $x = 4$.

$y = \dfrac{5}{2}x - 5$

$y = \dfrac{5}{2}(4) - 5$

$y = 10 - 5$

$y = 5$

The solution is $(5, 4)$.

54. How many solutions does the equation $y = -2x + 20$ have? Explain how to find them.

CHALLENGE

For the equation $5y + 2 = 0$, find:

55. y when $x = 4$

56. x when $y = -\dfrac{2}{5}$

57. Write an equation in two variables that has the pairs $(3, 0)$ and $(0, 3)$ as solutions.

58. Write an equation in two variables that has the pairs $(0, 6)$ and $(3, 0)$ as solutions.

59. Write an equation in two variables that has the pairs $(6, 0)$ and $(0, -3)$ as solutions.

MAINTAIN YOUR SKILLS (Sections 4.7, 6.6, 6.7)

60. Is $x = -5$ a solution of $3x - 45 - 2x = 50$?

61. Is $y = 0.05$ a solution of $2y - 13.5 = 12.6 - 16y$?

62. Change 4.75 feet to inches.

63. Find the value of t in the equation $t = -4c - 3d + 5$ if $c = 6$ and $d = -2$.

64. Find the value of p in the equation $p = -0.03s - 2.5v$ if $s = 4.5$ and $v = -0.75$.

8.3   Graphing Linear Equations in Two Variables

OBJECTIVE

Draw the graph of a linear equation in two variables.

HOW AND WHY
Objective

Draw the graph of a linear equation in two variables.

The graph of the linear equation, $y = \dfrac{2}{3}x - 2$, can be drawn by first finding several solutions of the equation and using these as coordinates of points. To make a table of values, select a value for x and solve for y. The easiest value to use is $x = 0$. Other values to use are those that are multiples of 3, such as 3, 6, 9, -3, -6, and -9.

Let $x = 0$.

$$y = \frac{2}{3}(0) - 2$$
$$y = 0 - 2$$
$$y = -2$$

Let $x = 3$.

$$y = \frac{2}{3}(3) - 2$$
$$y = 2 - 2$$
$$y = 0$$

Let $x = -3$.

$$y = \frac{2}{3}(-3) - 2$$
$$y = -2 - 2$$
$$y = -4$$

Let $x = -6$.

$$y = \frac{2}{3}(-6) - 2$$
$$y = -4 - 2$$
$$y = -6$$

List these solutions in a table.

x	y
0	-2
3	0
-3	-4
-6	-6

Now construct a rectangular coordinate system. Plot the ordered pairs $(0, -2)$, $(3, 0)$, $(-3, -4)$, and $(-6, -6)$, and draw the line joining the points.

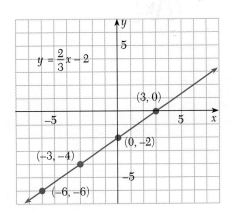

The significance of the graph is that every ordered pair of numbers that makes the equation true represents a point that lies on the line, and every point on the line has coordinates that make the equation true. We say that the line is the **graph of the equation.** Since two points determine a straight line, we only need two solutions to draw the graph; however, using three or more solutions acts as a check and helps avoid errors.

C A U T I O N   **Always use at least three points when graphing a linear equation to avoid errors.**

 To draw the graph of a linear equation in two variables

1. Find at least three solutions by choosing values for x (or y), and solving for y (or x).
2. Plot the solutions on a rectangular coordinate system.
3. Draw the straight line that passes through all of the plotted points.

Examples A–C   **Warm Ups A–C**

Directions: Graph the equation

Strategy: Find at least three solutions. Plot these points. Draw the straight line through them.

A. $y = \dfrac{1}{5}x - 1$.   A. $y = 2x - 6$

x	y
-0	-1
5	0
-5	-2

Find one point by letting $x = 0$, another by letting $y = 0$, and a third by letting $x = -5$.

Plot the points.

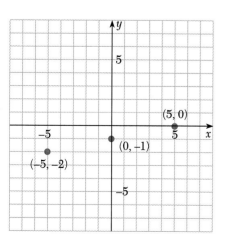

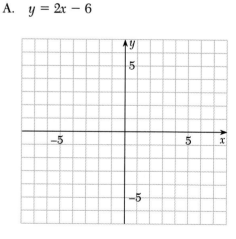

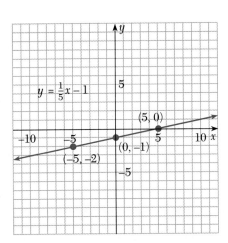

Use a straight edge to draw the line through the points.

B. Draw the graph of $y = \dfrac{6}{5}x - 6$.

x	y
5	0
0	−6
3	−2.4

Make a table of values.

Let $x = 5$ to find one point, $x = 0$ to find another, and the third by letting $x = 3$.

B. Draw the graph of $y = \dfrac{4}{5}x - 4$.

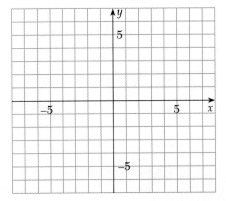

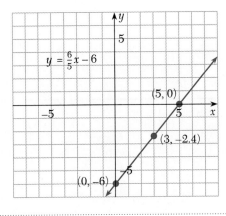

Plot the points and draw a straight line.

Answers to Warm Ups   **A.**

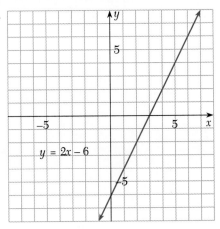

B.

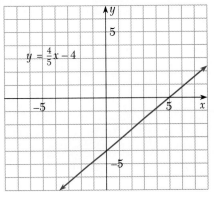

C. A company can depreciate the value of a small electronic control unit for tax purposes. One procedure is to use "straight-line depreciation." The procedure can be written as $y = -20x + 250$, where y represents the value of the unit in any given year and x represents the number of years the unit is in operation.

  One of these units costs $250 and has a scrap value of $50 at the end of 10 years. Graph the equation and list the coordinates that show the value of the unit each year.

Make a table of values and plot the points.

x	y
0	250
4	170
8	90

$y = -20(0) + 250 = 250$

$y = -20(4) + 250 = 170$

$y = -20(8) + 250 = 90$

In order to make it possible to draw the graph on one page, we use the following scale:

On the x-axis, 1 unit = 1 year.
On the y-axis, 1 unit = 10 dollars.

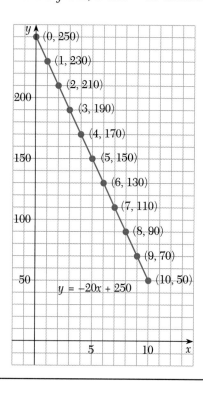

Plot the points with these coordinates, and draw the line joining them. Label the points that show the value at the end of each year.

C. A second electronic control unit costs $450 and has a straight-line depreciation given by $y = -50x + 450$. Graph the equation for $x = 0$ to $x = 9$, and list the coordinates that show the value of the unit each year.

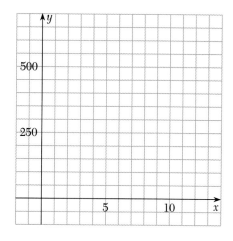

Answer to Warm Up   C.

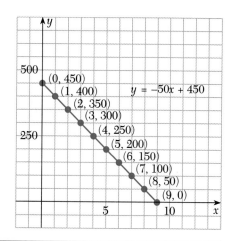

Exercises 8.3

OBJECTIVE: *Draw the graph of a linear equation.*

A.

Complete the table of values and graph each equation.

1. $y = \dfrac{1}{2}x$

x	y
-6	
0	
6	

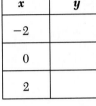

2. $y = 3x$

x	y
-2	
0	
2	

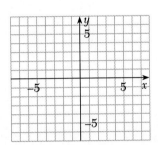

3. $y = -x - 2$

x	y
0	
	0
1	

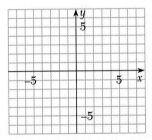

4. $y = -x + 4$

x	y
0	
	0
	-1

5. $y = x - 4$

x	y
0	
	0
2	

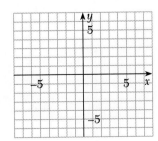

6. $y = x + 2$

x	y
0	
	0
1	

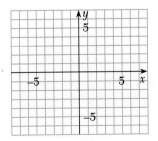

753

7. $y = -x - 9$

x	y
-3	
	-5
-6	

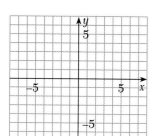

8. $y = x - 7$

x	y
1	
3	
	-1

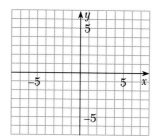

9. $y = 2x + 1$

x	y
4	
-2	
	5

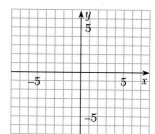

10. $y = -2x + 5$

x	y
0	
	-1
4	

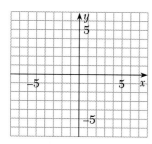

B.

Draw the graph of the equation

11. $y = -\dfrac{1}{2}x + 3$

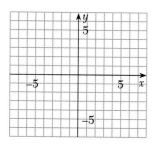

12. $y = -2x + 2$

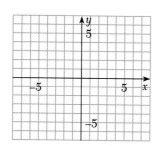

13. $y = 2x + 4$

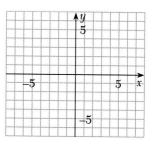

754

14. $y = -\dfrac{2}{3}x + 2$

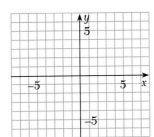

15. $y = -\dfrac{3}{4}x + 3$

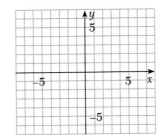

16. $y = \dfrac{5}{3}x - 5$

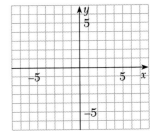

17. $y = \dfrac{4}{5}x - 4$

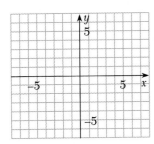

18. $y = -\dfrac{6}{5}x + 6$

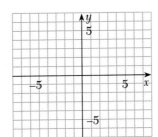

19. $y = -\dfrac{3}{5}x - 3$

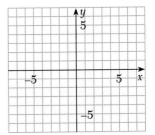

20. $y = \dfrac{2}{3}x - 2$

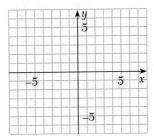

C.

Draw the graph of the equation.

21. $y = \dfrac{4}{3}x - 4$

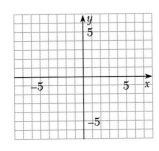

22. $y = \dfrac{1}{4}x - 2$

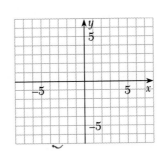

23. $y = \dfrac{1}{2}x + 3$

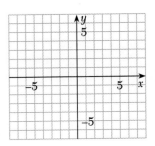

755

24. $y = -\dfrac{5}{2}x + 5$

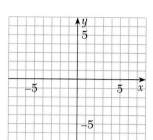

25. $y = \dfrac{5}{7}x - \dfrac{10}{7}$

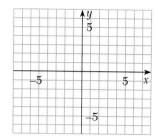

26. $y = \dfrac{4}{5}x - \dfrac{18}{5}$

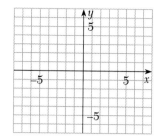

27. $y = 0.6x - 1.2$

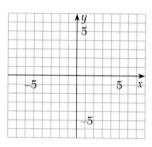

28. $y = -0.4x + 1.2$

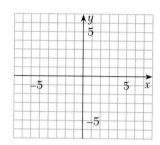

29. $y = -\dfrac{7}{4}x + 7$

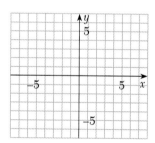

30. $y = \dfrac{5}{4}x - 5$

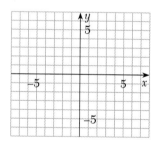

31. A business purchases an automobile for $8000 and will use straight-line depreciation over a period of 5 years to a value of $500. This can be described as $y = -1500x + 8000$, where y is the value of the automobile and x is the number of years from 0 to 5. Graph this equation.

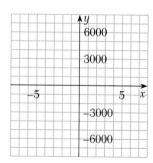

32. If a loan valued at $1000 including interest is being paid off at $75 a month, the balance B after n months is given by the formula $B = 1000 - 75n$. Graph this equation for the first 10 months. Ordered pairs are of the form (n, B).

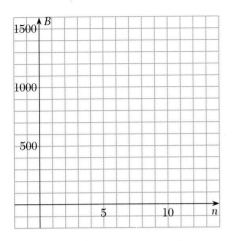

33. A service station makes a profit of 25¢ on each gallon of regular gasoline and 30¢ on each gallon of premium gasoline. If x gallons of regular and y gallons of premium are sold, the profit is $300. This is represented by the equation $0.25x + 0.30y = 300$. Graph this equation.

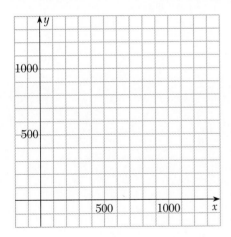

34. The measurement of the length, L, of a stretched spring can be shown by $L = \frac{1}{4}x + 3$, where x represents the number of pounds of force applied to stretch the spring. The spring is 3 inches long before it is stretched. Draw the graph of the equation, ignoring the negative values of x. Ordered pairs are of the form (x, L).

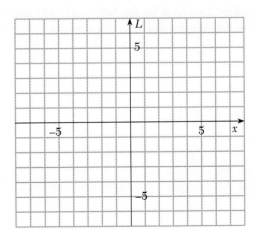

35. The speed of a supersonic aircraft in miles per hour, S, can be shown by the equation $S = 740M$, where M is the Mach number speed of the aircraft. (Under certain conditions the speed of sound is 740 miles per hour.) Draw the graph of the equation, ignoring the negative values of M. (*Hint:* Let each unit of scale on the S-axis represent 1000 and each unit on the M-axis represent 1.) Ordered pairs are of the form (M, S).

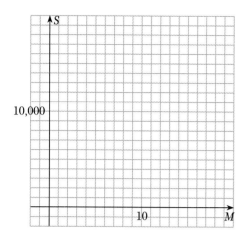

36. A cosmetic salesperson has to buy a sample kit for \$150. She receives \$15 for each makeup kit she sells. The equation $P = 15k - 150$ gives the amount of profit, P, the salesperson makes by selling k kits. Graph this equation letting k be the horizontal axis and P be the vertical axis. Ordered pairs are of the form (k, P).

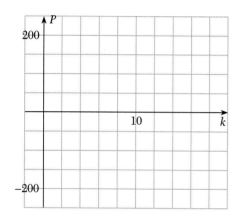

37. The equation $P = 40 + 2w$ gives the perimeter, P, of a rectangle with length 20 cm and width w cm. Graph this equation letting w be the horizontal axis and P be the vertical axis. Ordered pairs are of the form (w, P). Ignore negative values of w.

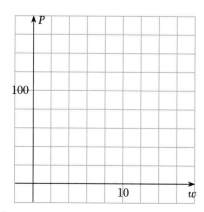

38. Graph the equation $I = 500t$, using the ordered pairs of the form (t, I).

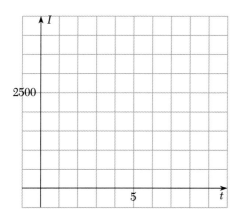

Exercises 39–42 refer to the Chapter 8 Application. See page 725.

39. Compare this graph with the one you drew in Section 8.1, Exercise 71. Devise a test to see if the equation $I = 500t$ describes the data in the savings bond table. Give a written description of your test results.

40. Graph the equation $V = 1000t$, using ordered pairs of the form (t, V).

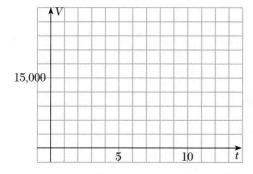

41. Compare this graph with the one you drew in Section 8.1, Exercise 71. Devise a test to see if the equation $V = 1000t$ describes the data in the savings bond table. Give a written description of your test and the results.

STATE YOUR UNDERSTANDING

42. What is another name for a first-degree equation in two variables, and why is it given this name?

43. Describe the relationship between a line graph and an equation.

44. Explain how to graph $3x + y = 18$.

CHALLENGE

Graph the following equations.

45. $2y - 12 = 0$

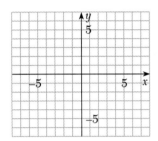

46. $5x + 10 = 0$

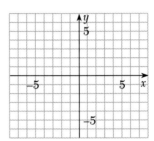

GROUP ACTIVITY

47. A line that is not parallel to the x-axis will always intersect the x-axis. Can your group identify a unique feature of the point of intersection?

MAINTAIN YOUR SKILLS (Sections 2.2, 2.3, 5.6, 6.9, 8.2)

48. Find the perimeter of a triangle with sides of 28 cm, 42 cm, and 16 cm.

49. Is a triangle with sides of 28 cm, 42 cm, and 16 cm a right triangle?

50. The cost of a high-grade carpet installed is $56 per square yard. What is the cost of carpeting a room that measures 15 feet by 21 feet?

51. Find the solution to the equation $4x - 7y = -30$ if $x = -4$.

52. Evaluate $\dfrac{a - c}{b - d}$ if $a = 6$, $b = -2$, $c = -4$, and $d = 5$.

8.4 Slopes and Intercepts (Optional)

OBJECTIVES

1. Find the *x*- and *y*-intercepts of a line.

2. Find the slope of a line given two points on the line.

VOCABULARY

The **x- and y-intercepts** of a graph are the points where the graph crosses (intersects) the *x*- and *y*-axes (Figure 8.5). The *x*-intercept is (*a*, 0) and the *y*-intercept is (0, *b*).

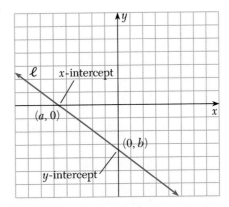

Figure 8.5

The **rise** is the change in the vertical direction, when you move from one point on a line to a second point. The **run** is the change in the horizontal direction.

The **slope of a line** (*m*) is the ratio of the rise to the run.

HOW AND WHY
Objective 1

Find the x- and y-intercepts of a line.

The *x*- and *y*-intercepts are the points where the line crosses the axes. They can be found by locating the points and reading the coordinates.

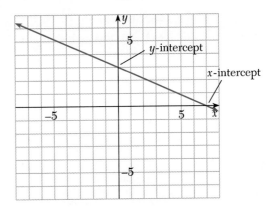

The *x*-intercept is at (7, 0) and the *y*-intercept is at (0, 3).

The intercepts can also be found from the equation of the line by using the fact that every point on the x-axis has a y-value of zero and that every point on the y-axis has an x-value of zero. So, for the equation

$$y = \frac{5}{7}x + 5$$

we find the y-intercept by letting x be zero and solving for y:

$$y = \frac{5}{7}(0) + 5$$

$$y = 5$$

So, the y-intercept is $(0, 5)$.

Likewise the x-intercept can be found by letting y be zero and solving for x:

$$0 = \frac{5}{7}x + 5$$

$$7(0) = 7(\frac{5}{7}x + 5) \qquad \textbf{Multiply both sides by 7.}$$

$$0 = 5x + 35$$

$$-35 = 5x$$

$$-7 = x$$

So, the x-intercept is $(-7, 0)$.

▶ ***To find the x- and y-intercepts of a line***

x-intercept:
1. Replace y with 0 in the equation.
2. Solve for x.
3. Write the ordered pair $(x, 0)$.

y-intercept:
1. Replace x with 0 in the equation.
2. Solve for y.
3. Write the ordered pair $(0, y)$.

Examples A–B   **Warm Ups A–B**

Directions: Find the x- and y-intercepts.

Strategy: From the graph read the coordinates of the points where the line crosses the axes. From the equation replace y with 0 to find the x-intercept. Replace x with 0 to find the y-intercept.

A.

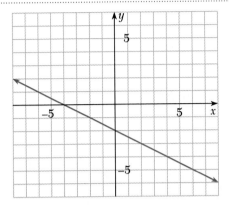

Read the coordinates where the line intersects (crosses) the axes.

A.

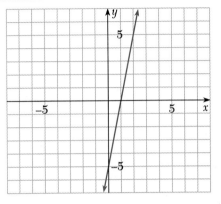

The x-intercept is $(-4, 0)$ and the y-intercept is $(0, -2)$.

B.   Find the intercepts for the graph of

$$y = \frac{5}{2}x - 7.$$

x-intercept:

$$0 = \frac{5}{2}x - 7 \qquad \textbf{Replace } y \textbf{ with 0.}$$

$$2(0) = 2(\frac{5}{2}x - 7) \qquad \textbf{Multiply both sides by 2.}$$

$$0 = 5x - 14$$

$$14 = 5x$$

$$\frac{14}{5} = x \qquad\qquad (\frac{14}{5}, 0) \textbf{ or } (2.8, 0)$$

y-intercept:

$$y = \frac{5}{2}(0) - 7 \qquad \textbf{Replace } x \textbf{ with 0.}$$

$$y = -7 \qquad\qquad (0, -7).$$

The x-intercept is $(\frac{14}{5}, 0)$ or $(2.8, 0)$ and the y-intercept is $(0, -7)$.

B.   Find the intercepts for the graph of

$$y = \frac{3}{2}x - 8.$$

Answers to Warm Ups   A. x-intercept: $(1, 0)$; y-intercept: $(0, -5)$ B. x-intercept: $(\frac{16}{3}, 0)$; y-intercept: $(0, -8)$

Find the slope of a line given two points on the line.

The slope of a line (m) is the ratio of the vertical change (rise) to the horizontal change (run) between any two points on a line (Figure 8.6). Following tradition, we use the symbol m to designate the slope of a line.

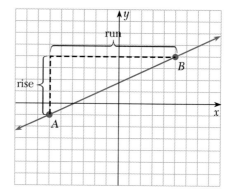

Figure 8.6

Since the slope is a ratio, we can write it in fraction form as

$$\text{slope} = m = \frac{\text{rise}}{\text{run}} = \frac{\text{change in } y}{\text{change in } x}$$

provided the change in x is not 0.

In Figure 8.6, we can find the slope of the line by counting the units of rise and run on the graph. Going from A to B, the rise is up 5 units or $+5$. The run is right 11 units or $+11$. So, the slope is

$$m = \frac{\text{rise}}{\text{run}} = \frac{5}{11}$$

The slope of a line describes the slant (pitch, grade) of a line and is constant regardless of the points chosen on the line.

Consider the line joining points A and B in Figure 8.7.

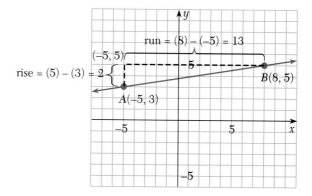

Figure 8.7

Going from A to B, the rise from y-value 3 to y-value 5 is the difference of 5 and 3, or $5 - 3$, or 2. The run from x-value -5 to x-value 8 is the difference of 8 and -5, or $8 - (-5)$, or 13. Therefore,

$$\text{slope} = m = \frac{\text{change in } y}{\text{change in } x} = \frac{5 - 3}{8 - (-5)} = \frac{2}{13}$$

Note that if the slope is calculated from B to A, the rise is actually a fall of 2, or -2, and the run is to the left, or -13. The slope is still $\dfrac{2}{13}$.

$$\text{slope} = m = \frac{\text{change in } y}{\text{change in } x} = \frac{3-5}{-5-8} = \frac{-2}{-13} = \frac{2}{13}$$

▶ **To find the slope of a line given two points, (x_1, y_1) and (x_2, y_2)**

Find the ratio of the change in y to the change in x:

$$\text{slope} = m = \frac{y_2 - y_1}{x_2 - x_1}$$

where $x_2 - x_1 \neq 0$.

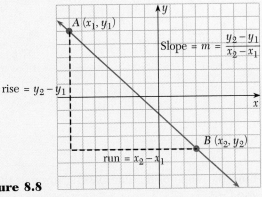

Figure 8.8

So, if a line contains the points $(6, -8)$ and $(4, 12)$, the slope is

$$m = \frac{(12) - (-8)}{(4) - (6)} = \frac{12 + 8}{-2} = \frac{20}{-2} = -10$$

Examples C–E   **Warm Ups C–E**

Directions: Find the slope of the line.

Strategy: Given the graph of the line, count the units of rise and run and then find the ratio of rise to run. Given two points on a line, substitute in the slope formula.

C.
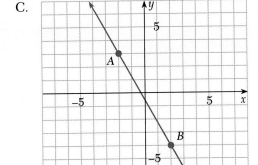

From A to B the rise is -7 and the run is 4.

C.
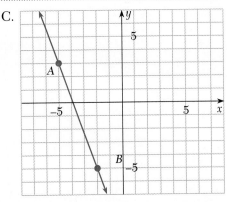

(continues next page)

Answer to Warm Up   C. $m = -\dfrac{8}{3}$

$$m = \frac{-7}{4} = -\frac{7}{4} = -1.75$$

The slope is $-\dfrac{7}{4}$ or -1.75.

D. Find the slope of the line that contains the points $(9, 5)$ and $(-10, 6)$.

Strategy: Use the formula: $m = \dfrac{y_2 - y_1}{x_2 - x_1}$

Let $P_1 = (9, 5)$ and $P_2 = (-10, 6)$.   Label the points.

$$m = \frac{y_2 - y_1}{x_2 - x_1} = \frac{6 - (5)}{-10 - (9)}$$

Substitute 6 for y_2, 5 for y_1, -10 for x_2, and 9 for x_1.

$$= \frac{1}{-19} = -\frac{1}{19}$$

So, the slope is $-\dfrac{1}{19}$.

D. Find the slope of the line that contains the points $(2, -5)$ and $(-3, 3)$.

E. Find the slope of the line passing through the points $(-5, -7)$ and $(-3, 7)$.

Let $P_1 = (-5, -7)$ and $P_2 = (-3, 7)$.   Label the points.

$$m = \frac{y_2 - y_1}{x_2 - x_1} = \frac{7 - (-7)}{-3 - (-5)}$$

Substitute 7 for y_2, -7 for y_1, -3 for x_2, and -5 for x_1.

$$= \frac{14}{2}$$

$$= 7$$

The slope is 7.

E. Find the slope of the line passing through the points $(1, 4)$ and $(3, -12)$.

Answers to Warm Ups   D. $m = -\dfrac{8}{5}$ E. $m = -8$

Exercises 8.4

OBJECTIVE 1: *Find the x- and y-intercepts of a line.*

A.

Find the *x*- and *y*-intercepts from the graph.

1.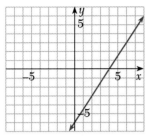

2.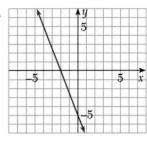

3.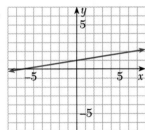

4.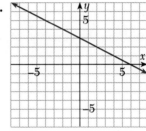

Find the *x*- and *y*-intercepts from the equation of the line.

5. $4x + y = 16$

6. $6x - y = 18$

7. $2x + 9y = 18$

8. $6x - 15y = 90$

9. $8x - 3y = 48$

10. $-5x + 7y = 35$

B.

Estimate the x- and y-intercepts from the graph.

11.
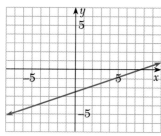

12.
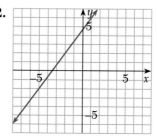

13.
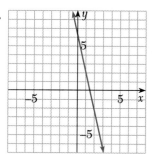

14.
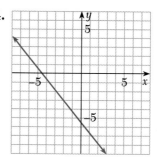

Find the x- and y-intercepts from the equation of the line.

15. $-3x + 2y = -15$

16. $-6x + 5y = -3$

17. $y = -\dfrac{1}{3}x + 7$

18. $y = \dfrac{5}{8}x + \dfrac{9}{8}$

19. $4y = -7x - 8$

20. $7y = 3x + 24$

OBJECTIVE 2: *Find the slope of a line.*

A.

Find the slope of the line given its graph.

21.
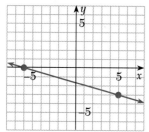

22.
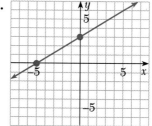

23.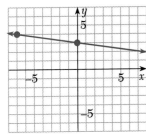

24.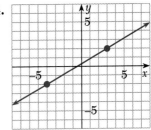

Find the slope of the line given two points.

25. $(0, 7), (3, 0)$

26. $(-4, 0), (0, 5)$

27. $(3, 7), (4, 6)$

28. $(4, 7), (9, 11)$

29. $(-10, 2), (6, -1)$

30. $(0, 7), (4, -6)$

B.

Estimate the slope of the line from the graph.

31.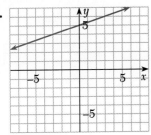

32.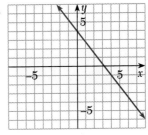

33.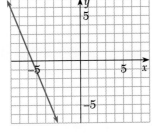

34.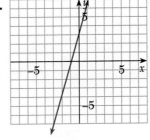

Find the slope of the line given two points.

35. $(5, -4), (-2, -1)$

36. $(-4, 0), (5, -6)$

37. $(-2, 5), (5, 1)$

38. $(-3, -1), (7, 2)$

39. $(-12, -2), (0, 3)$

40. $(-14, -14), (4, -11)$

C.

Find the x- and y-intercepts from the equation of the line. Find the slope of the line from the intercepts.

41. $3x + 7y = 14$

42. $6x + 5y = 25$

43. $-3x - 4y = 9$

44. $-3x + 8y = 24$

45. $8x - 2y = 7$

46. $8x + y = -4$

47. A home in Hibbing, Minnesota, has a steep roof. For every $2\frac{1}{2}$ feet of run it has $1\frac{1}{2}$ feet of rise. What is the pitch of the roof? (The pitch of the roof is the slope of the roof.)

48. A home in Aspen, Colorado, has a steep roof. For every 4 feet of run it has $5\frac{1}{4}$ feet of rise. What is the pitch of the roof?

49. The Timber Mountain water ride at Knott's Berry Farm has a final plunge from a height of 38 ft while moving 27 ft horizontally. Draw a graph that depicts this and calculate the slope of the ramp.

50. The Tidal Wave water ride at Magic Mountain has a final plunge from a height of 50 ft while moving 50 ft horizontally. Draw a graph that depicts this and calculate the slope of the ramp.

51. The Jurassic Park water ride at Universal Studios has a final plunge from a height of 84 ft while moving 66 ft horizontally. Draw a graph that depicts this and calculate the slope of the ramp.

52. Which of the rides in Exercises 49–51 is the steepest? Why?

Exercises 53–54 refer to the Chapter 8 Application. See page 725.

53. How much money do the grandparents invest each year into the college fund? Using the ordered pairs (1, 500) and (5, 2500), calculate the slope of the line between them. Consider the investment equation, $I = 500t$. What is the slope of this line? Indicate on the graph how this is determined.

54. By how much money does the face value of the college fund grow each year? Using the ordered pairs (1, 1000) and (10, 10,000), calculate the slope of the line between them. Consider the value equation $V = 1000t$. What is the slope of this line? Indicate on the graph how this is determined.

STATE YOUR UNDERSTANDING

55. What do you know about the slope of this line?

56. Explain how to find the x-intercept and the y-intercept. Indicate their location on the graph.

57. Explain how to use the graph of a line to calculate its slope.

CHALLENGE

Determine the slope of the line given two points.

58. $(-5, -4), (6, -4)$ **59.** $(7, -5), (7, 7)$ **60.** $\left(-\dfrac{12}{17}, -\dfrac{8}{21}\right), \left(\dfrac{9}{11}, -\dfrac{23}{33}\right)$

GROUP ACTIVITY

61. If a line has slope $m = \dfrac{2}{3}$ and contains the point $(-2, 6)$, can your group find two more points on the line? How? Be prepared to show how in class.

MAINTAIN YOUR SKILLS (Sections 7.4, 7.6)

Change to a percent.

62. 0.067   **63.** 3.45

Change to a decimal.

64. 18.6%   **65.** 113.8%

66. A certain spaghetti sauce has a 12.5% fat content. How many grams of fat are there in a serving of 90 grams?

8.5 Variation (Optional)

OBJECTIVE	**1.** Solve problems involving direct variation.
	2. Solve problems involving inverse variation.

VOCABULARY **Direct variation** is a relationship between two quantities such that their **ratio** is constant. This constant is called the **constant of variation.** We say that one quantity **varies directly as** another.

Inverse variation is a relationship between two quantities such that their **product** is a constant. We say that one quantity **varies inversely as** another.

HOW AND WHY
Objective 1   **Solve problems involving direct variation.**

In our physical world, most things are in a state of change or variation. For example, changes in temperature, rainfall, light, heat, and fuel are measured and recorded. Relationships between such measured changes can be expressed by equations. Here we are concerned about two kinds of related change: direct and inverse variation.

We begin by discussing direct variation. Here are two examples of quantities that vary directly.

1. When traveling at 35 mph, the distance, d, traveled varies directly as the time, t, traveled. This means that the ratio of d and t is constant. The constant of variation is 35. We write

$$\frac{d}{t} = 35 \text{ or } d = 35t$$

2. The circumference, C, of a circle varies directly as the radius, r. This means the ratio of C and r is a constant. The constant is 2π. We write

$$\frac{c}{r} = 2\pi \text{ or } C = 2\pi r$$

Notice in each example that as one variable increases, so does the other. If one gets smaller, so does the other. We call this direct variation because the value of the variables move in the same direction.

Given certain data, we can predict the value of a quantity that varies directly with another quantity by using the *equation of direct variation.*

Equation of
Direct Variation

If two quantities x and y vary directly, then there is some nonzero number k so that

$$\frac{y}{x} = k \quad \text{or} \quad y = kx$$

Examples A–B   **Warm Ups A–B**

....

Directions: Solve.

Strategy: Use the equation of direct variation and the given data to find the constant of variation. Use this value and the given data to find the unknown.

....

A. The weight, w, of a piece of aluminum varies directly as its volume v. If the volume of one piece is 1.5 cubic feet and its weight is 252 pounds, find the weight of a piece whose volume is 5 cubic feet.

Strategy: Write an equation of direct variation. Let k be the constant of variation. Since weight varies directly as volume, we write

$w = kv$

$252 = k(1.5)$ **Find the constant of variation by replacing w with 252 and v with 1.5.**

$\dfrac{252}{1.5} = k$ **Divide both sides by 1.5.**

$168 = k$

Knowing the value of k, we write the equation of direct variation again:

$w = 168v$ **Replace k with 168.**

Now find the weight of the second piece with volume 5 cubic feet.

$w = 168(5)$ **Replace v with 5.**

$w = 840$

The weight of the piece of aluminum with volume 5 cubic feet is 840 pounds.

A. Find the weight of a piece of aluminum whose volume is 7.5 cubic feet. Use the relationship expressed in Example A.

B. The cost of chicken varies directly with the weight of the chicken. If 5 pounds of chicken costs $14.45, what is the cost of 8 pounds?

Strategy: Find the constant of variation using the first 5 pounds of chicken at $14.45. Let C represent the cost and p the weight.

$C = kp$ **Equation of direct variation.**

$14.45 = k(5)$ **Replace C with 14.45 and p with 5.**

$2.89 = k$ **The constant of variation.**

Now use the constant of variation to find the cost of 8 pounds of chicken.

$C = 2.89p$ **Equation of direct variation.**

$C = 2.89(8)$ **Replace p with 8.**

$ = 23.12$

The cost of 8 pounds of chicken is $23.12.

B. The cost of Yukon Gold potatoes varies directly with weight of the potatoes. If 10 pounds of potatoes costs $8.37, what is the cost of 24 pounds?

Answers to Warm Ups   A. 1260 pounds   B. $20.09

HOW AND WHY

Objective 2 **Solve problems involving inverse variation.**

Inverse variation is recognized between quantities when their product is constant. Here is an example of quantities that vary inversely:

Traveling 70 miles, time, t, varies inversely as the speed, r.

This means the product of t and r is constant. The constant of variation is 70. We write:

$$rt = 70$$

Some possible values for r and t are listed in the table.

r	t
70 mph	1 hr
35 mph	2 hr
14 mph	5 hr
10 mph	7 hr

Notice that as the values for r get smaller, the values for t get larger. The values of the quantities are moving in the opposite direction, hence the name inverse variation.

Equation of Inverse Variation

If two quantities x and y vary inversely, then there is some nonzero number k so that

$$xy = k$$

Examples C–E   **Warm Ups C–E**

Directions: Solve.

Strategy Use the equation of inverse variation and the given data to find the constant of variation. Use this value and the data given to find the unknown value.

C. If y varies inversely as x and $y = 12$ when $x = 5$, what is the value of x when $y = 120$?

$yx = k$ **Equation of inverse variation.**

$12(5) = k$ **Substitute 12 for y and 5 for x.**

$60 = k$ **The constant of variation.**

Now use the constant of variation to find x when $y = 60$.

$yx = 60$ **Equation of inverse variation.**

$120x = 60$ **Substitute 120 for y.**

$x = \dfrac{1}{2}$

The value of x is $\dfrac{1}{2}$.

C. If y varies inversely as x and $y = 20$ when $x = 8$, what is the value of y when $x = 48$?

Answer to Warm Up   C. $y = \dfrac{10}{3}$

D. If a varies inversely as the square of b and $a = 9$ when $b = 4$, find the value of a when $b = 36$.

$ab^2 = k$ **Equation of inverse variation.**

$(9)(4)^2 = k$ **Substitute 9 for a and 4 for b.**

$k = 144$ **The constant of variation.**

Now use the constant of variation to find a when $b = 36$.

$ab^2 = 144$ **Equation of inverse variation.**

$a(36)^2 = 144$ **Substitute 36 for b.**

$1296a = 144$

$a = \dfrac{1}{9}$

The value of a is $\dfrac{1}{9}$.

D. If the square of x varies inversely as y and $x = 3$ when $y = 100$, find the value of y when $x = 5$.

E. The base b of a rectangle with constant area k varies inversely with its height. One rectangle has a base of 10 cm and a height of 6 cm. Find the height of another rectangle that has a base of 12 cm.

Use the formula $bh = k$ since the base and the height vary inversely. The product is a constant.

$bh = k$

$10(6) = k$ **Substitute the known values.**

$60 = k$ **The constant of variation.**

$bh = 60$ **Substitute the value for k.**

$12h = 60$ **The new rectangle has a base of 12.**

$h = 5$

Therefore, the height of the second rectangle is 5 cm.

E. Find the height of a rectangle with the same area ($k = 60$) as the one in Example E if the base is 8 cm.

Answers to Warm Ups   D. $y = 36$ E. 7.5 cm

Exercises 8.5

OBJECTIVE 1: *Solve problems involving direct variation.*

A.

Write an equation expressing the variation relationship given in each of the following. Use k as the constant of variation.

1. If time is held constant, the distance (d) varies directly as the rate.

2. The circumference (C) of a circle varies directly as the diameter (d).

3. The perimeter (P) of a square varies directly as the length of its side (s).

4. The volume of a cone (V) varies directly as the area of its base (B) if its height is a constant.

5. The total cost (C) of purchasing soup varies directly with the number of cans (n) bought if the price per can is constant.

6. The interest earned (I) on a savings account varies directly with the amount of savings (s) if the interest rate is fixed.

7. If r varies directly as m, as r goes from 5 to 15, m goes from 20 to _____.
8. If t varies directly as q, as t goes from 100 to 50, q goes from 10 to _____.

B.

Find the constant of variation for each of the stated conditions.

9. y varies directly as x, and $y = 24$ when $x = 16$.

10. y varies directly as x, and $y = 12$ when $x = 9$.

11. y varies directly as the square of x, and $y = 14$ when $x = 7$.

12. y varies directly as x, and $y = 120$ when $x = 480$.

Solve by first obtaining the constant of variation.

13. If a varies directly as b and $a = 14$ when $b = 6$, find a when $b = 16$.

14. If a varies directly as m and $a = 16$ when $m = 12$, find a when $m = 23$.

15. If m varies directly as the square of n and $m = 30$ when $n = 5$, find m when $n = 16$.

16. If z varies directly as the square of w and $z = 42$ when $w = 5$, find z when $w = 2$.

OBJECTIVE 2: *Solve problems involving inverse variation.*

A.

Write an equation expressing the variation relationship given in each of the following. Use k as the constant of variation.

17. If the area of a rectangle is held constant, the length (ℓ) varies inversely as its width (w).

18. If temperature is held constant, the volume (V) of a gas varies inversely as the pressure (P) applied.

19. If the total dollars spent is held constant, the number of cans (C) of fruit that can be bought varies inversely as the price (p) of a single can.

20. The number of pieces (n) that can be cut from a board of fixed length varies inversely as the length of the piece of board (t).

21. If books are of uniform size, the number of books (n) that will fit on a shelf of fixed length varies inversely with the width (s) of a spine.

22. The time (t) it takes to go from A to B varies inversely as the speed (r) that one travels.

23. If A varies inversely as B, as A goes from 10 to 5, B goes from 20 to _____.

24. If q varies inversely as w, as q goes from 11 to 33, w goes from 18 to _____.

B.

Find the constant of variation for each of the stated conditions.

25. y varies inversely as x, and $y = 108$ when $x = \dfrac{2}{3}$.

26. y varies inversely as x, and $y = 11$ when $x = 6$.

27. y varies inversely as x, and $y = 24$ when $x = 0.5$.

28. y varies inversely as x, and $y = 22$ when $x = 4$.

Solve by first finding the constant of variation.

29. If y varies inversely as x and $y = 42$ when $x = 16$, find y when $x = 48$.

30. If y varies inversely as x and $y = 36$ when $x = 24$, find y when $x = 12$.

31. If y varies inversely as x and $y = 22$ when $x = 4$, find y when $x = 8$.

32. If r varies inversely as t and $r = 31$ when $t = 2$, find r when $t = 16$.

C.

33. The weight of a metal ingot varies directly as the volume v. If an ingot containing $1\frac{2}{3}$ cubic feet weighs 350 pounds, what will an ingot of $2\frac{1}{3}$ cubic feet weigh?

34. The time it takes to make a certain trip varies inversely as the speed. If it takes 5 hours at 50 mph, how long will it take at 60 mph?

35. A car salesman's salary varies directly as his total sales. If he receives $372 for sales of $3100, how much will he receive for sales of $5400?

36. The amount of quarterly income a person receives varies directly as the amount of money invested. If Dan earns $40 on a $2000 investment, how much would he earn on a $3400 investment?

37. The number of amperes varies directly as the number of watts. For a reading of 50 watts, the number of amperes is $\frac{5}{11}$. What are the amperes when the watts are 75?

38. The weight of wire varies directly as its length. If 1000 ft of wire weighs 45 lb, what will 2200 feet of wire weigh?

39. The length of a rectangle with a constant area varies inversely as the width. If one rectangle has a length of 12′ and a width of 8′, what will be the length of a rectangle with a width of 6′?

40. The force needed to raise an object with a lever varies inversely as the length of the lever. If it takes 40 lb of force to lift a certain object with a 2-ft-long lever, what force will be necessary if you use a 3-ft lever?

41. As a rule of thumb, realtors suggest that the price you can afford to pay for a house varies directly as your annual salary. If a person earning $18,500 can purchase a $46,250 home, what price home can a person earning $24,000 annually afford?

42. Assuming that each person works at the same rate, the time it takes to complete a job varies inversely as the number of people assigned to it. If it takes five people 12 hours to do a job, how long will it take three people?

43. The amount of money invested by her grandparents in Emma's college fund is an example of direct variation. Explain why.

44. Emma will be 18 in 2004. Use the investment equation and the value equation to determine the total amount that the grandparents have invested, and the face value of the college fund. See Exercises 38–41, Section 8.3.

STATE YOUR UNDERSTANDING

45. As the surroundings get warmer, both the Fahrenheit and Celsius temperatures increase. Does either temperature vary directly as the other? Explain your answer using the meaning of direct variation.

CHALLENGE

46. The weight of cable varies directly as its length. If 560 feet of cable weighs 38 pounds, what will 1.5 miles of cable weigh, to the nearest half pound?

47. If distance is constant, time varies inversely as rate. A trip takes 2.5 hours at 55 miles per hour. How long, to the nearest tenth of an hour, will it take at 100 kilometers per hour?

GROUP ACTIVITY

48. In 2004, not all the bonds in the Chapter 8 Application will have matured to their face value, so Emma will not actually have $19,000 in her account. Assume that the interest rate is 6% (so the bonds will mature in 12 years). Calculate the actual amount of money in the account. You will need to use the equation $V = 500(1.06)^t$ that gives the value, V, of a bond bought for $500 after t years. Make a table that shows your strategy.

MAINTAIN YOUR SKILLS (Sections 6.3, 6.5, 6.7)

49. Find the quotient of $-21.06t^2$ and $-1.56t$.

50. Find the product of $-8.99w$ and 2.9.

51. Evaluate $\dfrac{9r^2}{0.5s}$ when $r = 2.3$ and $s = -1.8$.

52. Evaluate $r^2 - 34s^2$ when $r = 2.3$ and $s = -1.8$.

53. Is -3.5 a solution of $0.8 = 31.2z + 100$?

CHAPTER 8

OPTIONAL Group Project *(1–2 weeks)*

A part of the tax table for 1995 is given below. The figures are for a single or head of household taxpayer.

Taxable Income ($)	Tax Owned ($)
15,000	2254
16,000	2404
17,000	2554
18,000	2704
19,000	2854
20,000	3004
21,000	3154
22,000	3304
23,000	3454
24,000	3692
25,000	3972
26,000	4252
27,000	4532
28,000	4812
29,000	5092
30,000	5372

a. Graph the data carefully on graph paper. Label both axes, the scale on each, and each point. Describe the pattern made by the data.

b. Pick 10 pairs of data points and calculate the slope of the line between them. Summarize your results in a table. How do the slopes relate to your graph?

c. Of the following list of equations, one describes part of the data, another describes the rest of the data, and the others do not describe more than two data points. In each equation, I is the amount of income and T is the tax owed. Outline a decision strategy, show all calculations, and identify the two equations that together describe all the data.

$T = 0.172I - 436$

$T = 0.1936I - 868$

$T = 0.28I - 3028$

$T = 0.238I - 2020$

$T = 0.15I + 4$

$T = 0.2368I - 1732$

d. Using two different colors, graph your two selected equations on the same graph as your data. Label clearly.

e. Explain why more than one line is needed to describe the data. What significance does the slope of each equation have?

CHAPTER 8

True-False Concept Review

ANSWERS

Check your understanding of the language of algebra and arithmetic. Tell whether each of the following statements is True (always true) or False (not always true). For each statement you judge to be false, revise it to make a statement that is true.

1. _____

   1. Equations with two variables have more than one solution.

2. _____

   2. The ordered pair $(2, 3)$ is a solution of the equation $2y - x = 1$.

3. _____

   3. The solutions for equations with two variables are always ordered pairs.

4. _____

   4. The solution for equations with two variables can always be graphed on a rectangular coordinate system.

5. _____

   5. The x- and y-axes of a coordinate system are number lines.

6. _____

   6. The quadrants of the rectangular coordinate system are numbered clockwise.

7. _____

   7. The point where the x-axis and y-axis intersect is called the "starting point."

8. _____

   8. The x-value of an ordered pair shows the distance of a point from the x-axis.

9. _____

   9. If both the x- and y-values of an ordered pair are negative, then the point corresponding to the pair is in quadrant III.

10. _____

   10. If either x or y is zero in an ordered pair, then the point corresponding to the pair is in quadrant I.

11. _____

   11. For all values of a and b, the graph of $ax + by = c$ is a straight line.

12. _____

   12. At least three points are needed to determine the position of a line on a graph.

13. _____

13. A line graph is a picture of the solution of an equation of the form $ax + by = c$, a and b not both zero.

14. _____

14. Line graphs are useful in mathematics, but not for practical applications.

15. _____

15. The point $(-2, -1)$ is in quadrant II.

16. _____

16. The x-value of every y-intercept is 0.

17. _____

17. The x- and y-intercepts are the coordinates of the points where the line crosses the x- and y-axes.

18. _____

18. The slope of a line that goes up from right to left is positive.

19. _____

19. If $\dfrac{a}{b}$ always has the same value even if a and b both change, then a varies directly as b.

20. _____

20. If ab always has the same value even if a and b both change, then a varies inversely as b.

CHAPTER 8

Review

Section 8.1 Objective 1

Check whether the ordered pair is a solution of the equation.

1. $(4, -1)$; $3x - 8y = 20$
2. $(5, 0)$; $2x - 3y = -15$
3. $(-1, -1)$; $4x - 7y = 3$
4. $(0.5, 9)$; $8x - y = -5$
5. $(-3, -8)$; $7x + 5y = 19$

Section 8.1 Objective 2

Construct a rectangular coordinate system and plot the following points.

6. $(-4, -6)$
7. $(4, -1)$
8. $(0, -5)$
9. $(\frac{1}{2}, -6)$
10. $(-3, -5.5)$

Section 8.1 Objective 3

Identify the coordinates of the points on the following graph.

11. A
12. B
13. C
14. D
15. E

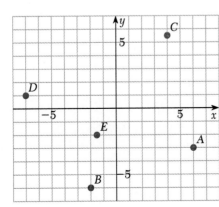

Section 8.2

Given the equation $y = \frac{4}{5}x - 4$, solve for y, given the following values of x. Write the solutions as ordered pairs:

16. $x = 5$
17. $x = 10$
18. $x = -4$

Given the equation $y = \frac{6}{7}x - 6$, solve for x, given the following values of y. Write the solutions as ordered pairs.

19. $y = 6$
20. $y = -\frac{3}{7}$

Section 8.3

Make a table of values and graph each of the following.

21. $y = 3x - 6$
22. $y = -\frac{1}{4}x + 1$
23. $y = -\frac{2}{3}x - 2$
24. $y = \frac{3}{5}x - \frac{9}{5}$
25. $y = \frac{1}{6}x - \frac{1}{2}$

Section 8.4 Objective 1

26. Find the x- and y-intercepts given the graph.

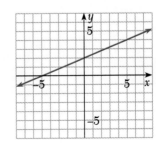

787

© 1997 Saunders College Publishing.

Find the x- and y-intercepts given the equation.

27. $3x - 8y = 24$
28. $2x - 5y = 20$
29. $6x + 7y = 15$
30. $y = 7x - 21$

Section 8.4   *Objective 2*

Find the slope of the line from the graph.

31.

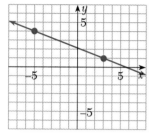

Find the slope of the line given two points on the line.

32. $(-5, 6)$ and $(3, -2)$
33. $(4, 3)$ and $(5, -2)$
34. $(-7, -3)$ and $(5, -2)$
35. $(9, -13)$ and $(21, 14)$

Section 8.5   *Objective 1*

36. Find the constant of variation if y varies directly as x, and $y = 40$ when $x = 8$.

37. If s varies directly as t and $s = 24$ when $t = 6$, find s when $t = 96$.

38. If q varies directly as r and $q = 5.25$ when $r = 8.4$, find q when $r = 10.7$.

39. The weight of a metal object varies directly as the volume. If an object containing 4.3 cubic feet weighs 1225.5 pounds, what will a 2.8-cubic-foot metal object weigh?

40. The amount of annual income received from an investment varies directly as the amount invested. If an investment of \$4800 returns \$504, what will be the return on \$13,500?

Section 8.5   *Objective 2*

41. Find the constant of variation if y varies inversely as x, and $y = 40$ when $x = 8$.

42. If y varies inversely as x and $y = 32$ when $x = 4$, find y when $x = 100$.

43. If a varies inversely as b and $a = 45$ when $b = 15$, find a when $b = 75$.

44. The time it takes to travel from point A to point B varies inversely as the speed. If it takes seven hours at 42 mph, how long will it take at 50 mph?

45. The average donation needed to raise a certain amount of money varies inversely as the number of donors. If an average gift of \$25 is needed with 250 donors, what average gift would be needed if there were only 200 donors?

CHAPTER 8

Test

ANSWERS

1. _____

2. _____

3. _____

1. Given the equation $y = \dfrac{3}{5}x - 9$, solve for y if $x = -25$.

2. Is $(-7, 4)$ a solution of $3x + 7y = 7$?

3. Locate the following points on the coordinate system: $A(-3, 3)$, $B\ (4, -3)$, $C\ (0, 3.5)$, $D\ (-1.5, -4)$, $E\ (2, 4)$.

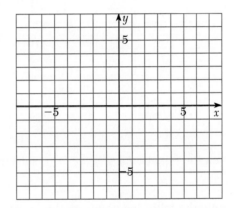

4. _____

4. Make a table of values and draw the graph of $y = 2x - 6$.

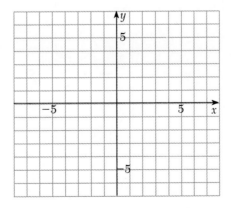

5. _____

5. Is $(-4, -8)$ a solution of $4y - 7x = 40$?

6. _____

6. Identify the coordinates of the points on the following graph.

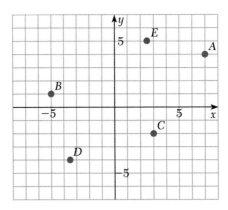

7. _____

7. Make a table of values and draw the graph of $y = -\dfrac{1}{2}x + 3$.

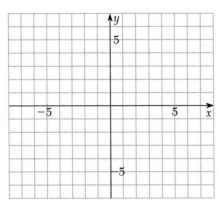

8. _____

8. Given the equation $y = \dfrac{1}{7}x - \dfrac{12}{7}$, solve for y if $x = -2$.

9. _____

9. Locate the following points on the coordinate system: $A\,(-5,\,-8)$, $B\,(-7,\,6)$, $C\,(0,\,-6)$, $D\,(-4.5,\,0)$, $E\,(4.5,\,4.5)$.

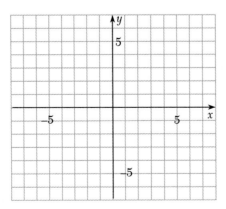

10. _____

10. Given the equation $y = -\dfrac{1}{11}x + 11$, solve for y if $x = 44$.

11. _____

11. Is $(-5.5,\,0.4)$ a solution of the equation $y = -3x - 16.1$?

790

12. _____

12. Make a table of values and draw the graph of $y = -4x + 6$.

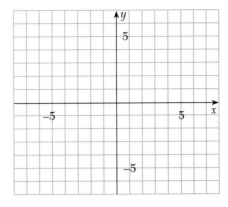

13. _____

13. Identify the x- and y-intercepts from the graph. Round to the nearest half unit.

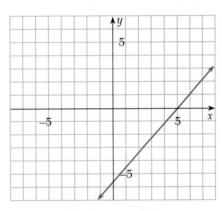

14. _____

14. Find the slope of the line given its graph.

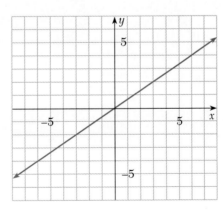

15. _____

15. Find the x- and y-intercepts of the graph of $3x - 7y = 18$.

16. _____

16. Find the slope of the graph containing the points $(-3, 5)$ and $(6, -7)$.

17. _____

17. If y varies directly as x and $y = 9$ when $x = 40$, find y when $x = 90$.

791

18. _____

18. If y varies inversely as x and $y = 18$ when $x = 27$, find y when $x = 45$.

19. _____

19. The cost of renting a car is $20 per day plus $0.25 per mile driven. The daily rental is expressed by $C = 20 + 0.25m$, where C is the total cost and m is the number of miles driven. Draw the graph of the relationship and find the cost if the car is driven 325 miles in one day.

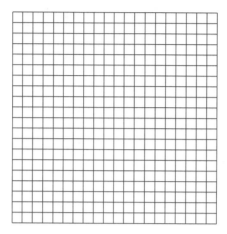

CHAPTER 1–8

Cumulative Review

---

1. Find the average of 3,456, 7,234, 8,110, 1,375, and 9,120.
2. Simplify:
   $9 \cdot 12 + 11(82 - 56) + 5^2 - (12 - 7)^2$
3. Use the property of exponents for raising a product to a power to write $(9 \cdot 2)^5$ in exponent form.
4. Use the property of exponents for raising a power to a power to write $(9^3)^4$ in exponent form.
5. Find the value of $297,987 \times 10^6$.
6. Translate to algebra, let n represent the number: A number multiplied by 10 and that product squared.
7. Evaluate $C = \pi D$ if $\pi = 3.14$ and $D = 8$.
8. Find the perimeter of a rectangular tract of land which has a length of 203 feet and a width of 93 feet.
9. Is 12 a solution of $8x - 15 = 81$?
10. Subtract: $28x - 12x - 17x$
11. Mustoffa bought a piece of real estate for $253,000. During the next couple of years the area was hard hit by a recession and he was only able to sell it for $215,000. Express his loss as a signed number.
12. A gambler lost $15 on each of eight consecutive rolls of the dice at the craps table. Express her total loss as a signed number.
13. Simplify: $\dfrac{4^2 - 4[6 + 3(-4)]^2}{(-14 + 16)^2}$
14. Simplify: $3xy(4x - 15y - 12)$
15. The sum of 6, -8, and four times a number is 10 plus two times the number. What is the number?
16. Find the LCM of 18, 28, and 48.
17. Add: $-\dfrac{2x}{9} + \dfrac{7x}{12} + \dfrac{x}{4}$
18. Find the average of $\dfrac{4}{9}, \dfrac{11}{12}, \dfrac{5}{6}, \dfrac{1}{2}$, and $\dfrac{5}{3}$.
19. Is $-\dfrac{4}{3}$ a solution of $\dfrac{5}{2} - \dfrac{3x}{8} = 3$?

20. Evaluate the formula $A = \dfrac{1}{2}h(B + b)$ if $h = 1\dfrac{2}{3}$, $B = 2\dfrac{1}{8}$, and $b = 1\dfrac{5}{8}$.
21. Solve: $0.8x - 0.024 = 16$
22. Solve: $9.82 - 0.5x = 3.72$
23. During the three years that Donna played basketball in high school, her team won 22 games the first year, 20 games the second year, and 18 games the third year. What was the average number of games won each year?
24. In 1 second a computer can do 8.9×10^5 calculations. Write this number of calculations in place value notation.
25. Mr. Mack's auto has a gasoline tank that holds $18\dfrac{1}{2}$ gallons of gasoline when full. He starts with a full tank and drives for 3 hours. He checks the gasoline gauge and sees that he has $\dfrac{1}{4}$ of a tank remaining. Assuming the gauge is correct, how many gallons still remain in the tank?
26. Solve the proportion: $\dfrac{15}{w} = \dfrac{0.74}{3.7}$
27. A watch gains 0.4 minute every 5 hours. How fast will the watch be in 3 days if it is not reset?
28. 165% of what number is 112.2?
29. To pass a written test to qualify for a job, Yien must score at least 70%. How many questions out of 148 must she get correct to qualify?
30. Grace left her waiter a $3.50 tip at dinner. If the dinner check was $14.70, what percent (to the nearest whole number) did she tip?

For Exercises 31–34, find the missing value for each of the following ordered pairs for the equation $7x - 2y = 14$.
31. $(-2, \quad)$
32. $(3, \quad)$
33. $(\quad , 0)$
34. $(\quad , -7)$
35. Graph the equation $3x - 2y = 2$.

The following table shows the number of automobile rentals at an agency for one month:

Compacts	2338
Standard	1147
Luxury	356
Limousines	23
Vans	108

36. What was the total number of rentals during the month?

37. What is the difference between the number of compact rentals and the total of all others?

38. What is the ratio of the number of standards to the number of compacts (to the nearest tenth)?

39. What percent of the rentals are compacts and standards (to the nearest percent)?

40. True or False: Luxury and limousine rentals make up less than 10% of the total rentals.

The following bar graph shows the occupancy rate of an apartment complex for a 4-year period:

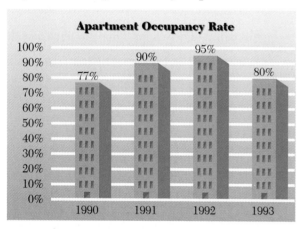

Apartment Occupancy Rate

41. What year was the occupancy rate the highest?

42. In what year were the apartments all occupied?

43. What will the occupancy rate be in 1994 if the occupancy rate increase is the same as between 1990 and 1991?

44. What will the occupancy rate be in 1994 if the occupancy rate decline is the same as between 1992 and 1993?

45. Assuming that in 1992 there were 1500 apartments available, how many were vacant?

46. Add: 6 ft 10 in.
            11 ft 7 in.
          + 8 ft 4 in.

47. Subtract: 3 hr 25 min 41 sec
                  − 1 hr 30 min 54 sec

48. A can of cherry pie filling contains 624 grams. How many grams of the pie filling are in 13 cans?

49. The In-Out Market sold 12 lb of hamburger on Monday, 22 lb on Tuesday, 35 lb on Wednesday, 28 lb on Thursday, 45 lb on Friday, and 60 lb on Saturday. What was the average number of pounds of hamburger sold during the 6 days?

50. LIKE International agrees to donate 1 cent for every foot that Kevin jogs during 1 week to the World Food Bank. Kevin jogged the following miles during the week: Monday, 3; Tuesday, 2.5; Wednesday, 4; Thursday, 1.5; Friday, 3; Saturday, 4.5; Sunday, 1. How much does LIKE donate to the World Food Bank?

51. Convert $\dfrac{320 \text{ miles}}{1 \text{ hour}}$ to $\dfrac{? \text{ feet}}{1 \text{ second}}$.

52. While in England for vacation, Rashi was advised to lose 42 kilograms of weight for health reasons. How many pounds does this represent? Find to the nearest pound.

53. Find the perimeter of the following figure.

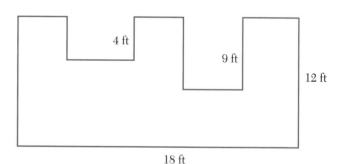

54. Audrey is planning to put a fringe around her circular rug. The rug has a diameter of 3 meters. If the fringe costs $0.45 per centimeter, find the cost of the fringe to the nearest dollar.

55. April's house has a large deck attached. The deck measures 16 feet by 21 ft. April plans to stain the deck. One gallon of the stain will cover 125 ft^2. How many gallons of stain must she purchase? (Assume the stain is sold only in gallon units.)

56. Find the area of the following figure:

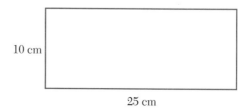

57. Find the volume of a cylinder that has a 21″ diameter and is 30″ tall. Round to the nearest cubic inch.

58. Given the equation $y = -\dfrac{3}{8}x - 2$, solve for y if $x = -16, -4$, and 8. Write the solutions as ordered pairs.

59. Identify the coordinates of the points on the graph.

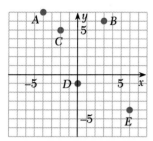

60. Draw the graph of $y = -\dfrac{5}{6}x + 3$.

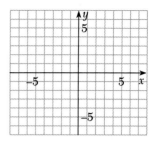

Formulas

Perimeter and Area

Square

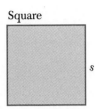

Perimeter: $P = 4s$
Area: $A = s^2$

Rectangle

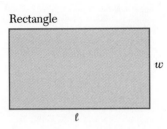

Perimeter: $P = 2\ell + 2w$
Area: $A = \ell w$

Triangle

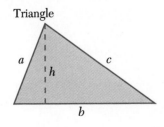

Perimeter: $P = a + b + c$
Area: $A = \dfrac{bh}{2}$

Parallelogram

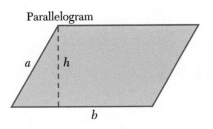

Perimeter: $P = 2a + 2b$
Area: $A = bh$

Trapezoid

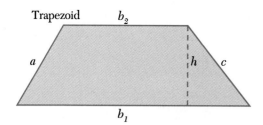

Perimeter: $P = a + b_1 + c + b_2$

Area: $A = \dfrac{1}{2}(b_1 + b_2) \cdot h$

Circle

Circumference: $C = 2\pi r$

$C = \pi d$

Area: $A = \pi r^2$

Volume

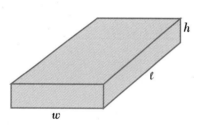

$V = \ell w h$

Cube

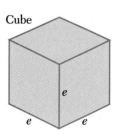

$V = e^3$

Sphere

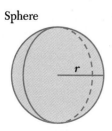

$V = \dfrac{4}{3}\pi r^3$

A2

Cylinder

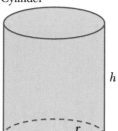

$$V = \pi r^2 h$$

Cone

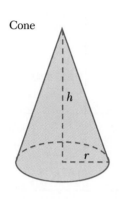

$$V = \frac{1}{3} \pi r^2 h$$

Miscellaneous

FORMULA	DESCRIPTION
$D = rt$	Distance equals rate (speed) times time.
$I = prt$	Interest equals principal times rate (%) times time.
$C = np$	Cost equals number (of items) times price (per item).
$S = C + M$	Selling price equals cost plus markup.
$s = S - D$	Sale price equals original selling price minus discount.
$V = v - D$	Current value equals original value minus depreciation.
$E = W + C$	Earnings equal base pay plus commission.
$F = \frac{9}{5} C + 32$	Fahrenheit temperature equals $\frac{9}{5}$ times Celsius temperature plus 32.
$S = v + gt$	Velocity (of object falling) equals initial velocity plus gravitational force times time.
$R \cdot B = A$	Rate (%) times base equals amount (percentage).
$a^2 + b^2 = c^2$	The square of the length of one leg of a right triangle plus the square of the length of the other leg is equal to the square of the length of the hypotenuse.

Measures (English, Metric, and Equivalents)

ENGLISH MEASURES AND EQUIVALENTS

Length	Time
12 inches (in.) = 1 foot (ft)	60 seconds (sec) = 1 minute (min)
3 feet (ft) = 1 yard (yd)	60 minutes (min) = 1 hour (hr)
5280 ft (ft) = 1 mile (mi)	24 hours (hr) = 1 day
	7 days = 1 week

Liquid Volume	Weight
3 teaspoons (tsp) = 1 tablespoon (tbs)	16 ounces (oz) = 1 pound (lb)
2 cups (c) = 1 pint (pt)	2000 pounds (lb) = 1 ton
2 pints (pt) = 1 quart (qt)	
4 quarts (qt) = 1 gallon (gal)	

METRIC MEASURES AND EQUIVALENTS

Length (Basic Unit is 1 Meter)

1 millimeter	(mm)	=		=	0.001	m
1 centimeter	(cm)	= 10 millimeters		=	0.01	m
1 decimeter	(dm)	= 10 centimeters		=	0.1	m
1 METER	(m)	= 10 decimeters		=	1	m
1 dekameter	(dam)	= 10 meters		=	10	m
1 hectometer	(hm)	= 10 dekameters		=	100	m
1 kilometer	(km)	= 10 hectometers		=	1000	m

Weight* (Basic Unit is 1 Gram)

1 milligram	(mg)	=		=	0.001	g
1 centigram	(cg)	= 10 milligrams		=	0.01	g
1 decigram	(dg)	= 10 centigrams		=	0.1	g
1 GRAM	(g)	= 10 decigrams		=	1	g
1 dekagram	(dag)	= 10 grams		=	10	g
1 hectogram	(hg)	= 10 dekagrams		=	100	g
1 kilogram	(kg)	= 10 hectograms		=	1000	g
1 metric ton		= 1000 kilograms				

METRIC MEASURES AND EQUIVALENTS

Liquid and Dry Measure (Basic Unit is 1 Liter)

1 milliliter	(mℓ)	=	=	0.001 ℓ
1 centiliter	(cℓ)	= 10 milliliters	=	0.01 ℓ
1 deciliter	(dℓ)	= 10 centiliters	=	0.1 ℓ
1 LITER	(ℓ)	= 10 deciliters	=	1 ℓ
1 dekaliter	(daℓ)	= 10 liters	=	10 ℓ
1 hectoliter	(hℓ)	= 10 dekaliters	=	100 ℓ
1 kiloliter	(kℓ)	= 10 hectoliters	=	1000 ℓ

°These units are technically reserved for measuring mass. We will use them for weight or mass. The difference between weight and mass is covered in science classes.

SYSTEM-TO-SYSTEM EQUIVALENTS

English-Metric Conversions	Metric-English Conversions
1 inch $\approx$ 2.54 centimeters	1 centimeter $\approx$ 0.3937 inch
1 foot $\approx$ 0.3048 meter	1 meter $\approx$ 3.281 feet
1 yard $\approx$ 0.9144 meter	1 meter $\approx$ 1.094 yards
1 mile $\approx$ 1.609 kilometers	1 kilometer $\approx$ 0.6214 mile
1 quart $\approx$ 0.946 liter	1 liter $\approx$ 1.057 quarts
1 gallon $\approx$ 3.785 liters	1 liter $\approx$ 0.2642 gallon
1 ounce $\approx$ 28.35 grams	1 gram $\approx$ 0.0353 ounce
1 pound $\approx$ 453.59 grams	1 gram $\approx$ 0.0022 pound

Calculators

The wide availability and economical price of current hand-held calculators make them ideal for doing time-consuming arithmetic operations. Even people who are very good at math use calculators under certain circumstances (for instance, when balancing their checkbooks). You are encouraged to use a calculator as you work through this text. Learning the proper and appropriate use of a calculator is a vital skill for today's math students.

As with all new skills, your instructor will give you guidance as to where and when to use it. Calculators are especially useful in the following instances.

1. For doing the fundamental operations of arithmetic (addition, subtraction, multiplication and division);

2. For finding powers or square roots of numbers;

3. For evaluating algebraic expressions; and

4. For checking solutions to equations.

There are several different kinds of calculators available.

- A basic 4-function calculator will add, subtract, multiply and divide. Sometimes these calculators also have a square root key. These calculators are not powerful enough to do all of the math in this text, and they are not recommended for math students at this level.

- A scientific calculator generally has about eight rows of keys on it, and is usually labeled "scientific." Look for keys labeled "sin," "tan," and "log." Scientific calculators also have power keys and parentheses keys, and the order of operations is built into them. These calculators are recommended for math students at this level.

- A graphing calculator also has about eight rows of keys, but it has a large, nearly square display screen. These calculators are very powerful, and you may be required to purchase them in later math courses. However, you will not need all that power to be successful in this course, and they are significantly more expensive than scientific calculators.

We will assume that you are operating a scientific calculator. (Some of the keystrokes are different on graphing calculators, so if you are using one of these calculators, please consult your owners manual.) Study the following table to discover how the basic keys are used.

Expression	Key Strokes	Display
$144 \div 3 - 7$	[144] [÷] [3] [−] [7] [=]	41.
$3(2) + 4(5)$	[3] [×] [2] [+] [4] [×] [5] [=]	26.
$13^2 - 2(-12 + 10)$	[13] [x²] [−] [2] [×] [(] [12] [+/−] [+] [10] [)] [=]	173.
$\dfrac{28 + 42}{10}$	[(] [28] [+] [42] [)] [÷] [10] [=]	7.
	or	
	[28] [+] [42] [=] [÷] [10] [=]	7.
$\dfrac{-288}{6 + 12}$	[288] [+/−] [÷] [(] [6] [+] [12] [)] [=]	−16.
$\sqrt{400} - 3^5$	[400] [√] [−] [3] [xʸ] [5] [=]	−223.
$2\dfrac{1}{3} + \dfrac{5}{6}$	[2] [aᵇ/c] [1] [aᵇ/c] [3] [+] [5] [aᵇ/c] [6] [=]	3 ⌐1 ⌐6

Notice that the calculator does calculations when you hit the = key. The calculator automatically uses the order of operations when you enter more than one operation before hitting =. Notice that if you begin a sequence with an operation sign, the calculator automatically uses the number currently displayed as part of the calculation. There are three operations which only require one number: squaring a number, square rooting a number, and taking the opposite of a number. In each case, enter the number first and then hit the appropriate operation key. Be especially careful with fractions. Remember that when there is addition or subtraction inside a fraction, the fraction bar acts as a grouping symbol. But the only way to convey this to your calculator is by using the grouping symbols (and). Notice that the fraction key is used between the numerator and denominator of a fraction and also between the whole number and fractional part of a mixed number. It automatically calculates the common denominator when necessary.

Model Problem Solving

Practice the following problems until you can get the results shown.

Answers

a. $47 + \dfrac{525}{105}$ — 52

b. $\dfrac{45 + 525}{38}$ — 15

c. $\dfrac{648}{17 + 15}$ — 20.25

d. $\dfrac{140 - 5(6)}{11}$ — 10

e. $\dfrac{3870}{9(7) + 23}$ — 45

f. $\dfrac{5(73) + 130}{33}$ — 15

g. $100 - 2^5$ — 68

h. $100 - (-2)^5$ — 132

i. $\sqrt{729} + \sqrt{57 + 24}$ — 36

j. $4\dfrac{2}{7} - 3\dfrac{3}{5}$ — $\dfrac{24}{35}$

Prime Factors of Numbers 1 through 100

	Prime Factors		Prime Factors		Prime Factors		Prime Factors
1	none	26	$2 \cdot 13$	51	$3 \cdot 17$	76	$2^2 \cdot 19$
2	2	27	3^3	52	$2^2 \cdot 13$	77	$7 \cdot 11$
3	3	28	$2^2 \cdot 7$	53	53	78	$2 \cdot 3 \cdot 13$
4	2^2	29	29	54	$2 \cdot 3^3$	79	79
5	5	30	$2 \cdot 3 \cdot 5$	55	$5 \cdot 11$	80	$2^4 \cdot 5$
6	$2 \cdot 3$	31	31	56	$2^3 \cdot 7$	81	3^4
7	7	32	2^5	57	$3 \cdot 19$	82	$2 \cdot 41$
8	2^3	33	$3 \cdot 11$	58	$2 \cdot 29$	83	83
9	3^2	34	$2 \cdot 17$	59	59	84	$2^2 \cdot 3 \cdot 7$
10	$2 \cdot 5$	35	$5 \cdot 7$	60	$2^2 \cdot 3 \cdot 5$	85	$5 \cdot 17$
11	11	36	$2^2 \cdot 3^2$	61	61	86	$2 \cdot 43$
12	$2^2 \cdot 3$	37	37	62	$2 \cdot 31$	87	$3 \cdot 29$
13	13	38	$2 \cdot 19$	63	$3^2 \cdot 7$	88	$2^3 \cdot 11$
14	$2 \cdot 7$	39	$3 \cdot 13$	64	2^6	89	89
15	$3 \cdot 5$	40	$2^3 \cdot 5$	65	$5 \cdot 13$	90	$2 \cdot 3^2 \cdot 5$
16	2^4	41	41	66	$2 \cdot 3 \cdot 11$	91	$7 \cdot 13$
17	17	42	$2 \cdot 3 \cdot 7$	67	67	92	$2^2 \cdot 23$
18	$2 \cdot 3^2$	43	43	68	$2^2 \cdot 17$	93	$3 \cdot 31$
19	19	44	$2^2 \cdot 11$	69	$3 \cdot 23$	94	$2 \cdot 47$
20	$2^2 \cdot 5$	45	$3^2 \cdot 5$	70	$2 \cdot 5 \cdot 7$	95	$5 \cdot 19$
21	$3 \cdot 7$	46	$2 \cdot 23$	71	71	96	$2^5 \cdot 3$
22	$2 \cdot 11$	47	47	72	$2^3 \cdot 3^2$	97	97
23	23	48	$2^4 \cdot 3$	73	73	98	$2 \cdot 7^2$
24	$2^3 \cdot 3$	49	7^2	74	$2 \cdot 37$	99	$3^2 \cdot 11$
25	5^2	50	$2 \cdot 5^2$	75	$3 \cdot 5^2$	100	$2^2 \cdot 5^2$

The Properties of Zero

Historically, the number zero is a recent development in mathematics. There is evidence that notched sticks or bones were used for counting as long as 30 million years ago, but the use of a symbol for zero began from about 1200 to 1800 years ago (the development was gradual and not a sudden decision).

Zero is *not* a counting number.

When counting a group of objects we usually count "one, two, three, four," and so on.

Zero is a whole number.

The classification of zero as a whole number is an arbitrary, although useful, one.

Zero is an integer.

All whole numbers are also integers.

Zero is a rational number.

Zero can be written as a fraction $\frac{a}{b}$ where a and b are integers, $b \neq 0$, since $0 = \frac{0}{1} = \frac{0}{2}$.

Zero is a real number.

All rational (and irrational) numbers are called real numbers.

The zero power of any nonzero real number is 1.
$x^0 = 1, x \neq 0$

Examples: $1^0 = 1$, $2^0 = 1$, $6^0 = 1$, $10^0 = 1$, and so on. This is an arbitrary definition in most arithmetic and algebra texts. It is a useful definition because it fits mathematical patterns such as these:

$3^3 = 1 \cdot 3 \cdot 3 \cdot 3 = 27$
$3^2 = 1 \cdot 3 \cdot 3 \quad\ = 9$
$3^1 = 1 \cdot 3 \qquad\ = 3$
$3^0 = 1 \qquad\qquad = 1$

and

10^3 has place value "thousand."
10^2 has place value "hundred."
10^1 has place value "ten."
10^0 has place value "one" or "units."

Any number plus zero is that number.
$a + 0 = a$

Examples: $0 + 0 = 0$, $1 + 0 = 1$, (also $0 + 1 = 1$), $2 + 0 = 2$ (also $0 + 2 = 2$), and so on. This is called the Addition Property of Zero or the Identity Property of Addition.

Any number times zero is zero.
$a \cdot 0 = 0$

Examples: $0 \cdot 0 = 0$, $0 \cdot 1 = 0$ (also $1 \cdot 0 = 0$), $0 \cdot 6 = 0$ (also $6 \cdot 0 = 0$), and so on. This is called the Multiplication Property of Zero.

Zero divided by any nonzero real number is zero.

$0 \div a = 0, a \neq 0$

Examples: $0 \div 1 = 0$, $0 \div 2 = 0$, $0 \div 5 = 0$, and so on. Division does not work in reverse (division is not commutative). Division is the inverse (a kind of opposite) of multiplication. The statement $0 \div 1 = 0$ is true because $1 \cdot 0 = 0$. The statement $0 \div 5 = 0$ is true because $5 \cdot 0 = 0$. Because of this, zero can be used as a numerator for a fraction if the denominator is a nonzero real number.

Division by zero is not defined.

$a \div 0$ has no value

Examples: $1 \div 0$, $2 \div 0$, and $6 \div 0$ have no value.

$1 \div 0$ is not 0 because $0 \cdot 0$ is not 1.

$1 \div 0$ is not 1 because $0 \cdot 1$ is not 1.

$6 \div 0$ is not 0 because $0 \cdot 0$ is not 6.

$6 \div 0$ is not 6 because $0 \cdot 6$ is not 6.

By the definition of division, if zero were not an exception, any answer would check.

$0 \div 0 = 17$ (?) because $0 \cdot 17 = 0$.

$0 \div 0 = 2$ (?) because $0 \cdot 2 = 0$.

$0 \div 0 = 0$ (?) because $0 \cdot 0 = 0$.

Such problems are of no use, so division by zero is not defined for *any* dividend. Because of this, zero cannot be used as the denominator of a fraction.

Plane Geometry

Geologists often use similar triangles to approximate the height of rock formations. The height of the large formation on the right can be approximated. Make the height one leg of a right triangle, and measure the angle to the base of the left face of the rock formation. (See Section G.6, Exercise 25.)

OBJECTIVE   Use inductive and deductive reasoning.

VOCABULARY   A **prediction** is the act of foretelling, or forecasting, a particular event, or happening. A **conclusion** is a decision or conviction that is reached as a consequence of an investigation.

Inductive reasoning is the name given to the process in which one draws conclusions based on a recognition of patterns. A study is made of the given information, and conclusions are based on the patterns and repetitions observed. Predictions are based on those patterns.

Deductive reasoning is the name given to the process through which one draws conclusions based on known facts, previously proven facts, basic assumptions, and given information.

HOW AND WHY   Inductive reasoning is used by scientists as they perform experiments repeatedly. If a certain event occurs each time the experiment is performed, the conclusion is that the same event will occur every time the experiment is performed. In this case, specific examples are used to state a general conclusion. For example, we conclude that each time we toss an object into the air it will return to the ground. The reason for this conclusion is that past experience has shown that the object has always returned to the ground.

Look at the sequence of numbers below to see if you can determine the number that will appear next.

3, 9, 15, 21, 27, . . .

Did you predict 33 would be the next number? If you did, then you made an observation and drew a conclusion. Perhaps you observed that

$$3 = 3 \cdot 1$$
$$9 = 3 \cdot 3$$
$$15 = 3 \cdot 5$$
$$21 = 3 \cdot 7$$
$$27 = 3 \cdot 9$$

so you predicted that the next number will be $3 \cdot 11$, or 33.

You could also use a different pattern. Since there is a difference of 6 between any two consecutive numbers in the sequence, the next term of the sequence can be found by adding 27 and 6 to get 33. Both patterns lead to the same conclusion. Each of the processes is based on the assumption that the sequence of numbers will continue in the same pattern.

Deductive reasoning, a second type of reasoning, is used in formal proof and involves using known facts and previously proven data to draw conclusions. For example:

It is known that if it rains, James will stay home.
It did rain. Therefore, we can conclude that James stayed home.

Mathematically this can be stated in three ways:

If A then B.
A occurred; therefore, B must occur.
B did not occur; therefore, A did not occur.

We must be careful to draw a valid, or correct, conclusion. This is shown in the following example.

It is known that if it rains, James will stay home.
James did stay home; therefore, we conclude that it did rain.

The fact that James stayed home does not mean that it rained. Perhaps he was ill or had homework to do. He could have stayed home for any of several reasons.
We need to use caution using either kind of reasoning. Make sure that enough information is given to draw the valid conclusion.

Examples A–F

A. Predict what numbers seem most likely to appear next in the following sequence.

2, 4, 6, 6, 4, 2, 2, 4, 6, _, _, _,   **Note that the digits 2, 4, and 6 are used. Their order of appearance keeps reversing each time they appear. Use this pattern to make the prediction. This is an example of inductive reasoning.**

The next three numbers in the sequence will probably be 6, 4, 2.

B. Predict what figure seems most likely to appear next in the following sequence.

The dot appears to be moving in a clockwise direction.

Because of the pattern, using inductive reasoning we conclude that the next figure is

C. Based on the first statement (assumption), use deductive reasoning to conclude whether the second statement is valid.

1. If Jill sleeps too late, she will be late for class.
2. Jill slept too late; therefore, Jill was late for class.

The statement is valid. The first statement specifies that if she did sleep too late, then she would be late for class.

Warm Ups A–F

A. Predict the next three numbers in the following sequence.

1, 3, 5, 7, 5, 3, 1, 1, 3, 5, 7, _, _, _

B. Predict what figure seems most likely to appear next in the following sequence.

C. Based on the first statement (assumption), use deductive reasoning to conclude whether the second statement is valid.

1. If the sun is shining, John will not wear a jacket.
2. The sun is shining; therefore, John does not wear a jacket.

D. Based on the first statement (assumption), use deductive reasoning to conclude whether the second statement is valid.

  1. If Jill sleeps too late, she will be late for class.

  2. Jill is late for class; therefore, she slept too late.

  There could be other reasons that she was late for class. For example, her auto may have run out of gasoline, or she may have met a friend and talked too long. Actually, there are any number of reasons that could have caused her to be late.

E. Predict what number seems most likely to appear next in the following sequence.

  2, 4, 12, 24, 72, _   **Observe: $2 \cdot 2 = 4$, $4 \cdot 3 = 12$, $12 \cdot 2 = 24$, and $24 \cdot 3 = 72$. We use this pattern to make the prediction. Other patterns are possible.**

  The pattern indicates that the next term probably will be $72 \cdot 2$, or 144; therefore, we say the next number will probably be 144.

F. Predict what number seems most likely to appear next in the following sequence.

  5, 11, 23, 47, 95, _   **Observe: $5 \cdot 2 + 1 = 11$, $11 \cdot 2 + 1 = 23$, $23 \cdot 2 + 1 = 47$, and $47 \cdot 2 + 1 = 95$.**

  We conclude that the next number in the sequence probably will be

  $95 \cdot 2 + 1 = 191$

D. Based on the first statement (assumption, use deductive reasoning to conclude whether the second statement is valid.

1. If the sun is shining, John will not wear a jacket.

2. John did not wear a jacket; therefore, the sun is shining.

E. Predict what number seems most likely to appear next in the following sequence.

  3, 9, 18, 54, 108, _

F. Predict what number seems most likely to appear next in the following sequence.

  4, 10, 22, 46, 94, _

Exercises G.1

OBJECTIVE: *Use inductive and deductive reasoning.*

A.

Predict what numbers or figures seem most likely to appear next in the following sequences.

1. 1, 4, 9, 16, 25, _

2. 6, 10, 14, 18, 22, _

3. 5, 9, 13, 17, 21, _

4. 1, 2, 3, 6, 5, 4, 7, 8, 9, 12, 11, 10, _

5. 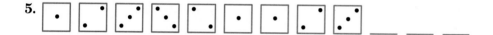 ___ ___ ___

6. 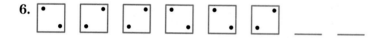 ___ ___

7. 0, 7, 14, 11, 18, 25, 22, 29, _, _

8. 1, 11, 20, 28, 35, _

Based on the first statement (assumption), use deductive reasoning to conclude whether the second statement is valid.

9. If a triangle has three equal angles, then it is equiangular. This triangle has three equal angles; therefore, it is equiangular.

10. If a triangle has three equal sides, then it is equilateral. This triangle has three equal sides; therefore, it is equilateral.

11. If Jill sleeps too late, she will be late for class. Jill is not late for class; therefore, she did not sleep too late.

12. If the sun is shining, then John will not wear a jacket. John did wear a jacket; therefore, the sun is not shining.

13. All high school students take math. Mary takes math; therefore, Mary is a high school student.

14. All entering freshmen must take a placement test. Greg took a placement test; therefore, Greg is a freshman.

15. All rectangles are parallelograms. This figure is not a parallelogram; therefore, it is not a rectangle.

16. All regular polygons have equal sides. This polygon does not have equal sides; therefore, it is not a regular polygon.

B.

Predict what numbers or figures seem most likely to appear next in the following sequences.

17.

18.

19.

20.

21. −3, 0, 5, 12, 21, _   **22.** 4, 7, 13, 25, 49, _

23. 6, 8, 12, 18, _   **24.** 3, 8, 18, 33, 53, _

A16

Based on the first statement (assumption), use deductive reasoning to conclude whether the second statement is valid.

25. All fishermen using worms for bait caught fish. Joe caught a fish; therefore, Joe used worms for bait.

26. All mountain climbers are physically fit. Mary is a mountain climber; therefore, Mary is physically fit.

27. All asparagus is grown in Washington. Marge loves asparagus; therefore, Marge lives in Washington.

28. All Olympic skiers train on Mt. Hood. John trains on Mt. Hood; therefore, John is an Olympic skier.

29. All corn grown in Kansas is at least six feet tall. Thelma's corn grows to a height of six feet; therefore, Thelma's corn was grown in Kansas.

30. All zinnias in Peter's garden are red. Lucy's zinnias came from Peter's garden; therefore, Lucy's zinnias are red.

31. All students must take writing to graduate. Jenny did graduate; therefore, Jenny took writing.

32. All students must take writing to graduate. Bill did not graduate; therefore, Bill did not take writing.

C.

Predict what numbers seem most likely to appear next in the following sequences.

33. 78, 63, 48, 33, _

34. 20, 10, 5, 2.5, 1.25, _

35. 1, 3, 7, 15, 31, _

36. 1, 5, 15, 75, 225, _

37. 1, 2, 3, 5, 8, 13, _

38. 3, 4, 7, 16, 43, _

Decide whether the reasoning in the following is a result of inductive or deductive reasoning.

39. All triangles have three sides. Polygon ABC has three sides; therefore, polygon ABC is a triangle.

40. Every circle is bisected by its diameter. A clock is circular in shape; therefore, the clock is bisected by its diameter.

41. A student observes that all prime numbers end in 1, 3, 5, 7, or 9 (that is 1, 3, 5, 7, or 9 is the unit's digit). She concludes that the number 1003 is a prime number.

42. A sailor observes the number of times that bad weather follows when the sky is very red at sunrise. He concludes, "Red sky in the morning, sailor take warning."

43. After measuring several swinging weights, Galileo concluded that the time of the swing of a pendulum is directly related to the square root of the length of the pendulum.

44. After washing his car on eight different occasions, Jorge observed that it rained the next day. Jorge decides that it will rain the next time he washes his car.

45. On various occasions, Kendall has burned his dinner while studying. He notices that chicken, carrots, beans, and other foods turn black if they are heated for a long time. He concludes that anything heated will eventually turn black.

46. Elan worked these exercises on her calculator.

$$0(9) + 1 = 1$$
$$1(9) + 2 = 11$$
$$12(9) + 3 = 111$$
$$123(9) + 4 = 1111$$
$$1234(9) + 5 = 11111$$

She concluded, without using her calculator, that

$$12345(9) + 6 = 111111$$

STATE YOUR UNDERSTANDING

47. Explain the difference between deductive and inductive reasoning.

48. Give an example of how you have used inductive reasoning in your life.

CHALLENGE

Predict what numbers seem most likely to appear next in the following sequences.

49. 6, 9, 18, 99, _, _

50. 8, 5, 1, 16, 13, 9, 32, _, _

MAINTAIN YOUR SKILLS (Sections 2.1, 2.2, 6.6)

51. Express 5.25 kilometers as meters.

52. Some drug dosages are measured in grains. For example, a common aspirin tablet contains 5 grains. If there are 15.43 grams in a grain, how many grams do two aspirin tablets contain?

53. Add: 7 g 88 mg
       + 2 g 95 mg
              = ? g

54. Multiply: (2.83 kg)(5.1) = ? kg

55. Divide: $\dfrac{116.9 \text{ cm}}{5.6}$ = ? cm

G.2 Geometry of Angles

OBJECTIVES

1. Classify angles according to size.

2. Find the supplement or the complement of a given angle.

VOCABULARY

In the geometry of this chapter, there are four undefined words: point, line, plane, and straight. These objects do, however, have certain properties. We generally say that a **point** has no dimensions. A **line** has no width or thickness, only length. A **plane** has only length and width, no thickness. **Straight** is usually thought of as not curving or bending.

Points are named (usually) by capital letters.

A
•

B   C
•   •

A, B, and *C* are points.

A **line** extends infinitely in both directions. A line can be named with a single lowercase letter or by two of the points on the line.

A   B   **This is called line ℓ, or line AB.**
•————•——→ ℓ

A **ray** has an endpoint and extends infinitely in one direction. A ray is named by the endpoint and one other point on the ray.

A   C   **This is called ray AC. We can also write this as $\overrightarrow{AC}$.**
•————•——→

A **line segment** is a portion of a line with two distinct endpoints. A line segment is named by the two endpoints.

X   Y   **This is line segment XY. We can also write this as $\overline{XY}$.**
•————————•

Equal line segments have the same length. A point or a line is said to **bisect** a line segment when it divides the segment into two equal parts.

Two lines **intersect** when they have one point in common.

An **angle** is formed by two rays or two line segments with a common endpoint. ∠ is a symbol used to denote an angle. The common endpoint is called the **vertex.** An angle can be named in several ways. It can be named by the letter at the vertex (common endpoint) if it is the only angle with that vertex. If can be named by a number placed inside the angle. It can also be named by three points, one on each side and the third at the vertex. (When this name is used, the vertex is in the middle.) An angle can also be thought of as the amount of rotation necessary to bring one of the rays (or line segments) into alignment with the other ray (or line segment) which formed the angle.

This angle can be called ∠A, ∠1, or ∠CAB or ∠BAC.

Adjacent angles have a common vertex and a common side between them.

∠**DAC and** ∠**CAB are adjacent angles.**

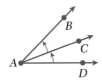

A comon unit of measure for angles is a **degree.** A degree is 1/360 of a complete revolution. The symbol for degree is °. **Equal angles** have the same degree measure.

The **bisector** of an angle is the line segment or ray with one endpoint at the vertex of the angle that divides the angle into two equal angles. The line segment or ray is said to **bisect** the angle.

Angles can be classified by their measure.

Acute angle   **This angle measures between 0° and 90°.**

Right angle   **This angle measures exactly 90° (one quarter of a complete rotation).**

A square at the vertex is used to symbolize a right angle.

Obtuse angle   **This angle measures between 90° and 180°.**

Straight angle   **This angle measures exactly 180° (one-half of a complete rotation).**

Reflex angle   **This angle measures greater than 180°.**

Two angles are **complementary** if the sum of their measures is 90°, or, if they are adjacent angles, they form a right angle. Two angles are **supplementary** if the sum of their measures is 180°, or, if they are adjacent angles, they form a straight angle.

Two lines or line segments are said to be **perpendicular** if they intersect so that the angles formed are equal (right angles). The symbol used to indicate that two lines or segments are perpendicular is ⊥.

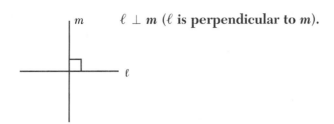

$\ell \perp m$ (ℓ is perpendicular to m).

HOW AND WHY A polygon is a figure bounded by straight sides. Examples of polygons are triangles, rectangles, parallelograms, squares, pentagons, and hexagons. Each pair of consecutive (adjacent) sides forms an angle. Thus, a polygon has as many angles as it has sides. For example, a triangle has three sides and three angles; a rectangle has four sides and four angles; and a hexagon has six sides and six angles.

Basic properties of mathematics are needed to solve the problems of geometry. Many of the properties used here are comparable to those used in algebra.

Here is a comparison of some of the properties of algebra and geometry:

ALGEBRA

Reflexive Property

$a = a$
"A number is equal to itself."

Transitive Property

If $x = y$ and $y = z$,
then $x = z$.

Division Property of Equality

If $2x = 4$, then $x = 2$.
Recall that division by
zero is not defined.

Addition or Subtraction

Property of Equality

If $x + 30 = 50$, then
$x + 30 - 30 = 50 - 30$.

Substitution Property

If $x + y = 75$ and $y = 30$,
then $x + 30 = 75$.

GEOMETRY

Identity

$\angle A = \angle A$, or $\overline{CE} = \overline{CE}$
"A quantity is equal to

Quantities equal to the
or equal quantities are
each other.

Equals divided by
are equal. Halves of
equals are equal,
of equals are equal,

Equals added to
equals subtracted
are equal.

Substitution

If $\angle A = \angle B$ and
$\angle A + \angle B + 50° = 180°$,
then $\angle A + \angle A + 50° = 180°$

Another property of geometry is that the whole is equal to the sum of its parts. For example:

$\angle$**CAB** and $\angle$**DAC** are adjacent angles.
$\angle$**CAB** + $\angle$**DAC** = $\angle$**DAB**

Since two angles are complementary when their sum is 90°, to find the complement of a given angle, subtract the given angle measure from 90°. Thus, the complement of an angle whose measure is 18° is 72° since

$$90° - 18° = 72°$$

Since two angles are supplementary when their sum is 180°, to find the supplement of a given angle, we subtract the given angle measure from 180°. Thus the supplement of an angle whose measure is 113° is 67° since

$$180° - 113° = 67°$$

Examples A–F

A. Find the complement of 16°.

$90° - 16° = 74°$ **Two angles are said to be complementary when their sum is 90°, so we subtract the given angle from 90°.**

The complement of 16° is 74°.

B. Find the supplement of 16°.

$180° - 16° = 164°$ **Two angles are said to be supplementary when their sum is 180°, so we subtract the given angle from 180°.**

The supplement of 16° is 164°.

C. What kind of an angle is one whose measure is 62°?

An acute angle **Since the measure of the angle is less than 90°, it is an acute angle.**

D. Given: ∠BAD = ∠FEH, AC bisects ∠BAD and EG bisects ∠FEH.

Does ∠1 = ∠2?

∠BAD = ∠FEH   **Given.**
∠1 is half of ∠BAD   **Definition of bisect.**
∠2 is half of ∠FEH   **Definition of bisect.**
∠1 = ∠2   **Halves of equals are equal.**

Warm Ups A–F

A. Find the complement of 39°.

B. Find the supplement of 116°.

C. What kind of an angle is one whose measure is 143°?

D. Given: ∠1 = ∠2, ∠2 = ∠3, and ∠3 = ∠4.

Does ∠1 = ∠4?

Answers to Warm Ups   A. 51°   B. 64°   C. Obtuse   D. Yes

E. Given: ∠EAC = ∠DAB

    Does ∠EAD = ∠CAB?

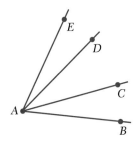

∠EAC = ∠DAB	**Given.**
∠DAC = ∠DAC	**Identity.**
∠EAC − ∠DAC = ∠DAB − ∠DAC	**Equals subtracted from equals are equal.**
∠EAD = ∠CAB	**Substitution in Step 3.** ∠EAC − ∠DAC = ∠EAD and ∠DAB − ∠DAC = ∠CAB

E. Given: $\overline{AC} = \overline{BD}$

    Does $\overline{AB} = \overline{CD}$?

```
├────┬──────────┬──────┤
A    B          C      D
```

F. Given: ∠1 + ∠2 + ∠3 = 150°

        ∠4 + ∠5 + ∠6 = 150°

               ∠1 = ∠4

               ∠2 = ∠5

Does ∠3 = ∠6?

∠1 + ∠2 + ∠3 = 150°	**Given.**
∠4 + ∠5 + ∠6 = 150°	**Given.**
∠1 + ∠2 + ∠3 = ∠4 + ∠5 + ∠6	**Quantities equal to the same quantity are equal to each other.**
∠1 = ∠4 and ∠2 = ∠5	**Given.**
∠1 + ∠2 + ∠3 − ∠1 − ∠2 = ∠4 + ∠5 + ∠6 − ∠4 − ∠5	**Equals subtracted from equals are equal.**
∠3 = ∠6	**Simplify.**

F. Given:

    ∠1 and ∠2 are supplementary

    ∠3 and ∠4 are supplementary

    ∠1 = ∠3

    Does ∠2 = ∠4?

Answers to Warm Ups E. Yes F. Yes

Exercises G.2

OBJECTIVE 1: *Classify angles according to size.*

A.

Classify each of the following angles as acute, right, obtuse, straight, or reflex.

1. 80°   **2.** 305°   **3.** 98°

4. 90°   **5.** 149°   **6.** 180°

7. 4°   **8.** 193°

9. 135°   **10.** 72°

B.

Use the following list of angle measures to answer Problems 11–15: 95°, 18°, 210°, 45°, 180°, 82°, 121°, 175°, 315°, 90°, and 282°.

11. Identify the acute angles.

12. Identify the straight angles.

13. Identify the reflex angles.

14. Identify the right angles.

15. Identify the obtuse angles.

16. If the sum of ∠C and ∠D is a right angle and ∠C = 32°, find ∠D.

OBJECTIVE 2: *Find the supplement or the complement of a given angle.*

A.

Find the complement of each of the following angles.

17. 40°   **18.** 25°   **19.** 85°

20. 36°   **21.** 15°   **22.** 1°

B.

Find the supplement of each of the following angles.

23. 80°   **24.** 125°   **25.** 25°

26. 115°   **27.** 55°   **28.** 147°

29. If ∠A and ∠B are supplementary and ∠B = 134°, find ∠A.

30. If ∠A and ∠B are supplementary and ∠B = 37°, find ∠A.

31. If ∠C and ∠D are complementary and ∠D = 15°, find ∠C.

32. If ∠C and ∠D are complementary and ∠D = 78°, find ∠C.

C.

Identify each of the following statements as either true (always true) or false (not always true).

33. The complement of an acute angle is an acute angle.

34. The supplement of an obtuse angle is an obtuse angle.

35. The supplement of an acute angle is a reflex angle.

36. The sum of a right angle and an acute angle is an obtuse angle.

37. The sum of two acute angles is an obtuse angle.

38. The sum of two obtuse angles is a reflex angle.

39. The sum of an obtuse angle and an acute angle is a reflex angle.

40. A straight angle minus an obtuse angle is an acute angle.

41. A line that bisects a straight angle forms two right angles.

42. A line that bisects an obtuse angle forms two acute angles.

43. Given: $\angle A + \angle B = 90°$
$\qquad \angle A + \angle C = 90°$
Does $\angle B = \angle C$?

44. Given: $\angle A + \angle B + \angle C = 140°$
$\qquad \angle B + \angle C = \angle A$
$\qquad\qquad \angle B = \angle C$
Does $\angle C = 35°$?

45. Given: $\angle A$ and $\angle B$ are supplementary
$\qquad \angle B$ and $\angle C$ are complementary
$\qquad \angle C = 36°$
Does $\angle A = 144°$?

46. Given: $\angle A$ is bisected to form $\angle 1$ and $\angle 2$
$\qquad \angle 1$ is a right angle
Is $\angle A$ straight angle?

STATE YOUR UNDERSTANDING

47. Describe an obtuse angle.

48. Explain what is meant by the phrase "two angles are supplementary."

CHALLENGE

49. Given: $\angle BAD = \angle ABD$

$$\angle BAD = \frac{1}{2}\angle BAC$$

$$\angle ABD = \frac{1}{2}\angle ABC$$

Does $\angle BAC = \angle ABC$?

50. Line segments $\overline{AB}$ and $\overline{BC}$ are bisected at D and E, respectively. Segments $\overline{BD}$ and $\overline{CE}$ are bisected at F and G, respectively. Finally, $\overline{BF} = \overline{CG}$. Does $\overline{AB} = \overline{BC}$?

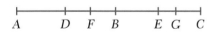

MAINTAIN YOUR SKILLS (Sections 2.2, 5.4, 6.6)

51. Write the rate comparing 450 miles to 2 hours and express the rate to the nearest foot per second.

52. Write the rate comparing 5 pounds to 3 feet and express the rate to the nearest ounce per inch.

53. Convert 17,824 inches to feet.

54. Mary Ann earns $9.80 per hour. Find her wage in cents per minute.

55. Find the perimeter, in meters, of a triangle with sides of 28 cm, 42 cm, and 16 cm.

OBJECTIVES

1. Identify triangles by the lengths of sides.

2. Identify triangles by the measures of the angles.

3. Find the missing angle in a triangle.

VOCABULARY

A **triangle** is a three-sided polygon. Each side is a line segment. The symbol for triangle is $\triangle$.

An **angle of a triangle** is formed by two consecutive sides of the triangle. The two sides of the triangle are the sides of the angle. A triangle has three angles.

Triangles are classified by their angles or by their sides. Triangles classified by their angles include the following:

Acute triangle	All three angles are acute.
Right triangle	One angle of the triangle is a right angle.
Obtuse triangle	One angle is obtuse.
Equiangular triangle	All angles have equal measure.

Triangles classified by their sides include the following:

Scalene triangle	All sides are of different length.
Isosceles triangle	At least two sides are equal in length.
Equilateral triangle	All sides are equal in length.

The **vertex angle** of an isosceles triangle is the angle formed by the two equal sides. The **base angles** of an isosceles triangle are the angles opposite the equal sides. The **legs** of an isosceles triangle are the two equal sides.

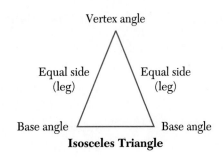

Isosceles Triangle

An angle is said to be **included** between two sides when it is formed by those two sides.

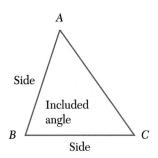

$\angle$B is included between $\overline{AB}$ and $\overline{BC}$.

A side is said to be an **included side** when it is included between two angles; that is, it is a side of each of the angles.

A

Side $\overline{BC}$ is included between $\angle B$ and $\angle C$.

Angle B Included side   Angle C

HOW AND WHY So that we may work with triangles in later applications, we need to become familiar with the different classifications. We must recognize that a triangle in which none of the sides are equal is called **scalene,** whereas a triangle with two sides equal is **isosceles.** If all three sides have the same length, the triangle is **equilateral.**

The sum of the measures of the angles of a triangle is 180°.

Examples A–F	**Warm Ups A–F**
A. Classify $\triangle ABC$ if $\overline{AB} = 12$, $\overline{AC} = 15$, and $\overline{BC} = 18$. $\triangle ABC$ is scalene.　**The length of each side is given. So we classify by the sides. None of the sides are equal.**	A. Classify $\triangle ABC$ if side $\overline{AB} = 16$, side $\overline{BC} = 11$, and side $\overline{AC} = 16$.
B. Classify $\triangle ABC$ if $\angle A = 60°$, $\angle C = 80°$, and $\angle B = 40°$. $\triangle ABC$ is acute.　**In this case, the measure of each angle is known. The measure of each angle is less than 90°.**	B. Classify $\triangle ABC$ if $\angle A = 48°$, $\angle B = 48°$, and $\angle C = 84°$.
C. Classify $\triangle ABC$ if $\angle C = 90°$, $\angle A = 42°$, and $\angle B = 48°$. $\triangle ABC$ is a right triangle.　**The triangle has a right angle.**	C. Classify $\triangle ABC$ if $\angle B = 134°$, $\angle A = 14°$, and $\angle C = 32°$.
D. Given: $\triangle ABC$ has $\angle A = 92°$ and $\angle C = 44°$. What is the measure of $\angle B$? $\angle A + \angle B + \angle C = 180°$　**The sum of the angles of a triangle is 180°.** $92° + \angle B + 44° = 180°$　**Substitution.** $\angle B + 136° = 180°$　**Simplify.** $\angle B + 136° - 136° = 180° - 136°$　**Equals subtracted from equals.** $\angle B = 44°$　**Simplify.**	D. Given $\triangle ABC$ with $\angle B = 99°$ and $\angle C = 44°$. What is the measure of $\angle A$?
E. Given: $\triangle ABC$ with $\angle C$ a right angle and $\angle A = \angle B$. What is the measure of $\angle A$ and $\angle B$? $\angle A + \angle B + \angle C = 180°$　**The sum of the angles of a triangle is 180°.**	E. Given: $\triangle ABC$ is isosceles with $\angle A$ and $\angle B$ the base angles. $\angle C$, the vertex angle, is 72°. What is the measure of each of the base angles?

$\angle C = 90°$	**Given.**
$\angle A = \angle B$	**Given.**
$\angle A + \angle A + 90° = 180°$	**Substitution.**
$2\angle A + 90° = 180°$	**Simplify.**
$2\angle A + 90° - 90° = 180° - 90°$	**Equals subtracted from equals.**
$2\angle A = 90°$	**Simplify.**
$\angle A = 45°$	**Division property of equality (halves of equals are equal).**
$\angle B = 45°$	**Substitution.**

F. Rosemary is installing a triangular dog run behind her house. One side of the dog run is formed by a side of the house, a second side of the run is at right angles to the first side, and the third side of the run forms an angle of 42° with the second side. Find the measure of the third angle of the run.

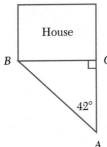

Given: $\angle A = 42°$

$\angle C$ is a right angle

Find: $\angle B = ?$

$\angle A = 42°$	**Given.**
$\angle C$ is a right angle	**Given.**
$\angle C = 90°$	**Definition of a right angle.**
$\angle A + \angle B + \angle C = 180°$	**The sum of the angles of a triangle is 180°.**
$42° + \angle B + 90° = 180°$	**Substitution.**
$\angle B + 132° = 180°$	**Simplify.**
$\angle B + 132° - 132° = 180° - 132°$	**Equals subtracted from equals.**
$\angle B = 48°$	**Simplify.**

The measure of the third angle is 48°.

F. Rosemary is installing a triangular dog run behind her house. One side of the dog run is formed by a side of the house, a second side of the run is at right angles to the first side, and the third side of the run forms an angle of 56° with the second side. Find the measure of the third angle of the run.

Exercises G.3

OBJECTIVE 1: *Identify triangles by the lengths of sides.*

A.

For Problems 1–4, classify the following triangles by sides.

1. $\triangle ABC$ with AB = 7, BC = 7, and AC = 15

2. $\triangle ABC$ with AB = 9.8, BC = 3.2, and AC = 9.7

3. $\triangle DEF$ with DE = 3.21, DF = 3.21, and EF = 3.21

4. $\triangle DEF$ with DE = 83, DF = 73, and EF = 63

OBJECTIVE 2: *Identify triangles by the measure of the angles.*

A.

For Problems 5–8, classify the following triangles by angles.

5. $\triangle ABC$ with $\angle A = 15°$, $\angle B = 15°$, and $\angle C = 150°$

6. $\triangle ABC$ with $\angle A = 35°$, $\angle B = 56°$, and $\angle C = 89°$

7. $\triangle DEF$ with $\angle D = 13°$, $\angle E = 90°$, and $\angle F = 77°$

8. $\triangle DEF$ with $\angle D = 91°$, $\angle E = 75°$, and $\angle F = 14°$

B.

For Problems 9–12, classify the following triangles by angles.

9. $\triangle PQR$ with $\angle P = 45°$ and $\angle Q = 30°$ 10. $\triangle PQR$ with $\angle P = 35°$ and $\angle Q = 55°$

11. $\triangle RST$ with $\angle R = 28°$ and $\angle S = 124°$ 12. $\triangle RST$ with $\angle R = 85°$ and $\angle S = 80°$

OBJECTIVE 3: *Find the missing angle in a triangle.*

A.

13. Find the measure of $\angle R$ in $\triangle PQR$ if $\angle P = 45°$ and $\angle Q = 30°$.

14. Find the measure of $\angle R$ in $\triangle PQR$ if $\angle P = 35°$ and $\angle Q = 55°$.

B.

15. Find the measure of $\angle T$ in $\triangle RST$ if $\angle R = 26°$ and $\angle S = 124°$.

16. Find the measure of $\angle T$ in $\triangle RST$ if $\angle R = 85°$ and $\angle S = 80°$.

17. Find the measure of $\angle C$ in $\triangle ABC$ if $\angle A = 20.3°$ and $\angle B = 94.7°$.

18. Find the measure of $\angle A$ in $\triangle ABC$ if $\angle B = 27.83°$ and $\angle C = 28.17°$.

19. Find the measure of $\angle D$ in $\triangle DEF$ if $\angle F = 129.6°$ and $\angle E = 37.5°$.

20. Find the measure of $\angle E$ in $\triangle DEF$ if $\angle D = 10.1°$ and $\angle F = 88.8°$.

C.

State whether the following are true (always true) or false (not always true).

21. A triangle can be both a right triangle and an isosceles triangle.

22. A triangle can be both an acute triangle and an equilateral triangle.

23. A triangle can be both an obtuse triangle and an equilateral triangle.

24. A triangle can be both an equiangular triangle and a scalene triangle.

25. Every acute triangle is also a scalene triangle.

26. Every equilateral triangle is also an isosceles triangle.

27. Some isosceles triangles are scalene triangles.

28. Some right triangles are obtuse triangles.

29. Every equilateral triangle is an acute triangle.

30. Every obtuse triangle is an isosceles triangle.

31. Some right triangles are isosceles triangles.

32. Some scalene triangles are right triangles.

STATE YOUR UNDERSTANDING

33. Describe the three different types of triangles based on the length of their sides.

34. Describe the four different types of triangles based on the measure of their angles.

CHALLENGE

35. Given: ∠2 and ∠3 are complementary

∠3 = ∠5

Name triangles ABC, CAE, ADE, ABE, and DBE by the measure of their angles.

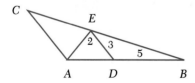

36. Given: $\overline{BF} = \overline{AF}$

$\overline{BD} = \overline{CD}$

∠2 = ∠4

∠12 = 80°

Find the measure of ∠6.

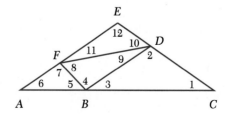

37. Given: $\angle CAB = \angle CBA$

$\qquad\angle 1 = \angle 2$

Does $\overline{AP} = \overline{BP}$?

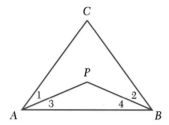

MAINTAIN YOUR SKILLS (Sections 2.3, 6.7, 6.9)

38. The cost of a high-grade carpet installed is $48 per square yard. What is the cost of carpeting a room that measures 18 feet by 24 feet?

39. Is a triangle with sides of 28 cm, 42 cm, and 16 cm a right triangle?

40. Find the area of a circle that has a radius of 19 in. (Let $\pi \approx 3.14$.)

41. Find the area of a square that is 23 m on each side.

42. Find the area of a rectangle that is 36 cm long and 24 cm wide.

G.4 Congruent Triangles

OBJECTIVES

1. Identify congruent triangles.

2. Use congruent triangles to identify angles and sides that are equal.

VOCABULARY

Axioms are statements (assumptions) that are accepted as true. **Postulates** are geometric statements (assumptions) that are accepted as true.

The **perpendicular bisector** of a line segment is the line (segment) that passes through the midpoint of the segment and is also perpendicular to the segment.

Congruent triangles are two or more triangles with exactly the same size and shape; that is, there is a correspondence between each vertex of one triangle with a vertex of the other triangle such that corresponding sides and angles are equal. The symbol used for congruent is $\cong$. If two triangles are congruent, corresponding parts are equal.

Corresponding parts of congruent triangles refers to pairs of angles or pairs of sides that must occupy the same relative position in their respective figures. The symbol $\therefore$ is an abbreviation of the word **therefore.**

HOW AND WHY

We accept the following three statements as postulates. Each is related to congruent triangles.

Congruent Triangle Postulates

SSS, SAS, ASA

If three sides of one triangle are equal to three sides of another, the two triangles are congruent. (Side, Side, Side, abbreviated SSS.)

If two sides and the included angle of one triangle are equal to two sides and the included angle of another, then the two triangles are congruent. (Side, Angle, Side, abbreviated SAS.)

If two angles and the included side of one triangle are equal to two angles and the included side of another, then the two triangles are congruent. (Angle, Side, Angle, abbreviated ASA.)

If two triangles have sides or combinations of sides and angles that satisfy any one of these three sets of conditions, then the two triangles are congruent. If they are congruent, corresponding parts are equal. Thus, if it is possible to show that two sides (line segments) or two angles are corresponding parts of congruent triangles, those line segments or angles are equal. This will be designated **corresponding parts of congruent triangles** or simply CPCT.

CAUTION **If three angles of one triangle are equal respectively to the three angles of another, the two triangles are not necessarily congruent.**

A common way to show the corresponding parts (sides and angles) in a figure is to mark them in the following way.

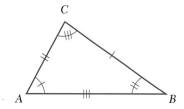 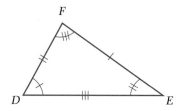

These corresponding parts of the two triangles are equal.

△ABC	△DEF

∠A, arc with one mark   ∠D, arc with one mark

∠B, arc with two marks   ∠E, arc with two marks

∠C, arc with three marks   ∠F, arc with three marks

$\overline{AB}$, side with three marks $\overline{DE}$, side with three marks

$\overline{BC}$, side with one mark $\overline{EF}$, side with one mark

$\overline{AC}$, side with two marks $\overline{DF}$, side with two marks

In figures marked in this way, we understand that angles or sides with the same marks are equal.

To say that △ABC ≅ △DEF is to say also that ∠A corresponds to ∠D, ∠B corresponds to ∠E, ∠C corresponds to ∠F, side $\overline{AB}$ corresponds to side $\overline{DE}$, side $\overline{BC}$ corresponds to side $\overline{EF}$, and side $\overline{AC}$ corresponds to side $\overline{DF}$.

Examples A–E

A. True or False: If △ABC ≅ △DEF, then ∠B = ∠F.

The statement is not necessarily true. ∠B corresponds to ∠E.

The order in which the letters are written shows that ∠B and ∠E are corresponding parts. It is possible that the angles are equal but not because the two triangles are congruent as stated.

Warm Ups A–E

A. True or False: If △ABC ≅ △DEF, then $\overline{AB}$ = $\overline{DE}$.

B. Is △ABC ≅ △DEF?

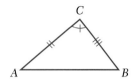

We see that $\overline{AC} = \overline{DF}$, $\overline{BC} = \overline{EF}$, and ∠C = ∠F. The angles are included between the pairs of equal sides. This satisfies the SAS postulate.

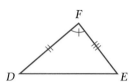

Yes, the two triangles are congruent.

C. Select the pairs of congruent triangles from the following group.

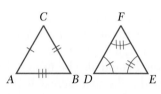

We look at the marks that indicate sides and angles are equal, keep-ing in mind the three postulates: SSS, SAS, and ASA.

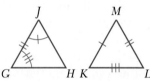

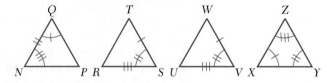

△ABC ≅ △KLM, SSS;

△GHJ ≅ △NPQ, ASA;

△RST ≅ △UVW, SAS;

△DEF and △XYZ are not necessarily congruent.

C A U T I O N **There is no congruence postulate for angle, angle, angle.**

D. Use the following figure to show that the base angles of an isosceles triangle are equal.

B. Is △ABC ≅ △DEF?

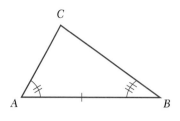

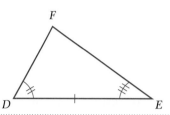

C. Select the pairs of congruent triangles from the following group.

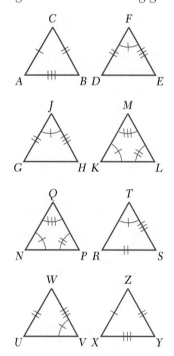

D. Using the figure from Example D, is $\overline{CD}$ perpendicular to $\overline{AB}$?

Answers to Warm Ups   B. Yes; ASA   C. △ABC ≅ △XYZ, SSS; △DEF ≅ △ GHJ, SAS; △KLM and △NPQ are not necessarily congruent. There is no AAA postulate for congruent triangles. △RST and △UVW are not necessarily congruent. The angle is not included between the sides.

A41

Given: Isosceles $\triangle ABC$ with $\overline{AC} = \overline{BC}$
where $\overline{CD}$ bisects side $\overline{AB}$

What postulate allows us to conclude that $\angle A = \angle B$?

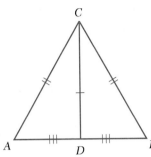

Recall that an isosceles triangle has two equal sides. The base angles of an isosceles triangle are opposite the equal sides. We first make a drawing to show the figure and label the equal parts.

We see that there are two triangles within the isosceles triangle. This pair of triangles ($\triangle ADC$ and $\triangle BDC$) is congruent by SSS. So $\angle A = \angle B$.

E. James want to construct a bridge across a pond on his farm. He needs to find the distance across the pond. To do this, he constructs a triangle in his field as shown in the following diagram. The triangle is constructed in the following way. He continues AC to the point E so that $\overline{CE} = \overline{AC}$. He continues BC to the point D so that $\overline{CD} = \overline{BC}$ and $\angle DCE = \angle ACB$. If $\overline{DE}$ is 75 ft, what is the distance across the pond?

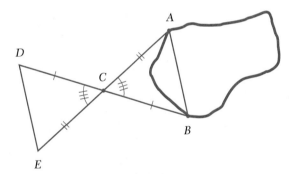

The figure is drawn and the equal parts are marked. We see that $\triangle DCE \cong \triangle BCA$ by the SAS postulate. Therefore, $\overline{DE}$ and $\overline{AB}$ are equal since they are corresponding parts of congruent triangles. Thus, Jamie will know that the distance across the pond is 75 ft.

E. If Jamie has a surveyor's instrument to measure angles in a vertical direction, show that the following method could also be used to compute the distance across his pond. He sets the instrument up on one side of the pond (perpendicular to the ground). He then sights through the instrument and adjusts it until he can see the far bank of the pond. He then turns the instrument and sights through it to a landmark on his side of the pond. He then measures the distance from the instrument to the landmark and claims that this is the distance across the pond. Is he correct? Why?

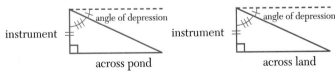

Exercises G.4

OBJECTIVE 1: *Identify the congruent triangles.*

A.

1. Name a third pair of parts needed to show that the two triangles shown are congruent by
   a. ASA   b. SAS   c. SSS

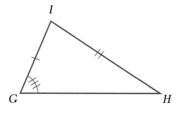
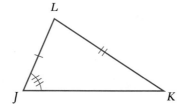

In Problems 2–4, there are two triangles that are congruent. Identify the pair of triangles and give the postulate that shows congruency.

2.

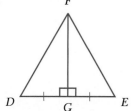

3.

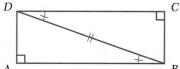

4.

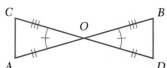

B.

Identify a pair of congruent triangles, and give the postulate that shows congruency.

5. Given: $\overline{CA} \perp \overline{AB}$
   $\overline{DB} \perp \overline{AB}$
   $\overline{CA} = \overline{DB}$
   $\overline{AO} = \overline{OB}$

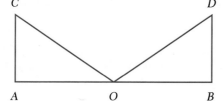

6. Given: $\overline{AB} = \overline{AD}$
$\overline{BC} = \overline{DC}$
$\angle 1 = \angle 3$
$\angle 2 = \angle 4$

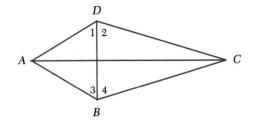

7. Given: $\overline{AC} = \overline{CB}$
E is the midpoint of $\overline{AC}$.
D is the midpoint of $\overline{CB}$.

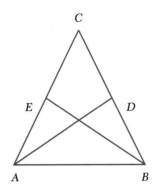

8. Given: $\overline{MR} = \overline{NS}$
$\angle MNP = \angle NMP$
$\angle RMP = \angle SNP$

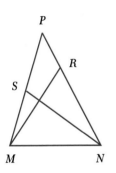

9. Given: $\angle A = \angle B$
$\overline{AC} = \overline{BC}$

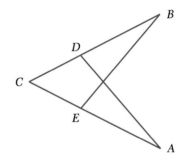

10. Given: $\angle 1 = \angle 2$
$\angle 3 = \angle 4$

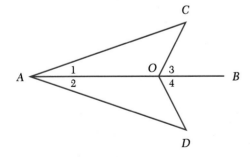

11. Given: $\overline{AD} = \overline{DC}$
$\angle 1 = \angle 2$

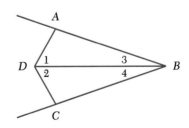

12. Given: $\overline{AB}$ bisects $\overline{CD}$ at E
   $\overline{CD}$ bisects $\overline{AB}$ at E

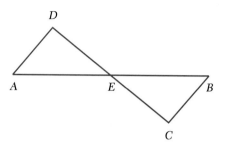

13. Given: $\overline{CD}$ bisects $\angle ACB$
   $\overline{AC} = \overline{BC}$

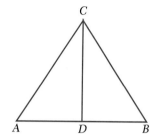

14. Given: $\angle EAB = \angle ABC = \angle BCD = \angle CDE = \angle DEA$
   $\overline{AB} = \overline{BC} = \overline{CD} = \overline{DE} = \overline{EA}$

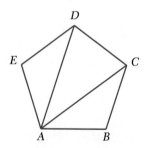

OBJECTIVE 2: *Use congruent triangles to identify angles and sides that are equal.*

A.

In Problems 15–20, assume $\triangle ABC \cong \triangle DEF$. Answer true or not necessarily true.

15. $\overline{AC} = \overline{DF}$

16. $\angle A = \angle D$

17. $\overline{AB} = \overline{EF}$

18. $\angle F = \angle A$

19. If $\overline{AC} = 12$ in., what side in $\triangle DEF$ is equal to 12 in.?

20. If $\angle E = 60°$, what angle in $\triangle ABC = 60°$?

B.

Identify a pair of congruent triangles and give the postulate that shows congruency.

21. Given: $\overline{DP} = \overline{PC}$
$\angle 1 = \angle 2$
$\overline{EP} \perp \overline{AB}$
$\overline{AP} = \overline{PB}$

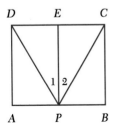

22. Given: $\overline{AB} = \overline{BC} = \overline{CD} = \overline{DA}$
$\angle 1 = \angle 2$

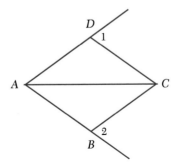

23. Given: $\overline{AC} = \overline{BC}$
$\overline{AD} = \overline{BE}$

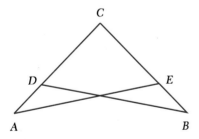

24. Given: $\overline{BD} \perp \overline{AC}$
$\overline{CD} = \overline{DE}$
$\angle AED = \angle C$

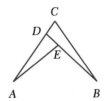

25. Given: $\overline{AF} = \overline{BE}$
$\overline{DE} = \overline{CF}$
$\overline{DE} \perp \overline{AB}$
$\overline{CF} \perp \overline{AB}$

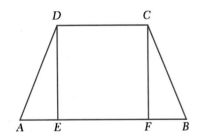

26. Given: $\overline{AC} = \overline{BC}$
$\overline{AE} = \overline{BD}$

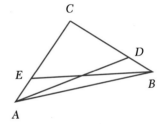

27. Given: $\overline{CE} \perp \overline{AB}$
$\overline{BF} \perp \overline{CD}$
$\overline{BE} = \overline{CF}$
$\overline{CE} = \overline{BF}$

28. Given: $\overline{DC} = \overline{AB}$
$\angle 1 = \angle 2$
$\angle 3 = \angle 4$

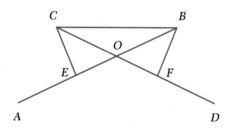

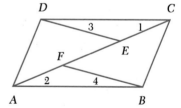

29. Given: Square ABCD
$\angle 1 = \angle 2$

30. Given: $\overline{BE} = \overline{DG}$
$\overline{CB} = \overline{FE}$
$\angle 1 = \angle 2$

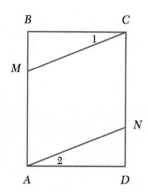

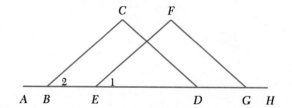

STATE YOUR UNDERSTANDING

31. Explain what is meant by "congruent triangles."

32. Explain what is meant by SAS when stating that two triangles are congruent.

CHALLENGE

33. Given: $\overline{OB} = \overline{OC}$
$\angle 1 = \angle 2$
Does $\angle OBD = \angle OCD$? Why?

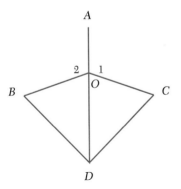

34. Given: $\overline{AD}$ bisects $\angle BAC$
$\overline{AD} \perp \overline{BC}$
Is $\triangle ABC$ isosceles? Why?

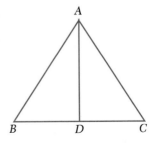

35. Given: $\angle 1 = \angle 2$
$\angle 3 = \angle 4$
Does $\overline{AD} = \overline{AE}$? Why?

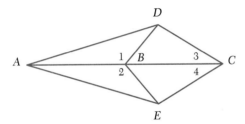

MAINTAIN YOUR SKILLS (Sections 2.3, 5.4, 6.7, 6.9)

36. Find the area of a triangle, in cm, that has a base of 3 m and a height of 3 m.

37. If a secretary can type 120 words per minute, how many words can she type in 8 seconds?

38. Find the area of a trapezoid that has a height of 1 meter and bases of 34 cm and 42 cm. Find the area in meters.

39. Find the area of a circle that has a radius of 6.75 km. (Let $\pi \approx 3.14$.)

40. Find the perimeter of a square that has an area of 655.36 square feet.

A48

G.5 Parallel Lines

OBJECTIVES

1. Classify angles formed by parallel lines cut by a transversal.

2. Find the measure of angles formed by parallel lines cut by a transversal.

VOCABULARY

Parallel lines are lines that are in the same plane and never intersect. The symbol $\parallel$ is used to indicate lines that are parallel.

$$\ell_1 \longleftrightarrow$$
$$\ell_2 \longleftrightarrow \qquad \ell_1 \parallel \ell_2$$

If lines are not parallel, they will intersect. When two lines intersect, they form four angles. These four angles are four pairs of adjacent angles. They also form two pairs of **vertical angles,** or **opposite angles.**

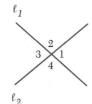

$\angle 1$ is adjacent to $\angle 2$ and $\angle 4$.
$\angle 2$ is adjacent to $\angle 1$ and $\angle 3$.
$\angle 2$ and $\angle 4$ are vertical angles.
$\angle 1$ and $\angle 3$ are vertical angles.

A **transversal** is a line that intersects two or more other lines. There are four angles formed at each intersection.

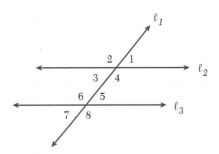

The following names are given to angles formed by a transversal intersecting two lines. In this case, transversal ℓ_1 intersects lines ℓ_2 and ℓ_3. $\angle 3$, $\angle 4$, $\angle 5$, and $\angle 6$ (those between the lines) are called **interior** angles. $\angle 1$, $\angle 2$, $\angle 7$, and $\angle 8$ (those outside the lines) are called **exterior** angles. $\angle 4$ and $\angle 6$ form a pair of **alternate interior** angles since they are on opposite sides of the transversal and are interior angles. Likewise, $\angle 3$ and $\angle 5$ are a pair of alternate interior angles. $\angle 1$ and $\angle 5$, $\angle 2$ and $\angle 6$, $\angle 3$ and $\angle 7$, and $\angle 4$ and $\angle 8$ are pairs of **corresponding** angles. Note that they are in corresponding positions at each intersection. $\angle 1$ and $\angle 2$, $\angle 2$ and $\angle 3$, $\angle 3$ and $\angle 4$, and $\angle 1$ and $\angle 4$ are supplementary, since together they form straight angles, $\angle 1$ and $\angle 3$, $\angle 2$ and $\angle 4$, $\angle 5$ and $\angle 7$, and $\angle 6$ and $\angle 8$ are pairs of vertical angles. $\angle 1$ and $\angle 7$ and $\angle 2$ and $\angle 8$ are pairs of **alternate exterior** angles.

HOW AND WHY

We will accept the following five statements about points and lines in the same plane as postulates.

Line	**Parallel, Perpendicular, Transversal**

Postulates

Through a point not on a line, one and only one line parallel to the given line can be drawn.

Through a point not on a line, one and only one line perpendicular to the given line can be drawn.

Two lines perpendicular to the same line are parallel.

If two lines are cut by a transversal so that the alternate interior angles are equal, the lines are parallel.

If two parallel lines are cut by a transversal, then the alternate interior angles are equal.

Examples A–F

A. Classify ∠3 and ∠8.

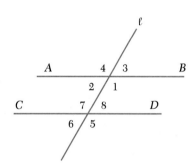

They are corresponding angles.   They are in corresponding positions at each intersection.

B. Find the measure of ∠1 if AB ∥ CD and ∠8 = 58°.

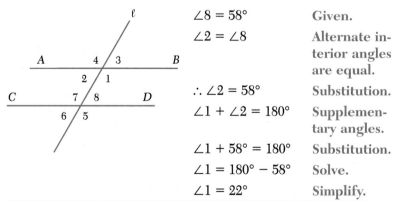

∠8 = 58°	Given.
∠2 = ∠8	Alternate interior angles are equal.
∴ ∠2 = 58°	Substitution.
∠1 + ∠2 = 180°	Supplementary angles.
∠1 + 58° = 180°	Substitution.
∠1 = 180° − 58°	Solve.
∠1 = 22°	Simplify.

C. If two straight lines intersect, are the vertical angles equal? We make a drawing and call the intersecting lines ℓ_1 and ℓ_2. The pairs of vertical angles formed are named ∠1 and ∠3 and ∠2 and ∠4.

Warm Ups A–F

A. Classify ∠1 and ∠7 in the figure in Example A.

B. Use the figure from Example B and find the measure of ∠3 if ∠7 = 132°.

C. If two lines intersect so that one angle formed is a right angle, are the lines perpendicular?

Answers to Warm Ups   A. Alternate interior angles.   B. 48°   C. Yes

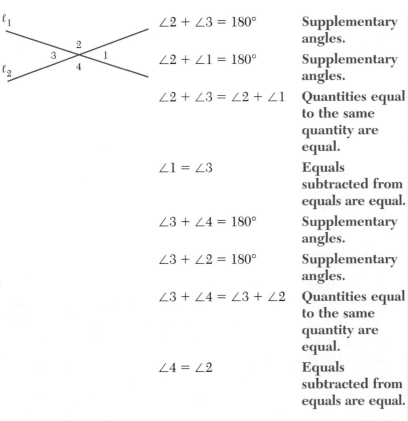

$\angle 2 + \angle 3 = 180°$	Supplementary angles.
$\angle 2 + \angle 1 = 180°$	Supplementary angles.
$\angle 2 + \angle 3 = \angle 2 + \angle 1$	Quantities equal to the same quantity are equal.
$\angle 1 = \angle 3$	Equals subtracted from equals are equal.
$\angle 3 + \angle 4 = 180°$	Supplementary angles.
$\angle 3 + \angle 2 = 180°$	Supplementary angles.
$\angle 3 + \angle 4 = \angle 3 + \angle 2$	Quantities equal to the same quantity are equal.
$\angle 4 = \angle 2$	Equals subtracted from equals are equal.

Since $\angle 1 = \angle 3$ and $\angle 2 = \angle 4$, the vertical angles are equal.

D. Given that AB ∥ CD and that they are cut by the transversal ℓ, are the alternate exterior angles equal? We make a drawing and label it as shown. Our job is to show $\angle 1 = \angle 7$ and $\angle 2 = \angle 8$.

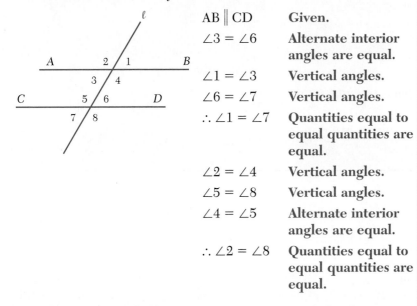

AB ∥ CD	Given.
$\angle 3 = \angle 6$	Alternate interior angles are equal.
$\angle 1 = \angle 3$	Vertical angles.
$\angle 6 = \angle 7$	Vertical angles.
$\therefore \angle 1 = \angle 7$	Quantities equal to equal quantities are equal.
$\angle 2 = \angle 4$	Vertical angles.
$\angle 5 = \angle 8$	Vertical angles.
$\angle 4 = \angle 5$	Alternate interior angles are equal.
$\therefore \angle 2 = \angle 8$	Quantities equal to equal quantities are equal.

Thus, if two parallel lines are cut by a transversal, the alternate exterior angles are equal.

D. If two lines are cut by a transversal so that the corresponding angles are equal, are the lines parallel?

E. If AB ∥ CD and they are cut by a transversal, are the interior angles on the same side of the transversal supplementary? We make a drawing and label it as shown. We must show that ∠1 + ∠3 = 180°.

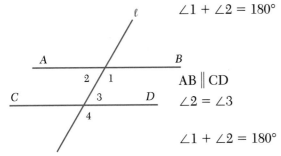

∠1 + ∠2 = 180° **Two adjacent angles that form a straight line.**

AB ∥ CD **Given.**

∠2 = ∠3 **Alternate interior angles.**

∠1 + ∠2 = 180° **Substitution.**

Therefore, the interior angles on the same side of the transversal are supplementary.

E. If two parallel lines are cut by a transversal, are the exterior angles on the same side of the transversal supplementary?

F. Louise constructs a table so that the supports bisect each other. Show that the table top is parallel to the floor.

We make a drawing and label it as shown. We also mark all of the known equal line segments. We will show that △COA ≅ △DOB.

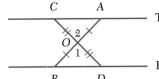

Given: AO = OB
       CO = OD
Show: CA ∥ BD

$\overline{AO} = \overline{OB}$ **Given.**

$\overline{CO} = \overline{OD}$ **Given.**

∠1 = ∠2 **Vertical angles.**

△COA ≅ △DOB **SAS.**

∠CAO = ∠DBO **Corresponding parts of congruent triangles.**

$\overline{CA} \parallel \overline{BD}$ **Alternate interior angles are equal.**

So the table top is parallel to the floor.

F. Show that two shelves that are built perpendicular to a vertical wall are parallel.

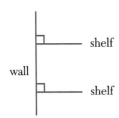

Exercises G.5

OBJECTIVE 1: *Classify angles formed by parallel lines cut by a transversal.*

A.

Use the following figure for Problems 1–8. It is known that the lines AB and CD are parallel.

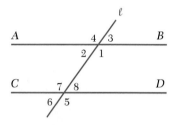

1. Specify all pairs of alternate interior angles.

2. Specify all pairs of corresponding angles.

3. Specify all interior angles.

4. Specify all exterior angles.

OBJECTIVE 2: *Find the measure of angles formed by parallel lines cut by a transversal.*

A.

Use the figure for problems 1–4.

5. Find the measure of ∠1 if ∠4 is 120°.   7. Find the measure of ∠1 if ∠6 is 65°.

6. Find the measure of ∠7 if ∠3 is 60°.   8. Find the measure of ∠5 if ∠4 is 146°.

B.

Use the following figure for Problems 9–12.

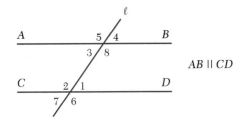

$AB \parallel CD$

9. What angle(s) in the diagram is (are) equal to ∠6?

10. What angle(s) in the diagram is (are) equal to ∠1?

11. What angle(s) in the diagram is (are) supplementary to ∠4?

12. What angle(s) in the diagram is (are) supplementary to ∠8?

C.

13. Use the following figure to show that the sum of the interior angles of a triangle is 180°.

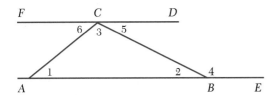

Given: △ABC with FD ∥ AE
Does ∠1 + ∠2 + ∠3 = 180°?

14. Given: AB ∥ CD and PQ ∥ RS
Does ∠1 = ∠2? Why?

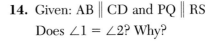
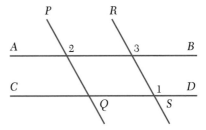

15. Given the following figure with ∠1 = ∠2, is AB ∥ CD? Why?

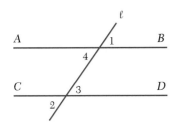

16. Given the following figure with ∠5 + ∠7 = 180°, is AB ∥ CD? Why?

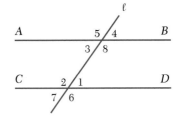

17. Given: EH ∥ AD
∠FBC = ∠GCB

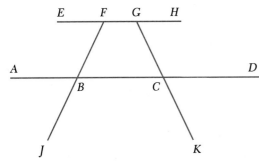

Does ∠EFB = ∠KCD?

18. Given: △ADC is isosceles with $\overline{CA} = \overline{CD}$
$\overline{BE} ∥ \overline{AD}$
Is △BEC an isosceles triangle? Why?

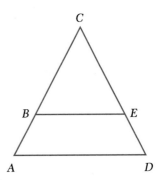

STATE YOUR UNDERSTANDING

19. Using an example from architecture or industry, explain what parallel lines are.

20. Explain what is meant by corresponding angles.

CHALLENGE

21. Given: ∠1 = ∠2
$\overline{OA} = \overline{OD}$
Does $\overline{AB} = \overline{CD}$? Why?

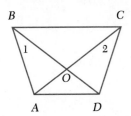

22. Given: $\overline{AB} ∥ \overline{CD}$
$\overline{EG} ∥ \overline{FH}$
Does ∠1 = ∠2? Why?

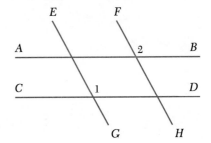

23. Given: △DEF is isosceles
$\overline{DE} = \overline{EF}$
∠1 = ∠2
Is $\overline{DG} ∥ \overline{EF}$? Why?

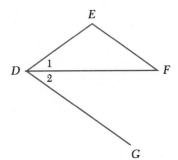

MAINTAIN YOUR SKILLS (Sections 2.4, 6.7)

24. Find the volume of a cube that is 7 inches on an edge.

25. Find the volume, in cubic meters, of a box that measures 5 cm by 8 cm by 6 cm.

26. Find the volume of a sphere that has a diameter of 9 inches. (Let $\pi \approx 3.14$.)

27. Find the volume of a cone that has a circular base with an 8-inch radius and a height of 1 foot. Find the volume to the nearest hundredth of a cubic foot.

28. Find the volume of a right circular cylinder with a radius of 5 feet and a height of 10 yards. Find the volume to the nearest tenth of a cubic yard. (Let $\pi \approx 3.14$.)

G.6   Similar Triangles

OBJECTIVES

1. Given two similar triangles, determine the measurements of designated sides and angles.

2. Show two triangles are similar.

VOCABULARY

When two or more polygons are **similar,** they have the same shape but are not necessarily the same size. Their corresponding sides are **proportional.** A **proportion** is a statement that says that two ratios are equal. Thus, if two figures are similar, their corresponding sides are in the same ratio. The symbol ~ is used to indicate similar.

HOW AND WHY

We accept the following three statements as postulates.

Similar Triangles Postulates

AAA, Proportional Sides

If three angles of a triangle are equal to three angles of another, the two triangles are similar.

If one angle of one triangle is equal to one angle of another and the two sides that include the angles are proportional, then the triangles are similar.

If the sides of two triangles are respectively proportional, the two triangles are similar.

We use these postulates together with deductive reasoning to find parts of similar triangles.

Examples A–E

A.   Given: $\triangle ABC \sim \triangle DEF$, $\overline{AC} = 10$, $\overline{AB} = 12$, $\overline{BC} = 14$, and $\overline{EF} = 7$
Find: The length of $\overline{DF}$ and $\overline{DE}$

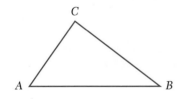

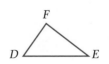

The ratio of $\overline{EF}$ to $\overline{BC}$ is

$\dfrac{7}{14} = \dfrac{1}{2}$

$\overline{BC}$ and $\overline{EF}$ are corresponding sides.

Warm Ups A–E

A.   Given: $\triangle ABC \sim \triangle DEF$, $\overline{AC} = 21$, $\overline{AB} = 27$, $\overline{BC} = 24$, and $\overline{EF} = 8$
Find: The length of $\overline{DF}$ and $\overline{DE}$

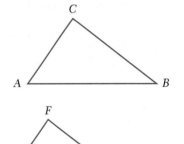

Answer to Warm Up   A. $\overline{DF} = 7$, $\overline{DE} = 9$

A57

$\overline{DF}$ corresponds to $\overline{AC}$ so, since $\overline{AC} = 10$, $\overline{DF} = 5$. $\overline{DE}$ corresponds to $\overline{AB}$ so, since $\overline{AB} = 12$, $\overline{DE} = 6$.

All corresponding sides must be in the same ratio. Each side in $\triangle DEF$ must be one half its corresponding side in $\triangle ABC$.

B. Show that a line parallel to one side of a triangle and intersecting the other two sides forms a triangle similar to the original triangle.

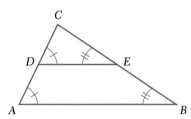

We draw a figure and label it as shown.

We know that $\overline{DE} \parallel \overline{AB}$, so $\angle CED$ and $\angle CBA$ are equal corresponding angles. For the same reason, $\angle CDE = \angle CAB$. $\angle C$ is an angle in each of the triangles, so the triangles are similar since the three angles of one are equal to the three angles of the other.

B. Show that a line parallel to one side of a triangle and intersecting the other two sides forms a triangle whose sides are proportional to the original triangle.

C. Given: Triangle ACD with $\overline{BE} \parallel \overline{CD}$
$\overline{ED} = 6$, $\overline{AE} = 8$, $\overline{BC} = 5$
Find: $\overline{AB} = ?$

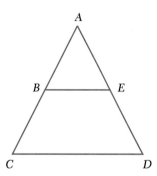

Since $\overline{BE} \parallel \overline{CD}$, we know that $\triangle ABE$ and $\triangle ACD$ are similar. That means that the corresponding sides are proportional. So
$\dfrac{\overline{AB}}{\overline{AC}} = \dfrac{\overline{AE}}{\overline{AD}}$; $\overline{AC} = \overline{AB} + \overline{BC} = \overline{AB} + 5$;
$\overline{AD} = \overline{AE} + \overline{ED} = 8 + 6 = 14$

C. Given: $\triangle ACD$ with $\overline{BE} \parallel \overline{CD}$
$\overline{ED} = 8$, $\overline{AE} = 12$, $\overline{BC} = 6$
Find: $\overline{AB} = ?$

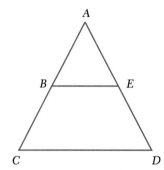

Answers to Warm Ups   B. Since the triangles are similar, corresponding sides are proportional.   C. $\overline{AB} = 9$

A58

Therefore, we solve the proportion

$$\frac{\overline{AB}}{\overline{AB} + 5} = \frac{8}{14}$$ **Substitute $\overline{AB} + 5$ for $\overline{AC}$, 8 for $\overline{AE}$, and 14 for $\overline{AD}$.**

$$14(\overline{AB}) = 8(\overline{AB} + 5)$$ **Cross multiply.**

$$14(\overline{AB}) = 8(\overline{AB}) + 40$$ **Simplify.**

$$6(\overline{AB}) = 40$$

$$\overline{AB} = 6\frac{2}{3}$$

D. Use corresponding sides to show that △ABC ∼ △DEF.

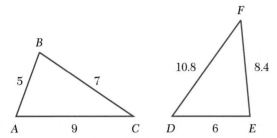

$$\frac{\overline{DE}}{\overline{AB}} = \frac{6}{5} = 1.2$$ **Find the ratios of corresponding sides.**

$$\frac{\overline{EF}}{\overline{BC}} = \frac{8.4}{7} = 1.2$$

$$\frac{\overline{DF}}{\overline{AC}} = \frac{10.8}{9} = 1.2$$

So △ABC ∼ △DEF   **Corresponding sides are proportional.**

D. Use corresponding sides to prove that △ABC ∼ △ECD.

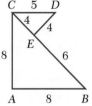

E. Florence decides to build a fence around a triangular field. Last year, she built a fence around a similar triangular field with sides of 300 ft, 200 ft, and 280 ft. How much fencing will she need if she knows that the length of the smallest side of the new field is 320 ft?

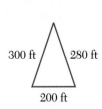 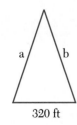

    Since the two triangles are similar, the corresponding sides must be proportional. We compute the ratio of the two smallest sides. The other corresponding sides have the same ratio.

The ratio of the smallest sides in the new triangle to the old triangle is

$$\frac{320}{200} = \frac{8}{5}$$

E. How much fencing would Florence need to fence her triangular field if the longest side measured 375 ft?

The other corresponding sides have the same ratio. For the longest side, we have

$$\frac{a}{300} = \frac{8}{5},$$

where a is the longest side in the new triangle.

$5a = 8 \cdot 300$ **Solve the proportion.**
$5a = 2400$ **Simplify.**
$a = 480$ **The longest side is 480 ft.**

For the remaining side, we have

$$\frac{b}{280} = \frac{8}{5},$$

where b is the remaining side in the new triangle.

$5b = 8 \cdot 280$ **Solve the proportion.**
$5b = 2240$ **Simplify.**
$b = 448$ **The remaining side is 448 ft.**

The distance around the new field is 320 ft + 448 ft + 480 ft = 1248 ft. She needs 1248 ft of fencing.

Exercises G.6

OBJECTIVE 1: *Given two similar triangles, determine the measurements of designated sides and angles.*

A.

Use the following data to answer Problems 1–5.
Given: $\triangle ABC \sim \triangle DEF$, $\angle A = 46°$, $\angle C = 76°$, $\overline{AC} = 12$, $\overline{CB} = 10$, $\overline{AB} = 13$, and $\overline{DF} = 15$

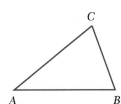 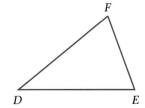

Find:

1. $\angle E$ **2.** $\angle D$

3. $\angle F$ **4.** $\overline{DE}$

5. $\overline{FE}$

Use the following to answer Problems 6–10.
Given: $\overline{ED} \parallel \overline{BA}$
$\overline{AB} = 20$, $\overline{EB} = 15$
$\overline{EC} = 10$, $\overline{AD} = 6$
$\angle CED = 40°$, $\angle C = 75°$

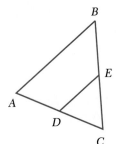

Find:

6. $\overline{ED}$ **7.** $\overline{DC}$

8. $\overline{AC}$ **9.** $\angle B$

10. $\angle A$

© 1997 Saunders College Publishing.

B.

Use the following figure to solve Problems 11–16.

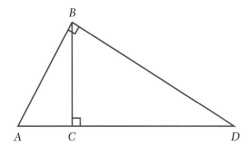

For Problems 11–13 it is given that $\overline{AB} = 13$, $\overline{AC} = 5$, $\overline{CB} = 12$, and $\triangle ABD \sim \triangle ACB \sim \triangle BCD$. Find the following:

11. $\overline{BD}$ **12.** $\overline{AD}$ **13.** $\overline{DC}$

For Problems 14–16 it is given that $\overline{BC} = 9$, $\overline{CD} = 12$, $\overline{BD} = 15$, and $\triangle ABD \sim \triangle ACB \sim \triangle BCD$. Find the following:

14. $\overline{AB}$ **15.** $\overline{AC}$ **16.** $\overline{AD}$

Use the following figure to solve Problems 17–20.

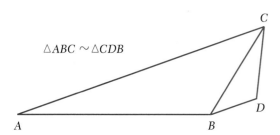

$\triangle ABC \sim \triangle CDB$

For Problems 17–18 it is given that $\angle ABC = \angle CDB$, $\overline{AB} = 21$, $\overline{CB} = 24$, and $\overline{CD} = 18$. Find the following:

17. $\overline{AC}$ **18.** $\overline{DB}$

For Problems 19–20 it is given that $\angle ABC = \angle CDB$, $\overline{BD} = 90$, $\overline{CB} = 120$, and $\overline{AB} = 100$. Find the following:

19. $\overline{AC}$ **20.** $\overline{CD}$

OBJECTIVE 2: *Show two triangles are similar.*

A.

Use corresponding sides to show that the following pairs of triangles are similar.

21.

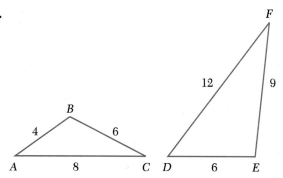

22.

B.

Use corresponding sides to show that the following pairs of triangles are similar.

23.

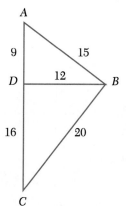

24.

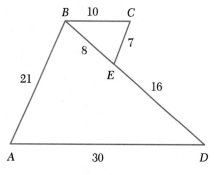

C.

25. To find the height of the rock formation a geologist uses similar triangles. She first forms a right triangle that has the height of the rock formation as one leg and then measures along the base of the rock until she comes to the lowest point of the left face. This is the other leg or the base of the triangle. This distance measures 75 ft. From the lowest point of the left face the angle to the top is 75°. She now draws a similar triangle with angles of 75°, 15°, and a right angle. The sides of this new triangle measure 5 in., 12 in., and 13 in. What is the approximate height of the rock formation?

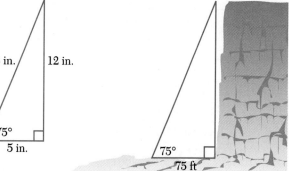

26. To find the distance across a pond, Marvin forms two similar triangles: $\triangle ABE \sim \triangle CDE$.

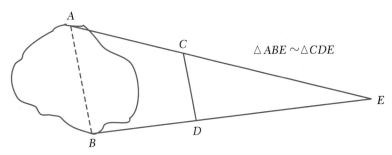

He also finds that $\overline{BE} = 450$ yds, $\overline{BD} = 180$ yds, and $\overline{CD} = 120$ yds. Find the distance across the pond.

27. A triangular plot of ground is surrounded by a sidewalk, as shown in the diagram, with $\triangle ABC \sim \triangle DEF$.

Given: $\overline{CB} = 40$ ft, $\overline{AC} = 34$ ft,
   $\overline{AB} = 52$ ft, and
   $\overline{FE} = 34$ ft

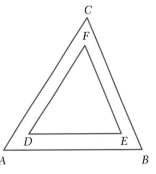

Find: The length of fence it will take to fence the perimeter of the plot of ground ($\triangle DEF$).

28. Given: ABCD is a trapezoid.
   Find x.

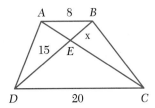

29. Given: ABCD is a parallelogram.
   Find x.

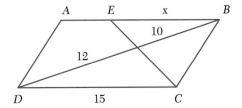

30. Given: $\overline{AC} = 18$, $\overline{DC} = 6$,
   $\overline{BC} = 24$, and $\overline{EC} = 8$
   Show that $\triangle CDE \sim \triangle CAB$.
   Is $\overline{DE} \parallel \overline{AB}$? Why?

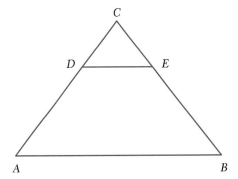

31. In △DEF, $\overline{DF} = 20$, $\overline{DE} = 16$, and $\overline{EF} = 24$. A point H is found on $\overline{DF}$ and a point K is found on $\overline{EF}$ such that $\overline{FH} = 15$ and $\overline{FK} = 18$. If line HK is drawn, find the length of $\overline{HK}$.

32. Given: Two squares with sides of 1 foot and 2 feet respectively with a diagonal drawn from one vertex to the opposite one in each square

Is △ABC ~ △DEF?

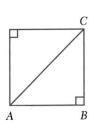
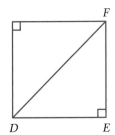

33. Two intersecting straight lines are cut by parallel lines, as shown in the figure below. Is △CDE ~ △ABC? Why?

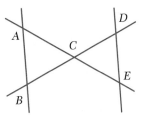

34. Two triangles are placed inside circles of different diameters so that each triangle has a vertex at the center of the circle. If ∠A = ∠D, are the triangles similar? Why?

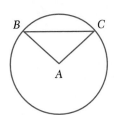
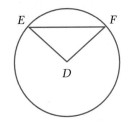

35. Given: A trapezoid with diagonals drawn

Are △CAB and △DAE similar? Why?

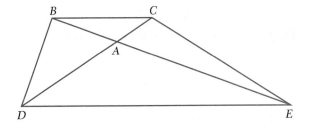

STATE YOUR UNDERSTANDING

36. Describe the differences between congruent triangles and similar triangles.

37. Explain what is meant when we say that the sides of two similar triangles are proportional.

CHALLENGE

38. Given: Rectangle ABCD

$\overline{DE} \perp \overline{AC}$

Is $\triangle CDE \sim \triangle ABC$? Why?

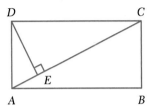

39. Given: $(\overline{AB})^2 = (\overline{BC})(\overline{BD})$

Is $\triangle ABC \sim \triangle ABD$? Why?

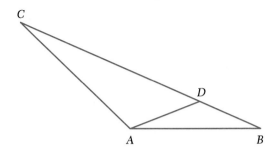

MAINTAIN YOUR SKILLS (Sections 4.8, 6.7)

40. Evaluate T in $T = 3a - 4b$, if $a = -5$ and $b = 9$.

41. Evaluate Q in $Q = -6x + 3y$, if $x = 3$ and $y = -1$.

42. Evaluate S in $S = -0.03s - 2.5t$, if $s = 4.5$ and $t = -0.75$.

43. Is $x = -5$ a solution of $3x - 45 - 2x = 50$?

44. Is $y = 0.05$ a solution of $2y - 13.5 = 3.5 - 16$?

PLANE GEOMETRY

True-False Concept Review

Check your understanding of the language of basic mathematics. Tell whether each of the following statements is True (always true) or False (not always true). For each statement you judge to be false, revise it to make a statement that is true.

ANSWERS

1. _____

    1. Jane opens four packages and each contains a glass tumbler. She concludes that a fifth package will also contain a glass tumbler. This is an example of inductive reasoning.

2. _____

    2. In the sequence, 3, 4, 6, 10, 18, _, _, the next two numbers are 34 and 66.

3. _____

    3. Angles with measures of 35° and 55° are complementary angles.

4. _____

    4. If ∠A is an obtuse angle and ∠B is an acute angle, the measure of ∠A is larger than the measure of ∠B.

5. _____

    5. The sum of two acute angles cannot be a right angle.

6. _____

    6. An isosceles triangle can never be an obtuse triangle.

7. _____

    7. If corresponding angles of two triangles are equal, the triangles are congruent.

8. _____

    8. If two sides and an angle of triangle #1 are equal to two sides and an angle of triangle #2, the triangles are congruent.

9. _____

    9. Alternate interior angles formed by a transversal intersecting two parallel lines are on the same side of the transversal.

10. _____

    10. Two lines are parallel if, when cut by a transversal, alternate interior angles are equal.

11. _____

    11. If two angles of a triangle are equal to two angles of a second triangle, the triangles are similar.

ANSWERS
12. _____

13. _____

14. _____

15. _____

16. _____

17. _____

18. _____

19. _____

20. _____

21. _____

22. _____

23. _____

24. _____

25. _____

12. If the sides of triangle #1 are three times the length of the sides of triangle #2, the triangles are congruent.

13. All similar triangles are also congruent triangles.

14. The following is an example of deductive reasoning: John eats only red tomatoes, therefore the tomato that John had for dinner was red.

15. All U.S. citizens over the age of 21 have a social security number. John has a social security number, so John is over 21 years old.

16. The supplement of an angle whose measure is 135° has a measure of 65°.

17. The sum of two right angles is a straight angle.

18. A scalene triangle has three equal angles.

19. A right triangle can also be a scalene triangle, an isosceles triangle, or an acute triangle.

20. If corresponding sides of two triangles are equal, the triangles are congruent.

21. If line AB is not parallel to line CD, then lines AB and CD will have a point of intersection.

22. Vertical angles formed by two intersecting lines are supplementary.

23. If two parallel lines are cut by a transversal, interior angles on the same side of the transversal are equal.

24. If two triangles are similar, the measure of the corresponding angles is equal.

25. All congruent triangles are similar triangles.

PLANE GEOMETRY

Review

Section G.1

Predict what numbers seem most likely to appear next in the following sequences.

1. 5, 11, 17, 23, _, _
2. 2, 5, 11, 23, _, _
3. 5, 3, 0, 7, 16, 14, 11, 18, 27, _, _

In the following examples, assume that the first statement is true. Then use deductive reasoning to conclude whether the second statement is valid.

4. All fourth graders wear blue shirts. José is wearing a blue shirt; therefore, José is a fourth grader.
5. All chickens from the Value Deli are grown in Oregon. Pete bought chicken for dinner from the Value Deli; therefore, Pete bought an Oregon-grown chicken.

Section G.2, *Objective 1*

Use the following list of angle measures to answer Problems 6–10: 103°, 37°, 210°, 86°, 346°, 90°, 2°, 98°, 180°, and 164°.

6. Identify the acute angles.
7. Identify the reflex angles.
8. Identify the obtuse angles.
9. Identify the straight angles.
10. Identify the right angles.

Section G.2, *Objective 2*

11. If ∠A and ∠B are complementary and ∠B = 55°, find ∠A.
12. If ∠A and ∠B are supplementary and ∠B = 55°, find ∠A.
13. If ∠R and ∠S are supplementary and ∠R = 178°, find ∠S.
14. If ∠R and ∠S are complementary and ∠R = 78°, find ∠S.
15. True or false: The supplement of a right angle is an acute angle.

Section G.3, *Objective 1*

16. Given △ABC with $\overline{AB}$ = 24, $\overline{BC}$ = 24, and $\overline{AC}$ = 36, what kind of triangle is △ABC?
17. Given △ABC with $\overline{AB}$ = 3.4, $\overline{BC}$ = 2.3, and $\overline{AC}$ = 4.4, what kind of triangle is △ABC?
18. Given △ABC with AB = 45, BC = 45, and AC = 45, what kind of triangle is △ABC?
19. True or false: Every isosceles triangle is an equilateral triangle.
20. True or false: Some scalene triangles are isosceles triangles.

Section G.3, *Objective 2*

21. Given △ABC with ∠A = 64° and ∠B = 47°, what kind of triangle is △ABC?
22. Given △ABC with ∠A = 52° and ∠B = 26°, what kind of triangle is △ABC?
23. Given △DEF with ∠D = 63° and ∠E = 27°, what kind of triangle is △DEF?

Section G.3, *Objective 3*

24. Find the measure of ∠R in △PQR if ∠P = 31° and ∠Q = 47°.
25. Find the measure of ∠Q in △PQR if ∠P = 110° and ∠R = 25°.

Section G.4, *Objective 1*

Use the following diagram to answer Problems 26–28.

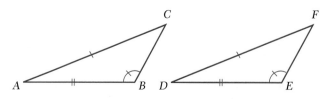

Name a third pair of parts needed to prove that the two triangles shown are congruent by the following:

26. ASA
27. SAS
28. SSS

In Problems 29 and 30, there are two triangles that are congruent. Identify the pair of triangles and give the postulate that shows congruency.

29.

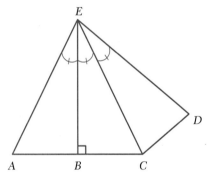

30.

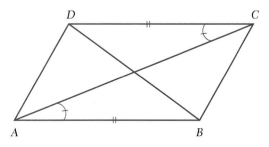

Section G.4, *Objective 2*

Given: △ABC = △RST

31. ∠A = ?
32. ∠T = ?
33. $\overline{BC}$ = ?
34. $\overline{AB}$ = ?
35. $\overline{RT}$ = ?

Section G.5, *Objective 1*

For Problems 36–40 use the following diagram.

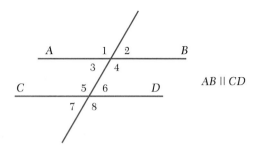

AB ∥ CD

36. What angle and ∠8 form a pair of alternate exterior angles?
37. What angle and ∠1 form a pair of vertical angles?
38. What angle and ∠5 form a pair of alternate interior angles?
39. What angle corresponds to ∠3?
40. Name two angles adjacent to ∠6.

Section G.5, *Objective 2*

Given: The diagram below

∠4 = 135°

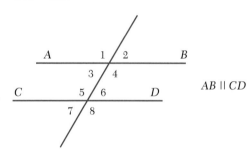

AB ∥ CD

Find:
41. ∠1
42. ∠8
43. ∠2
44. ∠5
45. ∠3

Section G.6, *Objective 1*

Use the following to answer Problems 46–48.

Given: △ACB ~ △ADC ~ △CDB

$\overline{AB}$ = 37, $\overline{AC}$ = 12, $\overline{CB}$ = 35

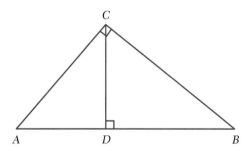

Find:
46. $\overline{BD}$
47. $\overline{CD}$
48. $\overline{AD}$

Use the following to answer Problems 49 and 50.

Given: △ACB ~ △AED

$\overline{AC}$ = 12 ft, $\overline{AB}$ = 16 ft, $\overline{BD}$ = 64 ft

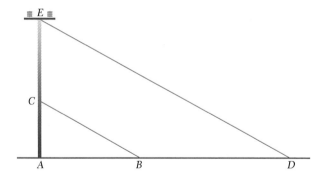

49. Find the height of the power pole ($\overline{AE}$).
50. Find the length of the guide wire attached to the top of the pole ($\overline{ED}$).

Section G.6, _Objective 2_

51. What two pairs of equal angles can be used to show that $\triangle BEC \sim \triangle AED$ in the diagram below?

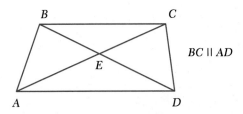

$BC \parallel AD$

52. What two pairs of equal angles can be used to show that $\triangle ACD \sim \triangle ACB$ in the diagram below?

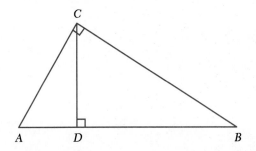

53. What two pairs of equal angles can be used to show that $\triangle AEC \sim \triangle CDB$ in the diagram below?

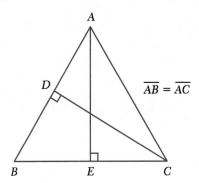

$\overline{AB} = \overline{AC}$

54. What pair of angles and what proportion of corresponding sides are needed to show that $\triangle AEB \sim \triangle DEC$ in the diagram below?

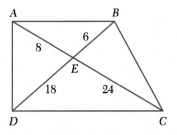

55. Show that $\triangle ABD \sim \triangle BDC$ by showing that corresponding sides are proportional.

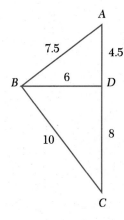

PLANE GEOMETRY

Test

ANSWERS

1. _____

1. Predict what figure seems most likely to appear next in the following sequence.

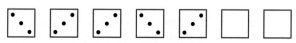

2. _____

2. True or false: The sum of two obtuse angles is a reflex angle.

3. _____

3. Given: AB ∥ CD in the diagram below, classify ∠2 and ∠7, ∠4 and ∠2, and ∠3 and ∠6.

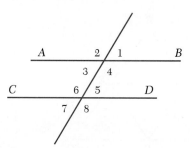

4. _____

4. In the diagram for Number 3, if ∠8 = 75°, find the measure of ∠3.

5. _____

5. Given △ABC with ∠A = 67° and ∠C = 75.8°, find ∠B.

6. _____

6. Given △DEF with $\overline{DE}$ = 23.5, $\overline{EF}$ = 20.7, and $\overline{DF}$ = 18.5, what kind of triangle is △DEF?

7. _____

7. In the following diagram, if ∠ABE = 90°, and ∠ABE = ∠C, is △ABE ∼ △ACD?

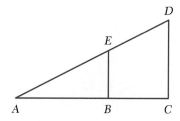

8. _____

9. _____

10. _____

11. _____

12. _____

13. _____

14. _____

15. _____

8. If ∠C and ∠D are complementary and ∠C = 48°, find ∠D.

9. True or false: If ∠A = 47° and ∠B = 23°, then △ABC is an acute triangle.

10. Given the angles 43°, 346°, 125°, 79°, 218°, 5°, 80°, 100°, identify the obtuse angles.

11. Given ∠A = ∠E, $\overline{EF} = \overline{AB}$, and $\overline{AC} = \overline{DE}$ in △ABC and △DEF, are the triangles congruent?

12. Is the following statement valid: All students must have a 2.00 or better GPA to graduate. June had a 3.76 GPA; therefore, June graduated.

13. If AB ∥ CD and cut by a transversal, are the interior angles on the same side of the transversal supplementary?

14. Given the following triangle, does $\overline{AF} = \overline{FC}$?

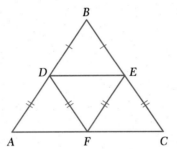

15. Given: △ABC ~ △DEF, $\overline{AB}$ = 60, $\overline{AC}$ = 75, $\overline{BC}$ = 100, and $\overline{DF}$ = 15, find DE.

Answers

Exercises 1.1

1. Five hundred forty-two   **3.** Eight hundred ninety
5. Seven thousand, fifteen   **7.** 57   **9.** 7500
11. 10,000,000   **13.** Twenty-five thousand,
three hundred ten   **15.** Two hundred five thousand,
three hundred ten   **17.** Forty-five million   **19.** 243,700
21. 23,470   **23.** 17,000,000   **25.** <   **27.** >   **29.** <
31. <   **33.** 690   **35.** 1700

	Number	Ten	Hundred	Thousand	Ten Thousand
37.	102,385	102,390	102,400	102,000	100,000
39.	7,250,978	7,250,980	7,251,000	7,251,000	7,250,000

41. 560,353,730   **43.** One million, two hundred thirty-five
thousand, nine hundred fifty-six   **45.** <   **47.** 1000
49. 43,780,000   **51.** 42,700; 42,800; rounding made the
tens digit approximate the second time, so rounding to the
hundreds place was inaccurate; the first method
53. Eighteen thousand, four hundred sixty-five   **55.** Three
hundred eighty-nine thousand, five hundred   **57.** $36,407
59. $686,000   **61.** 327,000,000 lb   **63.** Twenty thousand,
five hundred twenty-seven dollars   **65.** Maine
67. 93,000,000 mi   **69.** 500,000   **71.** Different round-off
places were used because of the population and severity of
the disease in the area   **73.** 52,000,000 sq mi

Challenge

77. Hundred billion   **79.** 1145, 1229, 1234, 1243, 1324,
1342, 1432   **81.** 10,000

Exercises 1.2

1. 97   **3.** 979   **5.** 992   **7.** 1   **9.** 4399   **11.** 69,543
13. 6571   **15.** 17,500   **17.** 1100   **19.** 11,000

21. 10,000   **23.** 19,000; 19,918   **25.** 160,000; 160,588
27. 201   **29.** 526   **31.** 451   **33.** Ten   **35.** 353
37. 158   **39.** 422   **41.** 4700   **43.** 2188   **45.** 300
47. 2000   **49.** 20,000   **51.** 500; 469   **53.** 4000; 3363
55. 3561   **57.** 163   **59.** 5320   **61.** 766   **63.** 8000;
9106   **65.** 42,400   **67.** 20,685   **69.** 40,000; 40,833
71. 0; 8233   **73.** 3,100,000   **75.** 10,100,000
77. $767,847   **79.** $323,725   **81.** 600,313
83. 264,271   **85.** $1,193,200   **87.** 167,210 gal
89. $13,100   **91.** Grand Canyon 5800 ft; Bryce Canyon
1900 ft; Zion 3500 ft; Greatest change is in Grand Canyon,
2300 ft more than Zion

Challenge

97. Four million, four hundred forty-three thousand, one
hundred forty-two   **99.** A = 3, B = 6, C = 8   **101.** A = 7,
B = 2, C = 3, D = 1

Exercises 1.3

1. 96   **3.** 660   **5.** 495   **7.** Thousands   **9.** 1596
11. 2848   **13.** 29,070   **15.** 13,300   **17.** 1600
19. 16,000   **21.** 8000   **23.** 32,000; 34,608   **25.** 35,000;
32,148   **27.** 71   **29.** 71   **31.** 2 R 5   **33.** Divisor
35. 3052   **37.** 24   **39.** 15 R 30   **41.** 58   **43.** 810
45. 10   **47.** 100   **49.** 50   **51.** 40   **53.** 600; 578 R 68
55. 2000; 1790 R 414   **57.** 251,490   **59.** 509
61. 1,200,000; 1,494,513   **63.** 20,000; 16,039 R 324
65. $600,000   **67.** $200,000; $228,550   **69.** $5000
71. $4100   **73.** 40,000; 38,880   **75.** 200; 188 R 759
77. 2278 salmon   **79.** 128 trees   **81.** 30,375 bacteria
83. $69,575   **85.** 9,934,000 gal   **87.** 2305 radios,
8 resistors   **89.** $35,100; $50,400; $15,300   **91.** 520 hrs
93. Latin America/Caribbean-North America; North
America-Western Europe; North Africa/Middle East-
Eastern Europe/Central Asia; North Africa/Middle East-
East Asia/Pacific; Eastern Europe/Central Asia-Australia;
East Asia/Pacific-Australia   **95.** $8060   **97.** Los Angeles;
the cost per passenger mile is about half that of Portland

Challenge

105. $23,439,000   **107.** A = 6, B = 3, C = 8, E = 1
109. A = 5, B = 1, C = 6

Exercises 1.4

1. 12^6 **3.** 81   **5.** 8   **7.** 1   **9.** Base; exponent; power or value   **11.** 216   **13.** 361   **15.** 10,000   **17.** 512
19. 6561   **21.** 4500   **23.** 70,000   **25.** 12   **27.** 340
29. Power or exponent   **31.** 4,350,000   **33.** 1200
35. 35,910,000   **37.** 302   **39.** 70,500,000,000
41. 9700   **43.** 10^{11} **45.** 38,416   **47.** 387,420,489
49. 3,350,000,000,000   **51.** 4380   **53.** $73,000,000
55. 32,000,000 shares   **57.** 10^7; 10^5 **59.** 5×10^5; 50×10^4; 500×10^3; 5000×10^2; $50,000 \times 10^1$; or $500,000 \times 10^0$ **61.** 5^9 dollars; $1,953,125

Challenge

65. $531,441; $797,160   **67.** 9736

Exercises 1.5

1. 6^6 **3.** 11^{10} **5.** 22^8 **7.** 14^8 **9.** 19^{15} **11.** 58^{26}
13. 6^{13} **15.** 8^4 **17.** 10^{12} **19.** 19^0 **21.** 4   **23.** 12^{18}
25. 10^{21} **27.** 18^{45} **29.** $2^2 \cdot 5^2$ **31.** $8^4 \cdot 9^4$
33. $13^7 \cdot 12^7$ **35.** $12^{12} \cdot 23^{12}$ **37.** $14^{19} \cdot 45^{19}$
39. $4^{11} \cdot 6^{11} \cdot 8^{11}$ **41.** 13^{28} **43.** 16^{18} **45.** 24^{63}
47. $3^{12} \cdot 7^8$ **49.** 5^{24} **51.** 125 cu in   **53.** $V = w^4$; 626 cu ft   **55.** $V = 729$ cu in   **57.** Sub-Saharan Africa; North Africa/Middle East — 10^5, SubSaharan Africa — 10^7, $10^2 \cdot 10^5 = 10^7$ — 100 (10^2) times North Africa/Middle East is Sub-Saharan

Challenge

61. 2   **63.** 3

Exercises 1.6

1. 53   **3.** 0   **5.** 27   **7.** 15   **9.** 25   **11.** 39   **13.** 145
15. 43   **17.** 31   **19.** 7   **21.** 35   **23.** 217   **25.** 24
27. 28   **29.** 5   **31.** 9   **33.** 13   **35.** 6   **37.** 12   **39.** 26
41. 40   **43.** 18   **45.** 31   **47.** 169   **49.** 110   **51.** 54
53. 45   **55.** 540   **57.** 596   **59.** 166   **61.** 716   **63.** 1
65. 160   **67.** 6010   **69.** 181,925   **71.** $900   **73.** 228
75. $61   **77.** 49 mph   **79.** $8159   **81.** 5,050,000; no, the actual cases in the two regions are too far apart to have the average make sense   **83.** 76 cents   **85.** 73

Challenge

91. 196   **93.** $1200

Exercises 1.7

1. 1992   **3.** $857 million   **5.** $248 million   **7.** $4700 million; $4707 million   **9.** Fish cakes   **11.** 48 mg   **13.** 72 g
15. 908 calories   **17.** Fish cakes, chicken dijon, and pepper steak   **19.** $106,930   **21.** $14,291   **23.** $210,478
25. 63 and 65   **27.** August   **29.** 3793   **31.** $11,855
33. 144   **35.** Fisher Zoo   **37.** 13,000,000
39. 1,251,036   **41.** $68,628,000

Challenge

45. 598,345

Exercises 1.8

1. 10–11   **3.** 250   **5.** 1425   **7.** 40   **9.** Full size
11. 400   **13.** 1996   **15.** 12,500   **17.** 20,500
19. $45,000   **21.** $15,000   **23.** $40,000

25.

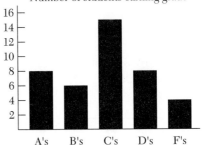

Number of students earning grade

27.

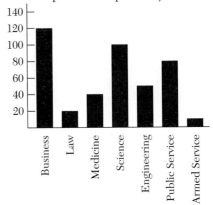

Career preference expressed by senior class

29.

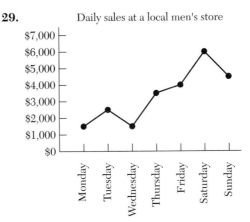

Daily sales at a local men's store

31.

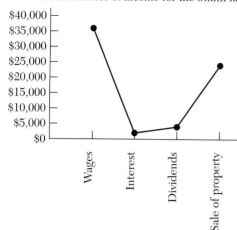

Annual sources of income for the Smith family

33.

Year	Cost of an average three-bedroom house 🏠 = $10,000 🏠 = $5000
1970	🏠 🏠 🏠 🏠 🏠 🏠
1975	🏠 🏠 🏠 🏠 🏠 🏠 🏠
1980	🏠 🏠 🏠 🏠 🏠 🏠 🏠 🏠 🏠
1985	🏠 🏠 🏠 🏠 🏠 🏠 🏠 🏠
1990	🏠 🏠 🏠 🏠 🏠 🏠 🏠 🏠
1995	🏠 🏠 🏠 🏠 🏠 🏠 🏠 🏠 🏠

35.

Year	Barrels produced 🛢 = 5000 🛢 = 2500
1991	🛢 🛢 🛢
1992	🛢 🛢 🛢 🛢 🛢
1993	🛢 🛢 🛢 🛢 🛢 🛢
1994	🛢 🛢 🛢 🛢 🛢 🛢 🛢
1995	🛢 🛢 🛢 🛢 🛢 🛢 🛢 🛢

Chapter 1 True–False Concept Review

1. True **2.** True **3.** False; the word name for 750 is seven hundred fifty **4.** False; 500 > 23 **5.** True
6. False; to the nearest ten, 7449 round to 7450 **7.** False; the rounded number may be the number itself; 750 rounded to the nearest ten is 750 **8.** True **9.** True
10. False; the sum of 8 and 5 is 13 **11.** True **12.** True
13. True **14.** True **15.** True **16.** False; the only factors of 15 are 1, 3, 5, and 15 **17.** True **18.** True

19. True **20.** True **21.** False; $\dfrac{22{,}000{,}000}{100{,}000} = 220$

22. False; in the order of operations, multiplication sometimes take precedence over division (if it appears before division when reading from left to right, and there are no grouping

symbols to affect the order) **23.** False; in the order of operations, multiplication is usually done before addition unless grouping symbols indicate otherwise
24. True **25.** True

Chapter 1 Review

1. Eight hundred ninety-two **3.** Six hundred eighty thousand, fifty-seven **5.** 208,025,608 **7.** < **9.** >
11. 4770 **13.** 67,300 **15.** 3,040,000 **17.** 2802
19. 1,498,060 **21.** 2500 **23.** 3700 **25.** 320,000
27. 87,843 **29.** 19,324 **31.** 200 **33.** 10,000
35. 40,000 **37.** 10,206 **39.** 49,644 **41.** 4000
43. 280,000 **45.** 2,800,000 **47.** 6834 **49.** 68 R 140
51. 60 **53.** 20 **55.** 1000 **57.** 1728 **59.** 625
61. 34,000,000 **63.** 2700 **65.** 39,000,000,000,000
67. 11^9 **69.** 31^8 **71.** 7^{10} **73.** 12^{12} **75.** 22^{42}
77. $9^6 \cdot 5^6$ **79.** $4^{10} \cdot 3^{10}$ **81.** 10 **83.** 100 **85.** 26
87. 109 **89.** 88 **91.** 1000 units **93.** Day shift
95. $404,000 **97.** Vans **99.** 50 two-door sedans

101.

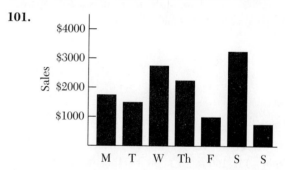

103.

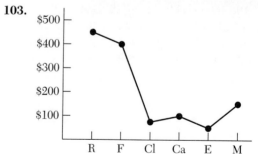

Chapter 1 Test

1. 109 **2.** 2279 **3.** 19 **4.** 40,704 **5.** > **6.** 760,000
7. 273,674 **8.** 450,082 **9.** 907 **10.** 73,300
11. 17,900 **12.** 160,000 **13.** 729 **14.** 5882 **15.** 43^{13}
16. 6453 **17.** Six thousand seven **18.** 30 **19.** 236,047
20. 60,500 **21.** 524,940,000 **22.** 600; 620 R 5 **23.** 17
24. 611 **25.** 666 **26.** 105 min **27.** $788,000; $39,400
28. Chevrolet; 700; 100 **29.** A and B; 225; 1200

30.

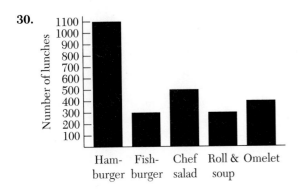

CHAPTER 2

Exercises 2.1

1. 24 ft **3.** 8 mℓ **5.** 4 gal **7.** 951 oz **9.** 50 hr
11. 208 lb **13.** 1120 sec **15.** 20 lb **17.** 17 yd
19. 20 g **21.** 205 kℓ **23.** 90 mm **25.** 104 yd
27. 553 gal **29.** 24 cm **31.** Answers vary **33.** 19 ft 7 in
35. 6 min 52 sec **37.** 3 hr 31 min 12 sec **39.** 6 yd 1 ft 2 in
41. 44 ft 6 in **43.** 2086 g **45.** 12° C **47.** 840 mg
49. 400 m **51.** The base of the fountain is a 4-ft square.
53. The patio is 16 ft long and 12 ft wide.

Maintain Your Skills

59. 832 **61.** $3910

Exercises 2.2

1. 19 in **3.** 28 m **5.** 34 mm **7.** 26 km
9. 83 mm **11.** 160 m **13.** 90 ft **15.** 100 cm
17. 78 yd **19.** 168 m **21.** 54 cm **23.** 348 mm
25. 13 ft 4 in **27.** 1 hr 48 min **29.** 92 in **31.** 117 ft
33. 19,376 yd **35.** The patio is an octagon. **37.** The
perimeter of the fountain base is 16 ft. **39.** The perimeter
of flower bed B is 36 ft.

Challenge

43. $240

Maintain Your Skills

45. 900,050 **47.** 32,570,000 **49.** 255,084

Exercises 2.3

1. 16 km^2 **3.** 20 yd^2 **5.** 60 ft^2 **7.** 52 m^2 **9.** 132 km^2
11. 238 yd^2 **13.** 396 cm^2 **15.** 135 in^2 **17.** 612 m^2
19. 101 ft^2 **21.** 6550 m^2 **23.** 5655 in^2 **25.** 9 ft^2 = 1
yd^2 **27.** No, she needs more than 3 gal **29.** 240 oz
31. 36 ft^2 **33.** 162 ft^2 **35.** 1332 in^2 **37.** 6 bags
39. The patio requires 2254 bricks.

Challenge

43. Joe needs 72 black 6-in and 64 white 6-in squares. He
needs 28 black and 28 white 12-in squares. **45.** $3348

Maintain Your Skills

49. 3050 **50.** $29

Exercises 2.4

1. 1000 m^3 **3.** 250 ft^3 **5.** 120 cm^3 **7.** 21,840 mℓ
9. 48 ft^3 **11.** 3060 in^3 **13.** 43,310 cm^3 **15.** 5880 in^3
17. 3 ft^3 **19.** 10 yd^3 **21.** 93,312 in^3 = 2 yd^3 **23.** 1200
in^3 **25.** 1320 yd^3 **27.** 177,120 cm^3 **29.** 16 truckloads
31. The bricklayer needs 20 ft^3 of mortar. **33.** The cement contractor needs 219 ft of lumber for the form.

Challenge

37. The pool holds 2400 ft^3. Seven loads of dirt were
hauled away.

Maintain Your Skills

41. 41 **43.** 65

Chapter 2 True–False Concept Review

1. False; metric system is the most commonly used
2. True **3.** False; a liter is a measure of volume **4.** True
5. False; volume is the measure of the inside of a solid
6. False; the volume of a cube is $V = s^3$ **7.** True **8.** True
9. True **10.** True **11.** True **12.** False; a parallelogram
has four sides **13.** False; volume is the number of cubes
an object will hold **14.** True **15.** True **16.** True
17. False; volume can be measured in cups or liters
18. True **19.** False; one square foot is 144 inches
20. False; "kilo" means 1000

Chapter 2 Review

1. 30 ft **3.** 24 ft **5.** 4750 ℓ **7.** 1650 mg **9.** 26 gal 3 qt
11. 12 ft **13.** 32 m **15.** $17,500 **17.** 189 ft^2
19. 648 in^2 **21.** 288 in^3 **23.** 1920 ft^3 **25.** 343 in^3

Chapter 2 Test

1. 7454 mm **2.** 136 cm **3.** 1728 in^3 **4.** 80 ft^2 **5.** 27 lb
6. 22 ft **7.** 30 in^2 **8.** 1 gal 3 qt 1 pt **9.** Possible answers: oz, lb, ton, mg, cg, g, kg **10.** 288 in^2 **11.** 46 ft
12. 264 m^3 **13.** 48 cm^3 **14.** 20 ft^3 **15.** 114 g
16. $132 **17.** Possible answers: cup, pint, qt, gal, mℓ, ℓ,
kℓ, in^3, ft^3, yd^3, mm^3, cm^3, m^3 **18.** 2736 mm^2
19. 1920 yd **20.** Answers vary

CHAPTER 3

Exercises 3.1

1. $x + 12$ **3.** $7x$ $20 - x$ **7.** $\dfrac{x}{5}$ **9.** $2y + 18$

11. $\dfrac{y}{2} - 15$ **13.** $8y - 2y$ **15.** $xy - 15$ **17.** $\dfrac{x}{y} + 20$

19. $15(w - 2)$ **21.** 38 **23.** 30 **25.** 12 **27.** 104
29. 36 **31.** 99 **33.** 42 **35.** 784 **37.** 42 **39.** 2
41. Substitution **43.** 84 **45.** $18n$; $72; $126; $216
47. $16n + 3$; $35; $83; $131 **49.** $11250x$; $56,250;
$123,750; $157,500 **51.** $ah + bh + ch$ or $(a + b + c)h$
gallons **53.** a) 58; b) $\dfrac{d_1 + d_2 + d_3 \ldots + d_{10}}{10}$ where A is

the average and $d_1, d_2, d_3 \ldots d_{10}$ are the number of acci-

dental drownings per year **55.** $B = \dfrac{x}{100}$

Challenge

61. 456 **63.** 10 **65.** 7757

Maintain Your Skills

69. 223 **71.** 832

Exercises 3.2

1. Yes **3.** Yes **5.** No **7.** Yes **9.** Yes **11.** Yes
13. No **15.** No **17.** Yes **19.** No **21.** $A = 3$ ft^2
23. $D = 110$ miles **25.** 25 in **27.** 2 cm **29.** 4096 m^3
31. 97 ft **33.** 720 in^3 **35.** 558 ft **37.** $x = 7$
39. $y = 15$ **41.** $w = 35$ **43.** $m = 11$ **45.** $x = 7$
48. $y = 12$ **49.** $w = 8$ **51.** $a = 11$ **53.** False
55. Two **57.** $S = 325$ **59.** $S = 10,100$ **61.** 450 ft^2
63. $577 **65.** s.g. = 5 **67.** IQ = 128 **69.** $E = 3I$;
$E = 6I$; $6900 to $13,800

71.

Number of guests	Hors d'oeuvres	Bartenders	Cocktail servers
x	$H = 24x$ or $H = 18x$	$B = \dfrac{x}{100}$	$C = \dfrac{x}{50}$
800	14,400 to 19,200	8	16
900	16,200 to 21,600	9	18
1000	18,000 to 24,000	10	20
1100	19,800 to 26,400	11	22

Challenge

75. Yes **77.** No

Maintain Your Skills

81. 11,148 **83.** 89 cm

Exercises 3.3

1. 75 **3.** $3xy$, $8xy$ **5.** $14m^2n$, $30m^2n$ **7.** 1

9. $17pq$, pq, $13pq$ **11.** $21a$ **13.** $27c$ **15.** $40x$ **17.** $6y$
19. $50w$ **21.** $76x$ **23.** $59y$ **25.** $244s$ **27.** $74w$
29. $8x + 7y$ **31.** $26r + 9t$ **33.** $19x^2 + 23x$
35. $4m^2 + 3s$ **37.** $3x + 8y$ **39.** $18xy + 22yz$
41. $94x^2 + 22x$ **43.** $28x^2 + 8x + 56$
45. $23xy + 43yz + 37wz$ **47.** $17x^2 + 34x + 37$
49. $8pq + 13qs$ **51.** $4abc + 6ab + 6c$ **53.** Like
55. $45x^2 + 2x$ cannot be simplified **57.** $8w + 12$
59. $T = 3i$; $36,000 **61.** $D = 4m$; $32
63. $L = y + y + 2y + 2y + 2y$; $L = 8y$; $L = 80$ ft
65. $6x + 28$ ft **67.** $1351P + 309Q$ **69.** $14B + 6C$;

$14\left(\dfrac{x}{100}\right) + 6\left(\dfrac{x}{50}\right)$ **71.** $R = 6x + 14\left(\dfrac{x}{100}\right) + 6\left(\dfrac{x}{50}\right)$

Challenge

75. $113a + 5b + 111c$ **77.** $42z$ **79.** $38z$

Maintain Your Skills

83. True **85.** 32 m^2

Exercises 3.4

1. $48w$ **3.** $63a^2b$ **5.** $8y$ **7.** $12y$ **9.** $320bc$ **11.** $13x$
13. $168xy^2$ **15.** $399s^2t^2$ **17.** $13ac$ **19.** $8d^2$
21. $4a + 3b + 5c$ **23.** $55z^2 + 10z$ **25.** $56xy^2z - 16x^2yz$
27. $30x^2yz - 44xy^2z + 50x^2y^2$ **29.** $4b + 3c + 5d$
31. $4z^2 + 5xz + 8yz$ **33.** $84a^2bc + 63ab^2c - 105abc^2$
35. $y^2 + 9y + 18$ **37.** $x^2 + 17x + 30$ **39.** $21x - 22w$
41. $18a - 13b + 9c$ **43.** $1748a^3$ **45.** $6x^2 + 29x + 35$
47. $24x^2 + 58x + 9$ **49.** x **51.** $85xy + 34x$
53. $A = 2w^2$; 242 sq ft **55.** $V = 6w^3$; 48 cu ft

57. $A = 2b^2$; 512 sq in **59.** $W = \dfrac{18c + 192}{12}$

61. Total number of chairs

63.

C = Chairs per Row	R = Number of Rows	Total Seating Capacity	Total Area of Hall
25	10	250	408,312 in.^{2} or 2835.5 ft^2
20	13	260	420,624 in.^{2} or 2921 ft^2
16	16	256	426,240 in.^{2} or 2960 ft^2
10	25	250	470,952 in.^{2} or 3270.5 ft^2

Challenge

69. $V = 24x^3 + 12x^2$ **71.** $V = 9r^3 + 12r^2$

Maintain Your Skills

73. 1425 **75.** $42 + 3n$ **77.** 1260 in^3

Exercises 3.5

1. $a = 27$ **3.** $c = 32$ **5.** $r = 8$ **7.** $t = 26$ **9.** $y = 0$
11. $m = 26$ **13.** $a = 14$ **15.** $c = 102$ **17.** $r = 39$
19. $t = 92$ **21.** $y = 28$ **23.** $m = 152$ **25.** $y = 32$
27. $t = 80$ **29.** $z = 108$ **31.** $a = 5$
33. $c = 148$ **35.** $u = 887$ **37.** $y = 231$
39. $p = 1020$ **41.** 59, answers vary   **43.** $x = 2$; answers
vary   **45.** 237 in   **47.** 237 m   **49.** a) $T =$
total money budgeted, $S =$ amount spent,
$R =$ amount remaining, $T = S + R$
b) Land acquisition, $R = \$1,581,172$
   Open space, $R = \$298,552$
   Pathways development, $R = \$153,443$
   Playfield improvements, $R = \$1,002,220$
51. $x - 44 = 56$; 100   **53.** $S = I + 4$ where $S =$ Saturn
rating and $I =$ Impreza rating; 24 mpg   **55.** $T = t + 26$
where $T =$ January temperature and $t =$ July temperature;
59°F   **57.** 308 double rooms   **59.** 52 more guests

Maintain Your Skills

65. 3000   **67.** $\dfrac{3n}{71}$

Exercises 3.6

1. $x = 4$ **3.** $u = 7$ **5.** $v = 9$ **7.** $y = 34$ **9.** $x = 20$
11. $y = 63$ **13.** $y = 36$ **15.** $c = 380$ **17.** $t = 13$
19. $w = 22$ **21.** $s = 11$ **23.** $s = 1584$ **25.** $u = 70$
27. $x = 714$ **29.** $w = 35$ **31.** $r = 2856$ **33.** $a = 25$
35. $y = 432$ **37.** $x = 7$ **39.** $x = 4032$ **41.** 4; Answers
vary   **43.** $x = 20$; answers vary   **45.** \$310   **47.** $N = 24c$;
17,352 cans   **49.** $N = 32d$; 6784 bricks   **51.** 25 m
53. $t = 2T$ where $t =$ July low temperature and $T =$
January high   **55.** $d = 2h$ where $d =$ distance to first row;
7 ft   **57.** The screen is too short for people in the first or
the last row; 9 ft high

Challenge

61. $x = 409$ **63.** $y = 17,675$ **65.** $t = 290$

Maintain Your Skills

67. $xy + 7$ **69.** $\dfrac{y}{17} - x$ **71.** $xy - \dfrac{x}{y}$

Exercises 3.7

1. $x = 2$ **3.** $y = 5$ **5.** $w = 8$ **7.** $t = 0$ **9.** $z = 7$
11. $a = 8$ **13.** $n = 7$ **15.** $p = 5$ **17.** $a = 19$
19. $c = 2$ **21.** $x = 6$ **23.** $y = 7$ **25.** $t = 25$
27. $a = 8$ **29.** $b = 4$ **31.** $c = 14$ **33.** $y = 55$
35. $w = 6$ **37.** $x = 23$ **39.** $z = 11$ **41.** 2;

answers vary   **43.** $x + 5 = 18$ or $7x = 91$; answers vary
45. $5x + 6 = 41$; 7   **47.** $92 = 4x - 12$; 26
49. $n + 3n = 1024$; 256 units; 768 units   **51.** 13 in

53. ATT, $C = 15m$; $m = 200$ min

Tone, $C = 490 + 10m$; $m = 251$ min

Pace, $C = 696 + 9m$; $m = 256$ min

Pace will give her the most value for her money.
55. $72 + 18x + 72 + 18x + 72 = 1200$ or
$36x + 216 = 1200$ **57.** Answers vary

Challenge

61. $x = 607$ **63.** $z = 5697$ **65.** $x = \dfrac{c - b}{a}$

Maintain Your Skills

69. 288   **71.** 36 ft   **73.** 528 ft^3

CHAPTER 3 TRUE–FALSE CONCEPT REVIEW

1. False; a *variable* is a placeholder for a number
2. True   **3.** True   **4.** True   **5.** True   **6.** True
7. True   **8.** True   **9.** False; $5ab$ and $5xy$ are unlike terms
because the variable factors are different   **10.** True
11. True   **12.** False; unlike terms cannot be combined
13. True   **14.** False; exponents of the same base are
added when terms are multiplied   **15.** False; only the coefficients are subtracted   **16.** False; equivalent equations
have the same solutions   **17.** True   **18.** False; this is the
procedure for checking an equation; to solve an equation,
we use the properties of equality to isolate the variable on
one side of the equation   **19.** True   **20.** False; an equation contains an equals sign; the expression $3x + 7$ has no
equals sign   **21.** False; the sum of 6 and 4 is 10; the product of 6 and 4 is 24   **22.** False; the square of 6 is 36
23. True   **24.** False; in general, $6x + 3y \neq 9xy$ **25.** True

CHAPTER 3 REVIEW

1. $x + 30$ **3.** $4x + 56$ **5.** $\dfrac{95}{7x}$ **7.** 22   **9.** 48   **11.** Yes
13. Yes   **15.** No   **17.** $V = 343$ **19.** $S = 533$
21. $A = 54$ **23.** $x = 14$ **25.** $y = 11$ **27.** $31t$
29. $23x^2 + 2x$ **31.** $7x + 4y$ **33.** $9x + 9y + 4z$
35. $a + 2b$ **37.** $216yz$ **39.** $7ab$ **41.** $12a^2 - 21a$
43. $2x^2 + 13x + 20$ **45.** $9x + 12y - 17$ **47.** $t = 125$
49. $s = 68$ **51.** $t = 7$ **53.** $y = 15$ **55.** $x = 98$
57. $w = 204$ **59.** $62c = 434$; \$7   **61.** $x = 7$ **63.** $w = 6$
65. $6x + 27 = 129$; 17

CHAPTER 3 TEST

1. $61z$ **2.** $x = 16$ **3.** 102 **4.** $S = 610$ **5.** $w = 1128$
6. $t = 126$ **7.** $12a$ **8.** $5c + 3 - 2d$ **9.** $42x^2y + 69xy^2$
$- 45xyz$ **10.** $227x$ **11.** 1763 **12.** $r = 28$ **13.** $64x^3$
14. $y = 11$ **15.** $x = 18$ **16.** $17y$ **17.** $x = 28$ **18.** No
19. $\dfrac{n + 18}{5}$ **20.** $c = 560$ **21.** $P = 76 + 21(24)$; $580
22. 83 ft **23.** $18c = 6264$; $348 **24.** $14y + 10 = 374$; 26

CHAPTERS 1–3 CUMULATIVE REVIEW

1. 8005 **3.** $>$ **5.** $11{,}348$ **7.** 5890 **9.** $722{,}000$
11. 521 **13.** 1728 **15.** $28{,}040$ **17.** 9^{12} **19.** 56
21. 361 **23.** 310 **25.** None **27.** 325

29.

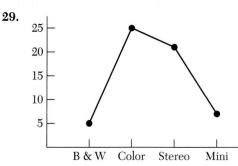

31. 1 hr 54 min 47 sec **33.** 34 lb **35.** 86 ft **37.** 3 gal
39. $11n - 9$ **41.** 62 **43.** Yes **45.** 42 **47.** $34x$
49. $4a$ **51.** $30x + 17y$ **53.** $6x^2 + 4xy - 6xz$
55. $432a^3$ **57.** $8ac$ **59.** $m = 1020$ **61.** $t = 980$
63. $x = 11$ **65.** $a = 2$ **67.** $18{,}300$ ft^2

CHAPTER 4

Exercises 4.1

1. 3 **3.** -14 **5.** 37 **7.** -17 **9.** -61 **11.** $75, -81$,
-95 **13.** $65, 82, 67$ **15.** -34 **17.** -75 **19.** 11
21. 33 **23.** 172 **25.** 144 **27.** 32 or -32 **29.** $75, 81$,
100 **31.** $99, 45, 115$ **33.** 6 **35.** 43 **37.** $<$ **39.** $>$
41. $>$ **43.** $>$ **45.** Greater **47.** $<$ **49.** $<$ **51.** $<$
53. $>$ **55.** 34 **57.** -30 **59.** True **61.** $-$2, a loss of
$2 **63.** $-12°C$ **65.** 1875 AD or $+1875$ **67.** -11 or 11
69. -101 or 101 **71.** $-115°C$ **73.** $-102°C$

75.

Africa	19,340	-512
Antarctica	16,864	-8327
Asia	29,028	-1312
Australia	7310	-52
Europe	18,510	-92
North America	20,320	-282
South America	22,834	-131

77. No, the Mariana Trench is farther away; absolute value
and greater than

Challenge

83. 4 **85.** Positive

Maintain Your Skills

89. 37^{20} **91.** 395

Exericses 4.2

1. -6 **3.** -1 **5.** -14 **7.** -34 **9.** -8 **11.** 5
13. -34 **15.** $5 + (-14) = -9$ **17.** 9 **19.** -95 **21.** 0
23. -4 **25.** 0 **27.** 1 **29.** -33 **31.** -110
33. $c = -37$ **35.** $a = 2$ **37.** -121 **39.** -189
41. -30 **43.** 81 **45.** -757 **47.** -114 **49.** $-349°C$
51. $x + (-67)$ **53.** 1 **55.** -113 lbs **57.** $29{,}826$ ft
59. $28{,}468$ volumes **61.** No; 2 yards short **63.** $-$13; a
loss of $13

Challenge

67. 15 **69.** 87 **71.** No solution

Maintain Your Skills

73. $978{,}900{,}000{,}000$ **75.** 26^{23} **77.** $\dfrac{65}{4x}$

Exercises 4.3

1. -10 **3.** 3 **5.** -17 **7.** 32 **9.** -26 **11.** 1 **13.** 0
15. -54 **17.** 87 **19.** -138 **21.** 4 **23.** 0 **25.** 177
27. 116 **29.** -677 **31.** Yes **33.** -10 **35.** 5
37. $b = -90$ **39.** $d = -12$ **41.** -87 **43.** -38
45. -57 **47.** -2 **49.** $-$2700 **51.** $97°C$ **53.** -43
55. Africa, 19,852 ft; Antarctica, 25,191 ft;
Asia, 30,340 ft; Australia, 7362 ft; Europe, 18,602 ft;
North America, 20,602 ft; South America, 22,965 ft;
Australia, it is a rather flat continent **57.** $-19{,}680$ ft or
19,680 ft below sea level **59.** $-$71 **61.** $-15°C$
63. $105°C$

Challenge

69. -45 **71.** $a = 3$ or $a = -25$

Maintain Your Skills

75. 38 **77.** 15

Exercises 4.4

1. $14x$ **3.** $-3x$ **5.** $2a$ **7.** $-30a$ **9.** $-18y$ **11.** $-3y$
13. $5a$ **15.** $-8x$ **17.** $2x^2$ **19.** $-a + 1$ **21.** $-4x^2 - 2x$

23. $-3x - 26$ **25.** $4xy$ **27.** $11y2 + 3y - 12$
29. $-4a - 5b + 20$ **31.** $9c - 3d + 3$ **33.** a
35. $-33b$ **37.** $2x^2 - 17x + 6$ **39.** $a - 16b - 5$
41. $-2a + 8b$ **43.** $5x + 17y - 15$ **45.** Ben, $3c + 4t$;
Josh, $5c + t$; $(5c + t) - (3c + 4t) = 2c - 3t$; cannot decide,
it depends upon the cost of the CDs and the tapes
47. $3x^2 - 15x + 10$

Challenge

53. $40a - 17b - 37$

Maintain Your Skills

57. Yes **59.** $24x - 96$

Exercises 4.5

1. Positive **3.** -6 **5.** 30 **7.** -48 **9.** 36 **11.** -11
13. -110 **15.** 144 **17.** -258 **19.** 210 **21.** -42
23. 48 **25.** 6 **27.** -4 **29.** -5 **31.** 7 **33.** -3
35. 6 **37.** -8 **39.** -5 **41.** -20 **43.** 25 **45.** -18
47. 102 **49.** -19 **51.** -33 **53.** -702 **55.** 16
57. 1300 **59.** -2496 **61.** 20 **63.** -63 **65.** $271{,}830$
67. -452 **69.** Yes **71.** No **73.** -24 lb **75.** $-\$87$
77. Asia **79.** Australia **81.** $-\$224$ **83.** \$12,350;
\$10,450; lost \$1900 or $-\$1900$ **85.** $-\$43{,}110$; $-\$28{,}740$

Challenge

91. -70 **93.** -9 **95.** 81

Maintain Your Skills

99. $-24s + 3t$ **101.** $y = 22$

Exercises 4.6

1. $-35c$ **3.** $40d$ **5.** $7ab$ **7.** $-6x + 15$ **9.** $18t - 42$
11. $-6s + 14$ **13.** $-6x^2$ **15.** $54b^2$ **17.** $132mn$
19. $80bc$ **21.** $-16x + 20$ **23.** $-28x - 8$
25. $-52x^2 + 91x$ **27.** $x^2 - 9x + 14$ **29.** $x^2 + x - 30$
31. $-3x$ **33.** $5y$ **35.** $-4b$ **37.** $-11b$ **39.** $-3n$
41. $4x - 3$ **43.** $8a$ **45.** $-8m$ **47.** $2b - 3c + 5d$
49. $x - 3$ **51.** $3a + 4d - 7e$ **53.** $3b + 2$
55. $b^2 + 19b + 90$ **57.** $3x - 5$ **59.** $y^2 - y - 42$
61. $c^2 + c - 72$ **63.** $15x^2 - 41x - 104$
65. $21ac - 14bc - 24a + 16b$ **67.** $10c^2 + 38c - 8$
69. $6x^2 + 15x - 9$ **71.** a) $15x$ or 15 of 20 parts b) x or 1
of 20 parts c) $9x$ or 9 of 20 parts **73.** $45x^2 - 65x - 24$

Challenge

77. $2x^2 + 2x - 61$ **79.** $4x^3 + 24x^2 - 9x - 54$

Maintain Your Skills

81. $P = 68$ **83.** $52w^2z$ **85.** $m = 115$

Exercises 4.7

1. -52 **3.** 55 **5.** -45 **7.** -3 **9.** -40
11. Exponents, $5^2 = 25$ **13.** -103 **15.** -33 **17.** 13
19. 22 **21.** 19 **23.** -6 **25.** -3 **27.** 1 **29.** -4
31. -1 **33.** -10 **35.** -3 **37.** -28 **39.** -7
41. -2 **43.** -16 **45.** 12 **47.** -12 **49.** $-56°C$
51. $-34°C$ **53.** 77 **55.** -41 **57.** -18 **59.** 1
61. -2065 **63.** -1594 **65.** -16 **67.** \$800 **69.** $-\$9$,
a loss **71.** Africa, 9414 ft; Antarctica, 4268 ft; Asia, 13,858
ft; Australia, 3629 ft; Europe, 9209 ft; North America,
10,019 ft; South America, 11,352 ft; largest, Asia; smallest,
Australia **73.** Athens, Bangkok, New Delhi **75.** \$3226

Challenge

83. -11

Maintain Your Skills

87. $6y$ **89.** $w = 43$

Exercises 4.8

1. -7 **3.** -12 **5.** -19 **7.** -10 **9.** $C = -2$
11. -108 **13.** 203 **15.** 7 **17.** Yes **19.** $S = -51$
21. -111 **23.** -180 **25.** 558 **27.** $-130°F$
29. $14°F$ **31.** No **33.** No **35.** No **37.** $10{,}080$
39. -6588 **41.** -469 **43.** -319
45. $3E = 2035 + (J + P)$; yes; answers vary

Challenge

51. 8

Maintain Your Skills

55. $t = 18$ **57.** $y = 11$

Exercises 4.9

1. $a = -23$ **3.** $x = -10$ **5.** $x = -9$ **7.** $c = 5$
9. $x = -12$ **11.** $y = 2$ **13.** $y = 7$ **15.** $x = 11$
17. $w = -5$ **19.** $a = 54$ **21.** $y = 0$ **23.** $x = -5$
25. $x = 1$ **27.** $k = -8$ **29.** $r = 15$ **31.** 4 sec **33.** 28
35. 8 **37.** $x = 14$ **39.** $p = 29$ **41.** $q = -5$

43.

°F	122°F	104°F	77°F	50°F	14°F	$-13°F$	$-40°F$
°C	50°C	40°C	25°C	10°C	$-10°C$	$-25°C$	$-40°C$

45. 14 payments   **47.** $7310 - 4L = 12558$, L represents the lowest point; Asia   **49.** $-35,840 - 2D = -2000$; Ionian Basin

Challenge

53. $x = 5$ **55.** $x = 5$

Maintain Your Skills

59. $z = 285$ **61.** $y = 23$

CHAPTER 4 TRUE–FALSE CONCEPT REVIEW

1. False; the opposite of -7 is 7   **2.** True   **3.** False; the difference of -3 and -10 is 7   **4.** True   **5.** True
6. True   **7.** False; the quotient is negative   **8.** True
9. True   **10.** False; the product of an odd number or negative factors is negative   **11.** False; $2x = -5$ does not have an integer solution   **12.** True   **13.** True   **14.** True
15. True   **16.** True   **17.** False; the commutative property does not hold for subtraction   **18.** True   **19.** False; the sum of two negative integers is negative, but the difference can be positive, $-3 - (-5) = 2$ **20.** False; $-5^2 = -25$

CHAPTER 4 REVIEW

1. 22   **3.** -32 **5.** 8   **7.** 87   **9.** 12   **11.** $<$ **13.** $<$
15. $<$ **17.** 41   **19.** $x = -206$ **21.** -132 **23.** 180
25. $k = -196$ **27.** $-23x$ **29.** $6a - 2b + 15$ **31.** 276
33. -348 **35.** $x = -567$ **37.** -21 **39.** 32   **41.** $132a$
43. $108z^2$ **45.** $2x^2 - 5x - 3$ **47.** $-15y$ **49.** $-9b$
51. 162   **53.** 29   **55.** 99   **57.** -7 **59.** -47 **61.** 37
63. -868 **65.** $y = 11$ **67.** $a = -9$ **69.** $x = 7$

CHAPTER 4 TEST

1. -35 **2.** $-165x^2$ **3.** 98   **4.** $-6x^2 + 21xy + 39x$
5. 88   **6.** -19 **7.** $-50b$ **8.** -17 **9.** 308   **10.** 170
11. $x = -11$ **12.** $>$ **13.** -90 **14.** 12   **15.** $-12x^2 + 15xy - 30x$ **16.** -35 **17.** $a = 1$ **18.** $-2b^2 - 3b - 56$
19. $-3a^2 - 19ab - 16b^2$ **20.** $-140ab^2c$ **21.** No
22. 135   **23.** $12m$ **24.** $d = -13$ **25.** -2772 **26.** 3
27. 2   **28.** $k = -9$ **29.** $120°F$ **30.** $-\$257$, a loss

CHAPTER 5

Exercises 5.1

1. $\dfrac{4}{7}$ **3.** $\dfrac{7}{10}$ **5.** $\dfrac{4}{3}$ **7.** $\dfrac{13}{10}$ **9.** $\dfrac{4}{6}, \dfrac{5}{6}$ **11.** $\dfrac{8}{12}, \dfrac{2}{14}, \dfrac{10}{12},$

$\dfrac{8}{14}, \dfrac{11}{22}$ **13.** $\dfrac{4}{5}, \dfrac{16}{17}, -\dfrac{99}{100}$ **15.** $-\dfrac{13}{19}, \dfrac{17}{14}, \dfrac{12}{17}$ **17.** $\dfrac{13}{17}$
19. $\dfrac{16}{19}, \dfrac{14}{3}, -\dfrac{16}{13}$ **21.** $\dfrac{14}{27}$ **23.** $1\dfrac{2}{3}$ **25.** $2\dfrac{1}{5}$ **27.** $3\dfrac{16}{25}$
29. $5\dfrac{9}{41}$ **31.** $\dfrac{13}{6}$ **33.** $\dfrac{9}{1}$ **35.** $\dfrac{143}{3}$ **37.** $\dfrac{855}{8}$ **39.** Any
whole number but zero   **41.** $\dfrac{7}{16}$; answers vary

43.

45. $\dfrac{25}{27}$ **47.** $\dfrac{843}{2147}$ **49.** 3 lb   **51.** 2495 sections   **53.** $\dfrac{1}{2}$
55. 70–79 years; answers vary   **57.** HH, HT, TH, TT; $\dfrac{1}{4}; \dfrac{1}{4}; \dfrac{2}{4}$

Challenge

63. $\dfrac{117}{9}; \dfrac{1521}{117}$ **65.** $1\dfrac{41}{6}; 2\dfrac{35}{6}; 3\dfrac{29}{6}; 4\dfrac{23}{6}; 5\dfrac{17}{6}$

Maintain Your Skills

71. 29   **73.** $>$

Exercises 5.2

1. 2, 4, 6, 8, 10   **3.** 5, 10, 15, 20, 25   **5.** 11, 22, 33, 44, 55
7. 25, 50, 75, 100, 125   **9.** 135, 270, 405, 540, 675   **11.** No
13. Yes   **15.** Yes   **17.** No   **19.** 72, 80, 88, 96, 104, 112
21. 2, 3, 5, 10   **23.** 3, 5   **25.** 2, 5, 10   **27.** 2, 5, 10   **29.** 5
31. 1, 2, 3, 4, 6, 12   **33.** 1, 13   **35.** 1, 2, 3, 4, 6, 9, 12, 18, 36   **37.** 1, 3, 37, 111   **39.** 1, 3, 5, 9, 15, 27, 45, 81, 135, 405   **41.** Prime   **43.** Composite   **45.** Composite
47. Prime   **49.** $3^2 \cdot 7$ **51.** $2^2 \cdot 11$ **53.** 11^2 **55.** $2 \cdot 5 \cdot 19$
57. $2^9 \cdot 3$ **59.** 1, 4, or 7   **61.** Six is a factor of 66; answers vary.   **63.** 4, 8, 12, 16, 20, 24, 28, 32, 36, 40, 44, 48, 52   **65.** Yes   **67.** 2025   **69.** 1999   **71.** 1009

Challenge

77. 1333   **79.** A number is divisible by 6 if it is divisible by 2 and 3   **81.** A number is divisible by 15 if it is divisible by 3 and 5

Maintain Your Skills

83. -3996 **85.** -84 **87.** $-78y$

Exercises 5.3

1. $\dfrac{4}{5}$ **3.** $\dfrac{1}{2}$ **5.** $\dfrac{7}{10}$ **7.** $\dfrac{3}{5}$ **9.** 14   **11.** $\dfrac{8w}{9}$ **13.** $\dfrac{3y^2}{5}$

15. $\dfrac{4}{6}, \dfrac{6}{9}, \dfrac{8}{12}, \dfrac{10}{15}$ **17.** $\dfrac{10c}{8}, \dfrac{15c}{12}, \dfrac{20c}{16}, \dfrac{25c}{20}$ **19.** $\dfrac{14}{32}, \dfrac{21}{48},$

$\dfrac{28}{64}, \dfrac{35}{80}$ **21.** $\dfrac{16}{30}, \dfrac{24}{45}, \dfrac{32}{60}, \dfrac{40}{75}$ **23.** 16 **25.** 24 **27.** $44x$

29. $72x^2$ **31.** $\dfrac{5}{14}$ **33.** $\dfrac{1}{6}$ **35.** $\dfrac{4}{5}$ **37.** $\dfrac{1}{2}$ **39.** 1

41. $\dfrac{1}{2}$ **43.** $43\dfrac{1}{8}$ **45.** $\dfrac{63ab^2}{125}$ **47.** $\dfrac{2}{3}$ **49.** $-\dfrac{16}{7}$

51. $\dfrac{1}{6}$ **53.** $-\dfrac{3}{4}$ **55.** $-\dfrac{200}{189}xy$ **57.** $\dfrac{125}{144}$ **59.** 5

61. $\dfrac{3}{10}$; answers vary **63.** $\dfrac{a}{15}$ **65.** $\dfrac{8s}{3}$ **67.** Yes

69. $\dfrac{12}{24}, \dfrac{16}{24}, \dfrac{4}{24}, \dfrac{15}{24}$ **71.** 960,625 lb **73.** $\dfrac{3}{7}$ **75.** $11\dfrac{1}{2}$

omelets **77.** $\dfrac{35}{16}$ or $2\dfrac{3}{16}$ in³ **79.** 1,386,000 in³

81. (1) card must be black; (2) card must be a seven;

$\dfrac{26}{52}, \dfrac{4}{52}$; answers vary **83.** $\dfrac{1}{2}$

Challenge

87. $\dfrac{6}{10}, \dfrac{12}{20}, \dfrac{18}{30}, \dfrac{36}{60}$

89.

91.

Maintain Your Skills

93. $\dfrac{11}{10}$ **95.** 177 **97.** 2

Exercises 5.4

1. 14 days **3.** 24 months **5.** 24 in **7.** 5280 ft
9. 200 cm **11.** 9 ft² **13.** 5 lb **15.** 150 mm **17.** 5 gal
19. 6 kg **21.** 10,000 g **23.** 3 yd **25.** 500 cm

27. 12 lb **29.** 1 m **31.** $2\dfrac{2}{3}$ ft **33.** 120 in

35. $1\dfrac{1}{2}$ tons **37.** 7 days **39.** 13,000 lb **41.** 66 ft/sec

43. 1500 lb/in **45.** 50 m/sec **47.** 24,000 kg/km²

49. $41\dfrac{2}{3}$ lb/min **51.** $1\dfrac{1}{2}$ words/sec **53.** 15 cc

55. $240,000 **57.** $\dfrac{24}{5}$ min or $4\dfrac{4}{5}$ min **59.** $\dfrac{1}{27}$ lb

61. 30 g; $4\dfrac{1}{2}$ or $4.50

Challenge

65. $2500

Maintain Your Skills

69. 312 **71.** Yes

Exercises 5.5

1. 28 **3.** 30 **5.** 6 **7.** 42 **9.** 60 **11.** 480 **13.** $204x^2$

15. $\dfrac{19}{21} < \dfrac{20}{21}$ **17.** $-\dfrac{3}{13} > -\dfrac{5}{13}$ **19.** $-\dfrac{4}{21} < \dfrac{3}{7}$

21. $\dfrac{13}{15} > \dfrac{21}{25}$ **23.** $\dfrac{20}{33} < \dfrac{8}{11}$ **25.** $-\dfrac{7}{30} < -\dfrac{2}{9}$ **27.** $\dfrac{3}{5}$

29. x **31.** $\dfrac{1}{48}$ **33.** $\dfrac{5}{12}$ **35.** y **37.** $-\dfrac{1}{14}$ **39.** $\dfrac{63}{80}$

41. $\dfrac{61}{60}w$ **43.** $\dfrac{1}{3}x - \dfrac{1}{8}$ **45.** $-\dfrac{3xy}{40}$ **47.** $\dfrac{1}{4}$ **49.** $\dfrac{1}{3}$

51. $\dfrac{1}{18}$ **53.** $\dfrac{4}{3}$ **55.** $\dfrac{1}{18}$ **57.** $-\dfrac{33}{20}$ **59.** $\dfrac{47}{140}st$

61. $-\dfrac{1}{6}x^2$ **63.** $6\dfrac{5}{7}$ **65.** $5\dfrac{2}{7}$ **67.** $14\dfrac{7}{12}$ **69.** $7\dfrac{3}{10}$

71. $4\dfrac{1}{2}$ **73.** $10\dfrac{5}{6}$ **75.** $9\dfrac{1}{3}$ **77.** $27\dfrac{2}{5}$ **79.** $28\dfrac{1}{3}$

81. $4\dfrac{25}{32}$ **83.** $19\dfrac{5}{8}$ **85.** 12, 24, 60, or 120

87. $2\dfrac{3}{4}$; answers vary **89.** $43\dfrac{13}{24}$ **91.** $94\dfrac{7}{72}$ **93.** $\dfrac{74}{75}$

95. $-\dfrac{23}{48}x^2$ **97.** $2x - 2y$ **99.** $\dfrac{5}{6}x + \dfrac{3}{8}y + \dfrac{3}{10}$

101. $3\dfrac{3}{8}$ points **103.** $1\dfrac{1}{8}$ in **105.** $1\dfrac{13}{16}$ in **107.** a) $3\dfrac{5}{8}$ in

b) $92\dfrac{5}{24}$ in c) $277\dfrac{1}{12}$ in d) $28\dfrac{1}{6}$ in **109.** $\dfrac{3}{4}$ in

111. Yes **113.** $100\dfrac{28}{33}$ bricks **115.** $\dfrac{1}{2}$ **117.** $\dfrac{9}{13}$

Challenge

123. 30, 60, 90, 120, 150; 30 **125.** $-\dfrac{11}{8}$

Maintain Your Skills

129. Yes **131.** $\dfrac{2}{3}$

Exercises 5.6

1. $\dfrac{5}{6}$ **3.** $-\dfrac{1}{2}$ **5.** $\dfrac{1}{4}$ **7.** $\dfrac{17}{36}$ **9.** $\dfrac{91}{240}$ **11.** $\dfrac{145}{36}$ or $4\dfrac{1}{36}$

13. $-\dfrac{8}{9}$ **15.** $-\dfrac{46}{61}$ **17.** $\dfrac{229}{225}$ **19.** $\dfrac{3}{8}$ **21.** $\dfrac{1}{2}$ **23.** $\dfrac{43}{64}$

25. $3\dfrac{13}{54}$ **27.** 44 cm **29.** 88 ft **31.** $\dfrac{110}{7}$ m or $15\dfrac{5}{7}$ m

33. 88 in^3 **35.** $\dfrac{360}{7}$ ft or $51\dfrac{3}{7}$ ft **37.** $254\dfrac{4}{7}$ in^2

39. $121\dfrac{23}{28}$ in^2 **41.** $6373\dfrac{5}{7}$ in^3 **43.** $\dfrac{14}{11}$ in^2 or $1\dfrac{3}{11}$ in^2

45. $2\dfrac{1}{4}$ or \$2.25 **47.** $\dfrac{49}{75}$ **49.** $\dfrac{43}{50}$ **51.** $33\dfrac{2}{3}$ in

53. $108\dfrac{1}{4}$ oz; 48 cents **55.** $10899\dfrac{3}{7}$ mm^2

57. $85\dfrac{5}{42}$ in^3 **59.** $\dfrac{4}{52}\cdot\dfrac{13}{52}=\dfrac{1}{52}$

Challenge

65. $14\dfrac{17}{72}$ **67.** \$58,331

Maintain Your Skills

69. $13\cdot 23$ **71.** $x=6$ **73.** $x=-4$

Exercises 5.7

1. $x=\dfrac{1}{5}$ **3.** $x=2$ **5.** $x=1$ **7.** $a=\dfrac{8}{3}$ **9.** $y=1$

11. $c=-\dfrac{32}{21}$ **13.** $y=\dfrac{13}{2}$ **15.** $x=1$ **17.** $y=\dfrac{23}{6}$

19. $a=-35$ **21.** $a=13$ **23.** $x=\dfrac{48}{43}$ **25.** $x=-\dfrac{20}{19}$

27. $c=\dfrac{308}{15}$ **29.** $b=-\dfrac{34}{11}$ **31.** $a=4$ **33.** $z=\dfrac{157}{105}$

35. $a=-\dfrac{67}{39}$ **37.** $y=-\dfrac{7}{4}$ **39.** $p=\dfrac{3}{4}$ **41.** $x=-\dfrac{37}{48}$

43. $x=-16$ **45.** $c=-\dfrac{38}{27}$ **47.** $x=-\dfrac{9}{10}$ **49.** $\dfrac{5}{6}$ in

51. 12 in **53.** 4 **55.** $3\dfrac{1}{2}$ in **57.** $\dfrac{49}{5}$ or $9\dfrac{4}{5}$ cm **59.** $\dfrac{1}{25}$

Challenge

63. $\dfrac{7}{32}$

Maintain Your Skills

67. $\dfrac{8y}{9}$ **69.** $-\dfrac{5y^2}{12}$

CHAPTER 5 TRUE – FALSE CONCEPT REVIEW

1. False; if the numerator and denominator of a fraction are both positive *and* the fraction is improper *and* the fraction is not equivalent to 1, then the denominator is less than the numerator **2.** True **3.** False; the proper fraction $\dfrac{5}{6}$ can be changed to the mixed number $0\dfrac{5}{6}$
4. False; the integer 7 is a factor of 35 **5.** True
6. False; every whole number except 0 and 1 is either a prime number or a composite number **7.** True **8.** False; when a fraction is reduced, its value remains the same
9. False; it may be possible to reduce the fraction part of a mixed number such as $5\dfrac{2}{4}$, but we do not reduce the mixed number **10.** True **11.** False; the numerator of the sum of two fractions is the sum of the numerators only when the fractions have a common denominator **12.** True
13. True **14.** True **15.** False; to find the reciprocal of a mixed number, first change it to an improper fraction
16. True **17.** False; a rational number *can be* written in fraction form **18.** True **19.** False; the opposite of a fraction may be positive or negative and hence may or may not have a negative numerator **20.** True **21.** True
22. False; building fractions is useful, at least for adding, subtracting, and comparing fractions **23.** False; a fraction renamed by building has the same value as the original fraction **24.** True **25.** False; the smallest divisor, other than 1, of the LCM of three numbers is the smallest prime number that is factor of one of the numbers **26.** True
27. False; the LCM of a list of numbers is the largest number in the list when it is a multiple of all the other numbers in the list **28.** False; the denominator of the sum of two fractions is the LCM of the denominators of the two fractions **29.** True **30.** True **31.** True
32. True **33.** False; mixed numbers are added by adding the whole number parts (which may involve "carrying") and the fraction parts (which may have a sum larger than one and should be renamed) **34.** True **35.** True
36. False; the order of operations is the same for every kind of number **37.** True **38.** False; the multiplication property of equality can be used to eliminate the fractions in an equation **39.** True **40.** False; some rational numbers are written in fraction form and some in whole number form (and some in decimal and other forms)
41. False; two fractions are equivalent if they represent the same number **42.** False; of two fractions with a positive common denominator, the one with the larger numerator has the larger value **43.** False; the commutative and associative properties are used to group the whole numbers and to group the fractions when adding mixed numbers
44. False; to add two fractions, build a common denominator, add the numerators, and keep tht common denominator **45.** False; the sum of two fractions may be larger or smaller than the fractions, depending on whether the fractions are positive or negative **46.** False; the product of two numbers is a common multiple of the numbers, but may not be the *least* common multiple **47.** True
48. True

CHAPTER 5 REVIEW

1. $\dfrac{3}{8}$ **3.** $\dfrac{7}{8}$ **5.** $\dfrac{11}{7}$ **7.** $\dfrac{7}{7}, \dfrac{27}{27}, \dfrac{149}{148}$ **9.** $\dfrac{8}{7}, -\dfrac{10}{9}, \dfrac{11}{10}, -\dfrac{12}{12}$

11. $-\dfrac{7}{9}$ **13.** $\dfrac{13}{7}$ **15.** $\dfrac{8}{3}$ **17.** $4\dfrac{3}{4}$ **19.** $20\dfrac{5}{7}$ **21.** $\dfrac{26}{3}$

23. $\dfrac{711}{7}$ **25.** $\dfrac{565}{11}$ **27.** 19, 38, 57, 76, 95 **29.** 33, 66, 99, 132, 165 **31.** Yes **33.** Yes **35.** Yes **37.** None **39.** None **41.** 1, 2, 4, 31, 62, 124 **43.** 1, 2, 3, 5, 6, 7, 9, 10, 14, 15, 18, 21, 30, 35, 42, 45, 63, 70, 90, 105, 126, 210, 315, 630 **45.** 1, 2, 3, 4, 6, 8, 12, 16, 24, 32, 48, 64, 96, 128, 192, 384 **47.** Composite **49.** Prime **51.** $2^6 \cdot 3$

53. $2 \cdot 5 \cdot 61$ **55.** $2^3 \cdot 3^2 \cdot 17$ **57.** $\dfrac{3}{4}$ **59.** $-\dfrac{2ab}{7}$ or $-\dfrac{2}{7}ab$ **61.** $\dfrac{12}{14}, \dfrac{18}{21}, \dfrac{24}{28}, \dfrac{30}{35}$ **63.** 40 **65.** $12yz$

67. $-\dfrac{3}{7}$ **69.** $\dfrac{129}{14}$ or $9\dfrac{3}{14}$ **71.** $\dfrac{6}{5}$ or $1\dfrac{1}{5}$ **73.** $\dfrac{4}{7}d$ or $\dfrac{4d}{7}$

75. 4 turns **77.** $15\dfrac{1}{2}$ gal **79.** $11\dfrac{1}{9}$ m/sec **81.** 40

83. 180 **85.** 2280 **87.** $\dfrac{8}{3} > \dfrac{9}{4}$ **89.** $\dfrac{15}{17} > \dfrac{29}{33}$ **91.** $\dfrac{16}{17}$

93. $\dfrac{3}{5}$ **95.** $\dfrac{1}{5}a$ **97.** $\dfrac{41}{36}$ or $1\dfrac{5}{36}$ **99.** $\dfrac{107}{90b}$ **101.** $\dfrac{4}{5}$

103. $-\dfrac{1}{2}$ **105.** $\dfrac{5}{12}y$ **107.** $22\dfrac{13}{18}$ **109.** $67\dfrac{31}{40}$ **111.** $3\dfrac{1}{4}$

113. $2\dfrac{11}{14}$ **115.** $117\dfrac{1}{90}$ **117.** $1\dfrac{5}{18}$ **119.** $A = 6630$

121. $\dfrac{25}{48}$ **123.** $\dfrac{49}{64}$ **125.** $6\dfrac{1}{16}$ **127.** $201\dfrac{1}{7}$ in^2

129. $8184\dfrac{11}{21}$ m^3 **131.** $x = 2$ **133.** $a = 2$ **135.** 5

CHAPTER 5 TEST

1. $\dfrac{3}{5}$ **2.** $5\dfrac{2}{5}$ **3.** $\dfrac{43}{8}$ **4.** 13, 26, 39, 52, 65 **5.** Yes

6. 1, 2, 3, 4, 6, 8, 9, 12, 18, 24, 36, 72 **7.** Prime

8. $2^2 \cdot 5 \cdot 13$ **9.** $\dfrac{2}{3b}$ or $\dfrac{2}{3}b$ **10.** $-\dfrac{5}{8}$ **11.** $\dfrac{3d^2}{17}$ or $\dfrac{3}{17}d^2$

12. $9\dfrac{9}{10}$ **13.** 3 **14.** $2\dfrac{2}{3}$ **15.** 40 **16.** 216 **17.** True

18. $\dfrac{1}{2}a$ **19.** $\dfrac{7}{9}$ **20.** $\dfrac{37}{42}y$ **21.** $\dfrac{2}{5}$ **22.** $\dfrac{23}{36}ab$

23. $9\dfrac{2}{15}$ **24.** $2\dfrac{23}{30}$ **25.** $\dfrac{27}{32}$ **26.** $y = \dfrac{11}{8}$ **27.** $4\dfrac{41}{64}$ in^2

28. $17\dfrac{1}{8}$ min **29.** $17\dfrac{23}{27}$ truckloads **30.** $2\dfrac{5}{8}$ ft

CHAPTER 6

Exercises 6.1

1. Twelve hundredths **3.** Two hundred sixty-seven thousandths **5.** Six and four ten-thousandths **7.** 0.11

9. 0.111 **11.** 2.019 **13.** Five hundred four thousandths **15.** Fifty and four hundredths **17.** Eighteen and two hundred five ten-thousandths **19.** 0.012 **21.** 700.096

23. 505.005 **25.** $\dfrac{33}{100}$ **27.** $\dfrac{3}{4}$ **29.** $\dfrac{111}{1000}$ **31.** $\dfrac{17}{50}$

33. $\dfrac{243}{500}$ **35.** $\dfrac{1}{5}$ **37.** 0.1, 0.6, 0.7 **39.** 0.05, 0.07, 0.6

41. 4.159, 4.16, 1.161 **43.** 0.072, 0.0729, 0.073, 0.073001, 0.073015 **45.** 0.88579, 0.88799, 0.888, 0.8881 **47.** 20.004, 20.039, 20.04, 20.093 **49.** False **51.** True

		Unit	Tenth	Hundredth
53.	15.888	16	15.9	15.89
54.	51.666	52	51.7	51.67
55.	477.774	478	477.8	477.77
56.	344.333	344	344.3	344.33
57.	0.7392	1	0.7	0.74
58.	0.92937	1	0.9	0.93

59. $33.54 **61.** $246.49

		Ten	Hundredth	Thousandth
63.	12.5532	10	12.55	12.553
64.	21.3578	20	21.36	21.348
65.	245.2454	250	245.25	245.245
66.	118.0752	120	118.08	118.075
67.	0.5536	0	0.55	0.554
68.	0.9695	0	0.97	0.970

69. $11 **71.** $1129 **73.** One and three tenths inches

75. $\dfrac{3}{50}$ in **77.** 1.1 in **79.** Week one, week two

81. Sixty-four and seventy-nine hundredths dollars

83. $\dfrac{1}{8}$ **85.** 1.62 **87.** 1.6 **89.** 98.35 cents;

Circle K Meats **91.** $1617.37 **93.** 127.7 mph
95. The digit after the round-off place is zero, so we chose the smaller number

Challenge

101. a) False b) True c) False d) True

Maintain Your Skills

107. $-\dfrac{1}{24}$ **109.** $-\dfrac{23}{12}$

Exercises 6.2

1. 0.7 **3.** 3.8 **5.** 7.7 **7.** 27.43 **9.** 3 **11.** 13.841
13. 23.40 **15.** 2.755 **17.** 9.0797 **19.** 82.593
21. 55.5363 **23.** 831.267 **25.** 26.981 **27.** 0.3
29. 3.4 **31.** 5.1 **33.** 6.41 **35.** 8.09 **37.** 0.457
39. 1.564 **41.** 2.609 **43.** 67.044 **45.** 2.76 **47.** 0.069
49. 0.17 **51.** 0.3 **53.** 6 **55.** 0.08 **57.** 1.6 **59.** 0.04;
0.04019 **61.** 0.11; 0.106794 **63.** -20; -15.3207
65. -0.1; -0.0653 **67.** -1.5; -1.3908 **69.** 92.105
71. $7.47t$ **73.** $-8.13x$ **75.** $-13.9a$ **77.** $11.797x - 4.692$
79. $-26.94b + 13.25c$ **81.** $-16.94x + 36.5y$
83. $-6.87a^2 - 3.8a$ **85.** 49.1 gal **87.** 471.9
89. 0.778 sec **91.** 38.95 sec **93.** 13.16 sec
95. \$18.7 billion **97.** \$1945.90 **99.** 0.35%
101. 75 ft **103.** 6.59375 in

Challenge

109. 23 **111.** 0.1877 **113.** 1.5725

Maintain Your Skills

117. $-\dfrac{7}{12}$ **119.** $-52x^4$

Exercises 6.3

1. 3.2 **3.** 9 **5.** 0.27 **7.** 0.2 **9.** 0.015 **11.** -0.064
13. Five **15.** 0.0149 **17.** 6.67 **19.** 4.1552 **21.** 2.6988
23. -30.1608 **25.** -0.117 **27.** $0.072x^2$ **29.** $-2.25t^4$
31. 24 **33.** 0.0035 **35.** 1.6 **37.** 0.0018 **39.** 10; 11.045
41. 0.028; 0.023852 **43.** 0.4; 0.3822 **45.** 0.004; 0.004455
47. 97.7 gal **49.** \$28.02 **51.** \$1.393 **53.** 400; 347.3184
55. 8000; 9576.252 **57.** 21,392.542 **59.** 835.15
61. $21.2b^2 + 16.43b$ **63.** $-9.45y^2 + 4.05y$ **65.** $1.2b^3 - 0.16b^2c$ **67.** \$556.10 **69.** \$219 **71.** Compact car; \$13.56
73. Store 3 **75.** 71.725 lb **77.** 66.99 points **79.** 1970,
528 lb; 1980, 505.6 lb; 1990, 449.2 lb **81.** \$194.42

Challenge

85. 84 **87.** 0.0216

Maintain Your Skills

91. 670 **93.** 230,000,000

Exercises 6.4

1. 0.425 **3.** 367 **5.** 628.33 **7.** 0.5692 **9.** 56.45
11. 8700 **13.** Right **15.** 6274 **17.** 0.185 **19.** 36,900
21. 0.68953 **23.** 1.478.000 **25.** 13.6794 **27.** 0.000458
29. 2.3×10^5 **31.** 3.5×10^{-4} **33.** 4.6795×10^2
35. 60,000 **37.** 0.008 **39.** 4780 **41.** 7.8×10^5
43. 3.45×10^{-5} **45.** 8.21×10^{-11} **47.** 3.567003×10^3
49. 0.000001345 **51.** 7,110,000,000 **53.** 0.000000444
55. 567.23 **57.** \$2229 **59.** \$98,500 **61.** 5.2×10^7 sq mi

63. Multiplying by 1000 **65.** 0.424
67. 3.3×10^{-6} sec **69.** 36,000,000
67,240,000
92,960,000
141,640,000
483,640,000
887,000,000
1,783,000,000
2,795,000,000
3,666,000,000
71. 0.0013 in **73.** 1.55×10^6 lbs fish; 6.36×10^6 lbs
poultry; 1.123×10^7 lbs red meat **75.** $\$1.582 \times 10^9$;
$\$1.582 \times 10^8$

Challenge

77. 1.92×10^{13} mi **79.** 6.5×10^6

Maintain Your Skills

83. 108 **85.** 600 R 23

Exercises 6.5

1. 0.5 **3.** 9.8 **5.** 183.1 **7.** 2020 **9.** Whole number
11. 12.6 **13.** -0.52 **15.** 1.3 **17.** 9.1 **19.** 15.555
21. 0.75 **23.** 0.375 **25.** 0.6875 **27.** 3.55 **29.** 11.024

		Tenth	Hundredth
31.	$\dfrac{3}{7}$	0.4	0.43
32.	$\dfrac{2}{9}$	0.2	0.22
33.	$\dfrac{5}{11}$	0.5	0.45
34.	$\dfrac{5}{6}$	0.8	0.83
35.	$\dfrac{2}{13}$	0.2	0.15
36.	$\dfrac{3}{14}$	0.2	0.21
37.	$\dfrac{8}{15}$	0.5	0.53
38.	$\dfrac{11}{19}$	0.6	0.58
39.	$5\dfrac{17}{18}$	5.9	5.94
40.	$11\dfrac{9}{17}$	11.5	11.53

41. 0.02 **43.** 10 **45.** 1 **47.** 400 **49.** 100; 97.06
51. 0.07; 0.08 **53.** 10; 17.14 **55.** $-21x$
57. $-0.375m$ **59.** -83.55 **61.** $1.473 per lb
63. $2.969 per lb **65.** $24.42 **67.** $27.08 **69.** $2a - 3$
71. $-9a - 0.04$

	Hundredth	Thousandth
72. $\frac{15}{43}$	0.35	0.349
73. $\frac{12}{31}$	0.39	0.387
74. $\frac{28}{57}$	0.49	0.491
75. $\frac{85}{91}$	0.93	0.934

77. $0.91\overline{6}$ **79.** $0.8\overline{846153}$ **81.** $4.15 **83.** 34 days
85. $1588.87 **87.** 24.1 ft **89.** $11.16 **91.** 2.93 ERA
93. 0.621

Challenge

99. $\frac{3}{2000}$

Maintain Your Skills

103. $3\frac{1}{8}$ **105.** $11\frac{2}{9}$ **107.** 139 books; $34

Exercises 6.6

1. 0.5 lbs **3.** 1.5 cm **5.** 0.55 ℓ **7.** 2.3 yd **9.** 36.24 in
11. 1830 mm **13.** 24076.8 ft **15.** 40.64 cm **17.** 4.76 qt
19. 14.92 in **21.** 0.07 sq ft **23.** 5.34 m² **25.** 4.5 m
27. 7.0 lb **29.** 5.1 mi **31.** 177.4 cℓ **33.** 1.3 m²
35. $\frac{2232.23 \text{ g}}{\text{m}}$ **37.** $\frac{2.54 \text{ g}}{\text{cm}^2}$ **39.** $\frac{1.10 \text{ lb}}{\text{qt}}$ **41.** 535.8 g
43. 19 cents/km **45.** 133.8 kg **47.** 29.0 m **49.** The
Canadian pack **51.** 68.1 cm **53.** $\frac{0.7 \text{ sec}}{1 \text{ yd}} = \frac{0.7655 \text{ sec}}{1m}$
55. 1 min 18.74 sec **57.** 1.852 km **59.** 0.869 nautical
miles = 1 statute mile

Challenge

65. 22.2 ℓ

Maintain Your Skills

65. $-\frac{49}{80}$ **67.** No.

Exercises 6.7

1. -1.5 **3.** 0.018 **5.** -10.67 **7.** 0.008 **9.** 0.4
11. No **13.** 0.367236 **15.** -30.18 **17.** -66
19. -52.4 **21.** 1000 **23.** 172.688 **25.** Yes
27. 12.675 **29.** 45.1 **31.** -0.003 **33.** 55.300
35. 102.656 **37.** Yes **39.** No **41.** 285.525 mi
43. 24,531 ft² or 24,544 ft² **45.** $4875.50 **47.** 904.6 ft²;
8 gal or 905.1 ft²; 8 gal **49.** 135 points
51. 17,000 farms lost/yr

53.

Time (hours)	1	2	3	4	5
Total distance of fastest cyclist	19	38	57	76	95
Total distance of slowest cyclist	16	32	48	64	80

55. 2:24 PM; 3.53 PM

Challenge

57. 12.8 horsepower **59.** $A \approx 3.14 \, (R^2 - r^2)$; 43.96 in²

Maintain Your Skills

61. $x = -4$ **63.** $y = -12$ **65.** $b = \frac{4}{5}$

Exercises 6.8

1. $x = 4.41$ **3.** $a = 5.347$ **5.** $x = -3.012$ **7.** $c = -10.18$
9. $b = 6.011$ **11.** $c = 170$ **13.** $b = 8.25$ **15.** $c = 56$
17. $x = -3.52$ **19.** $y = -8.33$ **21.** $w = -7.237$
23. $y = 251.05$ **25.** $b = -1.9$ **27.** $c = -0.65$
29. $463.35 **31.** 28 payments; $115.91 **33.** $210,404.80
35. $t \approx -12.2$ **37.** $a \approx 5.8$ **39.** $t \approx 24.1$ **41.** $z \approx 0.3$
43. 18.86 m **45.** 41.46 ft/sec **47.** 38.67 cm **49.** 185
bases **51.** 684 games **53.** 0°C

Challenge

59. $x = -3.6$ **61.** $y \approx 29.1$

Maintain Your Skills

65. 61 **67.** $-\frac{4}{9}$

Exercises 6.9

1. 9 **3.** 11 **5.** $\frac{2}{5}$ **7.** $\frac{11}{12}$ **9.** 22 **11.** 10.8
13. -8.83 **15.** 4.6 **17.** 0.82 **19.** 15 **21.** 13

23. 12 **25.** 50 **27.** 3.61 **29.** 5.86 **31.** 136.11
33. 14.38 **35.** 66.29 **37.** A right triangle **39.** Not a
right triangle **41.** A right triangle **43.** Not a right
triangle **45.** 1.8 sec
47. 46.3 ft

		Nearest Tenth	Nearest Hundredth	Nearest Thousandth	Nearest Ten-thousandth
49.	$\sqrt{365.96}$	19.1	19,13	19.130	19.1301
50.	$\sqrt{0.9682}$	1.0	0.98	0.984	0.9840
51.	$\sqrt{20.037}$	4.5	4.48	4.476	4.4763
52.	$\sqrt{0.03985}$	0.2	0.20	0.200	0.1996

53. $t = 7\frac{1}{3}$ **55.** 15.0 in **57.** 13.42 ft **59.** 61.0 ft
61. 206.2 mi **63.** 127.3 ft **65.** 10.2 mi

Challenge

69. $-\dfrac{157}{42}$ **71.** 16

Maintain Your Skills

75. 41 **77.** 29

CHAPTER 6 TRUE–FALSE CONCEPT REVIEW

1. False; it is common to write the decimal 6.7 in words:
six and seven tenths **2.** True **3.** True **4.** False; a decimal point is the separation of the ones and the tenths
place **5.** True **6.** False; to the nearest ten, 74.49 rounds
to 70 **7.** True **8.** False; the rounded value of a decimal
is usually smaller or larger than the original decimal (9.50
rounded to the nearest tenth is 9.5, which is equal to 9.50)
9. True **10.** True **11.** True **12.** False; it is possible to
subtract 9.4 from 6 $(6 - 9.4 = -3.4)$ **13.** True
14. False; the quotient of 0.6 and 0.5 is 1.2 **15.** False;
scientific notation is not only used by scientists who deal
with large numbers (astronomers) and small numbers
(physicists), but by engineers, mathematicians, and other
professionals who deal with large and small numbers
16. False; it is sometimes necessary to move the decimal
point when dividing decimals (when doing the division
problem $8.925 \div 5$, the decimal point does not need to be
moved) **17.** False; it is sometimes necessary to round off
when dividing decimals (in the example given in the answer
to Problem 16, no rounding is necessary; the answer is
1.785) **18.** True **19.** False; the procedure for dividing
rational numbers in decimal form is not the same as the
procedure for dividing rational numbers in fraction form (to
divide fractions, multiply by the reciprocal of the divisor; to
divide decimals, if necessary, move the decimal in both the
divisor and the dividend until the divisor is a whole number, then divide as in whole numbers **20.** True
21. False; the quotient of two decimals may be smaller
than either of the two decimals; $4 \div 0.2 = 20$ yields a larger
quotient than either of the two decimals, while $0.8 \div 1 =$
0.8 yields a number equal to one of the decimals
22. False; there are two basic rules for rounding; one rule
is called the "four-five" and the other is called "rounding by
truncation" **23.** False; the sum of 8.15 and 0.3 is 8.45
24. False; the product of two decimals may have more decimal places than either decimal; the product of 3.8 and 4.5
is 17.1; the product has the same number of places as either of the two factors; the product of 3.2 and 0.625 is 2,
which has no decimal places **25.** False; the smaller of two
decimals may have the greater number of decimal places
$(0.2 < 8.555$ is true; 0.2 has fewer decimal places)
26. True **27.** False; there are four decimal places in the
product of 7.61 and 4.63 **28.** False; $\sqrt{0.04} = 0.2$
29. False; the triangle must be a right triangle **30.** True

CHAPTER 6 REVIEW

1. 0.721 **3.** 18.0602 **5.** Three hundred forty-four and
eighty-two hundred-thousandths **7.** $\dfrac{17}{50}$ **9.** $\dfrac{7}{8}$ **11.** 0.2998,
0.3409, 0.3426, 0.345 **13.** 0.0907, 0.090733, 0.0909,
0.09099, 0.0976 **15.** True **17.** 0.04 **19.** 2632.938
21. 25.0693 **23.** 29.825954 **25.** $-0.885t$ **27.** 21.9646
29. 32.039v **31.** 0.16 **33.** 0.9 **35.** -0.9 **37.** 2.0328
39. 0.02555 **41.** 0.032 **43.** 0.000006 **45.** 0.036 **47.** 9.2
49. 0.000003467 **51.** 4×10^4 **53.** 7×10^{-8}
55. 6,780,000 **57.** 12.53 **59.** 2.336 **61.** 0.4375
63. 0.359375 **65.** 0.66 **67.** 0.3 **69.** 70 **71.** 0.000067
73. 133.245 g **75.** 911.92 m/min **77.** 119.8854
79. $P \approx 41.448$ **81.** $x = 0.1$ **83.** $z = 0.08$ **85.** $s = 20.5$
87. 30 **89.** 15.36 **91.** $a = 30$ **93.** $b \approx 28.72$
95. 625 ft **97.** Yes

CHAPTER 6 TEST

1. 0.826 **2.** 0.71 **3.** 0.067, 0.06729, 0.0673, 0.06893
4. One hundred six and four hundred eight hundred-
thousandths **5.** 27.2205 **6.** $S = 6.0288$ **7.** 5.00
8. $\dfrac{3}{125}$ **9.** 5.2166 **10.** $x = 1.8$ **11.** 7.1×10^{-6} **12.** No
13. $a = 2.19$ **14.** 6; 5.2839 **15.** 40.27 **16.** $4.382 \times$
10^4 **17.** False **18.** 0.6; 0.6287 **19.** -0.063; -0.06171
20. 40; 52.584 **21.** 0.00000026 **22.** $-12.77a - 7.14$
23. 244 mi **24.** 63,000 **25.** \$14.84 **26.** 240 m

CHAPTERS 1–6 CUMULATIVE REVIEW

1. 7974 **3.** 21,803,796 **5.** True **7.** $12x^2 + 9xy - 24xz$
9. $a = 4$ **11.** 7 yd 2 ft **13.** 72 bushels **15.** 190 sq in
17. Second **19.** 2 points **21.** 115 points **23.** -27
25. 300 **27.** $\dfrac{4}{9}$ **29.** 235, 470, 705, 940, 1175

31. Prime **33.** $\dfrac{2y}{3}$ **35.** $-\dfrac{100}{63}$ **37.** $20\dfrac{1}{20}$ **39.** $23\dfrac{19}{30}$

41. $2\dfrac{37}{40}$ yards **43.** $x = \dfrac{25}{4}$ **45.** $2\dfrac{1}{18}$ ft **47.** 16.012

49. $\dfrac{12}{25}$ **51.** 8.90 **53.** 1.9909 **55.** 100,250

57. 823.469 **59.** 9.3×10^7 **61.** 5.23 **63.** 0.265625
65. Approximately equal to 0.065 **67.** 1503
69. $570.08; $229.92

CHAPTER 7

Exercises 7.1

1. $\dfrac{1}{5}$ **3.** $\dfrac{6}{5}$ **5.** $\dfrac{8}{9}$ **7.** $\dfrac{1}{2}$ **9.** $\dfrac{2}{3}$ **11.** $\dfrac{1}{1500}$

13. $\dfrac{8 \text{ people}}{11 \text{ chairs}}$ **15.** $\dfrac{55 \text{ mi}}{1 \text{ hr}}$ **17.** $\dfrac{21 \text{ mi}}{1 \text{ gal}}$ **19.** $\dfrac{8 \text{ lbs}}{3 \text{ ft}}$

21. $\dfrac{2 \text{ trees}}{7 \text{ ft}}$ **23.** $\dfrac{2 \text{ books}}{5 \text{ students}}$ **25.** $\dfrac{85 \text{ people}}{3 \text{ rooms}}$

27. $\dfrac{15 \text{ pies}}{2 \text{ sales}}$ **29.** $\dfrac{25 \text{ mi}}{1 \text{ hr}}$ **31.** $\dfrac{4 \text{ ft}}{1 \text{ sec}}$ **33.** $\dfrac{66 \text{ yds}}{1 \text{ billboard}}$

35. $\dfrac{9 \text{ cents}}{1 \text{ lb potatoes}}$ **37.** $\dfrac{37.5 \text{ mi}}{1 \text{ gal}}$ **39.** $\dfrac{83.3 \text{ ft}}{1 \text{ sec}}$

41. $\dfrac{268.8 \text{ lbs}}{1 \text{ mi}^2}$ **43.** $\dfrac{16.1 \text{ gal}}{1 \text{ min}}$ **45.** $\dfrac{3}{4}; \dfrac{3}{7}$ **47.** No

49. $\dfrac{7}{12}$ **51.** $\dfrac{2}{3}$ **53.** $\dfrac{97.6 \text{ people}}{\text{mi}^2}$ **55.** Answers vary

57. $\dfrac{7}{20}; \dfrac{16}{3}$ **59.** 1.25 mg **61.** $\dfrac{\$0.89}{14 \text{ oz}} \approx \dfrac{\$0.06357}{1 \text{ oz}}; \dfrac{\$1.00}{16 \text{ oz}}$

$\approx \dfrac{\$0.0625}{1 \text{ oz}}$; the larger can **63.** The 5-lb bag **65.** No

CHALLENGE

69. $\dfrac{2}{7}$; not possible **71.** a) $\dfrac{21}{92}$ b) $\dfrac{24}{89}$ c) $\dfrac{117}{370}$ d) $\dfrac{18}{65}$

e) $\dfrac{243}{482}$ f) $\dfrac{171}{415}$ g) $\dfrac{30}{71}$ h) $\dfrac{63}{290}$

Maintain Your Skills

75. $x = \dfrac{17}{6}$ **77.** $y = -\dfrac{87}{5}$ or -17.4

EXERCISES 7.2

1. True **3.** True **5.** False **7.** True **9.** False
11. False **13.** True **15.** $a = 9$ **17.** $c = 6$ **19.** $a = 4$

21. $c = 10$ **23.** $x = 18$ **25.** $w = \dfrac{1}{2}$ or 0.5 **27.** $x = \dfrac{21}{2}$

or 10.5 **29.** $z = \dfrac{22}{5}$ or 4.4 **31.** $y = \dfrac{32}{3}$ **33.** $a = \dfrac{64}{5}$

or 12.8 **35.** $c = 0.6$ **37.** $b = \dfrac{3}{8}$ **39.** $x = 0.25$

41. $w = \dfrac{1}{2}$ or 0.5 **43.** $y = 216$ **45.** $t = 8$ **47.** $w \approx 1.4$

49. $y \approx 54.4$ **51.** $a \approx 0.22$ **53.** $c \approx 0.95$ **55.** 10;
answers vary **57.** $2(19) = 5x$; answers vary **59.** $3.11
61. $5.06

Challenge

65. $a = 10$ **67.** $w \approx 8.809$

Maintain Your Skills

71. $x = 3$ **73.** 48,350

Exercises 7.3

1. 6 **3.** 15 **5.** $\dfrac{6}{4} = \dfrac{15}{x}$ **7.** 30 **9.** x **11.** $\dfrac{30}{18} = \dfrac{x}{48}$

13. 8 **15.** x **17.** $\dfrac{8}{48} = \dfrac{x}{288}$ **19.** x **21.** $\dfrac{3}{65} = \dfrac{x}{910}$

23. 4 teachers **25.** $\dfrac{\frac{3}{2}}{30} = \dfrac{14}{X}$ **27.** 100 rows **29.** $26\dfrac{2}{3}$ lb

31.

	Case I	Case II
	12	x
	15	30

33. 24 games **35.** 4 weeks **37.** $145 **39.** $2.20
41. $9.48 **43.** 6 gal **45.** $350.90 **47.** 0.4 cubic centimeter **49.** $4.80 **51.** 12 lb cashews; 28 lb peanuts
53. 42 qt **55.** 70 lb **57.** £1495 **59.** $59.75
61. 2820 mg **63.** 9 cents

Challenge

67. 193 condors **69.** Santini, $802; Nguyen, $1103

Maintain Your Skills

73. 37.41 **75.** $-0.06 > -0.15$

Exercises 7.4

1. 19% **3.** 50% **5.** 70% **7.** 36% **9.** 40% **11.** 72%

13. $37\dfrac{1}{2}\%$ or 37.5% **15.** 37% **17.** 234% **19.** 140%

21. 200% **23.** 5.2% **25.** 10.25% **27.** 0.1% **29.** 0.01%
31. 66.7% **33.** 41.7% **35.** 57.1% **37.** 0.17 **39.** 0.0235

41. 3.15   **43.** 0.0012   **45.** 0.0025   **47.** 0.01   **49.** $\frac{37}{100}$

51. $\frac{1}{10}$ **53.** 0.04756   **55.** 0.038   **57.** 0.46375   **59.** $\frac{1}{3}$

61. $\frac{47}{300}$ **63.** $\frac{1}{5000}$ **65.** 70%   **67.** 2.2%   **69.** 18%

71. 37.5%   **73.** 0.075   **75.** 187%   **77.** 0.003   **79.** 0.006

81. 300%   **83.** 22%   **85.** $\frac{21}{50}$

Challenge

91. 0.24875; $\frac{199}{800}$

Maintain Your Skills

95. $5.79t^2 + 62.067t$ **97.** $y = 3.6$

Exercises 7.5

1. 18   **3.** 300%   **5.** 50%   **7.** 40   **9.** 36   **11.** 80%
13. 100   **15.** 1   **17.** 28   **19.** 40   **21.** 3.8   **23.** 1%
25. 120   **27.** 1%   **29.** 21.6   **31.** 600   **33.** 46
35. 52.8%   **37.** 10.8   **39.** 0.0288   **41.** 0.002
43. 172   **45.** 667   **47.** 15.6   **49.** 58.1   **51.** 10.7%

53. 136%   **55.** $\frac{26}{75}$

Challenge

59. $\frac{1}{6}$ **61.** $\frac{500}{9}$

Maintain Your Skills

65. -86.13072 **67.** 36.2   **69.** $t = -2.4$

Exercises 7.6

1. $5.20   **3.** $15,665   **5.** $394.91   **7.** 83%   **9.** 15%
11. 4761 men   **13.** 27.4%   **15.** 169,000   **17.** $160
19. Answers vary   **21.** 30%   **23.** $11.20   **25.** $245.24;
the competitor by $2.22   **27.** $567   **29.** 2.7%   **31.** 20%
33. $67.18; 52%   **35.** $181.97   **37.** $6.60   **39.** 1.65
tons   **41.** Hispanic   **43.** Salaries   **45.** $500,000

47.

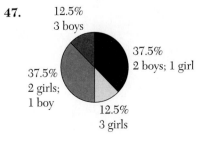

12.5%
3 boys

37.5%
2 boys; 1 girl

37.5%
2 girls;
1 boy

12.5%
3 girls

49.

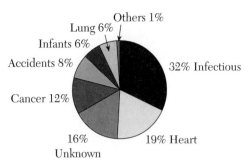

Others 1%
Lung 6%
Infants 6%
Accidents 8%
Cancer 12%
32% Infectious
19% Heart
16% Unknown

Challenge

53. $82.86   **55.** 42%

Maintain Your Skills

59. $t = -14.9$ **61.** $y = -1.0875$

CHAPTER 7 TRUE–FALSE CONCEPT REVIEW

1. True   **2.** True   **3.** False; the units are not the same
4. True   **5.** False; $t = \frac{28}{3}$ **6.** True
7. False; $\frac{7}{6} = 116\frac{2}{3}\%$ **8.** True
9. False; $7\frac{1}{2} = 750\%$ **10.** False; $0.009\% = 0.00009$
11. False; $B = 28{,}000$ **12.** False; 200% of 12 is 24
13. True   **14.** False; two decreases of 15% are the same
as a decrease of 27.75%   **15.** False; her salary is 1% less
16. True **17.** True   **18.** True   **19.** False; some ratios are
rates   **20.** True   **21.** False; $\frac{1}{2}\% = 0.005$ **22.** True
23. True   **24.** True   **25.** False; we can use the proportion $\frac{a}{b} = \frac{R}{100}$

CHAPTER 7 REVIEW

1. $\frac{7}{2}$ **3.** $\frac{1}{5}$ **5.** $\frac{9}{16}$ **7.** $\frac{949 \text{ cents}}{3 \text{ boxes}}$ **9.** $\frac{13 \text{ VCRs}}{2 \text{ CD players}}$
11. 0.925 per lb   **13.** 97.6 people per mi²
15. $\frac{17.5 \text{ nmiles}}{\text{gal}}$ **17.** False   **19.** False   **21.** $x = 26$
23. $x = 20.4$ **25.** $x = 0.012$ **27.** $10,120   **29.** 267 mi
31. 12.5%   **33.** 93.75%   **35.** 118.8%   **37.** 8.93%
39. 0.09%   **41.** 0.00375   **43.** 1.349   **45.** 0.0008642
47. 45%   **49.** 90.8   **51.** 20.5%   **53.** 31,500
55. $1522 **57.** $5016   **59.** $488   **61.** 20%   **63.** Auto
65. 50%

CHAPTER 7 TEST

1. $\frac{1}{5}$ **2.** True **3.** 160% **4.** 20% **5.** 0.78 **6.** $\frac{3}{4}$
7. $\frac{13}{40}$ **8.** $w = 2.4$ **9.** 95 **10.** 0.3% **11.** $\frac{2}{7}$
12. 0.0013 **13.** 24.38 **14.** $x = 10.5$ **15.** 30.4%
16. 39.1% **17.** 30 lb **18.** 4% **19.** 30%
20. $446.25

CHAPTER 8

Exercises 8.1

1. A solution **3.** Not a solution **5.** A solution
7. A solution **9.** A solution **11.** A solution **13.** Not a solution **15.** A solution **17.** Not a solution **19.** A solution

21–29.

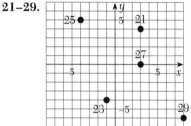

31–39.

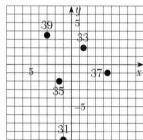

41. $(3, 4)$ **43.** $(2, -4)$ **45.** $(-1, 4)$ **47.** $(-3, 1)$
49. $(-7, 6)$ **51.** $(4, 1.5)$ **53.** $(-3, -0.5)$ **55.** $(4.5, -4)$
57. $(2, -6.5)$ **59.** $(0, 1.5)$ **61.** $(5, E)$ **63.** $(6, D)$
65. Longview **67.** $(5, D)$ **69.** $(5, B)$

71.

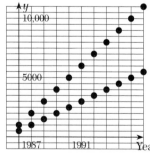

Challenge

75. Not a solution

Maintain Your Skills

79. $y = -7$ **81.** $a = 156.3$

Exercises 8.2

1. $(6, 12)$ **3.** $(13, 5)$ **5.** $(-3, 21)$ **7.** $(8, -9)$
9. $(0, -7)$ **11.** $(16, -11)$ **13.** $(0, 6)$ **15.** $(-3, \frac{36}{7})$
17. $(-15, \frac{12}{7})$ **19.** $(0, 10)$ **21.** $(-1.5, 8.8)$
23. $(-12.5, 0)$ **25.** 4 yrs, $30,600; 8 yrs, $16,200
27. 44.5 **29.** $(3, 12)$
31. $(-\frac{16}{3}, -\frac{1}{2})$ **33.** $(-4.2, 1.2)$ **35.** 11 six-lb boxes
37. $(3, 0)$ **39.** $(-5, \frac{56}{3})$ **41.** $(\frac{2}{7}, \frac{19}{3})$ **43.** 5 servings
45. 14 cm **47.** $174, $543.75, $21.75 **49.** $(1, 500)$,
$(2, 1000), (5, 2500), (10, 5000)$ **51.** $(1, 1000), (2, 2000)$,
$(3, 3000), (6, 6000), (10, 10,000)$

Challenge

55. $y = -\frac{2}{5}$ **57.** $y = -x + 3$ **59.** $y = \frac{1}{2}x - 3$

Maintain Your Skills

61. No **63.** $t = -13$

Exercises 8.3

1.

x	y
−6	−3
0	0
6	3

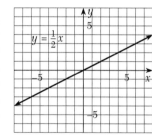

3.

x	y
0	−2
−2	0
1	−3

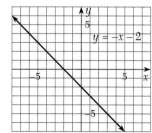

5.

x	y
0	−4
4	0
2	−2

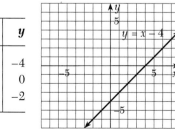

7.

x	y
-3	-6
-4	-5
-6	-3

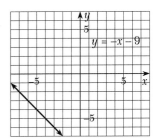

9.

x	y
4	9
-2	-3
2	5

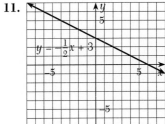

11.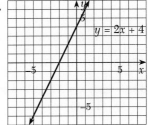

13.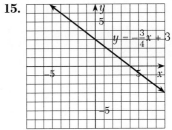

15.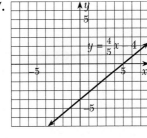

17.

19.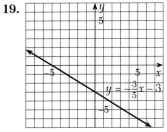

21.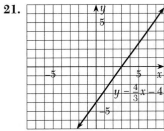

23.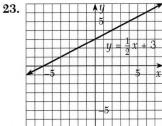

25.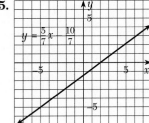

27.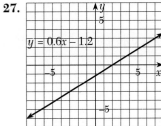

29.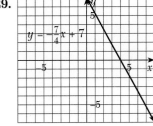

31.

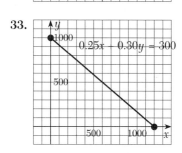

33.

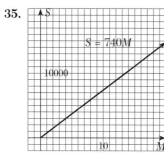

35.

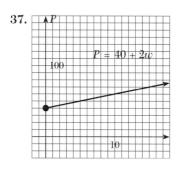

37.

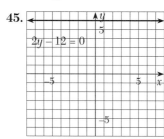

39. Answers vary   **41.** Answers vary

Challenge

45.

![graph with line 2y - 12 = 0 horizontal]

Maintain Your Skills

49. No   **51.** $(-4, 2)$

Exercises 8.4

1. $(4, 0), (0, -6)$ **3.** $(-7, 0), (0, 1)$ **5.** $(4, 0), (0, 16)$
7. $(9, 0), (0, 2)$ **9.** $(6, 0), (0, -16)$ **11.** $(7.5, 0), (0, -2.5)$
13. $(1.5, 0), (0, 6.5)$ **15.** $(5, 0), \left(0, -\dfrac{15}{2}\right)$ **17.** $(21, 0),$
$(0, 7)$ **19.** $(-\dfrac{8}{7}, 0), (0, -2)$ **21.** $m = -\dfrac{3}{11}$ **23.** $m = -\dfrac{1}{7}$
25. $m = -\dfrac{7}{3}$ **27.** $m = -1$ **29.** $m = -\dfrac{3}{16}$ **31.** $m = \dfrac{1}{3}$
33. $m = -\dfrac{12}{5}$ **35.** $m = -\dfrac{3}{7}$ **37.** $m = -\dfrac{4}{7}$ **39.** $m = \dfrac{5}{12}$
41. $\left(\dfrac{14}{3}, 0\right), (0, 2), m = -\dfrac{3}{7}$ **43.** $(-3, 0), \left(0, -\dfrac{9}{4}\right), m = -\dfrac{3}{4}$
45. $\left(\dfrac{7}{8}, 0\right), \left(0, -\dfrac{7}{2}\right), m = 4$ **47.** The pitch of the roof is $\dfrac{3}{5}$.
49. The slope of the ramp is $\dfrac{38}{27}$. **51.** The slope of
the ramp is $\dfrac{14}{11}$. **53.** $500, m = 500, m = 500$

Challenge

59. m is undefined

Maintain Your Skills

63. 345% **65.** 1.138

Exercises 8.5

1. $d = kr$ **3.** $P = ks$ **5.** $C = kn$ **7.** 60 **9.** $k = 1.5$
11. $k = \dfrac{2}{7}$ **13.** $a = \dfrac{112}{3}$ **15.** $m = 307.2$ **17.** $\ell w = k$
19. $CP = k$ **21.** $ns = k$ **23.** 40 **25.** $k = 72$ **27.** $k = 12$
29. $y = 14$ **31.** $y = 11$ **33.** 490 lb **35.** $648
37. $\dfrac{15}{22}$ amp **39.** 16 ft **41.** $60,000 **43.** The invest-
ment equation is $A = 500t$, where 500 is the constant of
variation.

Challenge

47. 2.2 hr

Maintain Your Skills

49. $13.5t$ **51.** -52.9 **53.** No

CHAPTER 8 TRUE – FALSE CONCEPT REVIEW

1. False; the equation $0x + 0y = 2$ has no solution
2. False; either of the pairs $(5, 3)$ or $(2, 1.5)$ is a solution of
$2y - x = 1$ **3.** True **4.** True **5.** True **6.** False; the

quadrants are numbered counterclockwise   **7.** False; the point where the axes intersect is called the origin
8. False; the x-value of an ordered pair shows the distance of a point from the y-axis   **9.** True   **10.** False; if either x or y is zero, then the point corresponding to the pair is on one of the axes   **11.** False; the graph of $0x + 0y = 3$ has no solution and no graph   **12.** False; two points are needed to determine the graph of a line   **13.** True
14. False; line graphs are useful in many practical applications including, for example, some cost and depreciation equations   **15.** False; the point $(-2, -1)$ is in quadrant III
16. True   **17.** True   **18.** False; a line that goes up from right to left has negative slope   **19.** True   **20.** True

25.

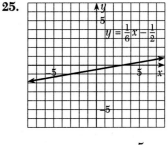

27. $(8, 0), (0, -3)$ **29.** $(\frac{5}{2}, 0), (0, \frac{15}{7})$ **31.** $m = -\frac{3}{8}$

33. $m = -5$ **35.** $m = \frac{9}{4}$ **37.** 384   **39.** 798 lb

41. 320   **43.** 9   **45.** \$31.25

CHAPTER 8 REVIEW

1. A solution   **3.** A solution   **5.** Not a solution

7.

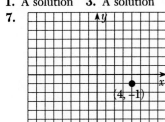

9.

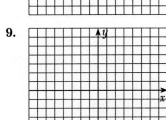

11. $(6, -3)$ **13.** $(4, 5.5)$ **15.** $(-1.5, -2)$ **17.** $(10, 4)$
19. $(14, 6)$

21.

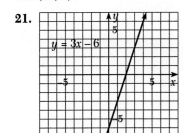

23.

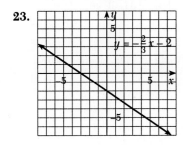

CHAPTER 8 TEST

1. $y = -24$ **2.** Yes

3.

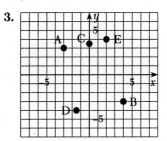

4.

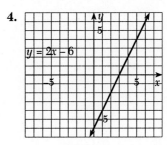

5. Not a solution   **6.** $A\,(7, 4)$, $B\,(-5, 1)$, $C\,(3, -2)$, $D\,(-3.5, -4)$, $E\,(2.5, 5)$

7.

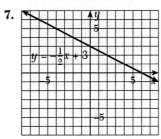

8. $y = -2$

9.

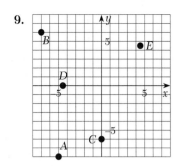

10. $y = 7$ **11.** Yes

12.

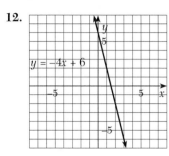

13. $(5, 0), (0, -5.5)$ **14.** $m = \dfrac{2}{3}$ **15.** $(6, 0), (0, -\dfrac{18}{7})$
16. $m = -\dfrac{4}{3}$ **17.** $y = 20.25$ **18.** $y = 18.0$

19.

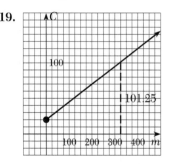

$101.25

CHAPTERS 1–8 CUMULATIVE REVIEW

1. 5859   **3.** $9^5 \cdot 2^5$ **5.** 297,987,000,000   **7.** $C = 25.12$
9. Yes   **11.** $-\$38,000$ **13.** -32 **15.** 6   **17.** $\dfrac{11x}{18}$
19. Yes   **21.** $x = 20.03$ **23.** 20 games   **25.** $4\dfrac{5}{8}$ gal
27. 5.76 min   **29.** 104   **31.** -14 **33.** 2

35.

---

37. 704   **39.** 88%   **41.** 1992   **43.** 93%   **45.** 75 vacancies
47. 1 hr 54 min 47 sec   **49.** $33\dfrac{2}{3}$ lb   **51.** $469\dfrac{1}{3}$ ft/sec
53. 86 ft   **55.** 3 gal   **57.** 10,386 in³   **59.** A $(-4, 7)$,
B $(3,6)$, C $(-2, 5)$, D $(0, -1)$, E $(6, -4)$

PLANE GEOMETRY

Exercises G.1

1. 36   **3.** 25   **5.**   **7.** 36, 33   **9.** True
11. True   **13.** False   **15.** True   **17.**   **19.**
21. 32   **23.** 26   **25.** False   **27.** False   **29.** False
31. True   **33.** 18   **35.** 63   **37.** 21   **39.** Deductive
41. Inductive   **43.** Inductive   **45.** Inductive

Challenge

49. 6660, 43,053,381

Maintain Your Skills

51. 5250 m   **53.** 9.183 g   **55.** 20.875 cm

Exercises G.2

1. Acute   **3.** Obtuse   **5.** Obtuse   **7.** Acute   **9.** Obtuse
11. 18°, 45°, 82°   **13.** 315°, 282°, 210°
15. 95°, 121°, 175°   **17.** 50°   **19.** 5°   **21.** 75°
23. 100°   **25.** 155°   **27.** 125°   **29.** $\angle A = 46°$
31. $\angle C = 75°$ **33.** True   **35.** False   **37.** False
39. False   **41.** True   **43.** Yes   **45.** No

Challenge

49. Yes

Maintain Your Skills

51. 330 ft/sec   **53.** $1485\dfrac{1}{3}$ ft

Exercises G.3

1. Isosceles   **3.** Equilateral   **5.** Obtuse   **7.** Right
9. Obtuse   **11.** Obtuse   **13.** $\angle R = 105°$ **15.** $\angle T = 30°$
17. $\angle C = 65°$ **19.** $\angle D = 12.9°$ **21.** True   **23.** False
25. False   **27.** False   **29.** True   **31.** True

Challenge

35. $\triangle ABC$, obtuse; $\triangle CAE$, right; $\triangle ADE$, acute; $\triangle ABE$,
right; $\triangle DBE$, obtuse   **37.** Yes

Maintain Your Skills

39. No   **41.** 529 m²

Exercises G.4

1. a) $\angle I = \angle L$; b) $GH = JK$ or $\angle I = \angle L$; c) $\overline{GH} = \overline{JK}$
3. $\Delta DAB \cong \Delta BCD$, ASA   **5.** $\Delta AOC \cong \Delta BOD$, SAS
7. $\Delta ADC \cong \Delta BEC$, SAS   **9.** $\Delta CEB \cong \Delta CAD$, ASA
11. $\Delta ADB \cong \Delta CDB$, SAS   **13.** $\Delta ADC \cong \Delta BDC$, SAS
15. True   **17.** Not necessarily true   **19.** $\overline{DF}$
21. $\Delta APD \cong \Delta BPC$, SAS or $\Delta PED \cong \Delta PEC$, SAS
23. $\Delta AEC \cong \Delta BDC$, SAS   **25.** $\Delta ADE \cong \Delta BCF$, SAS
27. $\Delta ECB \cong \Delta FBC$, SAS   **29.** $\Delta BCM \cong \Delta DAN$, ASA

Challenge

33. Yes, $\Delta OBD \cong OCD$, SAS   **35.** Yes, $\Delta ACD \cong \Delta ACE$
by SAS

Maintain Your Skills

37. 16 words   **39.** 143.06625 km^2

Exercises G.5

1. $\angle 1$ and $\angle 7$, $\angle 2$ and $\angle 8$ **3.** $\angle 1$, $\angle 2$, $\angle 7$, $\angle 8$
5. $\angle 1 = 120°$ **7.** $\angle 1 = 115°$ **9.** $\angle 2$, $\angle 5$, $\angle 8$ **11.** $\angle 2$,
$\angle 5$, $\angle 6$, $\angle 8$ **13.** Yes, $\angle 2 = \angle 5$, $\angle 1 = \angle 6$, $\angle 6 + \angle 3 + \angle 5 = 180°$ **15.** Yes, $\angle 2 = \angle 3$ so $\angle 3 = \angle 1$; therefore corresponding angles are equal.   **17.** Yes, $\angle EFB = \angle FBC$, $\angle FBC = \angle GCB$, so $\angle EFB = \angle GCB$ and $\angle GCB = \angle KCD$, so $\angle EFB = \angle KCD$.

Challenge

21. Yes, $\Delta ABD \cong \Delta DAC$, ASA   **23.** Yes, $\angle DFE = \angle 1$ (isosceles triangle); $\angle DFE = \angle 2$ (alternate interior angles).

Maintain Your Skills

25. 0.00024 m^3 **27.** 0.47 ft^3

Exercises G.6

1. 58°   **3.** 76°   **5.** 12.5   **7.** 4   **9.** 40°   **11.** 31.2
13. 28.8   **15.** 6.75   **17.** 28   **19.** 160
21. $\dfrac{DE}{DF} = \dfrac{AB}{AC} = \dfrac{1}{2}$; $\dfrac{BE}{EF} = \dfrac{AB}{BC} = \dfrac{2}{3}$; $\dfrac{FE}{DF} = \dfrac{BC}{AC} = \dfrac{3}{4}$
23. $\dfrac{AD}{DB} = \dfrac{DB}{DC} = \dfrac{3}{4}$; $\dfrac{AD}{AB} = \dfrac{DB}{BC} = \dfrac{3}{5}$; $\dfrac{DB}{AB} = \dfrac{DC}{BC} = \dfrac{4}{5}$
25. 180 ft

27. 107.1   **29.** $x = 12.5$ **31.** 12   **33.** Yes, corresponding angles are equal.   **35.** Yes, corresponding angles are equal.

Challenge

39. Yes, corresponding sides are proportional

Maintain Your Skills

41. $Q = -21$ **43.** No

Plane Geometry True–False Concept Review

1. True   **2.** True   **3.** True   **4.** True   **5.** False. $60° + 30° = 90°$ **6.** False; angles of $20°$, $20°$, and $140°$
7. False. They are similar.   **8.** False. They must be in corresponding positions.   **9.** False; on different sides
10. True   **11.** True   **12.** False. They are similar.
13. False. All congruent triangles are similar triangles.
14. True   **15.** False. People under 21 years old might have a social security number.   **16.** False; a measure of $45°$ **17.** True   **18.** False; no equal sides   **19.** False. It cannot be an acute triangle.   **20.** True   **21.** True
22. False. They are equal.   **23.** False. Alternate interior angles are equal.   **24.** True   **25.** True

Plane Geometry Review

1. 29, 35   **3.** 25, 22   **5.** True   **7.** 210°, 346°   **9.** 180°
11. 35°   **13.** 2°   **15.** False   **17.** Scalene   **19.** False
21. Acute   **23.** Right   **25.** 45°   **27.** $\angle A = \angle D$ or $\overline{BC} = \overline{FE}$ **29.** $\Delta AEB \cong \Delta CEB$, ASA   **31.** $\angle R$ **33.** $\overline{ST}$
35. $\overline{AC}$ **37.** $\angle 4$ **39.** $\angle 7$ **41.** 135°   **43.** 45°   **45.** 45
47. $11\dfrac{13}{37}$ **49.** 60 ft   **51.** $\angle ABC$ and $\angle ADB$; $\angle ACB$ and $\angle CAD$ **53.** $\angle ACE$ and $\angle CBD$; $\angle CEA$ and $\angle BDC$
55. $\dfrac{BA}{CB} = \dfrac{AD}{BD} = \dfrac{BD}{CD} = \dfrac{3}{4}$

Plane Geometry Test

1. [dice image]   **2.** True   **3.** Corresponding exterior angles, vertical angles, interior angles   **4.** 105°   **5.** 37.2°
6. Scalene   **7.** Yes, AAA   **8.** 42°   **9.** False   **10.** 125°, 100°   **11.** Yes, SAS   **12.** Not valid   **13.** Yes   **14.** Yes
15. 12

Index